Computing for Numerical Methods Using Visu[illegible]
Albert Y. Zomaya, and Sakhinah A. Bakar

Architecture-Independent Programming for Wireless Sensor Networks / Amol B. Bakshi and Viktor K. Prasanna

High-Performance Parallel Database Processing and Grid Databases / David Taniar, Clement Leung, Wenny Rahayu, and Sushant Goel

Algorithms and Protocols for Wireless and Mobile Ad Hoc Networks / Azzedine Boukerche (*Editor*)

Algorithms and Protocols for Wireless Sensor Networks / Azzedine Boukerche (*Editor*)

Optimization Techniques for Solving Complex Problems / Enrique Alba, Christian Blum, Pedro Isasi, Coromoto León, and Juan Antonio Gómez (*Editors*)

Emerging Wireless LANs, Wireless PANs, and Wireless MANs: IEEE 802.11, IEEE 802.15, IEEE 802.16 Wireless Standard Family / Yang Xiao and Yi Pan (*Editors*)

EMERGING WIRELESS LANs, WIRELESS PANs, AND WIRELESS MANs

EMERGING WIRELESS LANs, WIRELESS PANs, AND WIRELESS MANs

IEEE 802.11, IEEE 802.15, 802.16 WIRELESS STANDARD FAMILY

Edited by

Yang Xiao
Yi Pan

WILEY

A JOHN WILEY & SONS, INC., PUBLICATION

Published by John Wiley & Sons, Inc., Hoboken, New Jersey
Published simultaneously in Canada

Library of Congress Cataloging-in-Publication Data:

Xiao, Yang, 1966-
Emerging wireless LANs, wireless PANs, and wireless MANs : IEEE 802.11, IEEE 802.15, IEEE 802.16 wireless standard family/Yang Xiao and Yi Pan.
p. cm.
Includes bibliographical references and index.
ISBN 978-0-471-72069-0 (cloth)
1. IEEE 802.11 (Standard) 2. IEEE 802.16 (Standard) 3. Wireless LANs–Standards. 4. Personal communication service systems–Standards. 5. Wireless metropolitan area networks–Standards. I. Pan, Yi, 1960- II. Title.
TK5105.5668. X53 2008
004.6′8–dc22

2008021441

Printed in the United States of America

10 9 8 7 6 5 4 3 2 1

CONTENTS

PREFACE

The purpose of this book is to introduce current and emerging Institute of Electrical and Electronics Engineers (IEEE) 802 wireless standards to readers, including IEEE 802.11 wireless local area networks (WLANs)—WiFi, wireless personal area networks (WPANs) (IEEE 802.15.1 Bluetooth, IEEE 802.15.2 coexistence of WLANs and WPANs, IEEE 802.15.3 higher data rate WPANs, IEEE 802.15.4 sensor networks—Zigbee, and IEEE 802.15.5), and IEEE 802.16 wireless metropolitan area networks (WMANs)—WiMAX. The book introduces medium access control and physical layer protocols for all these standards as well as some research articles.

Experience has shown that reading the standards can be tedious and sometimes confusing since they are normally written in a way that is very detailed. Engineers and researchers in both industry and academia spend huge amounts of time trying to understand them. A good book can help them save this time and help them understand the standards and we hope this book does that.

About 10 standards are presented in this book. The actual text for each standard comprises about 300–600 pages. In our book all 10 standards are discussed in about 600 pages. The main purpose is to help readers understand the standards as well as related research issues.

The book is primarily written for scientists, researchers, engineers, developers, educators, and administrators of universities, industries, research institutes and laboratories, and government agencies working in the area of wireless networks. They will find this book a unique source of information on recent advances and future directions of WLANs, WPANs, and WMANs. We expect that the book will be an informative and useful reference in this new and fast-growing research area.

This book was made possible by the great efforts of our publishers and contributors. First, we are indebted to the contributors, who have sacrificed many days and nights to put together these excellent chapters for our readers. Second, we owe our special thanks to our publishers and staff members. Without their encouragement and quality work this book would not have been possible. Finally, we would like to thank our families for their support.

Yang Xiao
Yi Pan

Tuscaloosa, Alabama
Atlanta, Georgia
January 2009

CONTRIBUTORS

Usman A. Ali, Brunel University, West London, United Kingdom

Kaveh Ghaboosi, University of Oulu, Finland

Pin-Han Ho, University of Waterloo, Waterloo, Ontario, Canada

Ekram Hossain, University of Manitoba, Winnipeg, Manitoba, Canada

Ivan Howitt, University of North Caroling, Charlotte, North Carolina

Fei Hu, The University of Alabama, Tuscaloosa, Alabama

Yu-Kai Huang, National Taiwan University, Taipei, Taiwan, China

Hai Jiang, University of Alberta, Alberta, Canada

Babak H. Khalaj, Sharif University of Technology, Tehran, Iran

Sai Suhas Kolukula, Honeywell, Bangalore, India

Yu-Kwong Kwok, Colorado State University, Fort Collins, Colorado

Lutz Lampe, Department of Electrical and Computer Engineering, University of British Columbia, Vancouver, British Columbia, Canada

Matti Latva-aho, Department of Electrical Engineering, University of Oulu, Finland

Myung J. Lee, Department of Electrical Engineering, City University of New York, New York

Victor C. M. Leung, Department of Electrical and Computer Engineering, University of British Columbia, Vancouver, British Columbia, Canada

Jingli Li, University of Louisville, Louisville, Kentucky

Ming Li, California State University, Fresno, California

Tianji Li, Hamilton Institute, The National University of Ireland, Maynooth, County Kildare, Ireland

Xiaodong Lin, University of Ontario Institute of Technology, Ontario, Canada

Xiangqian Liu, University of Louisville, Louisville, Kentucky

Kuang-Hao Liu, University of Waterloo, Waterloo, Ontario, Canada

Rongxing Lu, University of Waterloo, Waterloo, Ontario, Canada

Jelena Mišić, University of Manitoba, Winnipeg, Manitoba, Canada

Qiang Ni, Brunel University, West London, United Kingdom

Dusit Niyato, University of Manitoba, Winnipeg, Manitoba, Canada

Yi Pan, Department of Computer Science, Georgia State University, Atlanta, Georgia

Ai-Chun Pang, National Taiwan University, Taipei, Taiwan, China

Tae Rim Park, Department of Electrical Engineering, City University of New York, New York

Michael J. Plyler, Freed-Hardeman University, Henderson, Tennessee

Anita Raja, University of North Carolina, Charlotte, North Carolina

Jeff J. Robertson, Department of Computer Science, The University of Memphis, Memphis, Tennessee

M. Sai Rupak, IBM, Bangalore, India

Robert Schoberg, Department of Electrical and Computer Engineering, University of British Columbia, Vancouver, British Columbia, Canada

Xuemin (Sherman) Shen, University of Waterloo, Waterloo, Ontario, Canada

Minghui Shi, University of Waterloo, Waterloo, Ontario, Canada

Krishna M. Sivalingam, University of Maryland, College Park, Maryland

Dionysios Skordoulis, Brunel University, West London, United Kingdom

Chris Snow, Department of Electrical and Computer Engineering, University of British Columbia, Vancouver, British Columbia, Canada

K. S. Sridharan, DMACS, Sri Sathya Sai University, Bangalore, India

Bo Sun, Lamar University, Beaumont, Texas

Hsueh-Wen Tseng, National Taiwan University, Taipei, Taiwan, China

Yang Xiao, Department of Computer Science, University of Alabama, Tuscaloosa, Alabama

Jiang Xie, University of North Caroling, Charlotte, North Carolina

Wenbing Yao, Brunel University, West London, United Kingdom

Zhanping Yin, Department of Electrical and Computer Engineering, University of British Columbia, Vancouver, British Columbia, Canada

Jianliang Zheng, Department of Electrical Engineering, City University of New York, New York

Chunhui Zhu, Samsung Electronics Corporation, San Jose, California

Haojin Zhu, University of Waterloo, Waterloo, Ontario, Canada

Weihua Zhuang, University of Waterloo, Waterloo, Ontario, Canada

PART I

IEEE 802.11 WIRELESS LANs

CHAPTER 1

IEEE 802.11 MEDIUM ACCESS CONTROL AND PHYSICAL LAYERS

KAVEH GHABOOSI, MATTI LATVA-AHO, and YANG XIAO

1.1 INTRODUCTION

A wireless local area network (WLAN) is an information system[1] intended to offer diverse location-independent network service access to portable wireless devices using radio waves instead of wired infrastructure. In corporate enterprises, WLANs are typically deployed as the ultimate connection between an existing cable infrastructure network and a cluster of mobile clients, giving them wireless access to the shared resources of the corporate network across a building or campus setting. Fundamentally, WLANs liberate customers from reliance on hard-wired access to the network backbone, giving them anywhere, anytime network services access. The pervasive approval of WLANs depends upon industry standardization to ensure product compatibility and reliability among various brands and manufacturers. Among existing system architectures, the IEEE 802.11 family is the most popular and accepted standard concerning medium access control (MAC) and physical (PHY) layers in WLANs; therefore, in this chapter, we briefly overview its basic features in both aforementioned layers. We start our investigation with the MAC layer and its fundamental components. Supported network types, different network services, and media access schemes are covered, accordingly. Subsequently, the physical layer and its basic characteristics are discussed. Different technologies,

[1] In telecommunications, an information system is any telecommunications and/or computer-related equipment or interconnected system or subsystems of equipment that are used in the acquisition, storage, manipulation, management, movement, control, display, switching, interchange, transmission, or reception of voice and/or data and includes software, firmware, and hardware (Federal Standard 1037C, MIL-STD-188, and National Information Systems Security Glossary).

Emerging Wireless LANs, Wireless PANs, and Wireless MANs. Edited by Y. Xiao and Y. Pan

including frequency hopping (FH), direct-sequence spread spectrum (DSSS) and its high rate (HR) counterpart (i.e., HR/DSSS), and orthogonal frequency division multiplexing (OFDM), recommended for the IEEE 802.11 physical layer are then explored. As a result, this chapter can be assumed as a comprehensive overview of the IEEE 802.11 standard.

1.2 IEEE 802.11 MAC PROTOCOL

In 1997, the IEEE 802.11 working group (WG) proposed the IEEE 802.11 WLAN standard and, subsequently, a revised version was released in 1999. The primary medium access scheme in IEEE 802.11 MAC is the distributed coordination function (DCF), a contention-based protocol which is based on the carrier sense multiple-access/collision avoidance (CSMA/CA) protocol. In the DCF, mobile terminals should contend for the shared wireless channel, and as a result, the medium access delay for each station (STA) cannot be bounded in heavy-traffic-load circumstances. Thus, the DCF is capable of offering only asynchronous data transmission on a best effort (BE) basis. In order to support real-time traffic such as voice and video, the point coordination function (PCF) scheme has been advised as a noncompulsory option. Basically, the PCF is based on a centralized polling scheme for which a point coordinator (PC) residing in an access point (AP) provides contention-free services to the associated stations in a polling list. In addition to the IEEE 802.11 standard [1], there is a well-known book by Gast [2] which is considered as a complete scientific review of 802.11 families. Due to the popularity of the aforementioned book, we will use it frequently throughout the section to refer the reader to more technical issues and discussions.

Recently, considerable interest in wireless networks supporting quality of service (QoS) has grown noticeably. The PCF is already available in IEEE 802.11 to offer QoS but has not yet been implemented in reality due to its numerous technical limitations and performance drawbacks. For that reason, the 802.11 WG initiated IEEE 802.11e activity to develop the existing 802.11 MAC to facilitate support of QoS. Regarding the 802.11e amendment, not only the IEEE 802.11e standard [3] but also recognized introductory and survey papers [4, 5] will be used repeatedly as key references. We cite many technical issues from these works and the references therein to more appropriately explain 802.11/802.11e-based system features.

1.2.1 Categories of 802.11 Networks

The key constructing component of an 802.11 network is the basic service set (BSS), a group of wireless terminals that communicate with each other over a common radio channel [1, 2]. Data transmission is accomplished within a *basic service area*, defined by radio propagation characteristics of wireless channel. BSSs come in two categories, as illustrated in Fig. 1.1.

The *infrastructure BSS* networks are primary for mobile stations to access the Internet via an AP so that, in most of cases, communications between two stations within the same service set do not happen. In communications between mobile stations in the same service set, the AP deployment acts as an intermediate node for all information exchanges comprising communications. In other words, any data communication between two wireless clients should take two successive hops, i.e., source STA to AP and AP to destination STA. Obviously, the exploitation of APs in infrastructure networks brings two major advantages. On the one hand, no restriction is placed on the physical distance between mobile stations. On the other hand, allowing straight communication between wireless terminals would apparently preserve system capacity[2] but at the cost of increased physical and MAC layer complexity. The most important functions of an AP are to assist stations in accessing the Internet and help save battery power in associated wireless stations. If a mobile terminal is in the power-saving (PS) mode, the AP buffers those frames destined to reach the station during the period it will be in PS status. When the terminal exits the PS mode, the AP forwards the cached data frames to the station one by one. Hence, APs evidently play a key role in infrastructure networks to make implementation of PS mechanisms possible [1].

In the independent BSS (IBSS), mobile stations are allowed to communicate directly. Characteristically, IBSSs are composed of a few stations configured

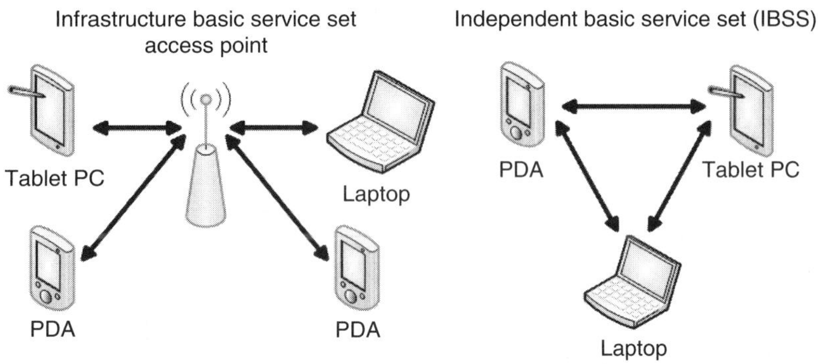

FIGURE 1.1 Infrastructure and independent basic service sets.

[2] In computer science, *channel capacity* is the amount of discrete information that can be reliably transmitted over a channel. By the noisy-channel coding theorem, the channel capacity of a given channel is the limiting information transport rate (in units of information per unit time) that can be achieved with vanishingly small error probability.

for a particular goal and for a short period of time. IBSSs are sometimes referred to as ad hoc BSSs or simply ad hoc networks.

IEEE 802.11 allows wireless networks of arbitrary size to be installed and utilized by introducing the extended service set (ESS) concept. Basically, the ESS is constructed by chaining neighboring BSSs and requires a backbone system that provides a particular set of services. Figure 1.2 illustrates an ESS as a combination of three neighboring BSSs. Switching between adjacent BSSs while being connected to the system is called a handoff.[3] Stations with the same ESS are able to communicate with each other even if they are not in the same BSS or are moving from one point to another. For associated stations within an ESS, a wireless network should behave as if it were a single layer 2 local area network (LAN). In such an architecture, APs are similar to layer 2 bridges; consequently, the backbone network should be a layer 2 network as well (e.g., Ethernet). Several APs in a single area may be connected to a single switch or can even use virtual LANs (VLANs) if the link layer connection spans a larger area. 802.11 supplies link layer mobility within an ESS, but only if the backbone network is a single link layer domain, such as a shared Ethernet or a VLAN [2].

Theoretically, extended service areas are the highestlevel abstraction supported by 802.11 wireless networks. In order to let non-802.11 network devices use the same MAC address to exchange data traffic with an associated station somewhere within the ESS, APs should mimic an absolute cooperative system. In Fig. 1.2, the illustrated gateway uses a single MAC address to deliver data frames to the targeted mobile stations in different BSSs. This is the MAC address of an AP with which the intended wireless station has been already associated. As a result, the gateway is unaware of the actual location of a tagged wireless terminal and relies only on the corresponding AP to forward data traffic [1, 2]. The backbone network to which APs are connected is called the distribution system (DS) since it makes delivery of information to and from the outside world possible.

It should be noted that technically different types of 802.11 networks may coexist at the same time. For instance, IBSSs might be constructed within the basic service area of an AP. Coexisting infrastructure BSSs and IBSSs should share the same radio channel capacity and, as a result, there may be adverse performance implications from colocated BSSs as well [2].

[3] In cellular telecommunications, the term *handoff* refers to the process of transferring an ongoing call or data session from one channel connected to the core network to another. In satellite communications, it is the process of transferring satellite control responsibility from one earth station to another without loss or interruption of service. The British term for transferring a cellular call is *handover*, which is the terminology standardized within European-originated technologies such as Global System for Mobile (GSM) communications and Universal Mobile Telecommunications System (UMTS). In telecommunications, there are two reasons why a handoff (handover) might be conducted: if the mobile terminal has moved out of range from one cell site, i.e., base transceiver station (BTS) or AP, and can get a better radio link from a stronger transmitter or, if one BTS/AP is full, the connection can be transferred to another nearby BTS/AP.

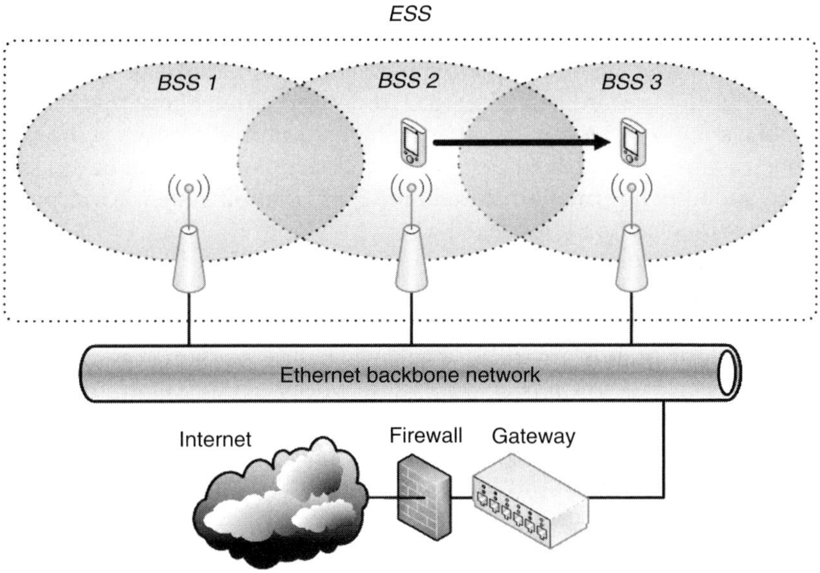

FIGURE 1.2 Extended service set.

1.2.2 IEEE 802.11 Networks Services

Generally the IEEE 802.11 standard provides nine dependent services: Three of these services are dedicated distinctively to data transfer purposes while the remaining ones are explicitly devoted to management operations enabling network systems to keep track of mobile stations and react in different circumstances accordingly.

The *distribution* service is exploited in infrastructure networks to exchange data frames. Principally, the AP, upon receiving a MAC protocol data unit (MPDU), uses this service to forward it to the intended destination station. Therefore, any communication with an AP should use a distribution service to be possible. *Integration* is a specific service provided by the distribution system that makes connection with a non-802.11 network possible. The integration function is not expressed technically by the standard, except in terms of the services it should offer.

MAC frame delivery to the associated terminals will not be possible unless the *association* service ensures that the AP and connected stations can work together and use the network services. Consequently, the distribution system is able to use the registration information to determine the AP with which a specific mobile station has been associated. In other words, unassociated wireless terminals are not permitted to obtain *any* service from the whole system. In an ESS, when a mobile station moves between different BSSs, there should be a set of handoffs to be accomplished in order to keep the station connected to the system. *Reassociation* is generally initiated by a wireless terminal once the signal strength indicates that a different association is necessary. This means that handoff and reassociation requests are never commenced by APs. Upon completion of reassociation, the distribution system renews its location records to reflect the latest information about reachability of the mobile station. To terminate an existing association, wireless stations may possibly use the so-called *disassociation* service. Upon invocation of disassociation, any mobility information stored in the distribution system corresponding to the requesting station is removed at once.

Authentication is an obligatory prerequisite to association due to the fact that only authenticated users are authorized to use the network resources. If the APs of a distribution system have been configured in such a way as to authenticate any station, then the system is called an "open system" or an "open network." These kinds of wireless networks can be found, for instance, at university campuses. *Deauthentication* terminates an authenticated relationship between an AP and a wireless station. Since authentication is required before system resources utilization, a side effect of deauthentication is termination of any existing association.

IEEE 802.11 offers a noncompulsory *privacy* service called wired equivalent privacy (WEP). WEP is not iron-clad security; in fact, it can be easily disabled. In response, the IEEE 802.11i task group (TG) is seeking an enhanced and stronger security scheme to be included in the next generation of 802.11 equipments. IEEE 802.11i, known as WiFi-protected access version 2 (WPA2), is an amendment to the 802.11 standard specifying security mechanisms for wireless networks. It makes use of the advanced encryption standard (AES) block cipher, while WEP and WPA (an earlier version) use the RC4 stream cipher. The 802.11i architecture contains the following components: 802.1X for authentication [entailing the use of extensible authentication protocol (EAP) and an authentication server], the robust security network (RSN) for keeping track of associations, and the AES-based counter mode with cipher block-chaining message authentication code protocol (CCMP) to provide confidentiality, integrity, and origin authentication. Another important element of the authentication process is an innovative four-way handshake.

The MPDU is a fancy name for 802.11 MAC frames. The MPDU does not, however, include PHY layer convergence procedure (PLCP) headers. On the other hand, the MAC service data units (MSDUs) are only composed of higher level data units [e.g., Internet protocol (IP) layer]. For instance, an 802.11 management frame does not contain an MSDU. Wireless stations provide the *MSDU delivery* service, which is responsible for getting the data to the actual recipient.

1.2.3 IEEE 802.11 Media Access Schemes

In what follows, we discuss the medium access rules defined in the 802.11 standard and its corresponding amendments. We begin the discussion with the contention-based *802.11 DCF* access scheme. Subsequently, a few paragraphs are dedicated to the *802.11 PCF*, which is a contention-free channel acquisition technique. Finally, the supplementary QoS-aware amendment of the IEEE 802.11 standard, i.e., the *802.11e hybrid coordination* function (*HCF*), is explored [1–3].

1.2.3.1 IEEE 802.11 DCF. The fundamental IEEE 802.11 access scheme is referred to as the DCF and operates based upon a listen-before-talk (LBT) approach and CSMA/CA.

As indicated, MSDUs are transmitted using MPDUs. If the wireless station chooses to fragment a long MSDU into a number of MPDUs, then it should send the long MSDU through more than one MPDU over the radio system. 802.11 stations deliver MSDUs following a media detection procedure dealing with an idle wireless channel that can be acquired for data transmission. If more than one station senses the communication channel as being idle at the same time, they might commence their frame transmissions simultaneously, and inevitably a collision occurs subsequently. To minimize the collision risk, the DCF uses carrier sense functions and a binary exponential backoff (BEB) mechanism. In particular, two carrier sense schemes, namely physical and virtual carrier sense functions, are employed to simultaneously resolve the state of the radio channel. The former is offered by the physical layer and the latter by the MAC layer, called network allocation vector (NAV). The NAV records the duration that the medium will be busy based upon information announced before the control/data frames are captured over the air interface. If either function indicates a busy medium, the medium is considered busy (i.e., reserved or occupied); if not, it is considered idle. Subsequent to detection of wireless medium as *idle*, for a so-called DCF interframe space (DIFS) time duration, stations continue sensing the channel for an extra random time period called a *backoff period*. The wireless station begins traffic delivery whenever the shared medium remains idle over this further random time interval. The backoff time is determined by each station as a multiple of a pre defined slot time chosen in a stochastic fashion. This means that a fresh independent random value is selected for every new transmission. In the BEB algorithm, each station chooses a random backoff timer uniformly distributed in an interval [0, CW – 1], where CW is the current *contention window* size. It decreases the backoff timer by 1 for every idle time slot. Transmission is started whenever the backoff timer reaches zero. When frame transmission fails due to any reason, the station doubles the CW until it reaches the maximum value CW_{max}. Afterward, the tagged station restarts the backoff procedure and retransmits the MAC frame when the backoff counter reaches zero. If the maximum transmission retry limit is reached, the retransmission should be stopped, the CW should be reset to the initial value CW_{min}, and the MAC frame is simply discarded. At the same time as a wireless station is counting down its backoff counter, if the radio channel becomes busy, it suspends its backoff counter decrement and defers from the media acquisition until the medium again becomes idle for a DIFS [1, 2, 4, 5].

Each MPDU requires the reception of an acknowledgment (ACK) frame to confirm its correct transmission over the wireless channel. If for any reason the intended ACK frame is not received right after the MPDU transmission, the source station concludes that the MPDU was not delivered successfully and may reiterate the transmission. Basically, the CW size of a contending station increases when the transmission fails. After an unsuccessful effort, the backoff procedure is restarted with a double-sized CW, up to a maximum value defined by CW_{max}. Alternatively, subsequent to a successful transmission, the tagged station exploits another random backoff, even if there is no further queued MSDU to be delivered over the air interface. In the literature, this extra backoff is referred to as *post-backoff* given that it is executed subsequent to the data frame departure. There is an exception to the above-mentioned rule: If an MSDU arrives from layer 3 when (1) the transmission queue is vacant, (2) the latest post-backoff has finished, and (3) the medium has been idle for at least one DIFS, then it may be delivered at once with no further backoff procedure [4].

To overcome the so-called *hidden terminal* problem, the IEEE 802.11 DCF media access scheme utilizes a request-to-send/clear-to-send (RTS/CTS) mechanism which can be exploited optionally prior to MPDU transmission. As illustrated in Figure 1.3, the source station sends an RTS control frame to its intended destination. Upon reception of the RTS by the receiver, it sends a CTS frame back to the source station. The RTS and CTS frames include information on how long it will take to deliver the upcoming data frame (in the fragmentation case, it indicates the duration of the first fragment) and the corresponding ACK over the radio link. Upon reception of either the RTS or CTS, wireless stations located in the radio range of the transmitting node, in addition to those hidden to the source node and located in the transmission range of the destination station, set their local timer NAV, with the duration announced within the RTS/CTS frames. The RTS and CTS frames protect the MPDU from interferences due to other neighboring wireless nodes. Stations that receive these control frames will not initiate transmission until the above-mentioned NAV timer expires. Between two consecutive frames in the sequence

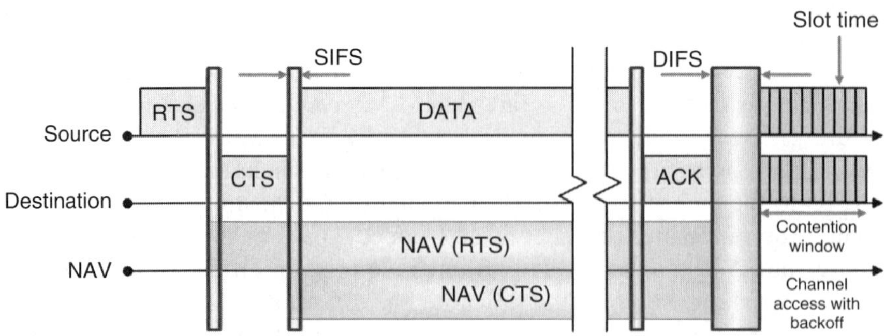

FIGURE 1.3 IEEE 802.11 RTS/CTS access scheme.

of RTS, CTS, MPDU, and ACK frames, a short interframe space (SIFS) gives transceivers time to switch (i.e., between transmitting and receiving modes). It is noteworthy that the SIFS is shorter than the DIFS, which gives the CTS and ACK the highest priority access to the wireless medium [1].

1.2.3.2 IEEE 802.11 PCF. IEEE 802.11 employs an optional PCF to support QoS for time-bound delay-sensitive services. The PCF offers techniques for prioritized access to the shared radio channel and is centrally coordinated by a PC station which is typically an AP. The PCF has higher priority than the DCF scheme. With the PCF, a contention-free period (CFP) and a contention period (CP) alternate periodically over time, where a CFP and the subsequent CP form an 802.11 superframe. The PCF is exploited throughout the CFP, while the DCF is used during the CP access phase. Each superframe is required to comprise a CP of a minimum length that allows at least one MSDU delivery (at least one frame exchange) of maximum size and at the slowest transmission rate under the DCF. A superframe is initiated by a beacon frame generated by the AP. The beacon frame is transmitted irrespective of whether or not the PCF is used. These frames are employed to preserve synchronization of the local timers in the associated stations and to deliver protocol-related parameters. The AP transmits these management frames at regular predefined intervals. Each station knows precisely when the subsequent beacon will arrive. These points in the time domain are referred to as target beacon transmission time (TBTT) and are announced in the previous beacon frame [1, 2].

During the CFP, there is no contention among wireless stations; instead, they are polled periodically by the AP. The PC polls a station requesting delivery of a pending data frame. Whenever the PC has a pending frame destined to an intended station, it utilizes a joint *data* and *poll* frame by piggybacking the CF-Poll frame onto the data frame. Upon reception of the so-called CF-Poll + Data, the polled station acknowledges the successful data reception and piggybacks an MPDU as well if it has any pending data frame targeted to the AP. If the PC does not receive a response from a polled station after waiting for a PCF interframe space (PIFS), it polls the next station or ends the CFP. Thus, no idle period longer than a PIFS occurs during a CFP. Bear in mind that a PIFS is longer than a SIFS but shorter than a DIFS. Since a PIFS is longer than an SIFS, a poll is never issued, e.g., between Data and ACK frames; hence a poll frame does not interrupt an ongoing frame exchange. The PC continues the aforementioned procedure until the CFP expires. A particular control frame, CF-End, is broadcast by the AP as the last frame within a CFP to indicate the end of the CFP (see Fig. 1.4) [1, 2, 4].

The PCF has many problems that have been reported in the literature [4]. Among many others, erratic beacon frame delay and indefinite transmission duration of the polled stations are the most important drawbacks. At the TBTT, the PC schedules the beacon as the next frame to be transmitted, but the beacon can only be transmitted when the medium has been determined to be

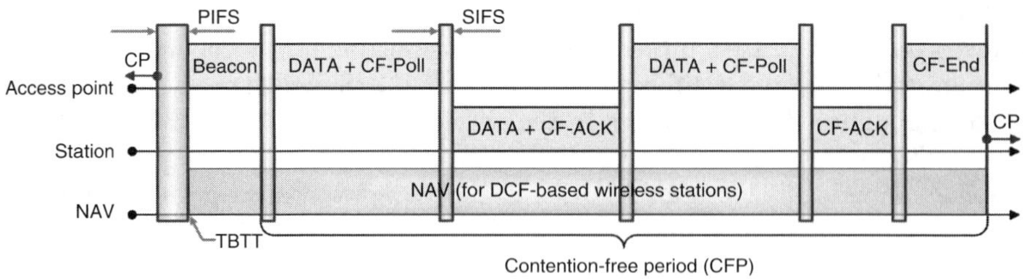

FIGURE 1.4 IEEE 802.11 PCF access scheme and TBTT.

idle for at least one PIFS. In IEEE 802.11, wireless stations are able to start their channel access even if the MSDU delivery is not finished before the upcoming TBTT. Depending upon whether the shared medium is idle or busy at the TBTT, a delay of the beacon frame might take place. The time the beacon frame is delayed from the TBTT determines the delay in a time-bounded MSDU transmission that has to be delivered in the CFP. This may rigorously influence the QoS as it introduces unpredictable time delays in each CFP. A further problem with the PCF is the unknown transmission duration of polled stations. A station that has been polled by the PC is allowed to deliver an MSDU that may be fragmented and of arbitrary length. In addition, different modulation and coding schemes are specified in the IEEE 802.11 family. Therefore, the duration of the MSDU is not under the control of the PC, which degrades the QoS offered to other stations polled during the rest of the CFP.

1.2.3.3 IEEE 802.11e: QoS Support in IEEE 802.11 MAC. The HCF, introduced in IEEE 802.11e, consists of two fundamental components: enhanced distributed-channel access (EDCA), an HCF contention-based channel access mechanism, and HCF controlled-channel access (HCCA). EDCA is the primary and mandatory access mechanism of IEEE 802.11e, while HCCA is optional and requires centralized polling and advanced scheduling schemes to distribute shared network resources among associated stations. According to the IEEE 802.11e, there can be two separate phases of operation within a superframe: CP and CFP. EDCA is used in the CP only, while HCCA is used in both phases. The HCF combines access methods of both the PCF and DCF, and this is why it is called *hybrid* [3].

The wireless station that operates as the central coordinator within a QoS-supporting basic service set (QBSS) is called a hybrid coordinator (HC). Similar to the PC, the HC resides within an 802.11e AP (i.e., QoS enabled access point (QAP)). There are *multiple* backoff entities operating in *parallel* within one QoS-aware 802.11e station (QSTA). A QSTA that is granted medium access opportunity should not occupy the radio resources for a time duration longer than a prespecified limit. This important characteristic of the 802.11e MAC protocol is referred to as transmission opportunity (TXOP). A TXOP is the time interval during which a backoff entity has the right to deliver MSDUs and is defined by its starting time and duration. TXOPs obtained throughout the contention-based phase are referred to as EDCA–TXOPs. Alternatively, a TXOP obtained via a controlled medium access scheme is called an HCCA–TXOP or *polled* TXOP. The duration of an EDCA–TXOP is limited by a QBSS-wide parameter referred to as the *TXOPlimit*. This parameter is distributed regularly by the HC within an information field of the beacon frame. A further enhancement is that backoff entities of QSTAs are totally forbidden from transmitting across the TBTT. That is, a frame transmission is commenced only if it can be completed ahead of the upcoming TBTT. This reduces the expected beacon delay, which gives the HC superior control over the wireless media, especially if the noncompulsory CFP is exploited after the beacon frame. Moreover, an 802.11e backoff entity is allowed to exchange data frames directly with another backoff entity in a QBSS without involving communication with the QAP. Whereas within an 802.11-based infrastructure BSS all data frames are either sent or received by the AP, an 802.11e QSTA can establish a direct link with another 802.11e QSTA using the direct-link protocol (DLP) prior to initiating direct frame transmissions. It should be noted that here the backoff entity deals with the local backoff entity of a tagged QSTA; therefore, they are used interchangeably [3–5].

1.2.3.3.1 IEEE 802.11e: EDCA. In EDCA, QSTAs have up to four distinct and parallel queues for incoming traffic. Each queue is coupled with a specific access category (AC) and contends for the radio channel independent of the others. Collisions among a tagged station's queues are resolved internally, allowing the higher priority queue to commence its transmission while forcing the lower priority queue(s) to perform a collision response.[4] Different levels of service are provided to each AC through a combination of three service differentiation mechanisms: arbitrary interframe space (AIFS), CW size, and TXOPlimit [3].

In contrast to the DCF access rules by which the backoff procedure is started after the DIFS from the end of the last indicated busy medium, EDCA backoff entities start at different intervals according to the corresponding AC of the traffic queue. As already pointed out, these time intervals are called AIFSs. The time duration of the interframe spaceAIFS[AC] is given by

$$\text{AIFS[AC]} = \text{SIFS} + \text{AIFSN[AC]} \times \text{aSlotTime}$$

where AIFSN[AC] $\geq$ 2. Note that AIFSN[AC] should be chosen by the HC such that the earliest access time of 802.11e stations to be the DIFS, equivalent to IEEE 802.11. Note that the parameter aSlotTime defines the duration of a

[4] In the literature, the internal collision between independent backoff entities is called *virtual collision*.

time slot. The smaller AIFSN[AC] corresponds to the higher medium access priority. The minimum size of CW, i.e., CW_{min}[AC], is another parameter which depends upon the AC. The initial value for the backoff counter is a random number taken from an interval defined by CW, which is exactly similar to the DCF case. Again, the smaller CW_{min}[AC] corresponds to a higher priority in acquiring a shared radio channel. An important difference between the DCF and 802.11e EDCA in terms of the backoff countdown rule is as follows: (1) The first backoff countdown takes place at the end of the AIFSN[AC] interval. (2) A frame transmission is initiated after a slot from the moment the backoff counter becomes zero. The CW increases upon unsuccessful frame exchanges but never exceeds the value of CW_{max}[AC]. This parameter is defined per the AC as well as part of the EDCA parameter set. Note that the smaller CW_{max}[AC] corresponds to a higher medium access priority. The aforementioned system configurations, in addition to the EDCA parameter set, are illustrated in Fig. 1.5.

As mentioned earlier, in addition to the backoff parameters, the TXOPlimit[AC] is defined per the AC as part of the EDCA parameter set. Apparently, the larger TXOPlimit[AC] is, the larger is the share of capacity for this AC. Once a TXOP is obtained, the backoff entity may keep transmitting more than one MSDU consecutively during the same TXOP up to a duration of TXOPlimit[AC]. This important concept in 802.11e MAC is referred to as *continuation* of an EDCA–TXOP (4).

As already explained, four self-governing backoff entities with different EDCA parameter sets exist inside an 802.11e QSTA. In a tagged QSTA when counters of two or more backoff entities reach zero simultaneously, they

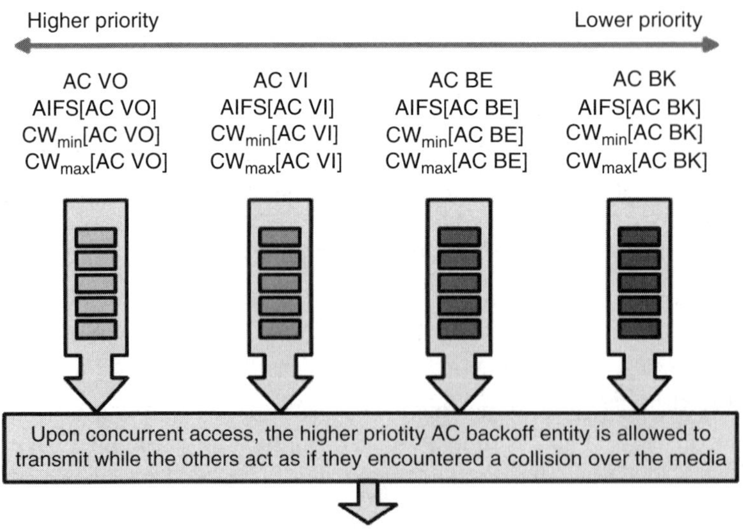

FIGURE 1.5 Four different access categories within a QSTA.

perform channel acquisition in the same time slot and consequently an internal *virtual collision* occurs. It should be noted that virtual collision is an abstract concept and there is no physical collision between contending backoff entities. When internal virtual collision occurs, the AC with the highest priority among collided entities is allowed to transmit, whereas all other backoff entities will act as if a collision has taken place on the shared radio channel.

1.2.3.3.2 IEEE 802.11e: HCCA. HCCA extends the EDCA medium access rules by assigning the uppermost precedence to the HC for the duration of both the CFP and CP. Basically, a TXOP can be attained by the HC through the controlled medium access stage. The HC may apportion TXOPs to itself in order to commence MSDU transactions whenever it requires, subsequent to detection of the shared wireless medium as being idle for PIFS, and without performing any further backoff procedure. To grant the HC a superior priority over legacy DCF and its QoS-aware counterpart, EDCA, AIFSN[AC] should be chosen such that the earliest channel acquisition for all EDCA stations can be the DIFS for any AC. During the CP, each TXOP of a QSTA begins either when the medium is determined to be available under the EDCA rules, i.e., after AIFS[AC] plus the random backoff time, or when a backoff entity receives a polling frame, the QoS CF-Poll, from the HC. The QoS CF-Poll is transmitted by the HC following a PIFS idle period and without any backoff procedure. On the other hand, for the duration of the CFP, the starting time and maximum duration of each TXOP is also specified by the HC, again by the use of QoS CF-Poll frames. In this phase, 802.11e backoff entities will not attempt to acquire the wireless media without being explicitly polled; hence, only the HC can allocate TXOPs by transmitting QoS CF-Poll frames or by immediately transmitting downlink data. Throughout a polled TXOP, the polled candidate mobile station can transmit multiple frames with a SIFS time space between two consecutive frames as long as the entire frame exchange duration does not exceed the dedicated maximum TXOPlimit. The HC controls the maximum duration of EDCA–TXOPs within its QBSS by the beacon frames. Thus, it is able to assign polled TXOPs at any time during the CP and the optional CFP [3].

Two supplementary schemes, namely block acknowledgement (BA) and DLP, which enhance the performance of the MAC protocol, have been taken into consideration in IEEE 802.11e [3, 4]. With the noncompulsory BA, the throughput efficiency of the protocol is improved. BA allows a backoff entity to send a number of MSDUs during one TXOP transmitted without individual ACK frames. The MPDUs delivered during the time of TXOP are referred to as a *block* of MPDUs in the literature and technical documents [4]. At the end of each block or in the next TXOP, all MPDUs are acknowledged at once by a bit pattern transmitted in the BA frame, and consequently the overhead of the control exchange sequences is reduced to a minimum of one ACK frame. On the other hand, each backoff entity is able to directly exchange information with any other backoff entity in the same QBSS without communicating through the QAP. For IEEE 802.11 and within a BSS, all data frames are sent to the AP and received from the AP. However, it should be obvious that this procedure consumes at least twice the channel capacity in comparison to direct communication. For that reason, DLP is defined to enable pairs of 802.11e backoff entities to establish direct links between each other.

1.3 IEEE 802.11 PHYSICAL LAYER FAMILIES

In this section we first introduce the concepts utilized in radio-based 802.11 physical layers and then present detailed explanations of these physical layers.

The IEEE 802.11 physical layer is divided into two sublayers: the PLCP and the physical medium dependent (PMD). The PLCP receives incoming MSDUs from the MAC layer, adds its own designated header, and then gives them to the PMD. It is mandatory for delivered information to have a *preamble*, which has a pattern that depends on the modulation technique deployed in the physical layer. The PMD is responsible for transmitting every bit it receives from the PLCP over the wireless medium. The physical layer also incorporates a clear-channel assessment (CCA) function to inform the MAC layer when a carrier is detected [2]. Figure 1.6 illustrates the logical structure of the physical layer.

Three different physical layers were standardized in the initial revision of 802.11: frequency-hopping spread spectrum (FHSS), DSSS, and infrared (IR) light. Consequently, supplementary amendments 802.11a, 802.11b, and 802.11 g were developed which are based on OFDM, high rate (HR)/DSSS, and the extended-rate PHY (ERP), respectively. Also, it is noteworthy to mention that 802.11n will be based on multi-input multi-output (MIMO) OFDM [6, 7].

In telecommunications, avoiding interference is a matter of law and the most imperative issue that should be taken into account. Thus, an official authority should impose rules on how the radio frequency (RF) spectrum is to be deployed. In the United States, the Federal Communications Commission (FCC) is responsible for regulating the use of the RF spectrum. European regulation is accomplished by the European Radio-communications Office (ERO) and the European Telecommunications Standards Institute (ETSI). The Ministry of Internal Communications (MIC) regulates radio exploitation in

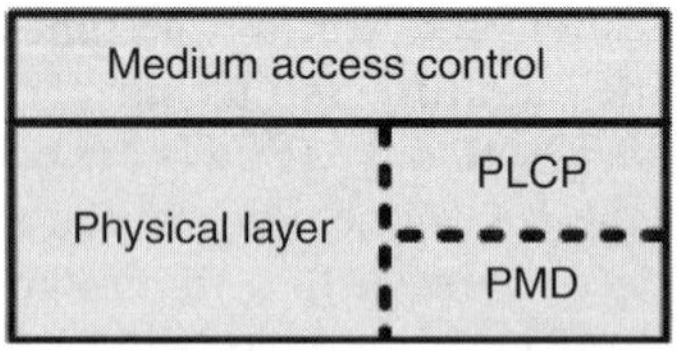

FIGURE 1.6 Physical layer logical structure.

Japan. Finally, worldwide regulation is done based upon recommendations of the International Telecommunications Union (ITU) [2].

The radio spectrum is partitioned into distinct frequency bands dedicated to particular applications. Among all frequency bands, the FCC and its counterparts in other countries designated particular frequency bands for the use of *industrial*, *scientific*, and *medical* (ISM) equipment. For instance, the 2.4-GHz band is available worldwide for unlicensed use. The use of RF equipment in the ISM bands is usually license free and RF devices operating in such frequency bands typically do not emit significant amounts of radiation. For example, microwave ovens are high-powered home/office devices, but they have extensive shielding to restrict interfering radio emissions. WLAN equipment, i.e., spread-spectrum-based IEEE 802.11b and OFDM-based IEEE 802.11 g, as well as Bluetooth, spread-spectrum cordless phones, and X10[5] communication system was developed for the 2.4-GHz ISM band [2]. On the other hand, in addition to 2.4 GHz, the 5-GHz frequency band is another ISM spectrum band dedicated to OFDM-based IEEE 802.11a; The United States was the first country to allow unlicensed device use in the 5-GHz range, though both Japan and Europe followed.

1.3.1 IEEE 802.11 Spread-Spectrum-Based Physical Layers

In this section, we start the discussion of the 802.11 SS–based physical layer with a short introduction of the system concept of different spread-spectrum architectures; then, FH 802.11 and the popular 802.11b extension based on DSSS and HR/DSSS will be covered separately.

1.3.1.1 Overview of Spread Spectrum. Spread-spectrum techniques are methods by which energy generated at one or more discrete frequencies is deliberately spread or distributed in either the frequency or time domain. This is accomplished for a variety of goals, including establishing secure communications, increasing resistance to natural interference and jamming, and preventing detection. Spread-spectrum telecommunications is a signal-structuring technique that utilizes direct sequence (DS), FH, or a hybrid of these and can be used for multiple access and/or multiple functions. This technique reduces the potential interference to other receivers while achieving *privacy*. Spread spectrum generally makes use of a sequential noiselike signal structure (i.e., spreading code) to broaden the narrowband information signal over relatively wideband radio frequencies. The intended receiver correlates the

[5] *X10* is an international and open industry standard for communication among devices used for home automation and domotic. It primarily uses power line wiring for signaling and control, where the signals involve brief RF bursts representing digital information. A radio-based transport is also defined. X10 was developed in 1975 by Pico Electronics of Glenrothes, Scotland, in order to allow remote control of home devices and appliances. It was the first *domotic* technology and remains the most widely available. *Domotics* is the application of computer and/or robotic technology to household appliances and buildings.

received signals to retrieve the original information signal. Originally, there were two motivations regarding the spread-spectrum concept: to resist efforts to jam radio communications and to hide the fact that communication is taking place, sometimes called low probability of intercept (LPI). Ultra wideband (UWB) is another modulation technique that accomplishes the same purpose based on transmitting short duration pulses. IEEE 802.11 uses either FHSS or DSSS in its radio interface. FH systems jump from one frequency to another in a random pattern, transmitting a short burst at each subchannel. The 2-Mbps FH physical layer is specified in clause 14. On the other hand, direct-sequence systems spread the power out over a wider frequency band using mathematical coding functions. Two DS structures were specified. The initial specification in clause 15 standardized a 2-Mbps physical layer, and 802.11b added clause 18 for the HR/DSSS physical layer.

1.3.1.1.1 Frequency-Hopping Spread Spectrum. FHSS is a method of transmitting radio signals by rapidly switching a carrier[6] among many frequency channels using a pseudorandom[7] sequence known to both transmitter and receiver. FH is similar to the well-known frequency division multiple access (FDMA) but with a key difference. In FDMA systems, devices are allocated fixed orthogonal nonoverlapping frequencies (i.e., fixed while totally distinct center frequencies and bandwidths). On the other hand, in FH-based systems, the frequency is time dependent; each frequency is used for a short portion of time, i.e., the so-called *dwell time*.

If two FH systems want to share the same frequency band, both of them can be configured such that they utilize different hopping sequences so that they do not interfere with each other. For the duration of each time slot, the aforementioned hopping sequences should be on dissimilar frequency slots. As long as the systems stay on different frequency slots, they do not encounter any interfere due to the other party. In general, orthogonal hopping sequences maximize wireless network throughput while increasing system complexity.

Beacon frames on FH networks include a timestamp and the so-called FH Parameter Set element. The FH Parameter Set element includes the hop pattern number and a hop index. By receiving a Beacon frame, a station knows everything it needs to synchronize its hopping pattern.

[6] A *carrier wave*, or simply *carrier*, is a waveform that is modulated to represent the information to be transmitted. This carrier wave is usually of much higher frequency than the baseband modulating signal, i.e., the signal which contains the information.

[7] A *pseudorandom* process is a process that appears stochastic but is not. Pseudorandom sequences typically exhibit statistical randomness while being generated by an entirely deterministic causal process. Such a process is easier to produce than a genuine random one and has the benefit that it can be used again and again to produce exactly the same numbers, useful for testing and fixing software. To date there is no known method to produce true randomness. The random-number generation functions provided in many software packages are pseudorandom.

Finally, it is noteworthy to mention that adaptive frequency hopping (AFH) spread spectrum, as used in Bluetooth, enhances system resistance to RF interference by avoiding using crowded frequencies in the hopping sequence. This sort of adaptive modulation is much easier to implement with FHSS than with DSSS.

1.3.1.1.2 Direct-Sequence Spread Spectrum. DS transmission is an alternative spread-spectrum technique that might be utilized to transmit a narrowband signal over a much wider frequency band. The fundamental approach of DS schemes is to cautiously spread the RF energy over a wide frequency band. The DS modulation scheme is accomplished by applying a chipping sequence (CS) to the information bit stream. A *chip* is a binary digit sequence employed by the DS system. These bits are higher level data while the chip signals are binary numbers used in encoding (transmitter side) and decoding (receiver side) procedures. Chipping streams, or the so-called pseudorandom noise (PN) codes, should have a much higher rate in comparison to the actual data stream.

For the PN code, IEEE 802.11 adopted an 11-bit Barker word meaning that each bit is encoded using the entire Barker word as a CS. Barker words have satisfactory autocorrelation, implying that the correlation function at the receiver operates as expected in a wide range of environments and is relatively tolerant to multipath delay spreads as incurred in multipath fading channels. The philosophy behind the deployment of exactly 11 bits is that most regulatory authorities usually necessitate a 10-dB processing gain in DS systems.

1.3.1.2 IEEE 802.11 FH Physical Layer. In 802.11 FH, the microwave ISM band is partitioned into a series of 1-MHz channels. Approximately 99% of the radio energy is confined to the channel. The modulation scheme employed by 802.11 encodes data bits as shifts in the transmission frequency from the channel center. Channels are defined by their center frequencies, which begin at 2.400 GHz for channel 0. Successive channels are derived by adding 1-MHz steps, meaning that channel 1 has a center frequency of 2.401 GHz, channel 2 has a center frequency of 2.402 GHz, etc., until channel 95, which has a center frequency at 2.495 GHz. Different regulatory authorities allow use of different parts of the ISM band. For example, the FCC in the United States and the ETSI in Europe (excluding France and Spain) allow channels 2–79 to be deployed while, in Japan, channels 73–95 might be utilized [1, 2].

The dwell time (see Section 1.3.1.1.1) in 802.11 FH systems is 390 time units, which is about 0.4 s. When an 802.11 FH physical layer hops between channels, the hopping process should take no longer than 224 μs. The frequency hops are subject to extensive regulation, in terms of both the size of each hop and the rate at which hops must occur [2].

1.3.1.3 IEEE 802.11b Physical Layer. The IEEE 802.11b amendment to the original standard was ratified in September 1999. 802.11b has a maximum raw data rate of 11 Mbps and uses the same CSMA/CA media access method defined in the original standard.

IEEE 802.11b products appeared on the market very quickly, since 802.11b is a direct extension of the DSSS modulation technique defined in the original standard. The 802.11b standard uses complementary code keying (CCK) as its modulation technique, which is a variation on code division multiple access (CDMA). Hence, chipsets and products were easily upgraded to support the 802.11b enhancements. The dramatic increase in throughput of 802.11b (compared to the original standard) along with substantial price reductions led to the rapid acceptance of 802.11b as the definitive WLAN technology.

Generally, 802.11b is used in point-to-multipoint configuration, wherein an AP communicates via an omnidirectional antenna with one or more clients that are located in a coverage area around the AP. Typical indoor range is 30 m (100 ft) at 11 Mbps and 90 m (300 ft) at 1 Mbps. With high gain external antennas, the protocol can also be used in fixed point-to-point arrangements, typically at ranges up to 8 km (5 miles), although some report success at ranges up to 80–120 km (50–75 miles) where line of sight (LOS) can be established. This is usually accomplished in place of costly leased lines or very cumbersome microwave communications equipment.

Channels for the 802.11b DS physical layer are much larger than the channels for the FH physical layer. The DS physical layer has 14 channels in the 2.4-GHz band, each 5 MHz wide. Channel 1 is placed at 2.412 GHz, channel 2 at 2.417 GHz, and so on, up to channel 13 at 2.472 GHz.Channel 14 was defined for operation in Japan and has a center frequency that is 12 MHz from the center frequency of channel 13 [2].

1.3.2 IEEE 802.11 OFDM-Based Physical Layers

In this section, we begin our discussion of 802.11 OFDM-based physical layers with a qualitative introduction to the basis of OFDM; subsequently, different 802.11 extensions, including 802.11a, g, h, and j, are covered separately.

1.3.2.1 Overview of OFDM. OFDM, essentially identical to coded OFDM (COFDM), is a digital multicarrier modulation scheme which deploys a large number of closely spaced orthogonal *subcarriers*. Each subcarrier is modulated with a conventional modulation scheme, e.g., quadrature amplitude modulation (QAM), at a low symbol rate, maintaining data rates similar to conventional *single-carrier* modulation schemes in the same bandwidth. In practice, OFDM signals are generated by the use of the fast Fourier transform (FFT) algorithm. The most important advantage of OFDM over single-carrier schemes is its ability to cope with severe channel conditions, e.g., multipath and narrowband interference, without complex equalization filters. Channel equalization is simplified due to the fact that OFDM may be viewed as using many slowly modulated narrowband signals rather than one rapidly modulated wideband signal. The orthogonality of the subcarriers results in zero cross-talk, even though they are so close that their spectra overlap. Low symbol rate helps manage time domain spreading of the signal by allowing the use of a guard interval between successive symbols. The guard interval eliminates the need for a pulse-shaping filter.

OFDM necessitates a high level of accuracy in frequency synchronization between the receiver and transmitter. In other words, any deviation causes the subcarriers to no longerbe orthogonal, resulting in intercarrier interference (ICI) and cross-talk between adjacent subcarriers. Frequency offsets are typically caused by mismatched transmitter and receiver oscillators or by *Doppler* shift due to mobile device movement. While Doppler shift alone may be compensated for by the receiver, the situation is worsened when combined with multipath, as reflections will appear at various frequency offsets, which is much harder to correct. This effect typically worsens as speed increases and is an important factor limiting the use of OFDM in high speed vehicles. Several techniques for ICI suppression have been suggested, but they may increase receiver complexity as well.

A key principle of OFDM is that, due to low symbol rate modulation, symbols are relatively longer than the channel time characteristics. As a result, it suffers less from intersymbol interference (ISI) caused by multipath. Since the duration of each symbol is long enough, it is feasible to insert a guard interval between the consecutive OFDM symbols, thus eliminating ISI. In addition, the guard interval also reduces the sensitivity to time synchronization problems. The *cyclic prefix*, which is transmitted during the guard interval, consists of the end of the OFDM symbol copied into the guard interval, and the guard interval is transmitted followed by the OFDM symbol. The guard interval consists of a copy of the end of the OFDM symbol so that the receiver will integrate over an integer number of sinusoid cycles for each multipath when it performs OFDM demodulation with FFT.

The stochastic effects due to channel frequency selectivity might be considered to be constant over an OFDM subchannel if the subchannel is sufficiently narrowbanded, i.e., if the number of subchannels is adequately large. This important feature makes equalization far simpler at the receiver side in OFDM systems than in conventional single-carrier modulation schemes. The equalizer simply has to multiply each subcarrier by a constant value, or even a rarely variable value. In addition, some subcarriers in some OFDM symbols might carry *pilot* signals for measurement of channel conditions, i.e., the equalizer gain for each subcarrier. In addition, pilot signals might be used for synchronization. If a differential modulation technique such as differential phase shift keying (DPSK) or differential quadrature phase shift keying (DQPSK) is applied to each subcarrier, equalization can be completely omitted, since these schemes are insensitive to slowly changing amplitude and phase distortion.

OFDM has being perpetually exploited in conjunction with channel coding schemes, i.e., forward error correction (FEC), and almost always uses frequency and/or time interleaving. On the one hand, frequency (subcarrier) interleaving increases resistance to channel frequency selectivity. On the other hand, time interleaving ensures that bits that are initially close together in the bit stream are transmitted far apart in time, thus mitigating against severe fading, as would happen when traveling at high speed. However, time interleaving is of little benefit in slowly fading channels while frequency interleaving offers little to no benefit for narrowband channels that suffer from flat fading. The reason interleaving is used in OFDM is to attempt to spread the errors out in the bit stream presented to the error correction decoder. A common type of error correction coding scheme used with OFDM-based systems is *convolutional* coding, which is often concatenated with *Reed–Solomon* coding. Convolutional coding is used as the inner code and Reed–Solomon coding is used for the outer code, usually with additional interleaving on top of the time and frequency interleaving and between the two layers of coding. The motivation for this combination of error correction coding is that the Viterbi decoder used for convolutional decoding produces short error bursts when there is a high concentration of errors, and Reed–Solomon codes are inherently well suited to correcting bursts of errors.

Finally, it should be noted that *windowing* is another technique which helps OFDM-based transceivers cope with real-world effects. Transitions can be abrupt at symbol boundaries, causing a large number of undesired high frequency components. To make OFDM transmitters robust against the aforementioned problems, it is widespread to add *padding* bits at the beginning and end of transmissions to allow transmitters to ramp up and down from full power. Padding bits are frequently required when error correction coding is deployed. In the literature, padding streams are usually referred to as *training* sequences.

1.3.2.2 IEEE 802.11a/h/j 5-GHz Physical Layer. The IEEE 802.11a amendment to the original standard was ratified in September 1999 [8]. Basically, it utilizes the same core protocol as the original standard, operates in the 5-GHz band, and uses a 52-subcarrier OFDM with a maximum raw data rate of 54 Mbps, which yields realistic net achievable throughput in the mid-20 Mbps. The data rate is reduced to 48, 36, 24, 18, 12, 9, and then 6 Mbps if required. 802.11a has 12 nonoverlapping channels, 8 dedicated to indoor and 4 to point to point. It is not interoperable with 802.11b, except if using equipment that implements both standards. Due to the fact that the 2.4-GHz frequency band has been heavily deployed, operating in the 5-GHz band gives 802.11a the advantage of less interference. However, this high carrier frequency also brings disadvantages. It restricts the use of 802.11a to almost LOS, necessitating more AP deployment [2, 9].

In IEEE 802.11a, out of 52 OFDM subcarriers, 48 are for data and 4 are pilot subcarriers with a carrier separation of 0.3125 MHz (20 MHz/64). Each of these subcarriers can be binary phase shift keying (BPSK), quadrature phase shift keying (QPSK), 16-QAM, or 64-QAM. The total bandwidth is 20 MHz with an occupied bandwidth of 16.6 MHz and symbol duration is 4 μs with a guard interval of 0.8 μs. The generation and decoding of orthogonal components are done in baseband using digital signal processing (DSP), which is then up converted to 5 GHz at the transmitter. Each subcarrier could be represented as a complex number. The time domain signal is generated by taking an inverse fast Fourier transform (IFFT). Correspondingly, the receiver down converts samples at 20 MHz and does an FFT to retrieve the original coefficients. The advantages of using OFDM include reduced multipath effects in reception and increased spectral efficiency [9].

In 802.11a, channels in the 5-GHz band are numbered starting every 5 MHz and each 20-MHz 802.11a channel occupies four channel numbers. Basically, 802.11a was originally designed for the United States. European channelization was added as part of 802.11 h in late 2003, and subsequently Japanese operation was appended with 802.11j in late 2004 [2].

1.3.2.3 IEEE 802.11 g 2.4-GHz Physical Layer. In June 2003, a third modulation standard was ratified: 802.11 g. This extension exploits the 2.4-GHz frequency band (similar to 802.11b) but operates at a maximum raw data rate of 54 Mbps, or about 24.7 Mbps net throughput, similar to 802.11a. IEEE 802.11 g hardware is compatible with its 802.11b counterpart. The modulation scheme used in 802.11 g is OFDM for the data rates of 6, 9, 12, 18, 24, 36, 48, and 54 Mbps, and, on the one hand, reverts to CCK to achieve 5.5 and 11 Mbps while, on the other hand, switches to DBPSK/DQPSK + DSSS for 1 and 2 Mbps. The maximum range of 802.11 g devices is slightly greater than that of 802.11b devices, but the range in which a client can achieve the full 54-Mbps data rate is much shorter than that which a 802.11b client can reach, 11 Mbps.

1.3.3 IEEE 802.11n Physical Layer

The emerging IEEE 802.11n specification differs from its predecessors in that it provides for a variety of optional modes and configurations that dictate different maximum raw data rates. This enables the standard to offer baseline performance parameters for all 802.11n devices while allowing manufacturers to enhance or tune capabilities to accommodate different applications and price points. With every possible option enabled, 802.11n could offer raw data rates up to 600 Mbps [6, 7].

In fact, the most widely celebrated component of 802.11n is the inclusion of MIMO technology. MIMO harnesses multipath transmission with a technique known as space division multiplexing. The wireless terminal basically splits a data stream into multiple concurrent divisions, called spatial streams, and delivers each one of them through separate antennas to corresponding antennas on the receiving end. The current 802.11n draft provides for up to four spatial streams, even though compliant hardware is not required to support that many. Doubling the number of spatial streams from one to two effectively doubles the raw data rate. There are trade-offs, however, such as increased power consumption and, to a lesser extent, cost. The 802.11n specification will include a MIMO power-saving mode which mitigates power consumption by using multiple paths only when communication would benefit from the additional performance. The MIMO power-saving mode is expected to be a compulsory feature in the ratified IEEE 802.11n final specifications. A non compulsory mode in 802.11n that effectively doubles the offered data rate is deployment of the communication channel of 40 MHz bandwidth. The main trade-off here is less channel availability for other mobile stations. In the case of the 2.4-GHz frequency band, there is enough room for three nonoverlapping 20-MHz channels. Needless to say, a 40-MHz channel does not leave much room for other devices to join the network or transmit in the same airspace. This means that dynamic radio resource management is critical to ensure that the 40-MHz channel option improves the overall system performance by balancing the high bandwidth demands of some clients with the needs of other clients to remain connected to the network [2].

Note that in the MAC layer IEEE 802.11n offers BAs similar to the concepts recommended in the 802.11e amendment. By removing the need for one ACK for every data frame, the amount of overhead required for the ACK frames, as well as preamble and framing, is considerably reduced. BAs are helpful, but only if all the frames in a burst can be delivered without any problem. Missing one frame in the block or losing the ACK itself carries a steep penalty in protocol operations since the entire block must be retransmitted again. In addition, in 802.11n MAC, frame aggregation is also expected to be a mandatory MAC entity component. Combining several small layer 3 packets into a single relatively large frame improves the data-to-overhead ratio. Frame aggregation is often used with MAC header compression, since the MAC header on multiple frames to the same destination is quite similar.

1.4 SUMMARY AND CONCLUDING REMARKS

In this chapter, a brief overview of existing MAC and physical layers in WLANs was provided. Starting from the MAC layer, we reviewed the well-known IEEE 802.11 and its QoS-aware amendment 802.11e. We pursued our discussion of the IEEE 802.11 physical layer, and different types of physical layers with diverse system architectures and modulation schemes were explored.

ACKNOWLEDGMENTS

This work was partially supported by Nokia Foundation and Elisa Foundation.

REFERENCES

1. IEEE 802.11 WG, Part 11: "Wireless LAN medium access control (MAC) and physical layer (PHY) specification," IEEE, New York, Aug. 1999.
2. M. Gast, *802.11 Wireless Networks: The Definitive Guide*, 2nd ed., O'Reilly Media Inc., 2005.
3. IEEE 802.11e, Part 11: "Wireless LAN medium access control (MAC) and physical layer (PHY) specifications: amendment 8: Medium access control (MAC) quality of service enhancements," supplement to IEEE 802.11, IEEE, New York, Nov. 2005.
4. S. Mangold, S. Choi, G. R. Hiertz, O. Klein, and B. Walke, "Analysis of IEEE 802.11e for QoS support in wireless LANs," *IEEE Wireless Commun.* 10(6), 40–50 (2003).
5. Y. Xiao, "IEEE 802.11e: a QoS provisioning at the MAC layer," *IEEE Wireless Commun.* 11(3), 72–79 (2004).
6. Y. Xiao, "IEEE 802.11N: enhancements for higher throughput in wireless LAN," *IEEE Wireless Commun.* 12(6), 82–91 (2005).
7. Y. Xiao, "Efficient MAC strategies for the IEEE 802.11n wireless LANs," *Wireless Commun. Mobile Comput.* 6(4), 453–466 (2006).
8. IEEE 802.11b, Part 11: "Wireless LAN medium access control (MAC) and physical layer (PHY) specifications: High-speed physical layer extension in the 2.4 GHz band," supplement to IEEE 802.11, IEEE, New York, Sept. 1999.
9. IEEE 802.11a WG, Part 11: "Wireless LAN medium access control (MAC) and physical layer (PHY) specification: high-speed physical layer in the 5 GHz band," IEEE, New York, Sept. 1999.

CHAPTER 2

FRAMEWORK FOR DECENTRALIZED WIRELESS LAN RESOURCE MANAGEMENT

JIANG XIE, IVAN HOWITT, and ANITA RAJA

The proliferation of wireless local area network (WLAN) deployments in enterprises, public areas, and homes will cause frequent geographical coverage overlap among multiple networks. A recent growing interest is the coordination among WLAN providers for efficient resource management over a large coverage area. While radio resource management for a single WLAN has been studied extensively, little research work addresses cooperative resource management over multiple WLANs. The lack of cooperative resource management can cause significant performance degradation due to inter-WLAN interference. Moreover, unbalanced loads among multiple networks can incur congestion in a few WLANs while foregoing unused excess resources in others. Hence, resource management among multiple WLANs can make the best use of available radio resources and accommodate more users systemwide. In this chapter, a fully decentralized cooperative resource management framework using multiagent systems for multiple WLANs in interference environments is explained that incorporates the predictability of network states and decentralized control through multiagent systems. The proposed framework emphasizes the underlying predictability of network conditions and promotes management solutions tailored to different interference environments. The impacts of both inter-WLAN cochannel interference and colocated interference sources from wireless personal area networks (WPANs) are considered.

2.1 INTRODUCTION

There has recently been a remarkable increase in the usage of IEEE 802.11–based WLANs [1] due to low cost, installation simplicity, and high data rates.

Emerging Wireless LANs, Wireless PANs, and Wireless MANs. Edited by Y. Xiao and Y. Pan

Many hot spots are emerging and multiple WLANs are being deployed within a small geographic vicinity such as office buildings, multitenant residential complexes, city downtown areas, and university campuses. Different WLANs in a particular area may be deployed by different operators. In such an environment with multiple WLANs coexisting, a growing interest is that WLAN providers may set up reciprocal agreements and coordinations so that mobile users may share the usage of multiple WLANs.

A direct benefit of resource sharing among multiple WLANs is the expansion of network coverage. A WLAN usually suffers limited communication range. By integrating different overlapping WLANs, WLAN providers can offer value-added inter-WLAN roaming services to subscribers who need wider roaming areas. Meanwhile, mobile users can roam among multiple networks, enjoying the wide-area wireless access. Another benefit of this integration is the cooperation of resource management among multiple WLANs. Radio resources of each network are usually managed independently. The lack of cooperative resource management can cause significant performance degradation due to *inter-WLAN interference* [2]. Moreover, unbalanced loads among multiple networks can incur congestion in a few WLANs while foregoing unused excess resources in others. Hence, resource sharing among multiple WLANs can make the best use of available radio resources and accommodate more users systemwide. At the same time, WLAN providers may also benefit from being able to improve service quality and network utilization through cooperative WLAN resource management.

WLANs and WPANs often operate in a shared spectrum, the 2.4-GHz unlicensed industrial, scientific, and medical (ISM) band. When a WLAN such as the IEEE 802.11b is colocated with a WPAN such as Bluetooth or the IEEE 802.15.4 low rate WPAN (LR-WPAN), the issue of coexistence between different wireless networks needs to be considered to ensure their performance requirements are maintained [3–6]. Colocation of wireless services may occur under a number of different scenarios. For example, a WPAN can be deployed to support a sensor array within the same location as an established WLAN. Alternatively, a hierarchical network structure based on the strengths of each wireless service is straightforward to envision, where WPANs support local connectivity and WLANs provide the backbone for multiple WPANs. Therefore, resource management of WLANs will operate in a dynamic radio frequency (RF) environment often involving a diverse set of colocated wireless services.

Colocated wireless networks operating in the same unlicensed frequency band can cause interference to each other because of spectral overlap. Interference sources will impact mobile stations differently due to variations in RF path loss. Even if stations are at fixed locations, dynamics in the environment will significantly impact the RF propagation characteristics. These variations make it difficult and costly, in terms of radio resources, to maintain performance requirements. Hence, it is imperative that the dynamic effects of interference be incorporated into network management and control decision making.

Network management can be implemented through a centralized or distributed control method or a hybrid one. A centralized method involves a centralized controller and is able to provide the global optimal solution but requires periodic global information gathering on network states at the centralized controller and has the weakness of scalability. On the other hand, a distributed method has the advantages of scalability and easy collection of local inputs but may lead to local optimal solutions and longer convergence time of solution finding. A recent research effort on distributed control is to apply the agent technology to intelligent network management and data harvesting. Agents are autonomous entities that receive sensory inputs from the environment and then act on it using their effectors based on the knowledge they have of the environment [7]. A multiagent system (MAS) allows for the distribution of knowledge, data, and resources among individual agents and its modularity supports the development and maintenance of complex highly reliable systems [8]. Multiagent systems are also easier to scale up as they can speed up computation due to concurrent processing; they have less communication bandwidth requirements since processing is located nearer the source of information and facilitate real-time responsiveness as processing, sensing, and effecting can be colocated. Several distributed control algorithms based on MASs are proposed to solve centralized control problems efficiently and proved to converge to the global optimal solution [9–12]. Various MASs have been deployed on wireless sensor networks (WSNs) and other distributed networks for data processing and energy conservation in an intelligent fashion [13–15]. However, no existing work has addressed using MASs for cooperative network management for multiple WLANs and incorporated the dynamics of the interference environment into control decision making.

Although a considerable amount of research on radio resource management in a single WLAN has been proposed [16–19], cooperative resource management for multiple WLANs remains largely unexplored. Moreover, few protocols and algorithms incorporate the prediction of dynamic RF operational statistics from interference environments and investigate the application of MASs for distributed intelligent network management. Therefore, predictability-based cooperative resource management for multiple WLANs using MASs is proposed in this chapter. Predictability-based approaches may capture the effects of the time-varying nature of network links. It also helps determine the degree to which the state of the network can be reliably observed. By using predictability-based management approaches with the help of MASs, the changing operating conditions of multiple networks and the potential interference to WLANs from diverse colocated devices can be captured in advance, and this information can be distributed timely through multiple agents, which may help the decision making of resource management.

In this chapter, we focus on how to adaptively manage shared systemwide resources under time-varying network conditions among multiple WLANs in WLAN/WPAN interference environments. The impacts of both inter-WLAN cochannel interference and colocated interference sources from WPANs are considered. A fully decentralized resource management framework that incorporates the predictability of network states and the coordination between physical environment modeling and network management using MASs is proposed.

The rest of this chapter is organized as follows. In Section 2.2, existing work on WLAN resource management is described. In Section 2.3, a centralized multidomain WLAN resource management approach is first introduced. The goal of this centralized approach investigation is to get design insights and performance benchmark for the proposed decentralized approach. In Section 2.4, the proposed framework for decentralized WLAN resource management based on MASs is explained in detail, followed by the conclusions in Section 2.5.

2.2 EXISTING WORK ON WLAN RESOURCE MANAGEMENT

Resource management for WLANs includes dynamic channel assignment, dynamic transmit power control, and load balancing [16]. In this work, we focus on resource management for load balancing.

Resource management for load balancing in wireless networks has been extensively studied. In cellular networks, load balancing is usually achieved through dynamic channel allocation [20, 21]. This technique is not as suitable in WLANs where each access point (AP) normally uses one channel. Another approach is to use cell overlapping to reduce the blocking probability of calls and maximize the network utilization [22, 23]. In [24, 25], load balancing integrated with coordinated scheduling techniques for multicell packet data networks is proposed. However, these techniques consider different objective functions such as call blocking probability, which is not applicable for the load-balancing issue in the WLAN context [18].

Approaches for load balancing in a single WLAN can be classified into two categories. One is association control through which the network redistributes client associations among APs more or less uniformly so that no one AP is unduly overloaded [16]. The other is capacity control through which the network adjusts the maximum allowable throughput of each AP so that heavy-loaded APs can have more capacity to support users [2]. Three different techniques are proposed for association control. The *explicit channel switching* algorithm requests client stations to explicitly change their association from an overloaded AP to a less loaded neighboring AP [18, 19]. This algorithm trades off received signal strength with load by forcing stations to switch from an overloaded AP with a stronger signal to a lightly loaded AP with a possibly weaker signal within the radio range of the stations. In [17], an algorithm incorporating *transmit power control* and channel switching is proposed. Load balancing is achieved by adjusting transmit power of neighboring APs to change their radio coverage pattern. In this way, the coverage area of the overloaded AP is reduced, causing some of its client stations to handoff to lightly loaded APs with enlarged coverage areas. The third technique is *network-directed roaming* under which the network balances load by providing explicit feedback to users about where to roam to get the services they require [19]. This technique can achieve global load balancing over the entire WLAN, while the first two try to distribute load among neighboring APs.

All the above load-balancing schemes are designed for a single WLAN. They cannot be directly applied to multidomain WLANs because they cannot provide systemwide fair resource allocation among multiple networks. Colocated WLANs often use the same limited number of orthogonal channels (e.g., three orthogonal channels are available in IEEE 802.11b/g networks). Hence, the effect of interdomain cochannel interference becomes severe as the number of colocated WLANs increases. The load of a cell in one domain determines the level of its interference on other cochannel cells both inside and outside the domain. Therefore, achieving systemwide fair resource allocation should not be restricted to only independent load balancing per domain without global cooperation. It should incorporate the load and interference interactions between different domains. In [2], an interdomain radio resource management scheme for WLANs is proposed. This work provides an exciting insight into the optimization of resource sharing among multiple domains. However, the interactive effects of cochannel interference among multiple domains are not considered in the optimization process.

To the best of our knowledge, very little research work addresses cooperative resource management over multiple WLANs. Moreover, very little work on resource management for either single or multiple WLANs has considered the interference from possible colocated WPANs in the operational environment. As stated previously, WPAN interference sources require the WLAN to consume additional network resources in order to maintain performance requirements. Therefore, it is imperative that the dynamic effects of colocated WPAN interference be incorporated into network management.

2.3 THIRD-PARTY-BASED CENTRALIZED WLAN RESOURCE MANAGEMENT

Multidomain WLAN resource sharing and management can be implemented through either a centralized approach or a decentralized approach. Under the centralized approach, a centralized controller collects estimates of resource utilization and interference level from all APs in multiple WLANs and generates global optimal control decisions to feed back to each AP. The resource optimization under the centralized architecture can achieve the global optimal performance, which can be used as a performance benchmark for the decentralized approach.

Therefore, we first conduct research on using a centralized resource optimization approach for multidomain WLANs. Then, the performance results from this centralized approach will be used as a benchmark for our proposed decentralized approach to achieve the global optimal performance.

2.3.1 Third-Party-Based Centralized Architecture

We first propose a third-party-based centralized resource management architecture for the integration of multiple WLANs. A new entity, *local network controller* (*LNC*), is connected to all the APs of multiple WLANs, as shown in Fig. 2.1. WLANs under the control of an LNC form a *WLAN cluster*. The LNC acts as a radio resource coordinator across domains and takes care of issues related to interdomain roaming and resource sharing within a WLAN cluster. The LNC can integrate any number of WLANs belonging to different providers. As the number of domains in a WLAN cluster increases, the LNC can be built in a hierarchical structure to make it more scalable. As shown in Fig. 2.1 a *global network controller* (*GNC*) is connected to all LNCs supporting inter-WLAN cluster roaming and resource sharing. When a mobile station (MS) roams between WLANs in different WLAN clusters or when load balancing over multiple WLAN clusters needs to be addressed, the GNC is involved for resource management. A third-party agent can be the operator of the LNCs and GNC. It is responsible for the design, implementation, and maintenance of the control functions provided by the LNCs and GNC.

Providers of different WLANs in a WLAN cluster set up service-level agreements with the LNC. The operator of the LNC generates revenue from WLAN providers who agree to share their network resources with others. A similar business model is used by iPass [26] to provide global remote access services. Through the coordination of the LNC, providers may offer inter-WLAN roaming services to their subscribers as a value-added service feature. They can also support communications with better quality signals since the impact of interactive interference is globally balanced across the WLAN cluster through the control of the LNC. The functions related to user authentication, billing, security and privacy, and mobility management can be implemented in the LNC using similar models proposed in [27]. Here, we focus on how to fairly balance systemwide resources in order to accommodate more users with the least amount of cost.

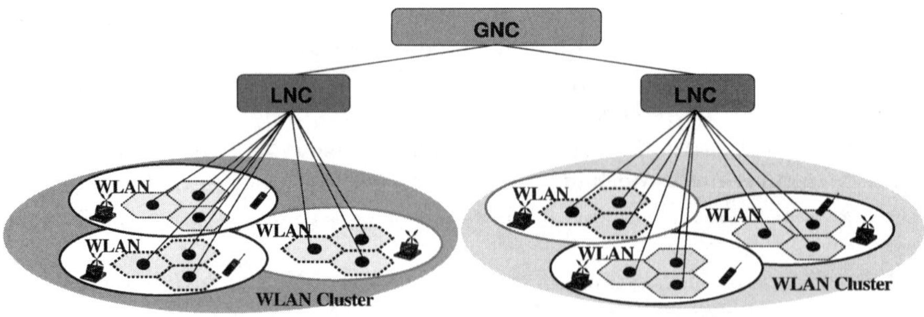

FIGURE 2.1 Third-party-based multidomain WLAN resource management architecture.

The LNC gathers the measured resource usage statistics from all the APs via simple network management protocol (SNMP) [28]. SNMP also provides security-related functions such as user authentication and message encryption [29]. Most enterprise-class APs can support SNMP [2]. APs collect signal characteristics from client stations in each domain. The IEEE 802.11k task group [30] is developing a radio resource measurement extension to the IEEE 802.11 WLAN standard. As suggested by the IEEE 802.11k task group, a portion of the signal characteristics are obtained directly from the WLAN cluster. The data can be augmented by an additional sensing network, potentially located at each AP, to provide additional data specifically associated with WPAN interference sources in the environment. The measured data can then be used by the LNC to generate the control decisions to optimize the performance of the entire WLAN cluster.

The LNC and APs periodically collect the required information for resource management. The LNC calculates the optimal resource allocation across domains and applies control decisions to APs. The decision making is updated periodically in order to address changes in the traffic load and interference environment. Note that the load at APs does not vary frequently if stations are not highly mobile. Previous studies on WLAN measurement and user behavior show that users have a quasi-static mobility pattern [31–33], which means users are free to move from place to place, but they tend to stay in the same physical locations for long time periods [18]. In addition, it is important to remark that the periodic data-collecting from stations does not imply measuring instantaneous small-scale multipath signal characteristics, which are very time sensive. Instead, measurements should be targeted at capturing large-scale changes in signal characteristics due to variations in traffic pattern, station mobility, interference sources, and interference mobility. In other words, the measurement is based on the factors that influence the LNC management of the WLAN performance. Therefore, control decisions need not be updated frequently and they should target long-term performance improvement. Thus, the control overhead resulting from the periodic updating can be kept at a reasonable level.

2.3.2 Resource Utilization Modeling and Optimization

In this section, we introduce the resource management scheme for multidomain WLANs under the third-party-based centralized architecture.

2.3.2.1 Motivation. We first use the example shown in Fig. 2.2 to explain the interactions between different WLAN domains due to cell load and cochannel interference.

In Fig. 2.2, two IEEE 802.11b WLANs, A and B, are colocated in the considered region. Each circle represents an AP: 13 APs for WLAN A and 12 APs for WLAN B. The locations of APs and the channels used by APs are shown in the figure. Since only three nonoverlapping channels are available, the

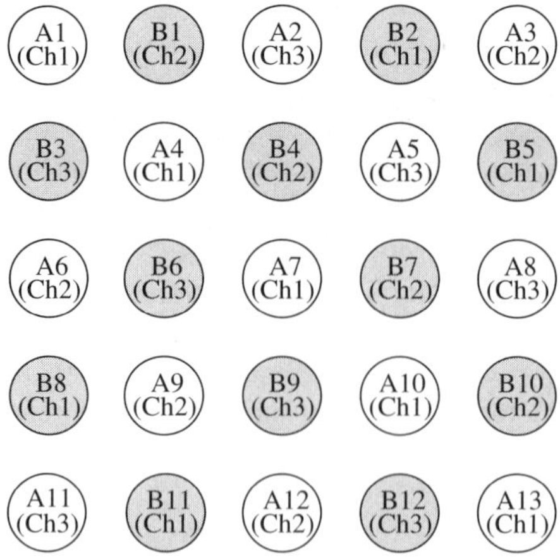

FIGURE 2.2 Cell layout of two colocated WLANs A and B.

channels are selected so that the APs using the same channel are geographically separated the furthest.

When multiple WLANs colocate, the traffic load carried by one WLAN will impact the resource utilization of other WLANs due to cochannel interference. This is illustrated in Fig. 2.2 by considering cell 4 in WLAN A, which operates on frequency channel 1. If offered traffic load is increased in this cell, then the impact will be felt in closeby cells sharing the same frequency channel, that is, cells 1 and 7 in WLAN A and cells 2 and 8 in WLAN B. Depending on the RF propagation characteristics, it is possible for additional cells to be impacted by interference signals that are of sufficient strength to cause contention within nearby cells. Therefore, stations in the impacted area contend for the channel based on the additional interference traffic. From the figure, it can be observed that optimizing the load in domain A independent of the resource requirement of domain B is likely to impact domain B's performance. As a result, load-balancing techniques designed for one domain may not be suitable for multiple colocated WLANs if the interrelationship between colocated domains is not taken into consideration. This is the motivation for developing a new resource management scheme in a multidomain environment.

2.3.2.2 Overview of the Resource Management Scheme. The goal of the resource management scheme is to minimize the total system cost by adjusting resource allocation in each domain. The cost is what the system needs to pay to support all the client stations to achieve performance requirements. It is related to the available radio resources for supporting the offered load in each domain and mitigating interference from the operational environment. The LNC manages resource sharing across domains by controlling the maximum allowable throughput of each AP. When the maximum allowable throughput at an AP changes, the available radio resources of the cell is limited. Consequently, the cell utilization changes, which leads to a different system cost. Therefore, minimizing the overall system cost is equivalent to finding the optimal allowable throughput at each AP. In addition, WPAN interference can adversely affect the WLAN performance by changing its resource utilization requirements and thereby needs to be considered. Moreover, due to the dynamics in the RF environment, signal characteristics, traffic load, and interference intensity are time variant. As a result, the optimal resource allocation decision should be dynamically adjusted to reflect the influences of the time-varying environment.

The resource management scheme under the third-party-based centralized architecture includes three steps. First, based on the overall traffic load distribution at all the APs in a WLAN cluster, the impact of cochannel interference at each cell can be calculated. Then, by incorporating the impact of interference from other sources in the operational environment, the communication cost of the overall system can be derived, which is a function of cell load, cochannel interference, and interference from other wireless services. Second, the LNC finds the optimal pattern of maximum allowable throughput at each AP in multiple domains. In other words, the LNC decides which AP should provide how much capacity to its users. This optimal throughput pattern results in the minimum system cost. Finally, the LNC sends control signals to APs to instruct them on how to update their allowable resources for users based on the calculated optimal throughput. Therefore, by the global control of the LNC on restricting the maximum allowable throughput at APs, the effect of cochannel interference is balanced across multiple domains and thereby the overall system resource utilization is minimized. Note that it is up to the individual WLANs to determine which method should be used to achieve the optimal throughput, for example, load balancing within the domain, limiting average throughput, and the like.

2.3.3 Problem Formulation

There are four possible resource management scenarios in multidomain WLANs as follows:

1. *Intradomain* resource optimization *without* the consideration of potential interference from colocated WPANs
2. *Intradomain* resource optimization *with* the consideration of potential interference from colocated WPANs
3. *Interdomain* resource optimization *without* the consideration of potential interference from colocated WPANs
4. *Interdomain* resource optimization *with* the consideration of potential interference from colocated WPANs

Scenarios 1 and 2 are the cases in which each domain optimizes radio resource usage independently. The already proposed load-balancing techniques designed for a single WLAN can be applied to scenario 1. Scenarios 3 and 4 involve the LNC to help control the resource allocation in each domain.

The intra- and interdomain load-balancing issue can be formulated as an optimization problem. The LNC controls the maximum allowable throughput of each AP. It periodically optimizes the resource usage in each domain by minimizing the overall system cost function $F(\cdot)$ where $F(\cdot)$ is the total communication cost the system needs to pay for supporting all the client stations. More specifically, assume there are M domains in a WLAN cluster. Let $\mathbf{N}^{\mathrm{T}} = [N_1, N_2, \ldots, N_M]$ be the number of APs in each domain. Hence, for a particular domain j, there are N_j APs in the network. Let $\mathbf{C}_j^{\mathrm{T}} = [C_{1j}, C_{2j}, \ldots, C_{N_j j}]$ be the maximum allowable throughput of each AP in domain j. The objective function and the constraints for intradomain resource optimization in domain j are

$$\begin{aligned} &\text{Minimize } F_{\text{intra}}(\mathbf{C}_j) = \sum_{i=1}^{N_j} \Phi\big(f_{ij}(C_{ij}, I_{\text{co\_ij}}, I_{\text{e\_ij}})\big) \\ &\text{Subject to } f_{ij}(\cdot) \le 1 \\ &\qquad \mathbf{e}^{\mathrm{T}}\mathbf{C}_j \le T_{D_j} \\ &\qquad C_{\min} \le C_{ij} \le C_{\max} \end{aligned} \tag{2.1}$$

where $f_{ij}(C_{ij}, I_{\text{co\_ij}}, I_{\text{e\_ij}})$ is the normalized resource usage at cell i in domain j, $I_{\text{co\_ij}}$ is the cochannel interference from other cells in the domain to cell i, and $I_{\text{e\_ij}}$ is the interference from other colocated WPANs to cell i. The resource usage at a cell is determined by the cell utilization U, cochannel utilization I_{co}, and environmental interference I_e. The cell utilization U can be approximated by calculating the ratio of the measured load to the maximum allowable throughput [2], that is, $U_{ij}(I_{\text{e\_ij}}) = \rho_{ij}(I_{\text{e\_ij}})/C_{ij}$, where ρ_{ij} is the time-varying measured load at cell i in domain j. Thus, $f_{ij}(C_{ij}, I_{\text{co\_ij}}, I_{\text{e\_ij}})$ is

$$\begin{aligned} f_{ij}(C_{ij}, I_{\text{co\_ij}}, I_{\text{e\_ij}}) &= U_{ij}(I_{\text{e\_ij}}) + I_{\text{co\_ij}}(I_{\text{e\_ij}}) \\ &= \frac{\rho_{ij}(I_{\text{e\_ij}})}{C_{ij}} + I_{\text{co\_ij}}(I_{\text{e\_ij}}) \end{aligned} \tag{2.2}$$

where $U_{ij}(\cdot)$, $\rho_{ij}(\cdot)$, and $I_{\text{co\_ij}}(\cdot)$ are functions of the WPAN interference as presented in Section 2.3.3.2. $\Phi(\cdot)$ is a mapping function to map the resource usage to cost. It should be chosen as a convex function to get an effective strategy to facilitate optimization [34]. $\mathbf{e}^{\mathrm{T}}$ is a unit vector with adequate dimension and $\mathbf{T}_D^{\mathrm{T}} = [T_{D_1}, T_{D_2}, \ldots, T_{D_M}]$ is the maximum capacity of each domain. If no interdomain resource cooperation, the maximum domain throughput is the maximum data rate that can be supported by WLAN products. $C_{\min}$ and $C_{\max}$ are the minimum and maximum allowable throughput in each cell, respectively. Equation (2.1) shows that for intradomain resource optimization, the LNC finds the maximum allowable throughput for each AP in the domain, which results in the minimum communication cost.

For interdomain resource optimization, the LNC not only finds the optimal throughput pattern for all the APs but also determines the optimal capacity for each domain. In other words, the LNC optimizes the global resource sharing among multiple domains by finding the optimal $\mathbf{T}_D^{\mathrm{T}} = [T_{D_1}, T_{D_2}, \ldots, T_{D_M}]$, which leads to the minimum overall system cost. The objective function and the constraints for interdomain resource optimization in domain j are

$$\begin{aligned} \text{Minimizing} \quad & F_{\text{inter}}(\mathbf{C}_j) = \sum_{i=1}^{N_j} \Phi\left(\tilde{f}_{ij}(C_{ij}, \tilde{I}_{\text{co\_ij}}, I_{\text{e\_ij}})\right) \\ \text{Subject to} \quad & \tilde{f}_{ij}(\cdot) \leq 1 \\ & \mathbf{e}^{\mathrm{T}}\mathbf{C}_j \leq T_{D_j} \\ & C_{\min} \leq C_{ij} \leq C_{\max} \end{aligned} \tag{2.3}$$

Here, $\tilde{f}_{ij}$ is different from that in (2.1) since $\tilde{I}_{\text{co\_ij}}$ includes interference from cochannel cells both inside and outside domain j. The optimal allowable throughput $\mathbf{C}_j^*$ is given by

$$\mathbf{C}_j^* = \arg\min_{\mathbf{C}_j} F_{\text{intra}}(\mathbf{C}_j) \tag{2.4}$$

or

$$\mathbf{C}_j^* = \arg\min_{\mathbf{C}_j} F_{\text{inter}}(\mathbf{C}_j) \tag{2.5}$$

After the LNC finds the optimal resource allocation, resources at each domain should be updated. Therefore, at each time step, T_{D_j} is updated to optimize the global resource sharing. Assume $T_{D_j}^{(m)}$ is the maximum capacity of domain j after time step m. Then at time step $m+1$, T_{D_j} should be updated based on

$$T_{D_j}^{(m+1)} = T_{D_j}^{(m)} + \alpha(\psi_{D_j} - \widehat{E}[\Psi_D]) \tag{2.6}$$

where α is a constant controlling the update rate, ψ_{D_j} is the cross-domain impairment for domain j, which is obtained by

$$\psi_{D_j} = F_{\text{inter}}(\mathbf{C}_j) - F_{\text{intra}}(\mathbf{C}_j) \tag{2.7}$$

and $\widehat{E}[\Psi_D]$ is the sample mean over $\Psi_D = [\psi_{D_1}, \ldots, \psi_{D_j}, \ldots, \psi_{D_M}]$. Hence, $\widehat{E}[\Psi_D] = (1/M)\sum_{j=1}^{M} \psi_{D_j}$. The LNC periodically finds the optimal resource allocation and instructs APs to update resources based on (2.6).

2.3.3.1 *Interference from Colocated WLANs.* The method used for deriving $\tilde{I}_{\text{co\_ij}}$ is the same as presented in [2], which is

$$\tilde{I}_{\text{co\_ij}} = \sum_{d=1}^{M} \sum_{\substack{\ell=1 \\ \ell \neq i}}^{N_d} \frac{\delta_{\ell d|ij}\, U_{\ell d}\, S_{\ell d|ij}}{S_{ij}} \tag{2.8}$$

where

$$\delta_{\ell d|ij} = \begin{cases} 1 & \text{AP frequency channels are the same for} \\ & \text{cell } \ell \text{ in domain } d \text{ and cell } i \text{ in domain } j \\ 0 & \text{otherwise} \end{cases} \tag{2.9}$$

and N_d is the number of APs in the dth domain, S_{ij} is the coverage area for cell i in domain j, and $S_{\ell d|ij}$ is the overlap region between cell i in domain j coverage area and the interference area of cell ℓ in domain d. Every AP is assigned a single frequency channel and MSs are assumed to be uniformly distributed within the AP's coverage area. MSs are associated with their nearest AP. The coverage area and interference area are approximated by circles. As in [2], the ITU-R P.1238-2 indoor path loss model was used in evaluating $S_{\ell d|ij}$, which is expressed as

$$P_L = 20 \log_{10} f + 10n \log_{10} d - 28 \quad \text{(dB)} \tag{2.10}$$

where P_L is the RF signal propagation path loss based on distance d between the AP and the MS, f is the carrier frequency in megahertz, and n is the path loss exponent.

In a similar fashion to (2.8), $I_{\text{co\_ij}}$ is defined based on the cochannel interference within domain j only:

$$I_{\text{co\_ij}} = \sum_{\substack{\ell=1 \\ \ell \neq i}}^{N_d} \frac{\delta_{\ell j|ij}\, U_{\ell j}\, S_{\ell j|ij}}{S_{ij}} \tag{2.11}$$

2.3.3.2 *Interference from Colocated WPANs.* Next, we explain how to obtain the interference from colocated WPANs, I_e. Figure 2.3 illustrates a general scenario in which cell i in domain j (located at $x_{\text{AP\_ij}}$) and an associated MS (located at $x_{\text{STA\_ij}}$) are colocated with the kth WPAN (located at $x_{\text{BT\_k}}$). We use Bluetooth technology as an example of WPANs to derive I_e here. Due to the WPAN interference, packet retransmissions can be required in order to successfully transmit a packet between the AP and MS. The packet retransmission, in essence, increases the traffic load within the cell and thereby increases

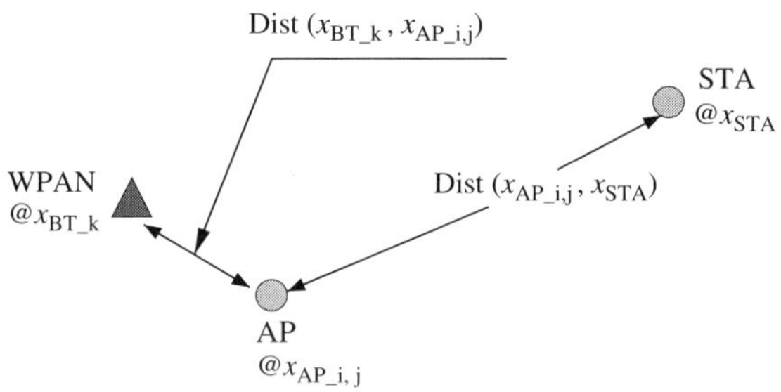

FIGURE 2.3 Cell layout with colocated WPAN interferers.

the utilization of the cell, that is,

$$U_{ij}(I_{e\_ij}) = \frac{\rho_{ij}(I_{e\_ij})}{C_{ij}} = \frac{\rho_{ij}\overline{N}_{Tx}(i,j)}{C_{ij}} \tag{2.12}$$

where $\overline{N}_{Tx}(i,j)$ is the expected number of transmissions required to successfully transmit a packet within cell i in domain j based on the local WPAN interference environment. The expected number of transmissions can be evaluated by

$$\overline{N}_{Tx}(i,j) = 1 + \frac{\Pr[C|i,j]}{1 - \Pr[C|i,j]} = \frac{1}{1 - \Pr[C|i,j]} \tag{2.13}$$

where $\Pr[C|i,j]$ is the probability of requiring a packet retransmission due to interference from one or more WPAN interference sources, that is, the probability of collision is given by

$$\Pr[C|i,j] = \sum_{k=1}^{W} \Pr[C_k|i,j] - \sum_{l=1}^{W} \sum_{\substack{k=1 \\ k \neq l}}^{W} \Pr[C_k|i,j] \Pr[C_l|i,j] + \cdots \tag{2.14}$$

where W is the number of colocated WPAN interferers. Equation (2.14) assumes the collision probabilities, $\Pr[C_k|i,j]$, for each of the active interference sources are independent. $\Pr[C_k|i,j]$ takes into account the dynamics between the interference signal's characteristics and the desired signal's characteristics at the intended receiver.

In order to evaluate $\overline{N}_{Tx}(i,j)$ based on (2.13), the collision probability $\Pr[C_k|i,j]$ needs to model the specific interference scenario. For the scenario where WPANs based on Bluetooth technology are operated in an IEEE 802.11b WLAN environment, the collision probability is associated with the likelihood that a Bluetooth packet and an IEEE 802.11b packet are time and frequency coincident, and the interference signals have sufficient power to cause an error. In evaluating $\Pr[C_k|i,j]$, the likelihood that the interference and desired signals are both time and frequency coincidence needs to be determined such that the interference signal has sufficient power to cause an error in the desired packet's reception. Based on the relative timing between the IEEE 802.11b packet transmission with the Bluetooth packet frame timing, the number of Bluetooth packets time coincident with the 802.11b packet transmission is either n_r or n_r-1 with corresponding probabilities $\Pr[n_r]$ and $\Pr[n_r-1]$. Based on typical packet lengths for each of the wireless standards, n_r is either 1 or 2 [3]. These two events are independent. Therefore,

$$\Pr[C_k|i,j] = \Pr[n_r]\ \Pr[C_k|i,j,n_r] + \Pr[n_r-1]\ \Pr[C_k|i,j,n_r-1] \tag{2.15}$$

where $\Pr[C_k|i,j,n_r]$ is the probability of collision given the IEEE 802.11b packet overlaps in time with n_r Bluetooth packets. Since Bluetooth transmissions are based on frequency hopping, the likelihood of frequency coincidence for each of the n_r packets is independent. Therefore,

$$\Pr[C_k|i,j,n_r] = 1 - (1 - L_{\mathrm{BT}} \Pr[C_f|i,j,\Omega_{\mathrm{I/S}}(i,j,x_{\mathrm{BT\_k}})])^{n_r} \tag{2.16}$$

where the parameter L_{BT} models the loading factor for a given Bluetooth piconet, that is, the percentage of time slots utilized by both the master and the slaves. $\Pr[C_f|i,j,\Omega_{\mathrm{I/S}}(i,j,x_{\mathrm{BT\_k}})]$ is the probability the interfering signal is frequency coincident with sufficient power to cause interference within cell i in domain j. The term $\Omega_{\mathrm{I/S}}(i,j,x_{\mathrm{BT\_k}})$ represents the interference-to-signal power ratio (I/S) in decibels at the receiver. For this study, the I/S is characterized for each cell by evaluating the received signal power at the AP from a typical MS located within its coverage area. The I/S expressed in decibels is given by

$$\Omega_{\mathrm{I/S}}(i,j,x_{\mathrm{BT\_k}}) = \Omega_{\mathrm{BT\_k}} - \Omega_{\mathrm{AP}} - 10n\log_{10}\left(\frac{\mathrm{Dist}_E(x_{\mathrm{AP}},x_{\mathrm{BT\_k}})}{\overline{\mathrm{Dist}_E(x_{\mathrm{AP}},x_{\mathrm{STA}})}}\right) \tag{2.17}$$

where Ω_{AP} and $\Omega_{\mathrm{BT\_k}}$ are typical IEEE 802.11b and Bluetooth transmit powers, respectively, expressed as the power ratio in decibels referenced to 1 mW (dBm), n is the path loss exponent, $\mathrm{Dist}_E(x,y)$ is the Euclidean distance between x and y, and $\overline{\mathrm{Dist}_E(x_{\mathrm{AP}},x_{\mathrm{STA}})}$ is the expected distance between the AP and an MS within its coverage area. Based on the Bluetooth hopping sequence uniformly covering the ISM band with bandwidth B_{UL}, $\Pr[C_f|i,j,\Omega_{\mathrm{I/S}}(i,j,x_{\mathrm{BT\_k}})]$ is expressed as

$$\Pr\left[C_f|i,j,\Omega_{\mathrm{I/S}}(i,j,x_{\mathrm{BT\_k}})\right] = \frac{2}{B_{\mathrm{UL}}}\int_0^{B_{\mathrm{UL}}/2} \Pr[\Omega_{\mathrm{I/S}}(i,j,x_{\mathrm{BT\_k}}) \geq \gamma(f_{\mathrm{offset}})|f_{\mathrm{offset}}]\, df_{\mathrm{offset}} \tag{2.18}$$

where $\gamma(f_{\text{offset}})$ is a random variable representing the susceptibility of the 802.11b receiver to Bluetooth interference based on the frequency offset, (f_{offset}), between the two signals' carrier frequencies. $\gamma(f_{\text{offset}})$ is modeled as a Gaussian random variable based on analysis with empirical data.

2.3.3.3 Summary. The procedure of the multidomain WLAN resource management scheme under the third-party-based centralized architecture is summarized in Fig. 2.4.

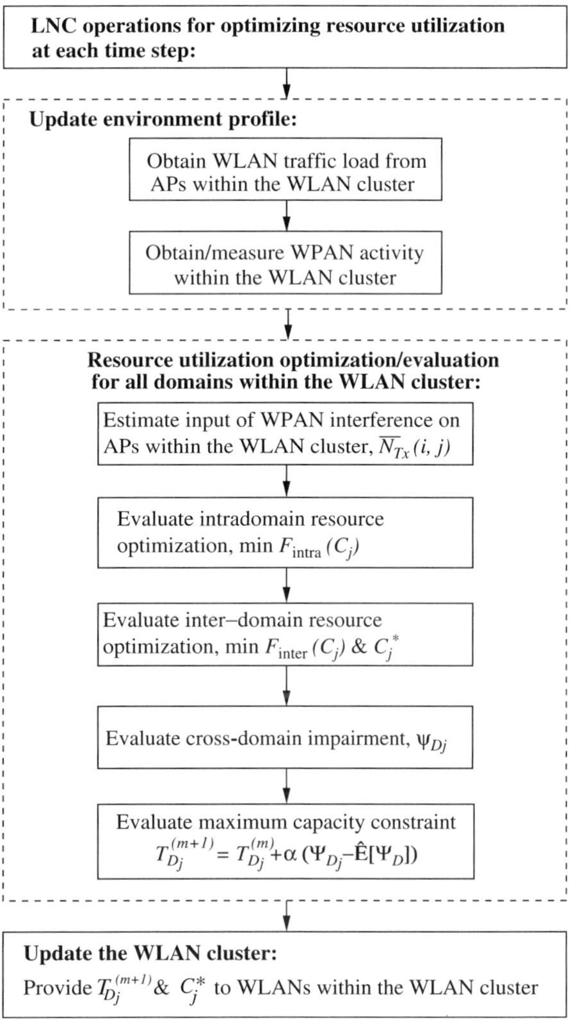

FIGURE 2.4 Procedure of the interdomain WLAN resource management scheme.

2.3.4 Performance Evaluation

In this section, we conduct simulation to demonstrate the performance improvement of the interdomain resource management scheme, compared with the intradomain schemes and the interdomain scheme but without the consideration of interference from colocated WPANs.

2.3.4.1 Simulation Environment. We simulate a two-domain WLAN environment with WLAN A and B colocated. Both WLANs are IEEE 802.11b–compliant networks. The locations of APs and the channels used by APs are the same as shown in Fig. 2.2. Multiple Bluetooth nodes are also colocated with the two WLANs. Their communications interfere with each other.

We define the *X* and *Y* axes as shown in Fig. 2.5. AP1 of WLAN A is located at the origin (0, 0). Each AP is separated by 30 m. APs of domain A and domain B are placed alternately: 13 APs of domain A and 12 APs of domain B. Their channels are selected so that APs using the same frequency channel are separated geographically the furthest. The coverage area of each AP is approximated as a circle with radius of 30 m. Two Bluetooth nodes are located

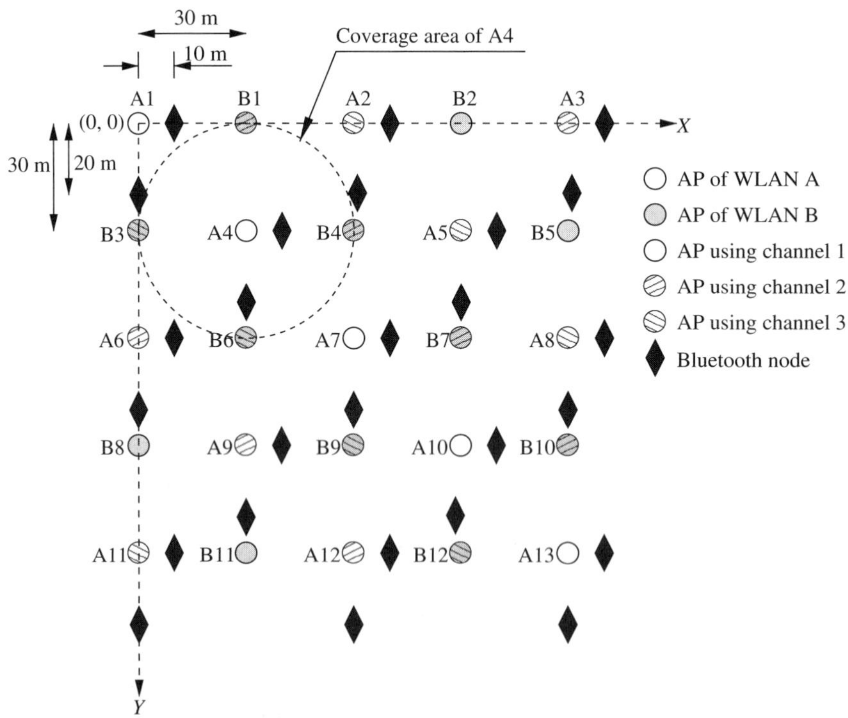

FIGURE 2.5 Simulation environment with colocated 802.11b and Bluetooth.

TABLE 2.1 Simulation parameters

	IEEE 802.11b WLAN		
	Domain A		Domain B
Number of APs	13		12
Radio frequency		2.4 GHz	
Number of channels		3	
Transmit power		15 dBm	
Cell radius		30 m	
Carrier sense threshold		−82 dBm	
	Bluetooth		
Radio frequency		2.4GHz	
Transmit power		10 dBm	
	Other Parameters		
Path loss exponent		3	
Measurement interval		5 min	

10 m away in the X direction and 20 m away in the Y direction to each AP of domain A, respectively. An example of the simulated WLAN and WPAN coexistence environment is an office building with two WLANs deployed. Each office room may have a Bluetooth device, for example, a Bluetooth-enabled palm pilot, a laptop with Bluetooth interface, or a Bluetooth headset, causing interference to WLAN communications.

The simulation parameters for WLANs and the Bluetooth are listed in Table 2.1. Using the WLAN outlined in Table 2.1 with (2.10), the WLAN cochannel interference radius is $\sim 82m$. This is the radius within which one AP will impact another AP's performance given they are cochannel. The impact of Bluetooth interference is based on evaluating (2.13), using (2.14) through (2.18) and evaluating the $\Pr[C_k| \cdot]$ versus I/S. In evaluating I/S, $\overline{\text{Dist}_E}(x_{\text{AP}}, x_{\text{STA}}) = (\text{cell radius})/\sqrt{2}$.

2.3.4.2 *Traffic Load Characterization.* For 802.11b-compliant systems, the maximum data rate at each cell is 11 Mbps. However, due to the physical (PHY) and media access layer (MAC) layer overhead, the net throughput is approximately 6 Mbps [35]. Based on the study on a campus WLAN shown in [2], the traffic load at an AP possesses characteristics of the truncated Pareto distribution with cutoff values equal to the upper limit of the MAC layer throughput, that is, 6 Mbps for IEEE 802.11b WLAN. The cumulative distribution function (cdf) of a generalized Pareto distribution is

$$P(x) = 1 - \left(\frac{\alpha}{x}\right)^{\beta} \tag{2.19}$$

where α and β are the location and scale parameters, respectively [36]. In addition, the traffic burst duration at an AP also follows a Pareto distribution.

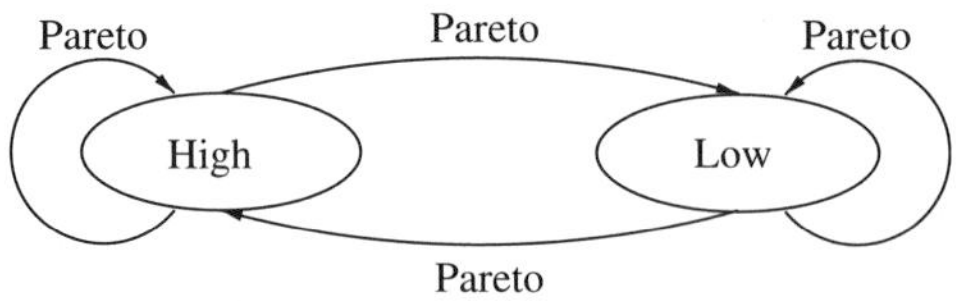

FIGURE 2.6 Traffic model for simulation.

Therefore, we use the two-state Markov traffic model shown in Figure 2.6 for our simulation. There are two Pareto distributions involved in the model: one for the traffic load with a cutoff value at 6 Mbps and the other for the HIGH/LOW state duration. The traffic is generated at both states with a burst threshold of 100 kbps, which means, when the generated traffic load is less than 100 kbps, we assume the AP is at the LOW state. The parameters used for generating the two Pareto distributions are listed in Table 2.2.

The Bluetooth traffic model is also based on a Markov model. Activity is checked on a 5-min interval and the traffic switches from an ON to an OFF state with probability given in Table 2.2. While in the ON state, the traffic load within the Bluetooth piconet corresponds to the $\Pr[C_k| \cdot]$. Details of the traffic load model can be found in [3].

2.3.4.3 Simulation Results. In the following, we demonstrate the performance of the four different resource management schemes, that is, *intradomain* resource optimization *without* the consideration of interference from colocated WPANs [in short, intra|(no WPAN)], *intradomain* resource optimization *with* the consideration of interference from colocated WPANs (in short, intra|WPAN), *interdomain* resource optimization *without* the consideration of interference from colocated WPANs [in short, inter|(no WPAN)], and our

TABLE 2.2 Traffic parameters

	IEEE 802.11b WLAN		
	Domain A		Domain B
HIGH-state load α	0.0096 Mbps		0.0054 Mbps
HIGH-state load β	0.61		0.75
HIGH-state load cutoff		6 Mbps	
HIGH-state duration α		2.1 min	
HIGH-state duration β		0.89	
LOW-state duration α		5.3 min	
LOW-state duration β		0.51	
	Bluetooth		
Probability of node active		0.6	

proposed *interdomain* resource optimization *with* the consideration of interference from colocated WPANs (in short, inter|WPAN).

2.3.4.3.1 Intradomain Optimization without I_e. Figure 2.7 shows the total system cost of the intra|(no WPAN) scheme. The solid line represents the sum of $F_{\mathrm{intra}}(\cdot|(\text{No WPAN}))$ for each domain, where $F_{\mathrm{intra}}(\cdot|(\text{No WPAN}))$ is obtained through (2.1) with $I_e = 0$, that is, $N_{Tx}(i,j) = 1\,\forall i, j$. Since this scheme does not consider the impact of interdomain cochannel interference and interference from colocated WPANs, we reevaluate the results by evaluating the optimal throughput $(\mathbf{C}_j^*)_{\mathrm{intra}|(\text{no WPAN})}$ using the interdomain cost function with WPAN interference, that is,

$$\left(\mathbf{C}_j^*\right)_{\mathrm{intra}|(\text{no WPAN})} = \arg\min_{\mathbf{C}_j} F_{\mathrm{intra}}\left(\mathbf{C}_j | I_e = 0\right) \tag{2.20}$$

and results reevaluated with

$$F_{\mathrm{inter}}\left((\mathbf{C}_j^*)_{\mathrm{intra}|(\text{no WPAN})} | I_e \text{ based on WPAN interference}\right) \tag{2.21}$$

The new cost is illustrated by the dashed line in the figure. It is shown in the figure that the new cost is alway higher or equal to the optimal system cost supported by the intra|(no WPAN) scheme. The gap between the two lines indicates the extra cost the intra|(no WPAN) scheme should pay for

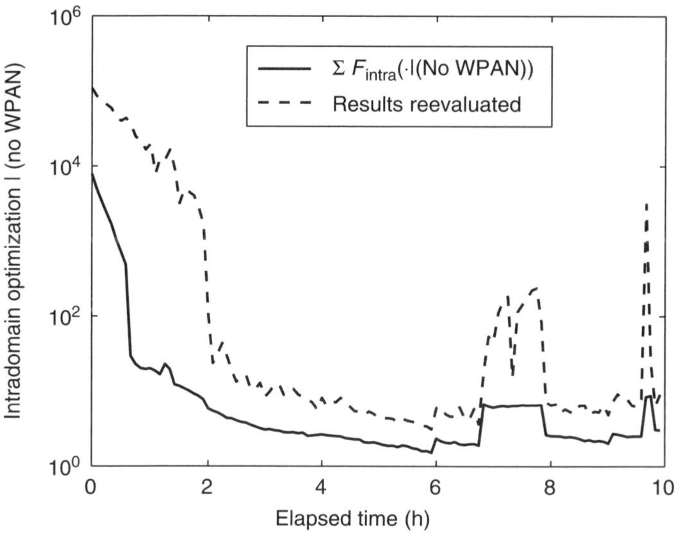

FIGURE 2.7 Comparison of total cost for intradomain optimization without consideration of WPAN interference.

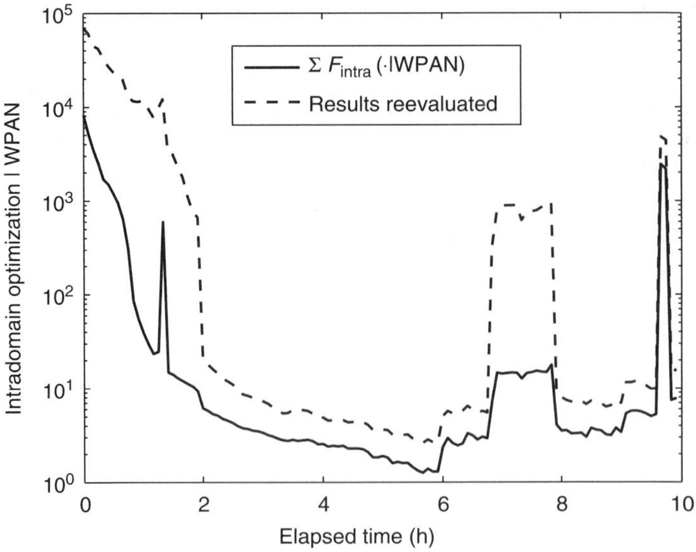

FIGURE 2.8 Comparison of total cost for intradomain optimization with consideration of WPAN interference.

interference mitigation in order to achieve the performance requirements expected by the scheme.

2.3.4.3.2 Intradomain Optimization with I_e. Figure 2.8 presents the total system cost of the intra|WPAN scheme. The solid line represents the sum of $F_{\text{intra}}(\cdot\,|\text{WPAN})$ for each domain, where $F_{\text{intra}}(\cdot\,|(\text{No WPAN}))$ is obtained through (2.1). Similar to the above case, we incorporate the impact of interdomain interference into the optimal cost provided by the intra|WPAN scheme and show through the dashed line the actual cost the system should pay in order to achieve the expected performance. Figures 2.7 and 2.8 tell us that independent resource optimization inside each domain cannot handle interdomain interactions. If the system does not spend extra resources to reduce the effect from interdomain interference, the communication quality will be lowered and the overall system resource utilization is not minimized for the offered traffic

2.3.4.3.3 Interdomain Optimization without I_e. Next, we present the performance of resource management when taking into account the cochannel interference from other colocated WLANs. Figure 2.9 plots the total system cost of the inter|(No WPAN) scheme in the solid line and the reevaluated data by incorporating the impact of interference from colocated WPANs in the dashed line. It is observed from the figure that the gap between the two lines is smaller than those in Figs. 2.7 and 2.8. The optimization control decisions made by the inter|(No WPAN) scheme are based on not only the cell utilization

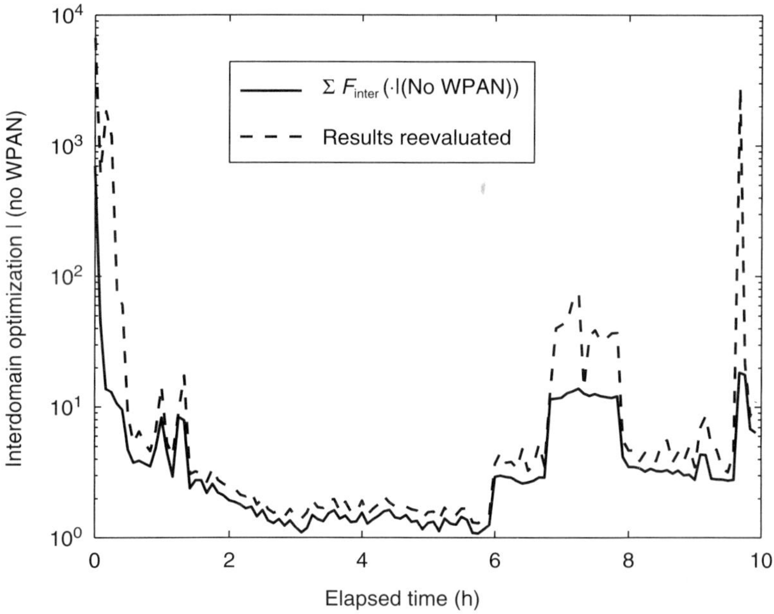

FIGURE 2.9 Comparison of total cost for interdomain optimization without consideration of WPAN interference

caused by the stations communicating in the cochannel cells inside each domain but also the interference caused by the stations communicating in other colocated WLANs using the same frequency channel. Therefore, comparing with the intra|(No WPAN) and intra|WPAN schemes, the inter|(No WPAN) scheme results in lower system cost to maintain the same performance quality.

2.3.4.3.4 Interdomain Optimization with I_e. Finally, we demonstrate the performance of our proposed interdomain resource management with the consideration of environmental interference, that is, the inter|WPAN scheme. Figure 2.10 plots the performance comparison of three schemes. The solid line is the cost difference of the intra|(No WPAN) scheme to the proposed inter|WPAN scheme, that is, $\sum F_{\text{intra}}(\cdot|(\text{No WPAN})) - \sum F_{\text{inter}}(\cdot|\text{WPAN})$, while the dashed line is the cost difference of the inter|(No WPAN) scheme to the inter|WPAN scheme, that is, $\sum F_{\text{inter}}(\cdot|(\text{No WPAN})) - \sum F_{\text{inter}}(\cdot|\text{WPAN})$. The figure shows that the other two schemes always pay higher cost than the proposed inter|WPAN scheme since the cost difference is always larger than zero. The proposed interdomain scheme can save up to 99.8% and 47.3% cost compared to the intra|(no WPAN) scheme and the inter|(No WPAN) scheme, respectively. The results indicate that the interdomain cooperative resource management scheme is more cost efficient for a WLAN/WPAN interference environment.

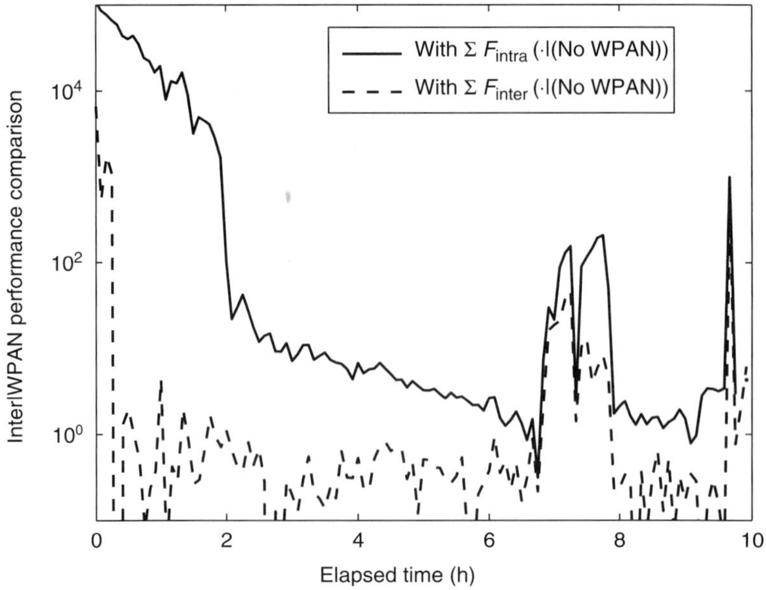

FIGURE 2.10 Cost difference of intradomain optimization without I_e to interdomain optimization with I_e as well as interdomain optimization without I_e to interdomain optimization with I_e.

2.4 DECENTRALIZED WLAN RESOURCE MANAGEMENT USING MULTIAGENT SYSTEMS

The centralized architecture often has the single-point failure and scalability problem. In order to achieve managing resources fairly among multiple WLANs through a fully decentralized way, a MAS-based approach is proposed to achieve information sharing and decision distribution among multiple WLANs in a distributed manner. WLAN providers may set up service-level agreements among themselves on how much data can be exchanged among agents. Compared to using a centralized controller, a MAS-based approach is more scalable.

2.4.1 Multiagent-Based Architecture

We propose a resource management architecture for multiple WLANs using multiagent systems, as shown in Fig. 2.11. Multiple WLANs are colocated within a particular geographic area. Communications inside the surrounding WPANs such as Bluetooth networks and WSNs generate interference to WLAN activities. An agent is located inside each AP within each WLAN and interacts with agents within its neighborhood. An agent's neighborhood

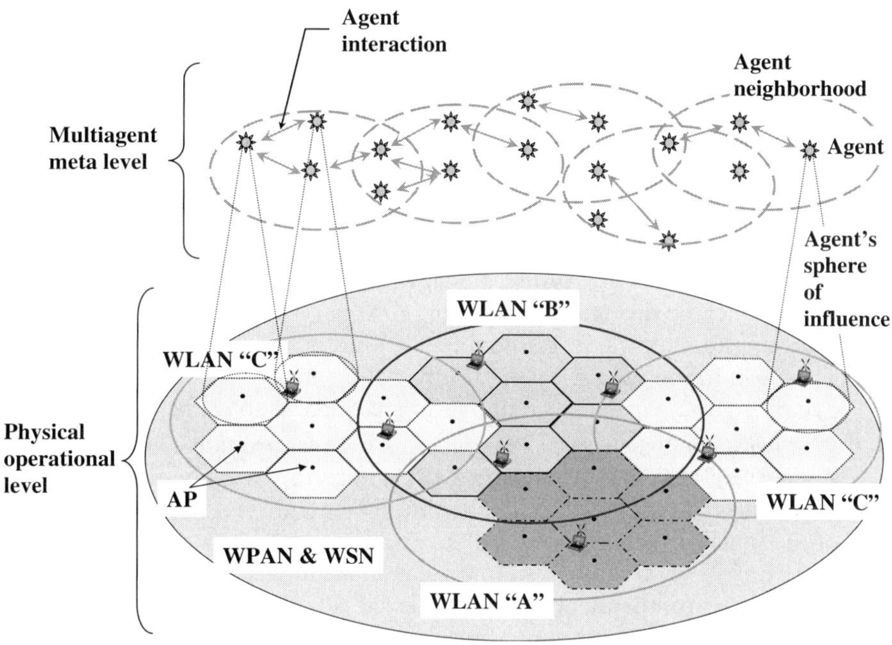

FIGURE 2.11 Architecture of decentralized WLAN resource management using multiagent systems.

consists of those agents with whom it has frequent interactions. These interactions include sharing of data and negotiating about resource assignments. Individual agents act as radio resource coordinators and cooperate with agents in their neighborhood to take care of resource management across multiple WLANs.

The agent at each AP collects the statistics from the measured operational environment as well as its neighborhood and estimates the required parameters for optimizing system performance based on predictive models. The agents use the measured data to generate local control decisions and try to optimize the performance of the entire WLAN system in a distributed fashion through agent interaction and coordination.

Agent interaction is an essential aspect of this architecture. Agent interaction occurs on the backbone network connecting all the APs. Therefore, the bandwidth requirement for agent interaction is not a critical issue. However, since multiple agents contribute to the control of optimal resource allocation across WLANs, they need to decide what information should be exchanged among neighbors, how often to exchange this information, and which neighbors should act as relay nodes for the data. When a control decision is made, an agent also needs to decide what actions its effector should take and how the control decision should be distributed to the desired area.

2.4.2 Framework of Predictability-Based Resource Management

Figure 2.12 presents a block diagram of a general framework for physical environment prediction and resource management using agent technologies. The major functional blocks are WLAN and WPAN cluster, RF environment sensing (RES), and agent operations that include predictive parameter estimation (PPE) and resource management optimization. They are explained in detail as follows.

- *WLAN and WPAN Cluster*. Each MS in WLANs operates within a dynamic RF environment comprising time-varying cochannel interference sources and time-varying interference sources from colocated WPANs. The agent inside each AP periodically collects measured statistics from the dynamic RF environment required for resource management.
- *RF Environment Sensing*. This block is used to provide estimates of the signal characteristics from both MSs within the WLAN cluster as well as potential interference sources within the operational environment. Part of the functions defined in this block can be provided by the specifications of IEEE 802.11k radio resource measurement [30]. Statistics related to WPAN environmental interference levels should be provided from an additional sensing component inside each AP. Measurements would be

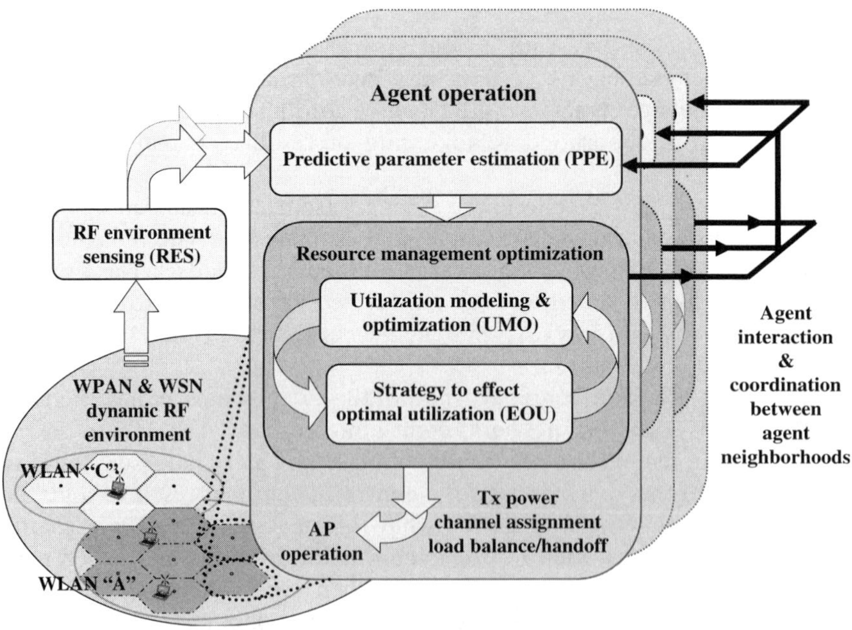

FIGURE 2.12 Block diagram of physical environment prediction and agent operations.

targeted at capturing large-scale changes in signal characteristics due to variations in shadowing, MS mobility, interference sources, and interference locations. In other words, the RES needs to measure the factors that influence the resource management of the WLAN performance.

- *Agent Operation—Predictive Models for Parameter Estimation.* Estimates of signal characteristics are input to the agent inside each AP. An agent also receives data from its neighborhood through agent interaction and coordination. The general concept for the PPE block is to use predictive models to generate parameter estimates required by the resource management optimization. The parameters to be estimated include:
  a. *Link Quality.* Link quality between each MS and its AP
  b. *Link Quality Rate.* Rate of changes in the expected link quality between each MS and its AP
  c. *Throughput.* Throughput for each WLAN cell based on the operational environment characteristics, current offered traffic, and projected offered traffic
  d. *Transmission Latency.* Expected time delay and the variance in the transmission time delay between each MS and its AP
  e. *Handoff Latency.* Expected probability distribution of the time delay required for an MS to be handed off from one AP to another
- *Agent Operation—Resource Management Optimization.* This block analyzes the parameter estimations and makes instructional decisions to optimize the overall WLAN performance based on designed optimization models. Instructional decisions include the optimal transmit power at APs, the optimal channel APs should operate in order to minimize interference levels and make the best use of overall resources, whether or not to accept association requests from specific MSs, whether to direct specific MSs to be associated to another AP for load balancing, and so on. These decisions are updated periodically in order to address changes in the traffic load and interference environment. They should target long-term performance improvement. The operational changes are downloaded to the WLAN cluster with the help of agent effectors and distributed to the neighborhood of agents through agent interaction and coordination.

Resource management optimization includes two components:

1. *Utilization Modeling and Optimization (UMO):* This block finds the optimal utilization, that is, the maximum allowable throughput, of each AP based on the environmental information agents possess. The decision of the optimal utilization is used by the effect optimal utilization block (which is explained in the following) to generate specific strategies to achieve the optimal utilization at each AP.
2. *Strategy to Effect Optimal Utilization (EOU):* Given the optimal utilization of each AP, instructional decisions are generated to achieve

the optimal utilization while minimizing interference to the environment. Operational changes are negotiated within the agent's neighborhood and applied to the WLAN cluster. They are also fed back to the UMO block to update the optimal utilization decision.

2.4.3 Predictive Models for Parameter Estimation

The PPE models provide an intelligent interface between the operational environment and the resource management optimization algorithm. The PPE uses observations from both the RES and APs within an agent's neighborhood. These observations are used in conjunction with a fundamental understanding of WLAN operations to extract necessary information concerning the time-varying signal characteristics and the impact of interference on WLAN operational parameters. The outputs of these predictive models are then utilized in the development of resource management schemes described in the next section.

A conceptual approach for implementing the PPE is illustrated in Fig. 2.13. The approach is based on using a set of integrated analytical models that utilize input from the RES and the agent interaction to estimate the dynamically varying control signals. The PPE includes analytical models to predict the current and near-term impact of the channel, MS trajectory, handoff latency, interference, and network traffic. In order to ensure that the control signals are being adequately estimated, the initial set of parameters used to optimize the control signals are intentionally selected to be extensive and inclusive. As introduced in Section 2.4.2, the initial set of parameters is given by $\mathbf{U}_{ij}^{\mathrm{T}} = [L_{ij}\; R_{ij}\; S_{ij}\; T_{ij}\; H_{ij}]$, where $\mathbf{U}_{ij}$ is the MS profile for the ith MS and the jth AP, and it is defined for all MSs and APs within an agent's neighborhood,

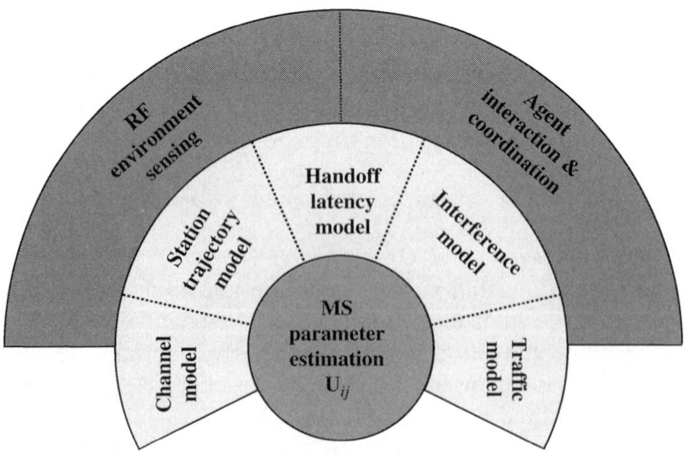

FIGURE 2.13 Conceptual diagram for PPE approach.

including the AP with which the MS is associated. The parameters in the profile are L_{ij} (link quality), R_{ij} (link quality rate), S_{ij} (throughput), T_{ij} (transmission latency), and H_{ij} (handoff latency).

The analytical models required for the PPE are built upon established models reported in the literature for the channel model [37–41], the station trajectory model [42], and the traffic model [19, 32, 33]. The interference model is one of the critical components of the PPE. As derived in [3], the parameters in $\mathbf{U}_{ij}$ are dependent on the interference environment where the WLAN is operating. According to [3–6, 43–48], coexistence analysis is based on evaluating the probability of collision, Pr[C]. Analytical models derived in [3–6, 43–48] are central to evaluating Pr[C] and are also partially explained in Section 2.3.3.2 when deriving the interference from Bluetooth to WLANs under the third-party-based centralized architecture.

The building layout in Fig. 2.14 is used to illustrate the operation of the PPE and the interaction between the various analytical models. Based on the RF propagation characteristics within the building, both APs depicted in the figure can provide service to an MS located at almost any point within the building layout. In addition to the WLAN, Bluetooth piconets are located throughout the building and will impact the WLAN activities depending on the Pr[C]. The shading in the figure represents the likelihood that the AP_1's signal received at

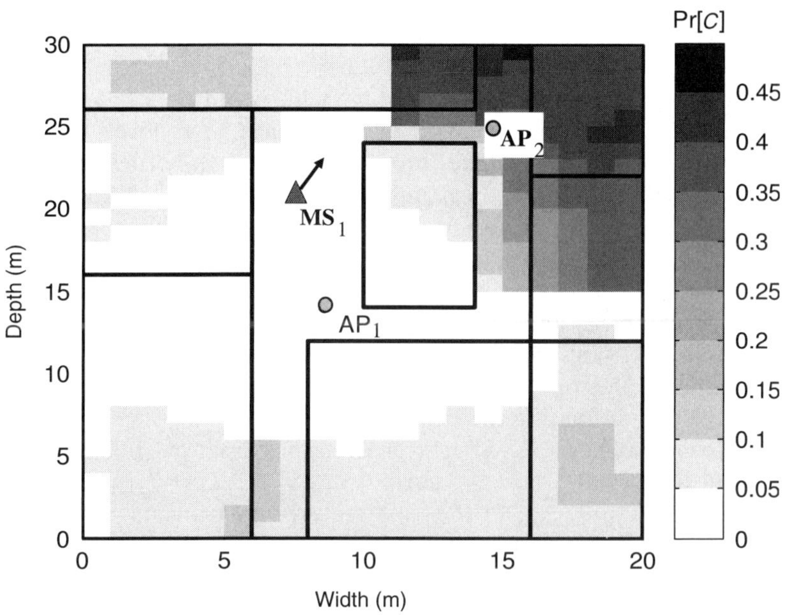

FIGURE 2.14 Pr[C] analysis results overlaid onto floor layout based on specific Bluetooth interference profile.

an MS will be corrupted by Bluetooth interference. The Pr[C] varies from 0 to 0.50. For example, at Pr[C] = 0.25, on average every fourth packet needs to be retransmitted in order to successfully transmit the packet. The agent within AP_1, estimates the MS's Pr[C] based on observations on Bluetooth piconet activities provided by the RES and based on an estimate of the MS's location. As illustrated in the figure, using AP_1's PPE Pr[C] estimate for MS_1, the set of parameters $\mathbf{U}_{11}$ can be predicted. In addition, due to the proximity of AP_2, $\mathbf{U}_{12}$ can be provided to AP_1 through agent interaction. Furthermore, using the directional estimate for MS_1, a time sequence for $\mathbf{U}_{11}$ can be estimated by AP_1's PPE with a corresponding confidence interval provided for each estimate. The PPE can therefore provide a powerful tool for enhancing the resource management optimization process.

The estimations of $\mathbf{U}_{ij}$ based on the PPE approach is outlined as follows:

Link quality, L_{ij}—estimated based on the expected packet error rate, E[PER], of the link between the ith MS and the jth AP. As derived in [3], in an interference environment, E[PER] = Pr[C]. Given multiple interference sources with corresponding Pr[C_i] and assuming the interferences are independent, $E[\text{PER}] = \sum \Pr[C_i] - \sum \sum_{i \neq j} \Pr[C_i]\Pr[C_j] + \cdots$.

Link quality rate, R_{ij}—estimated based on the expected rate of change in L_{ij}. As illustrated in the example above, the station trajectory model, channel model, and interference model are used in estimating the rate at which the link quality changes. The station trajectory model is used to estimate of the direction and rate of movement for the MS, which can be estimated using a similar approach presented in [42].

Throughput, S_{ij}—detailed analytical models for evaluating the throughput for IEEE 802.11 CSMA/CA-based MAC have been developed in [49, 50]. In [49], the proposed analytical throughput model is based on a two-dimensional Markov chain that takes into account the probability of collision for MSs. We have extended this model to take into consideration the effect of cochannel interference on the analytical throughput in WLANs with multiple cochannel cells [51]. In addition, we can also extend the throughput model to include Pr[C] due to interference from sources other than 802.11 transceivers. It is similar to the Markov model developed in [52] to evaluate packet transmission latency. Therefore, a complete analytical representation of WLAN cell throughput considering different interference sources can be developed. In addition to the channel, station trajectory, and interference models, the traffic model will also play a key role in estimating the throughput.

Transmission latency, T_{ij}—a first-order approximation for the expected packet transmission latency is derived in [3], $E[T] = T_{\text{normal}}(1 + \alpha \Pr[C])(1 - \Pr[C])^{-1}$, where T_{normal} is the time required to transmit a packet given no interference and α is a proportionality constant relating the T_{normal} to the time required to retransmit a packet, αT_{normal}. In addition, a more detailed analytical expression for expected packet transmission latency has been derived in [52], which is based on a more accurate model of the IEEE 802.11 backoff algorithm.

Handoff latency, H_{ij}—we have conducted research on setting up an analytical model for handoff latency analysis based on IEEE 802.11b MAC scheme [53]. This model considers comprehensive factors that influence the WLAN handoff latency such as MAC probability, binary exponential backoff latency, packet transmission delay, queuing delay at APs, and so on as well as the range these factors affect the handoff latency. The outcome of this research is the probability distribution of the handoff latency in a certain range based on the offered traffic load and network conditions in the WLAN. Therefore, from this research, we can predict the likelihood the handoff can be finished at a certain moment.

2.4.4 Resource Optimization Using Multiagent Systems

2.4.4.1 Overview of Resource Management Optimization. The goal of the UMO block is to adjust resource allocation in each WLAN in order to minimize the total system cost. Based on the optimal utilization for each AP derived from the UMO, the strategy to the EOU block inside each agent generates corresponding strategies to satisfy the optimal utilization requirement for each cell. These strategies include instructing the APs on which channels they should operate, which transmit powers they should use, whether or not to accept association requests from specific MSs, and so on. These actions are needed to make dynamic channel allocation, dynamic transmit power control, and load balancing possible, which can be expected to significantly improve the performance of multiple WLANs [16]. In this research, we focus on dynamic load balancing, that is, finding the optimal set of MSs under each AP and instructing specific MSs with which AP they should be associated. The association control from the EOU helps redistribute loads across neighboring APs by requesting MSs to explicitly change their association from an overloaded AP to a less loaded neighboring AP so that no one AP is unduly overloaded.

In a distributed implementation, the multiagent system directs the APs to redistribute associations of MSs. MSs that are redistributed perform handoffs to a new AP. The redistribution process considers the optimal allowable throughput S^* of each AP, which is calculated in order to minimize the overall cost in the agent neighborhood. Each agent negotiates with other agents in its neighborhood to decide which MS should be handed off to a neighboring AP, when to initiate the handoff, and when to complete the handoff. This coordination among neighboring agents will result in event triggers that indicate the need to balance the traffic load based on a distributed constraint optimization algorithm. The result of applying such an optimization algorithm is an optimal handoff strategy, that is, the EOU strategy.

Distributed-constraint optimization problems (DCOPs) have been used as fully distributed algorithms to solve existing centralized problems efficiently [9, 10]. In the following discussion, we first present a model based on a multiagent DCOP to illustrate our distributed approach to dynamic load balancing.

2.4.4.2 Scenario of WLAN Handoffs for Load Balancing Using DCOP Algorithm.

A discrete multiagent DCOP [11] is a tuple $\langle A, X, D, R \rangle$, where

- $A = \{A_1, \ldots, A_n\}$ is the set of agents interested in the solution; in the WLAN context, each access point AP_i is assigned an agent.
- $X = \{X_1, \ldots, X_m\}$ is the set of variables; in the WLAN context, each AP_i has a variable X_i for MS_i, which represents the new associated AP_j after a handoff.
- $D = \{d_1, \ldots, d_m\}$ is a set of domains of the variables, where each domain d_i is the set of APs in AP_i's neighborhood.
- $R = \{r_1, \ldots, r_p\}$ is a set of relations where a relation r_i is a utility function that provides a measure of the value associated with a given combination of variables. In WLANs, R represents objective functions, which are provided by the UMO block in Fig. 2.12. They are similar to (2.3) shown in the centralized implementation.

We describe a simple WLAN scenario, as shown in Fig. 2.15 to explain how to use the DCOP algorithm for WLAN load balancing. In this scenario, three APs are depicted in the figure at (x, y) locations AP_1: (0, 0); AP_2: (45, 90); AP_3:

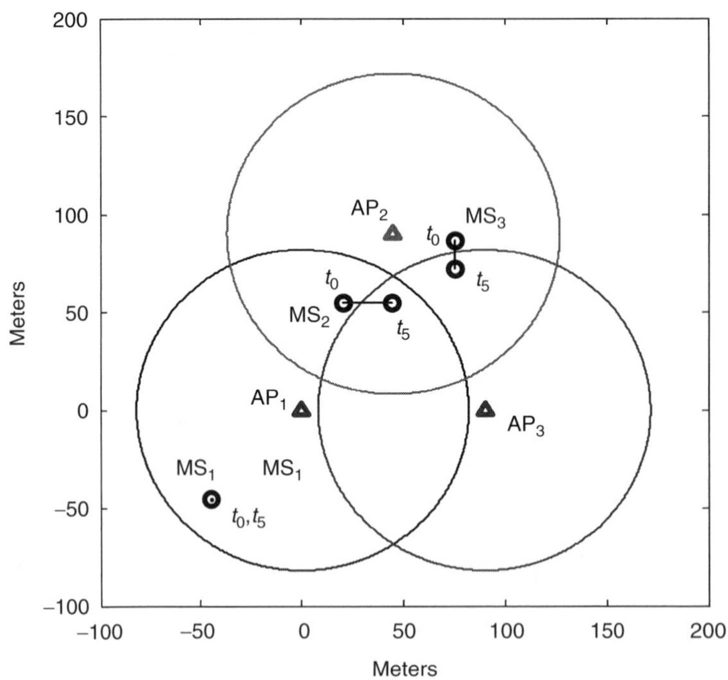

FIGURE 2.15 Scenario for illustrating DCOP algorithm for WLAN load balancing.

(90, 0). In addition, there are three MSs depicted in the figure. The MS_1 remains stationary at location (−45, −45) during the 5-s duration of the simulation from t_0 to t_5. MS_2's location at t_0 is at (20, 55) and moves in the x direction at 5 m/s. MS_3's location at t_0 is at (75, 87) and moves in the negative y direction at 3 m/s.

The goal of the DCOP algorithm is to dynamically assess the MS associations with the APs at time t_k based on the estimate of the state of the MSs at time t_{k+1} where $t_H = t_{k+1} - t_k$ is the fixed time required to handoff an IEEE 802.11 MS from one AP to a neighboring AP. For simplicity, for the simulation shown below, the handoff latency, t_H, is an expected value, and $t_H = 350\,\text{ms}$ corresponding to the expected handoff latency associated with the standard IEEE 802.11 protocol [54].

For the purpose of this example, the state of the MSs within a WLAN is defined by two parameters: the AP utilization and the link quality between the MS and AP. The AP utilization is based on the number of MSs associated with it during time interval t_H. Based on the current IEEE 802.11 protocol, each MS can only be associated with a single AP. Therefore, the association between AP_j and MS_i at time t_k is given by

$$\rho_{ij}(t_k) = \begin{cases} 1 & AP_j \text{ and } MS_i \text{ are associated at time } t_k \\ 0 & AP_j \text{ and } MS_i \text{ are not associated at time } t_k \end{cases} \tag{2.22}$$

Assuming the traffic offered by each MS is on average the same, then the AP utilization is estimated by

$$\rho_j(t_k) = \sum_{i=1}^{M} \rho_{ij}(t_k) \tag{2.23}$$

where M is the total number of MSs within the WLAN.

The link quality is based on the expected received power over a transmission distance of $d_{ij}(t_k)$ between AP_j and MS_i at time t_k given by

$$P_R(d_{ij}(t_k)) = P_T - [20 \log_{10} f_c + 10n \log_{10}(d_{ij}(t_k)) - 28] \quad (\text{dBm}) \tag{2.24}$$

where P_T is the WLAN transmit power, $P_T = 20\,\text{dBm}$; f_c is the WLAN carrier frequency, $f_c = 2.4\,\text{GHz}$; n is the path loss exponent, $n = 3$; $d_{ij}(t_k)$ is the Euclidean distance between AP_j and MS_i at time t_k. In the figure, the boundary for each AP's coverage range is depicted as a circle around the AP based on the power received threshold, γ, of −82 dBm.

The received power by MS_1 from the three APs remains constant over the duration of the simulation and are [−78.7, −90.9, −89.2], respectively, for [AP_1, AP_2, AP_3]. The received power by MS_2 and MS_3 from each of the APs changes continuously over the 5-s scenario based on the mobility profile for each MS.

The corresponding received power versus the scenario time is illustrated in Figs. 2.16 and 2.17.

For the scenario described above, suppose each access point AP_i is represented by an agent A_i. Each MS_j is initially associated with an AP_i at time t_k. Each AP_i has a variable X_i for MS_i, which represents the new association AP_j at time t_{k+1}. Hence, X_i is equivalent to $\rho_{ij}(t_{k+1})$ defined in (2.22). At any point in time, only a subset of the agents will be involved in the resource allocation process, which means that the multiagent system is constructed dynamically. Periodically, each agent listens to handoff event triggers. The event triggers for handoffs are usually requested from MSs need to be handed off or when an AP is overloaded. They initiate agent local computations and communications with other agents in the neighborhood to handle the resource allocation problem.

The DCOP is viewed as an optimization problem with the following criteria:

Criterion I: Maximize the utilization of each AP in order to accommodate more users, that is,

$$\begin{aligned} &\text{Maximize} \quad \rho_j(t_{k+1}) \\ &\text{Subject to} \quad P_R(d_{ij}(t_{k+1})) > \gamma \quad \forall i, j | \rho_{ij}(t_{k+1}) = 1 \end{aligned} \tag{2.25}$$

Criterion II: Maximize the minimum received power by each MS in order to improve the link quality and minimize the likelihood of packet loss, that is,

$$\text{Maximize} \min_{i,j|\rho_{ij}(t_{k+1}=1)} (P_R(d_{ij}(t_k))) \tag{2.26}$$

Criterion III: Distribute the load among viable APs in order to increase fairness as well as increase the overall systemwide utilization, that is,

$$\text{Minimize} \max_{j,l} \left(\left| \rho_j(t_{k+1}) - \rho_l(t_{k+1}) \right| \right) \tag{2.27}$$

Criterion IV: Each MS can only be assigned to one AP at any time according to the current IEEE 802.11 standard, that is,

$$\sum_j \rho_{ij} = 1 \tag{2.28}$$

We describe a simple distributed algorithm to solve the DCOP, which will specifically consider the handoff process at three time slots: $t_k = 0$, the start time; $t_k = 1.75$, when MS_2 crosses the boundary of AP_3; and $t_k = 2.45$, when MS_3 crosses the boundary of AP_3. We assume that for each of the three time slots, the MS_i's have the following initial association: MS_1 is associated with

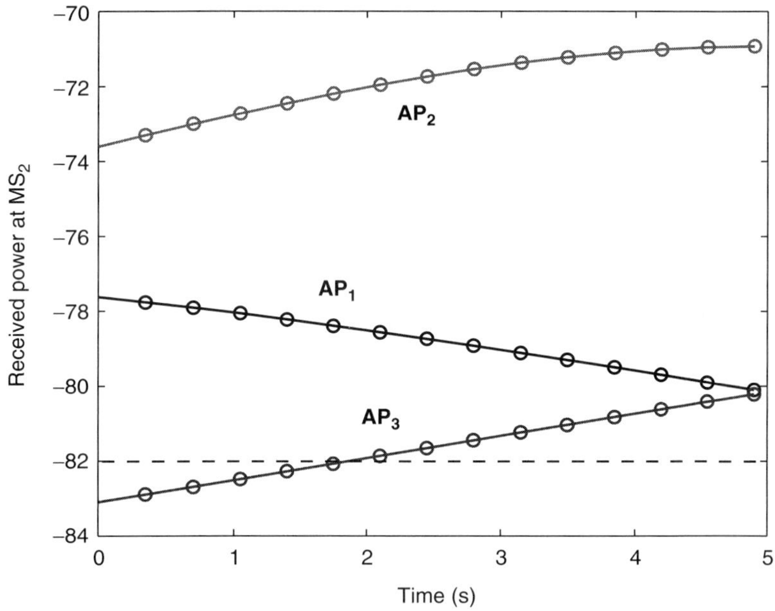

FIGURE 2.16 Received power at MS_2.

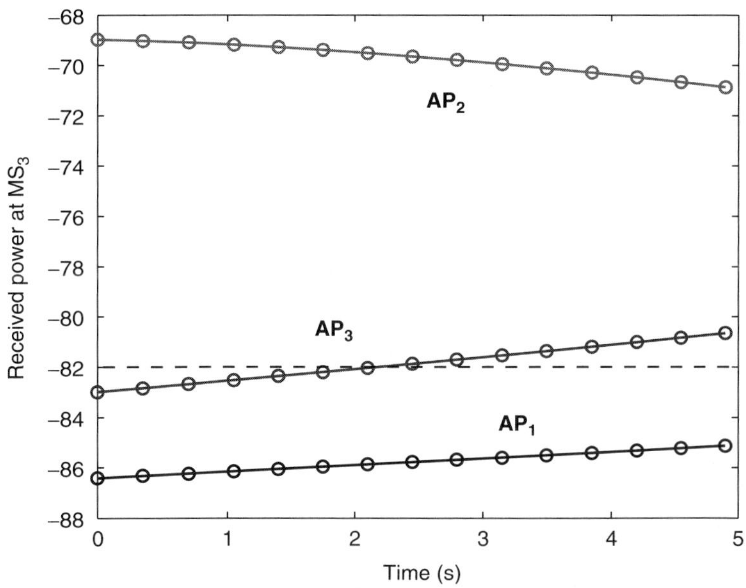

FIGURE 2.17 Received power at MS_3.

AP_1, MS_2 and MS_3 are both associated with AP_2. The destination APs computed by the decision process described below are independent of the agent performing the computation.

2.4.4.3 *Time t_k = 0.* Each agent A_i inside AP_i will identify all values for the variable X_i that satisfy criterion I, II, III, and IV. Criterion I and II can be verified locally within an agent, while criterion III requires communications with other agents about their respective assignments in order to determine the load of the APs.

Agent A_1 (at AP_1) computes the value assignment for X_1 to be AP_1, as it is the only AP from which the power received is greater than γ (−82 dBm). Criterion II is consistent with this value. Agent A_2 (at AP_2) computes the value assignment for X_2 to be $\{AP_1, AP_2\}$ as the power received from each of the two APs is greater than γ. Criterion II would lead to X_2 being assigned AP_2 as it maximizes the minimum power received. Agent A_3 (at AP_3) computes the value assignment for X_3 to be AP_2 as it is the only AP from which the power received is more than γ. Criterion II is consistent with this value. In order to verify criterion III, the agents exchange their assignment information and independently compute the load information. Although AP_2 has both MS_2 and MS_3 associated with it and AP_3 has none associated with it, this is determined to be a fair load since a reassignment of either MS_2 or MS_3 to AP_3 would cause criterion I to be violated. Criterion IV is consistent with this assignment.

Hence the final assignment at time $t_k = 0$ is $X_1 = AP_1$; $X_2 = AP_2$; $X_3 = AP_2$.

2.4.4.4 *Time t_k = 1.75.* Agent A_1 computes the value assignment for X_1 to be AP_1, as it is the only AP from which the power received is greater than γ (−82 dBm). Criterion II is consistent with this value. Agent A_2 computes the value assignment for X_2 to be $\{AP_1, AP_2\}$ as the power received from each of the two APs is greater than γ. Criterion II would lead to X_2 being assigned AP_2 as it maximizes the minimum received power. Agent A_3 computes the value assignment for X_3 to be AP_2 as it is the only AP from which the power received is more than γ. Criterion II is consistent with this value. In order to verify criterion III, the agents exchange their assignment information and independently compute the load information. Although AP_2 has both MS_2 and MS_3 associated with it and AP_3 has none associated with it, this is determined to be a fair load since a reassignment of either MS_2 or MS_3 to AP_3 would cause criterion I to be violated. Constraint IV is consistent with this assignment.

Hence the final assignment at time $t_k = 1.75$ is $X_1 = AP_1$; $X_2 = AP_2$; $X_3 = AP_2$.

2.4.4.5 *Time t_k = 2.45.* Agent A_1 computes the value assignment for X_1 to be AP_1, as it is the only AP from which the power received is greater than γ (−82 dBm). Constraint II is consistent with this value. Agent A_2 computes the value assignment for X_2 to be $\{AP_1, AP_2, AP_3\}$ as the power received from each of the three APs is greater than γ. Criterion II would lead to X_2 being assigned

AP_2 as it maximizes the minimum received power. Agent A_3 computes the value assignment for X_3 to be $\{AP_2, AP_3\}$ as the power received from both APs is more than γ. Criterion II would lead to X_2 being assigned AP_2 as it maximizes the minimum received power. In order to verify criterion III, the agents exchange their assignment information and independently compute the load information. AP_2 has both MS_2 and MS_3 associated with it and AP_3 has none associated with it. Using simple backtracking, it is determined that the AP utilization is best optimized for all three criteria when X_2 is reassigned to AP_3 (power received is $-69.6505\,dBm$) and X_3 retains its association with AP_2 (power received is $-81.6561\,dBm$). Criterion IV is consistent with this assignment.

Hence the final assignment at time $t_k = 2.45$ is $X_1 = AP_1$; $X_2 = AP_3$; $X_3 = AP_2$.

2.5 CONCLUSION

In this chapter, a framework for resource management across multiple WLANs in interference environments is introduced. The framework is based on multiagent systems for decentralized information sharing and network management decision making. It emphasizes the predictability of the time-varying network states using predictive models and incorporates the impact of interference into the resource optimization. A centralized resource optimization approach under a third-party-based architecture is first explained. The performance of the centralized approach is used as the performance benchmark for the proposed decentralized approach. Then, the functional details of each component under the MAS-based decentralized architecture are introduced. A handoff scenario is used to explain how to use a DCOP algorithm, a fully distributed algorithm based on the multiagent system, to make the handoff decisions for load balancing. This chapter is aimed at conveying to the research community the importance of cooperative network management for multiple WLANs and introducing a novel decentralized network control framework suitable for large-scale networks in interference environments.

REFERENCES

1. IEEE 802.11, "Wireless LAN medium access control (MAC) and physical layer (PHY) specifications," IEEE, New York, 1999.
2. Y. Matsunaga and R. H. Katz, "Inter-domain radio resource management for wireless LANs," in *Proc. IEEE Wireless Communications and Networking Conference* (*WCNC 2004*), Vol. 4, Atlanta, Georgia, 2004, pp. 2183–2188.
3. I. Howitt, "WLAN and WPAN coexistence in UL band," *IEEE Trans. Vehic. Technol.* 50(4), 1114–1124 (2001).
4. I. Howitt and S. Y. Ham, "Site specific WLAN and WPAN coexistence evaluation," in *Proc. IEEE Wireless Communications and Networking Conference* (*WCNC 2003*), Vol. 3, New Orleans, Louisiana, 2003, pp. 1487–1491.

5. I. Howitt, V. Mitter, and J. A. Gutierrez, "Empirical study for IEEE 802.11 and Bluetooth interoperability," in *Proc. IEEE Vehicular Technology Conference* (*VTC 2001*), Vol. 2, 2001, pp. 1109–1113.
6. I. Howitt, "IEEE 802.11 and Bluetooth coexistence analysis methodology," in *Proc. IEEE Vehicular Technology Conference* (*VTC 2001*), Vol. 2, 2001, pp. 1114–1118.
7. S. Russell and P. Norvig, *Artificial Intelligence: A Modern Approach*, 2nd ed., Prentice-Hall, Upper Saddle River, NJ, 2003.
8. M. Wooldridge, *An Introduction to Multiagent Systems*, Wiley, Hoboken, NJ, 2002.
9. P. J. Modi, P. Scerri, W.-M. Shen, and M. Tambe, "Distributed resource allocation: A distributed constraint reasoning approach," in *Distributed Sensor Networks: A Multiagent Perspective*, V. Lesser, C. L. Ortiz, M. Tambe, Eds. Kluwer Academic, Norwell, MA, 2003.
10. R. Mailler and V. Lesser, "Solving distributed constraint optimization problems using cooperative mediation," in *Proc. Third International Joint Conference on Autonomous Agents and Multiagent Systems* (*AAMAS 2004*), Vol. 1, 2004, New York, pp. 438–445.
11. A. Petcu and B. Faltings, "A distributed, complete method for multi-agent constraint optimization," in *Proc. Fifth International Workshop on Distributed* Constraint *Reasoning* (*DCR 2004*), Sept. Toronto, Canada, 2004.
12. A. Petcu and B. Faltings, "A scalable method for multiagent constraint optimization," in *Proc. International Joint Conference on Artificial Intelligence* (*IJCAI*), Edinburgh, Scotland, 2005, pp. 266–271.
13. D. Marsh, R. Tynan, D. O'Kane, and G. M. P. O'Hare, "Autonomic wireless sensor networks," in *Engineering Applications of Artificial Intelligence*, Elsevier, Amsterdam, 2004, pp. 741–748.
14. J. S. Sandhu, A. M. Agogino, and A. K. Agogino, "Wireless sensor networks for commercial lighting control: Decision making with multi-agent systems," in *Proc. American Association for Artificial Intelligence Workshop on Sensor Networks*, 2004.
15. R. Tynan, A. Ruzzelli, and G. M. P. O'Hare, "A methodology for the deployment of multi-agent systems on wireless sensor networks," in *Proc. International Conference on Software Engineering and Knowledge Engineering* (*SEKE 2005*), Taipei, Taiwan, 2005.
16. A. Hills and B. Friday, "Radio resource management in wireless LANs," *IEEE Commun. Mag.* 42(12), S9–S14 (2004).
17. Y. Wang, L. G. Cuthbert, and J. Bigham, "Intelligent radio resource management for IEEE 802.11 WLAN," in *Proc. IEEE Wireless Communications and Networking Conference* (*WCNC 2004*), Vol. 3, Atlanta, Georgia, 2004, pp. 1365–1370.
18. Y. Bejerano, S.-J. Han, and L. Li, "Fairness and load balancing in wireless LANs using association control," in *Proc. ACM MOBICOM 2004*, Philadelphia, PA, 2004, pp. 315–329.
19. A. Balachandran, P. Bahl, and G. M. Voelker, "Hot-spot congestion relief in public-area wireless networks," in *Proc. IEEE Workshop on Mobile Computing Systems and Applications* (*WMCSA 2002*), Callicoon, NY, 2002, pp. 70–80.
20. I. Katzela and M. Nagshineh, "Channel assignment schemes for cellular mobile telecommunication systems: A comprehensive survey," *IEEE Personal Commun.* 3(3), 10–31 (1996).

21. I. Tinnirello and G. Bianchi, "A simulation study of load balancing algorithms in cellular packet networks," in *Proc. ACM/IEEE MSWiM*, Rome, Italy, 2001, pp. 73–78.
22. T. P. Chu and S. R. Rappaport, "Overlapping coverage with reuse partitioning in cellular communication systems," *IEEE Trans. Vehic. Technol.* 46(1), 41–54 (1997).
23. X. Lagrange and B. Jabbari, "Fairness in wireless microcellular networks," *IEEE Trans. Vehic. Technol.* 47(2), 472–479 (1998).
24. S. Das, H. Viswanathan, and G. Rittenhouse, "Dynamic load balancing through coordinated scheduling in packet data systems," in *Proc. IEEE INFOCOM 2003*, Vol. 2, San Francisco, CA, 2003, pp. 786–796.
25. A. Sang, X. Wang, M. Madihian, and R. D. Gitlin, "Coordinated load balancing, handoff/cell-site selection, and scheduling in multi-cell packet data systems," in *Proc. ACM MOBICOM 2004*, Philadelphia, PA, 2004, pp. 302–314.
26. iPass, http://www.ipass.com.
27. I. F. Akyildiz, S. Mohanty, and J. Xie, "A ubiquitous mobile communication architecture for next-generation heterogeneous wireless systems," *IEEE Commun. Mag.* 43(6), S29–S36 (2005).
28. D. Harrington, R. Presuhn, and B. Wijnen, "An architecture for describing simple network management protocol (SNMP) management frameworks," *Request for Comments* (RFC) 3411, Internet Engineering Task Force, Dec. 2002.
29. U. Blumenthal and B. Wijnen, "User-based security model (usm) for version 3 of the simple network management protocol (SNMPv3)," *Request for Comments* (RFC) 3414, Internet Engineering Task Force, Dec. 2002.
30. IEEE Std 802.11k, "Radio resource measurement of wireless LANs," June 2008, Amendment to IEEE Std 802.11-2007, IEEE, New York.
31. A. Balachandran, G. M. Voelker, P. Bahl, and P. V. Rangan, "Characterizing user behavior and network performance in a public wireless LAN," in *Proc. ACM SIGMETRICS*, Marina del Rey, CA, 2002, pp. 195–205.
32. D. Kotz and K. Essien, "Analysis of a campus-wide wireless network," in *Proc. ACM MOBICOM*, Atlanta, Georgia, 2002, pp. 107–118.
33. M. Balazinska and P. Castro, "Characterizing mobility and network usage in a corporate wireless local-area network," in *Proc. ACM MobiSys*, San Francisco, CA, 2003, pp. 303–316.
34. M. Alanyali and B. Hajek, "On simple algorithms for dynamic load balancing," in *Proc. IEEE INFOCOM*, Vol. 1, Boston, MA, 1995, pp. 230–238.
35. A. Kamerman and G. Aben, "Throughput performance of wireless LANs operating at 2.4 and 5 GHz," in *Proc. IEEE International Symposium on Personal, Indoor and Mobile Radio Communications (PIMRC)*, Vol. 1, London, UK, 2000, pp. 190–195.
36. M. O. Lorenz, "Methods of measuring the concentration of wealth," *Publ. Am. Statist. Assoc.* London, UK, 9, 209–219 (1905).
37. S. Y. Seidel and T. S. Rappaport, "914 MHz path loss prediction models for indoor wireless communications in multifloored buildings," *IEEE Trans. Antennas Propagation* 40(2), 207–217 (1992).
38. M. Hassan-Ali and K. Pahlavan, "A new statistical model for site-specific indoor radio propagation prediction based on geometric optics and geometric probability," *IEEE Trans. Wireless Commun.* 1(1), 112–124 (2002).

39. C. Perez-vega, "Simple approach to a statistical path loss model for indoor communications," in *Proc. 27th European Conference and Exhibition: Bridging the Gap between Industry and Academia*, 1997, pp. 617–623.
40. P. T. Kuruganti and J. Nutaro, "A comparative study of wireless propagation simulation methodologies: Ray tracing, FDTD, and event based TLM," in *Proc. Huntsville Simulation Conference*, Huntsville, Alabama, Oct. 2006.
41. J. Nutaro, "A discrete event method for wave simulation," *ACM Trans. Modeling Computer Simulation* 16(2), 174–195 (2006).
42. R. Hsieh and A. Seneviratne, "A comparison of mechanisms for improving Mobile IP handoff latency for end-to-end TCP," in *Proc. ACM MOBICOM 2003*, San Diego, CA, 2003, pp. 29–41.
43. I. Howitt and V. Mitter, "Analytical tools for evaluating coexistence in UL band," 802.15 TG2, IEEE, New York, 2001.
44. V. Mitter, I. Howitt, and J. Gutierrez, "Empirical study for coexistence of WLAN and Bluetooth in ISM band," 802.15 TG2, IEEE, New York, 2001.
45. I. Howitt and J. A. Gutierrez, "A tool for evaluating Bluetooth co-existence with other 2.4 ghz ISM devices," in *Proc. Bluetooth Congress 2001*, 2001.
46. I. Howitt, "Bluetooth performance in the presence of 802.11b WLAN," *IEEE Trans. Vehic. Technol.* 51(6), 1640–1651 (2002).
47. I. Howitt and J. A. Gutierrez, "IEEE 802.15.4 low rate wireless personal area network coexistence issues," in *Proc. IEEE WCNC 2003*, Vol. 3, New Orleans, Louisiana, 2003, pp. 1481–1486.
48. I. Howitt, "Mutual interference between independent bluetooth piconets," *IEEE Trans. Vehic. Technol.* 52(3), 708–718 (2003).
49. G. Bianchi, "Performance analysis of the IEEE 802.11 distributed coordination function," *IEEE J. Sel. Areas Commun.* (*JSAC*) 18(3), 535–547 (2000).
50. P. Ferre, A. Doufexi, A. Nix, and D. Bull, "Throughput analysis of IEEE 802.11 and IEEE 802.11e MAC," in *Proc. IEEE WCNC 2004*, Atlanta, Georgia, 2004, pp. 783–788.
51. C. Nie and J. Xie, "An analytical model for the IEEE 802.11 DCF WLAN with multiple co-channel cells," in *Proc. IEEE Global Communications Conference* (*GLOBECOM 2006*), San Francisco, CA, 2006.
52. I. Howitt and F. Awad, "Optimizing IEEE 802.11b packet fragmentation in collocated Bluetooth interference," *IEEE Trans. Commun.* 53(6), 936–938 (2005).
53. J. Xie, I. Howitt, and I. Shibeika, "IEEE 802.11-based Mobile IP fast handoff latency analysis," in *Proc. IEEE International Conference on Communications* (*ICC 2007*), Glasgow, Scotland, 2007.
54. I. Ramani and S. Savage, "SyncScan: Practical fast handoff for 802.11 infrastructure networks," in *Proc. IEEE INFOCOM*, Vol. 1, Miami, Florida, 2005, pp. 675–684.

CHAPTER 3

INCENTIVE ISSUES IN IEEE 802.11*X* WIRELESS NETWORKS

YU-KWONG KWOK

3.1 INTRODUCTION

Wireless local area networks (WLANs), based on the IEEE 802.11*x* (here, *x* could be substituted by a, b, e, i, g, etc.) technologies, are quickly proliferating, both in household environments and in workplaces. Indeed, many handheld gadgets are now equipped with WLAN interfaces in the hope that the gadgets can be hooked onto a WLAN if one is available nearby. However, as many WLANs are deployed independently by different autonomous organizations, roaming or shared access are not supported, making it inconvenient for nomadic and mobile WLAN users. Surely, through administrative agreements such as those in cellular networks, roaming access can be provided. Yet the overheads involved may not be justified as IEEE 802.11*x*–based wireless networking is usually of a low cost or even free of charge.

It is now clear to many researchers that providing shared WLAN access needs a novel technical solution rather than an administrative solution. The key issue here is that the WLANs are owned and operated by different people who might not be cooperative. As such, proper *incentives* are needed in order to entice cooperation. The goal of this chapter is to survey such incentive techniques, as detailed in Section 3.3.

Apart from wireless link access sharing, WLAN devices, when together in close proximity, could form an ad hoc network, which is an impromptu network without centralized base-station or access point support. However, establishing a multihop path in such an ad hoc network is mandatory but challenging again due to the possible noncooperativeness. There are also a plethora of novel approaches suggested for encouraging cooperation in a multihop ad hoc network. We survey these approaches in Section 3.4.

In the next section, we first delineate different possible incentive techniques.

Emerging Wireless LANs, Wireless PANs, and Wireless MANs. Edited by Y. Xiao and Y. Pan

3.2 OVERVIEW OF INCENTIVE TECHNIQUES

To provide incentives in a wireless network with autonomous devices, there are basically six different classes of techniques.

1. *Payment-Based Mechanisms.* Users taking cooperative actions (e.g., sharing wireless bandwidth) would obtain payments in return. The payment may be real monetary units (in cash) or virtual (i.e., some tokens that can be redeemed for other services). Thus, two important components are needed: (1) currency and (2) accounting and clearing mechanism. Obviously, if the currency is in the form of real cash, we need a centralized authority, in the form of an electronic bank, that is external to the wireless system. If the currency is in the form of some virtual token, then it might be possible to have a peer-to-peer clearing mechanism. In both cases, the major objective is to avoid fraud at the expense of usually significant overhead. Proper pricing of cooperative actions is also important—overpriced actions would make the system economically inefficient while underpriced actions would not be able to entice cooperation.
2. *Auction-Based Mechanisms.* In some situations, in order to come up with an optimal pricing, an auction is an effective mechanism. In simple terms, an auction involves bidding from the participating users so that the user with the highest bid get the opportunity to serve (or to be served, depending on the context). An important issue in auction-based systems is the valuation problem—how much a user should set in the bid? If every user sets a bid higher than its true cost in providing a service, then the recipient of the service would pay too much. On the other hand, if the bids are too low, the service providers may suffer. Fortunately, in some form of auctions, we can design proper mechanisms to induce bidders to bid at their true costs.
3. *Exchange-Based Mechanisms.* Compared to payment-based and auction-based systems, exchange (or barter)-based techniques manifest as a purer peer-to-peer interaction. Specifically, in an exchange-based environment, a pair of users (or, sometimes, a circular list of users) serve each other in a rendezvous manner. That is, service is exchanged in a synchronous and stateless transaction. For example, a pair of users meet each other and exchange files. After the transaction, the two users can forget about each other in the sense that any future transaction between them is unaffected by the current transaction. Thus, an advantage is that very little overhead is involved. Most importantly, peers can interact with each other without the need of intervention or mediation by a centralized external entity (e.g., a bank). Furthermore, free-riding is impractical. Of course, the downside is that service discovery and peer selection (according to price and/or quality of service) could be difficult.

4. *Reciprocity-Based Mechanisms.* While pure barter-based interactions are stateless, reciprocity generally refers to stateful and history-based interactions. Specifically, a peer A may serve another peer B at time t_1 and does not get an immediate return. However, the transaction is recorded in some history database (centralized in some external entity or distributed in both A and B). At a later time $t_2 > t_1$, peer B serves peer A, possibly because peer B selects peer A as the client due to the earlier favor from A. That is, as peer A has served peer B before, peer B would give a higher preference to serve peer A. A critical problem is how we can tackle a special form of free-riding behavior, namely the "whitewashing" action (i.e., a user leaves the system and rejoins with a different identity), which enables the free-rider to forget about his/her obligations.
5. *Reputation-Based Mechanisms.* A reputation-based mechanism is a generalized form of reciprocity. Specifically, while a reciprocity record is induced by a pair of peers (or a circular list of more than two peers), a reputation system records a score for each peer based on the assessments made by many peers. Each service provider (or consumer, depending on the application) can then consult the reputation system in order to judge whether it is worthwhile or safe to provide service to a particular client. Reputation-based mechanism is by nature globally accessible and, thus, peer selection can be done easily. However, the reputation scores must be securely stored and computed, or otherwise, the scores cannot truly reflect the quality of peers. In some electronic market place such as eBay, the reputation scores are centrally administered. But such an arrangement would again need an external entity and some significant overhead. On the other hand, storing the scores in a distributed manner at the peers would induce problems of fraud. Finally, similar to reciprocity-based mechanisms, whitewashing is a low cost technique employed by selfish users to avoid being identified as a low quality users, which would be excluded from the system.
6. *Game-Theoretic Strategies.* Using a game-theoretic approach, "best strategies" are derived for each user. Specifically, an important concept is the *Nash equilibrium* [1] at which state there will not be any user (called player) unilaterally deviate from his/her strategy (i.e., the chosen action) because no gain (in terms of utility) can be obtained from such deviation. The key parts in a game-theoretic modeling approach are: (a) definition of a utility function (e.g., achieved network throughput) and (b) definition of possible actions (e.g., forwarding data or not). An important point to note in a game-theoretic analysis is that a Nash equilibrium state, at which all users use their respective "best strategies," may not constitute a system optimal point. This is common because of the fact that selfish actions, albeit locally are the best strategies, could actually hurt the whole community. Indeed, a classic situation is the "tragedy of the commons" in which selfish actions taken individually eventually inflict harm on the community as well as the individuals themselves.

3.3 SINGLE-HOP NETWORKS

In this section, incentive approaches suggested for handling wireless link access in a single-hop WLAN are briefly discussed. Specifically, in present-day WLAN deployment, there is very little, if any, capacity and coverage planning. As a result, there are usually nearby WLAN access points (APs) providing overlapping coverage. Indeed, it is not uncommon that a user belonging to one AP roams into the coverage area of another nearby AP, as illustrated in Fig. 3.1.

Kawade et al. [2] reported their theoretical and practical study on the viability of a wide-area shared wireless access system called urban residential WiFi (UR-WiFi). Their study showed that a well-planned WiFi system is highly attractive.

Akella et al. [3] studied the performance degradation of IEEE 802.11 users in unplanned WLANs. The key problem is the potentially severe interference resulting in the unplanned deployment of WLANs. Accordingly, they proposed

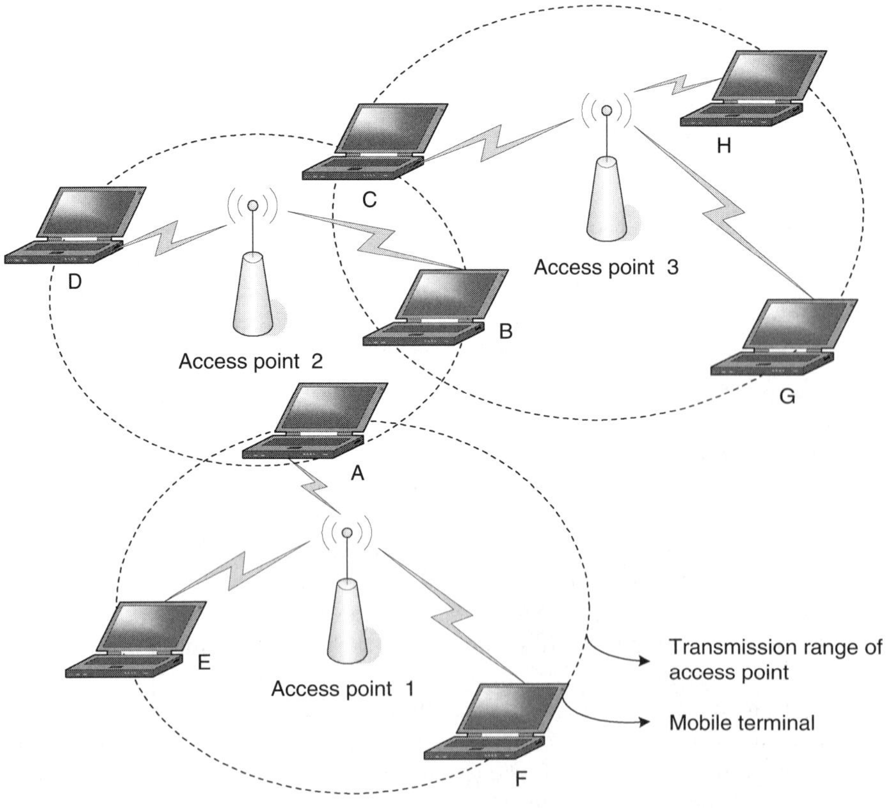

FIGURE 3.1 Sharing wireless link access (e.g., device G actually belongs to access point 1 but roams to access point 3).

a "self-managing" wireless access protocol for the WLAN users. In the proposed protocol, each user adaptively chooses the transmit power and transmission rate. The adaptation is shown to be able to combat the inter-WLAN interference. Siris et al. [4] also performed a theoretical study on the congestion prices involved in a shared WLAN environment.

3.3.1 Perils in the Common Good

Damsgaard et al. [5] illustrated the concept of "the tragedy of the commons" [6] by delineating five different selfish behaviors in sharing WLAN access among users of several nearby networks. Specifically, when some users behave selfishly in sharing some common resource so as to get short-term gain, other users may copy the selfish behaviors. Thus, such selfish behaviors become more wide-spread and eventually lead to destruction of the common resource. Consequently, these users suffer from their selfish behaviors. The five possible selfish behaviors in sharing WLAN access identified by Damsgaard et al. [5] are:

- *Overgrazing*. This destructive phenomenon occurs when there are too many WLAN devices sharing the available bandwidth. The policy usually used for deterring this is to impose limits on the wireless access. For example, access time can be limited for each device. Data rate could also be limited.
- *Stealing*. This occurs when some nomadic WLAN devices make use of the bandwidth originally allocated to legitimate devices of a WLAN without asking for permission. This can happen through impersonation or cracking of access keys. A common method to control the stealing problem is to use intrusion detection systems (IDSs) such as Airsnort to detect and subsequently exclude some nomadic users.
- *Poaching*. In contrast to stealing, poaching is even more detrimental because it involves blocking legitimate WLAN devices from accessing the shared bandwidth, in the hope that the malicious poachers can obtain more bandwidth. Again, to deal with the poaching problem, an IDS is needed.
- *Tainting*. This malicious behavior occurs when an adversary installs viruses or worms into legitimate users' devices. Another tainting example is that an adversary installs a rogue access point in a shared environment. To avoid this, every new device or access point has to be carefully checked (e.g., scanned for known viruses) before granting wireless bandwidth access.
- *Contamination*. This phenomenon manifests itself as the shutting down of a WLAN access point so that other users cannot use it. Another possible contaminating behavior is that a malicious user intentionally brings an IEEE 802.11b device into an IEEE 802.11 g network, thereby lowering the signal quality of the latter. To tackle this problem, a judicious malicious behavior detection system has to be used.

In summary, as the cost of being selfish is low but the potential gain could be high, it is difficult to deter the above selfish behaviors. Perhaps a penalty scheme is needed to help shape the behaviors of nomadic WLAN users.

3.3.2 P2PWNC

Antoniadis et al. [7] and Efstathiou and Polyzos [8] discussed general issues about providing roaming support among several wireless Internet service providers (WISPs). Specifically, they proposed an architecture, called Peer-to-Peer Wireless Network Confederation (P2PWNC), to support roaming. In the proposed architecture, a key component is the domain agents (DAs), which are specialized WLAN devices that are responsible for provisioning of wireless access rights and the monitoring of the access usages.

The incentive scheme used in the proposed P2PWNC roaming architecture is based on reciprocity. When a foreign WLAN user is allocated wireless bandwidth, its home network's DA will then send tokens to the visited WLAN's DA. Consequently, a key condition for this scheme to work is that each DA must judiciously maintain a certain amount of tokens so that its users can roam to other networks. A DA achieves this by providing access to foreign users. In essence, using DAs in such an autonomous manner is targeted to avoid complicated administrative agreements among multiple WISPs that are currently used in the wired Internet. However, while Efstathiou and Polyzos [8] provided a very detailed account of the major issues in implementing and deploying such a roaming architecture, a major drawback is that the question of how to provide secure exchange of unforgeable tokens is not addressed.

Frangoudis and Polyzos [9] then extended the P2PWNC system to provide secure decentralized voice and video communications. Using their experimental testbed consisting of 14 desktop PCs, they demonstrated the viability of their approach.

3.3.3 Game Theoretic Random Backoff

Čagalj et al. [10] studied selfishness in the implementation of random backoff algorithm in the carrier sense multiple-access/collision avoidance (CSMA/CA) protocol, which is the basis of the IEEE 802.11 media access control (MAC). Specifically, they observed that a rational WLAN user could selfishly tune the contention window in order to gain a larger share of wireless bandwidth. They formulated the WLAN channel access mechanism as a game in which the WLAN users are players. Each player i's strategy is to determine the value of the contention window W_i. The utility function for each player is the achieved data throughput r_i. A smaller value of W_i leads to a larger value of r_i.

Čagalj et al. [10] first considered a static game where all the players make deterministic decisions simultaneously. They determined two kinds of Nash equilibria. In the first kind, there is exactly one selfish device that sets its contention window as one so that this device obtains all the bandwidth and the remaining users obtain zero throughput. In the second kind, there are multiple selfish devices, all of which set their contention windows to 1. Consequently, the "tragedy of the commons" occurs—all users, selfish or not, obtain zero throughput.

Čagalj et al. [10] then considered a dynamic game in which the users repeatedly make their decisions in multiple rounds. In this dynamic game, the Nash equilibria are of the form $W_i = \tau_i$, where $\tau_i \in (0,1)$ for all selfish user i. In essence, the selfish users cooperate to avoid system collapse and, thus, they set their contention windows judiciously. However, there is a possibility that some selfish users would deviate from this cooperation so as to gain more bandwidth. Thus, cooperative detection and penalty schemes are needed. Specifically, each selfish user proactively checks whether some other selfish users deviate from the Nash equilibrium point. If such a user is detected, all remaining selfish users penalize the deviated user by jamming the latter's transmission for a certain period of time. This penalty scheme provides an incentive for each selfish user not to deviate from the Nash equilibrium point.

Konorski [11] also worked on a game-theoretic investigation of the selfish random backoff problem. Unlike previous approaches, Konorski focused on the idea that devices might have different quality of service (QoS) sensitivity, that is, different valuation of the same offered throughput. A detailed mathematical analysis demonstrated that there is a trade-off between bandwidth utilization and fairness, pretty much like the problem of channel quality adaptation.

3.3.4 Reputation-Based Wireless Link Access

Salem et al. [12] proposed an architecture of wireless link access in which multiple WISPs provide WLAN access to users. Each user can select from among these WISPs for connecting to the Internet. In the proposed architecture, a key component is called the trusted central authority (TCA), which is a universally trusted machine that is responsible for maintaining and updating the reputation values of the various WISPs registered with the TCA. Specifically, a user selects the WISP that has the highest reputation and can provide the required level of QoS. After using the wireless service provided by the selected WISP, the user will pay the WISP and also send its feedback to the TCA. The TCA then updates the selected WISP's reputation value according to the user's feedback.

Salem et al. [11] showed that their scheme can handle various kinds of security attacks. Furthermore, the scheme is found to be robust and can provide a higher overall throughput under a wide range of simulation configurations.

3.3.5 Auction-Based Incentive Compatible MAC

BenAmmar and Baras [13] also considered the problem with the random backoff mechanism in the CSMA/CA protocol. In their study, they first formulated the noncooperative backoff process as a Bayesian game, in view of the fact that it is impractical for WLAN devices to have complete information about other players in the system. Assuming that all the devices have the same utility function under the same level of throughput, BenAmmar and Baras showed that the symmetric Nash equilibrium of the game results in a very low system throughput.

In view of the negative result in the symmetric Bayesian game, BenAmmar and Baras [13] considered the second-price Vickrey auction. Specifically, they formulated the WLAN link access process between the client devices and the AP as a Vickrey auction. Accordingly, they designed the link access protocol called incentive-compatible MAC (ICMAC). With an auction-based approach, heterogeneous valuation of wireless link throughput is modeled. That is, different devices could have different valuations of the same level of throughput. Consequently, the devices would submit different bids to the AP for the wireless link access. The AP then assigns time slots to the competing devices according to their bids.

Using optimized network engineering tools (OPNET), BenAmmar and Baras [13] obtained promising simulation results, indicating that the proposed ICMAC is practical in utilizing the wireless bandwidth efficiently even in the presence of selfish users. A major deficiency in their study is that the payment scheme is not clearly specified and demonstrated to be practicable.

3.4 MULTIHOP NETWORKS

In a wireless environment, however, the mere action of sending a request message from a client peer to a server peer would probably need several intermediate peers to help in forwarding the message because the server and client peers may be out of each other's transmission range. Consequently, incentives have to be provided to encourage such forwarding actions, as illustrated in Fig. 3.2.

Indeed, in a wireless computing system, the connectivity among peers is itself a bootstrap sharing problem. Specifically, if wireless users are unwilling to cooperate in performing routing and data forwarding, the wireless network can be partitioned so that service providers cannot be reached by potential service consumers. In view of this critical challenge, there has been a plethora of important research results related to incentive issues for ad hoc routing and data forwarding. In the following, we briefly survey several recently suggested approaches.

3.4.1 CONFIDANT Ad Hoc Routing Protocol

Buchegger and Le Boudec [14] proposed a cooperative routing system called CONFIDANT (cooperation of nodes: fairness in dynamic ad hoc networks). The major design objective of CONFIDANT is to detect and isolate

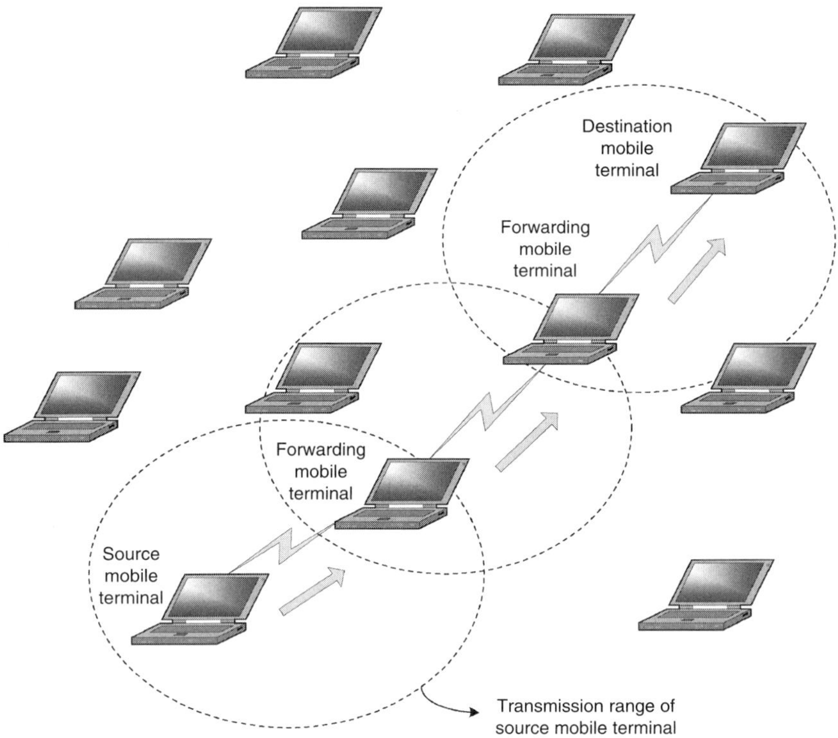

FIGURE 3.2 Cooperative data forwarding in wireless ad hoc network.

misbehaving devices in an ad hoc network. Such "expulsion" action therefore provides incentive for devices to cooperate.

The CONFIDANT system consists of four major components:

- *Montior*. Each device has a monitor keeping track of the behaviors of neighboring devices. Specifically, the monitor is designed to detect data integrity attacks (i.e., modifying the contents of a message) and non-forwarding attacks. If such malicious behaviors are observed, the device will notify the reputation system by sending out alarm messages.
- *Trust Manager*. This component is responsible for maintaining and managing the trust data in a device. The trust manager is composed of three subsystems: an alarm table, a trust table, and a friends list. The alarm table is a database of alarm messages received by the device. Each alarm message signifies the existence of a malicious user. The trust table is a database of the trust levels for known devices. The trust value of a device is used for judging whether an alarm message received from such a device is trustworthy. If an alarm message is generated by the device itself, such an alarm message will be forwarded to those devices on the friends list.

- *Reputation System*. This component is responsible for recording and updating the reputation ratings (different from trust levels) of other devices. The rating of a device is changed only when there is enough data showing that the device has exhibited malicious behaviors. The rating of a device is updated according to a weighted function in that a heavier weight is used for local observation of malicious behaviors, while a lighter weight is used for alarm messages received from other devices.
- *Path Manager*. This component manages the operations related to data-forwarding paths. Specifically, the path manager in a device performs reranking of paths according to the reputation ratings of devices on the path. Moreover, the path manager would remove paths that contain malicious devices. Furthermore, the path manager would also handle incoming path-related messages. If such messages are from a malicious device, the path manager will ignore them.

Buchegger and Le Boudec [14] conducted simulations to evaluate the performance of the proposed CONFIDANT system on top of the dynamic source routing (DSR) protocol [15]. Their study indicated that CONFIDANT is robust in that the overhead is small. Furthermore, CONFIDANT is scalable in terms of the total number of devices in the system.

Nevertheless, a key problem remains to be solved for the CONFIDANT system is that an incentive scheme has to be designed so that devices can become "friends" of each other.

Caituiro-Monge et al. [16] recently also proposed a similar system called Friend Relay.

3.4.2 Issues in Incentive-Based Multihop Forwarding

Mahajan et al. [17] presented a detailed account on the possible incentive techniques for multihop data forwarding and the problems expected when using game-theoretic methods for modeling and design. Specifically, they pointed out the challenges in each of the following possible incentive schemes:

- *Barter*. Simply put, a device would forward data for a neighboring device if the latter also forwards data for the former in return. However, the challenge is that barter is not suitable for asymmetry in time and space dimensions. That is, first of all, the two devices might not need the service of each other *simultaneously*. Apart from this time dimension problem, there is a space dimension problem—two remotely located devices would not help each other.
- *Currency*. In this scheme, each device pays a helping device with a virtual currency. Apart from the security aspects of implementation of the currency scheme, there are two major problems [17]. First, devices located at the edge of the network naturally would not be involved in many data-forwarding processes, and thus, would be unable to accumulate enough

"money" for its own "spending." The second problem, an even more subtle one, is that those "rich" devices may have a *low incentive* to participate further in data forwarding because they have already had more than enough money. This is in some sense contrary to the currency in the real world but is probable in a wireless network if the virtual currency has no "real value" apart from buying data-forwarding actions.

- *Game-Theoretic Forwarding*. Usually, with a game-theoretic modeling, a Nash equilibrium could be reached and all devices would participate in the data-forwarding process with some rational strategies. However, a key assumption, commonly used in game-theoretic modeling, is that all devices have symmetric workload. As such, the devices are mutually dependent on each other. However, this assumption may not be valid in a real environment because devices at the edge need more service from other devices, but the reverse is usually not true.
- *Evolutionary Scheme*. This is essentially a "repeated game" approach in that the devices' interactions are partitioned into multiple rounds. In each round, devices could use different strategies based on their "past experience." The rational goal of each device is to maximize its payoff in the current round through using a strategy that is considered the best based on past experience. However, it is still difficult to use this scheme to accurately model a real-world data-forwarding situation [17] because there are too many possible *combinations* of strategies to consider.

Mahajan et al. [17] also identified three problems in implementing a game-theoretic wireless data-forwarding system. These problems are not insurmountable but may reduce the practical effectiveness of a game-theoretic solution.

- *Identity*. It is common that a game-theoretic modeling requires an identity scheme that binds each device with a permanent, unforgeable identity. However, it is very difficult to bind a permanent identity to a user, especially in a wireless environment. Although hardware addresses might be used for such purposes, it is difficult to impose the requirement that the hardware address of a user be truthfully reported.
- *Cost Effectiveness*. Mahajan et al. [17] also insightfully pointed out that cooperation among devices in a wireless network may not be the most important goal. Indeed, routing, power management, and MAC may be more important from a system's perspective. Thus, the "cooperation cost" implied by a game-theoretic approach should be small.
- *Uncertainty*. There are inevitably uncertain events that can occur in a real wireless environment. For instance, due to time-varying channel quality, a packet might need to be retransmitted several times before successfully received by the destination device. Such retransmissions represent a significant cost to the forwarding device and are unpredictable in advance. A related uncertainty is then the accuracy of cheating detection. As a

receiver might rightfully claim that a packet is not successfully received, a sender might mistakenly treat this claim as cheating or a noncooperative behavior that induces punishment, resulting in unfair penalty. Thus, the uncertainty in cheating detection also impairs a game-theoretic approach relying on penalty-induced incentive.

3.4.3 Multiqueue-Based Fair Relaying

Casetti et al. [18] proposed a practical approach for fair relaying in a multihop IEEE 802.11–based ad hoc network. Specifically, in their proposed approach, there are two components. In the first component, two LLC (link layer control) queues are used. One LLC queue is designated for local traffic sending, while the other is for relaying other devices' traffic. With these two queues, a split-queue (SQ) scheduling algorithm is used:

- The relaying LLC queue is served if the local LLC queue was served in the previous round.
- The relaying LLC queue is also served if the local LLC queue length is smaller than a threshold.
- The local LLC queue is served if the relaying LLC queue was served in the previous round, provided that the local LLC queue length is larger than or equal to a threshold.

The second component is based on the IEEE 802.11e QoS facilities. Specifically, four access categories (AC) (in IEEE 802.11e) are defined, in decreasing priority:

- AC [0] for local user datagram protocol (UDP) traffic
- AC [1] for local transmission control protocol (TCP) traffic
- AC [2] for relaying UDP traffic
- AC [3] for relaying TCP traffic

Consequently, with these two components, one for the link layer and the other for the transport layer, the proposed approach is practical for a real wireless environment. Simulation results show that the proposed approach is effective.

3.4.4 Channel Adaptive Scheduling with Incentive

Wei and Gitlin [19] considered the data-relaying/forwarding problem in a hybrid wireless ad hoc network, as shown in Fig. 3.3. In such a hybrid wireless network, a dual-mode device could use its WLAN interface to relay traffic from the base station (BS) to another client device who is suffering from a poor third-generation (3G) channel quality. The problem here is that if there is no incentive provided by the BS for the dual-mode relaying device, the latter would not cooperate and help the device with a poor 3G channel quality.

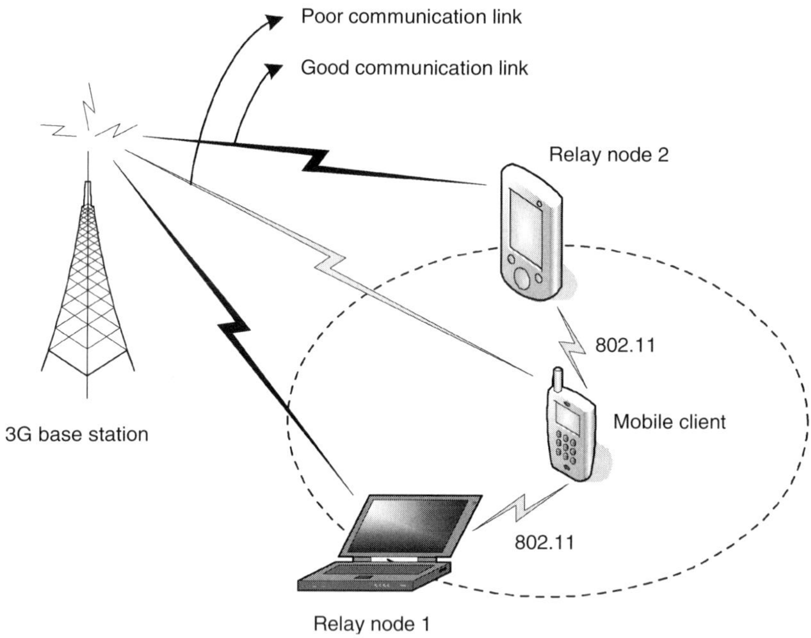

FIGURE 3.3 Hybrid wireless network in which some dual-mode (with two interfaces: 3G and IEEE 802.11*x* WLAN) devices can help other devices, suffering from poor 3G channel quality, by relaying their data from the BS to them.

They examined several classic scheduling algorithms to be used by the BS:

- *Round Robin.* This simple scheduler just allocates time slots to each user without regard to channel quality.
- *Maximum Rate.* This scheduler just allocates a time slot to user k given by

$$k = \arg\max_i \{r_i(t)\}$$

  where $r_i(t)$ is the instantaneous data rate of user i at time slot t.
- *Proportional Fair.* This scheduler, which is used in CDMA2000 HDR system, allocates a time slot to user k given by

$$k = \arg\max_i \left\{ \frac{r_i(t)}{\mu_i(t)} \right\}$$

  where $\mu_i(t)$ is the average rate allocated up to time slot t.

In summary, the above three schedulers do not provide special incentive for a dual-mode relaying device.

Using a game-theoretic analysis, Wei and Gitlin [19] showed that all the above three schedulers could result in a highly inefficient Nash equilbirum—no dual-mode relaying device in the system would relay traffic for other devices.

In view of the poor Nash equilibrium that could happen, Wei and Gitlin [19] proposed an enhanced scheduling algorithm that takes into account incentive. Specifically, a time slot is allocated to user k given by

$$k = \arg\max_i \left\{ \frac{c_i r_i(t)}{\mu_i(t)} \right\}$$

where c_i is the incentive parameter that equals to c_R ($c_R = 1.5$, is used for a device that has previously relayed traffic); otherwise, $c_i = 1$ for other devices.

Effectively, a device that previously relayed traffic for others would have a higher chance of getting a time slot, with everything else equal.

Wei and Gitlin [19] analytically showed that an efficient Nash equilibrium exists for the incentive-enhanced scheduler. Simulation results also indicated that the scheduler is effective.

3.4.5 Repeated Game Multihop Data Forwarding

Milan et al. [20] considered the modeling of the relaying actions among neighboring devices in an ad hoc network as a repeated game. In this model, the classical solution is the tit-for-tat (TFT) strategy:

- Device A relays data for a neighbor device B if device B has also relayed data for device A in the previous round.
- Device A punishes device B by refusing to relay data for it if device B is found to be selfish (i.e., refused to relay data for device A) in the previous round.
- Punishment is only for one round and cooperation is resumed in the next round.

However, as discussed earlier, channel errors in device B might be mistakenly perceived by device A as a selfish action in that the data forwarded by device B is corrupted by poor channel conditions. Thus, the TFT strategy could be enhanced by adding a tolerance of misbehaviors—a limited number of nonforwarding actions are not considered as selfish and will not be punished. The resulting strategy is called generous TFT (GTFT).

Milan et al. [20] also considered two schemes with more severe punishment strategies. The first one, called one-step trigger (OT), is a slight variation of the GTFT strategy. Specifically, in OT, even a small perceived deviation from cooperation triggers a punishment round. The second one, called grim trigger (GT), imposes permanent punishment (i.e., forever noncooperation) once a deviation is detected.

Milan et al. [20] then proved analytically that increasing punishment severity increases the network capacity because a heavy perceived punishment deters noncooperation.

3.5 CONCLUDING REMARKS

In this chapter, we have briefly surveyed recently suggested approaches in providing incentives for cooperation in a IEEE 802.11*x*–based wireless environment. While all of the six different types of incentive techniques have been employed in the surveyed approaches, much work still has to be done in designing a practicable system. Indeed, many simplifying assumptions are made in the surveyed approaches (e.g., permanent identity, symmetric workload, etc.) so that the designed algorithms may not work well in a real environment.

ACKNOWLEDGMENTS

This research was supported by the Hong Kong Research Grants Council (under project number HKU7157/04E). Thanks are due to Mr. Tyrone Kwok for his kind assistance in compiling the figures in this chapter. The author would also like to thank Professor Yang Xiao for his professional advice in preparing this chapter.

REFERENCES

1. Martin J. Osborne, *An Introduction to Game Theory*, Oxford University Press, New York, USA, 2004.
2. S. Kawade, J.-W. Van Bloem, V. S. Abhayawardhana, and D. Wisely, "Sharing your urban residential WiFi (UR-WiFi)," in *Proc. of the IEEE 63rd Vehicular Technology Conference* (*VTC 2006–Spring*), Vol. 1, Melbourne, Australia, 7–10 May 2006, pp. 162–166.
3. A. Akella, G. Judd, S. Seshan, and P. Steenkiste, "Self-management in chaotic wireless deployments," in *Proc. of the 11th Annual International Conference on Mobile Computing and Networking* (*MOBICOM 2005*), Cologne, Germany, 28 Aug.–2 Sept. 2005, pp. 185–199.
4. V. A. Siris and C. Courcoubetis, "Resource control for the EDCA mechanism in multi-rate IEEE 802.11e networks," in *Proc. of the 2006 International Symposium on World of Wireless, Mobile and Multimedia Networks* (*WoWMoM 2006*), Buffalo, New York, USA, 26–29 June 2006, pp. 419–428.
5. J. Damsgaard, M. A. Parikh, and B. Rao, "Wireless commons: Perils in the common good," *Commun. ACM*, 49(2), 104–109 (2006).
6. G. Hardin, "The tragedy of the commons," *Science*, 162, Dec. 1968.
7. P. Antoniadis, C. Courcoubetis, E. C. Efstathiou, and G. C. Polyzos, "Peer-to-Peer Wireless LAN Consortia: Economic modeling and architecture," in *Proc. of the 3rd International Conference on Peer-to-Peer Computing* (*P2P 2003*), Eds. Nahid Shahmehri, Ross Lee Graham, and Germano Carroni, Linköping, Sweden, 1–3 Sept. 2003, pp. 198–199.
8. E. C. Efstathiou and G. C. Polyzos, "A peer-to-peer approach to wireless LAN roaming," in *Proc. of the 1st ACM International Workshop on Wireless Mobile*

Applications and Services on WLAN Hotspots (*WMASH 2003*), San Diego, CA, USA, 19 Sept. 2003, pp. 10–18, ACM, New York, USA.

9. P. A. Frangoudis and G. C. Polyzos, "Peer-to-peer secure and private community based multimedia communications," in *Proc. of the 8th IEEE International Symposium on Multimedia* (*ISM 2006*), San Diego, CA, USA, 11–13 Dec. 2006, pp. 1004–1010.
10. M. Čagalj, S. Ganeriwal, I. Aad, and J.-P. Hubaux, "On selfish behavior in CSMA/CA networks," in *Proc. of the 24th Annual Joint Conference of the IEEE Computer and Communications Societies* (*INFOCOM 2005*), Eds. Kia Makki and Edward Knightly, Vol. 4, Miami, Florida, USA, 13–17 Mar. 2005, pp. 2513–2524.
11. J. Konorski, "Quality of service games in an IEEE 802.11 ad hoc wireless LAN," in *Proc. of the 9th ACM International Symposium on Modeling Analysis and Simulation of Wireless and Mobile Systems* (*MSWiM 2006*), Eds. Enrique Alba, Carla–Fabiana Chiasserini, Nael Abu-Ghazaleh, and Renato Lo Cigno, Torremolinos, Malaga, Spain, 2–6 Oct. 2006, pp. 265–272, ACM, New York, USA.
12. N. B. Salem, J.-P. Hubaux, and M. Jakobsson, "Reputation-based Wi-Fi deployment," *Mobile Comput. Commun. Rev.*, 9(3), 69–81.
13. N. BenAmmar and J. S. Baras, "Incentive compatible medium access control in wireless networks," *Proc. of the 2006 Workshop on Game Theory for Communications and Networks* (*GameNets 2006*), Article 5, Pisa, Italy, 14 Oct. 2006.
14. S. Buchegger and J.-Y. Le Boudec, "Performance analysis of the CONFIDANT protocol (Cooperation of Nodes: Fairness in Dynamic Ad Hoc Networks)," in *Proc. of the 3rd ACM International Symposium on Mobile Ad Hoc Networking and Computing* (*MOBIHOC 2002*), Lausanne, Switzerland, 9–11 June 2002, pp. 226–236, ACM, New York, USA.
15. D. B. Johnson and D. A. Maltz, "The dynamic source routing protocol for mobile ad hoc networks," Internet draft, Mobile Ad Hoc Network (MANET) Working Group, Internet Engineering Task Force, Oct. 1999.
16. H. Caituiro-Monge, K. Almeroth, and M. del Mar Alvarez-Rohena, "Friend relay: A resource sharing framework for mobile wireless devices," in *Proc. of the 4th International Workshop on Wireless Mobile Applications and Services on WLAN Hotspots* (*WMASH 2006*), Los Angeles, CA, USA, 29 Sept. 2006, pp. 20–29.
17. R. Mahajan, M. Rodrig, D. Wetherall, and J. Zahorjan, "Experiences applying game theory to system design," in *Proc. of the ACM SIGCOMM Workshop on Practice and Theory of Incentives in Networked Systems*, Portland, Oregon, USA, 3 Sept. 2004, pp. 183–190, ACM, New York, USA.
18. C. Casetti, C.-F. Chiasserini, and L. Previtera, "Fair relaying and cooperation in multi-rate 802.11 networks," in *Proc. of the IEEE 61st Vehicular Technology Conference* (*VTC 2005–Spring*), Vol. 3, Stockholm, Sweden, 30 May–1 June 2005, pp. 2033–2036.
19. H.-Y. Wei and R. D. Gitlin, "Incentive mechanism design for selfish hybrid wireless relay networks," *Mobile Networks Appl.*, 10, 929–937 (2005).
20. F. Milan, J. J. Jaramillo, and R. Srikant, "Achieving cooperation in multihop wireless networks of selfish nodes," in *Proc. of the 2006 Workshop on Game Theory for Communications and Networks* (*GameNets 2006*), Article 3, Pisa, Italy, 14 Oct. 2006.

CHAPTER 4

CAPACITY AND RATE ADAPTATION IN IEEE 802.11 WIRELESS LANs

MING LI and YANG XIAO

High capacity is critical for the success of the IEEE 802.11 wireless local area networks (WLANs) on providing high speed Internet access. However, due to *operating inefficiency* and *protocol inefficiency*, the actual achievable overall throughput in 802.11 WLANs has been shown to be much lower than its theoretical limit. This chapter reviews the recent advances in capacity analysis models of the IEEE 802.11 distributed coordination function (DCF) and various throughput improvement techniques. Furthermore, protocols on exploiting the multirate capability are also discussed.

4.1 BACKGROUND

WLANs are becoming more popular and increasingly relied on. The IEEE 802.11 WLAN is accepted as a complementary technology to high speed IEEE 802.3 (Ethernet) for portable and mobile devices. One reason for such success is that it keeps increasing data transmission rates while maintaining a relatively low price. The IEEE 802.11, 802.11b, and 802.11a/g specifications provide up to 2-, 11-, and 54-Mbps data rates [1, 2], respectively. Furthermore, the IEEE 802.11 working group is pursuing IEEE 802.11n, an amendment for higher throughput and higher speed enhancements. Different from the goal of IEEE 802.11b/.11a/.11 g, i.e., to provide higher speed data rates with different physical (PHY) layer specifications, IEEE 802.11n aims at higher throughput instead of higher data rates with PHY and medium access control (MAC) enhancements.

To provide better quality of service (QoS), especially for multimedia applications, increasing data rates is also highly desirable. The rationale is the same as Ethernet, which dramatically increases data rates from 10/100

Emerging Wireless LANs, Wireless PANs, and Wireless MANs. Edited by Y. Xiao and Y. Pan

Mbps to 10 Gbps. Data-rate-intensive applications exist, such as multimedia conferencing, Moving Pictures Experts Group (MPEG) video streaming, consumer applications, network storage, file transfer, and simultaneous transmission of multiple high definition TV (HDTV) signals, audio, and online gaming. Furthermore, there is a great demand for higher capacity WLAN networks in the market in such areas as hotspots, service providers, and wireless backhaul.

However, recent research results show that the overall throughput of IEEE 802.11 WLANs neither reaches optimal throughput under most network conditions [3–5] nor increases linearly with the increase of physical channel data rate; i.e., the maximum throughput (MT) is bounded even with infinitely high data rate [2]. These are two main reasons for this. First, the fixed contention window (CW) setting in 802.11 DCF does not adapt well under either low traffic load or congestion channel status due to a large number of competing stations. Using one single CW_{min}, no matter big or small, yields very low throughput in most scenarios. Second, the overhead of request-to-send/clear-to-send/acknowledgment (RTS/CTS/ACK) and backoff per channel access consumes a significant portion of channel time, especially with high physical channel data rates and small frame sizes. With the existing overhead, simply increasing the data rate cannot improve the throughput beyond a certain limit [2, 6]. These two types of issues are called operating inefficiency and protocol inefficiency, respectively.

To further improve network capacity, multiple rate capability has been enabled and supported by most commercial wireless adapters. For example, IEEE 802.11b supports 2, 5.5, and 11 Mbps, depending on the channel condition and signal strength. Generally, given a certain bit error rate (BER) and maximum channel data rate, the higher the data rate used, the higher the loss ratio. With higher loss ratio, data retransmission and collision probability significantly increase, yielding lower throughput. To handle the trade-off between data rate and frame loss ratio, many adaptation protocols have been proposed to determine the best possible rate based on channel loss ratio [7–9], signal-to-noise ratio (SNR) [10, 11], or some channel quality measures [8].

In this chapter, we provide a comprehensive study of the capacity analysis models and rate adaptation protocols, with an emphasis on various techniques that improve the network capacity. Section 4.2 reviews representative performance models on network capacity, optimal throughput, and theoretical throughput limit. Section 4.3 describes existing rate adaptation schemes such as autorate fallback (ARF) [7], receiver-based autorate (RBAR) [10], Onoe [9], and SampleRate [8] in detail. Section 4.4 introduces throughput-improving techniques such as adaptive parameter tuning, frame concatenation, frame piggybacking, and integration of frame concatenation and rate adaptation. Finally, Section 4.5 concludes this chapter with a discussion of future research trends on this topic.

4.2 CAPACITY ANALYSIS OF IEEE 802.11 DCF

The capacity of IEEE 802.11 wireless networks has been investigated extensively [2–6, 12–16]. Despite the fact that the maximum channel bandwidth of the IEEE 802.11a and 802.11b standards are 54 and 11 Mbps, respectively, it is found that the overall network capacity can be much lower than the specification. Basically, with the existing contention-based carrier sense multiple-access/collision avoidance (CSMA/CA) MAC access, several factors/parameters may lead to inefficient channel utilization and thus low capacity:

- *RTS/CTS Handshaking*. With RTS/CTS, each channel access yields higher overhead on transmitting these MAC frames. However, RTS/CTS will be very effective when network traffic is high, where collision occurs frequently and network capacity decreases due to large frames being dropped.
- *Minimum Contention Window* (CW_{min}). With a larger CW_{min}, a bigger backoff counter is chosen in general when the channel is sensed busy or collision occurs. In this case, more channel time may be wasted. However, a small CW_{min} will lead to identical backoff counters being chosen by different stations and thus increases collision probability and frame retransmissions.
- *Data Frame Size*. If the data frame size is very small, e.g., voice frames, a high portion of the channel time is used for backoff process, the DCF interframe space (DIFS), short interframe space (SIFS), and control frames such as RTS/CTS/ACK.
- *Network Traffic*. With heavy traffic load (e.g., large number of contending stations and/or high bit rate at the sender), the collision probability and frame retransmission increase significantly, leading to a larger backoff counter. In this case, more channel time is wasted due to a frame drop or backoff procedure.

Therefore, it is important to consider the above factors/parameters in order to fully understand the issue of network capacity in 802.11-based WLANs.

4.2.1 *p*-Persistent Model

Cali et al. [4] proposed one of the first analytical models, the *p-persistent IEEE 802.11 protocol*, as an approximation of the IEEE 802.11 DCF. In this protocol, it is assumed that for each transmission attempt a station uses a backoff interval sampled from a geometric distribution with parameter p, where $p = 1/(E[B]+1)$ and $E[B]$ is the average backoff time. Also, it is assumed that frame size follows the geometric distribution with parameter q. Then, with the geometric distribution assumption, it is easy to obtain that

$E[B] = (\mathrm{E[CW]} - 1)/2$ and $\rho_{\max}$, the normalized capacity, can be calculated as

$$\rho_{\max} = \frac{\overline{m}}{\frac{1-p}{Mp} t_{\text{slot}} + \frac{1-(1-p)^M - Mp(1-p)^{M-1}}{Mp(1-p)^{M-1}} [E[\text{Coll}] + \tau + \text{DIFS}] + E[S]}$$

where $\overline{m}$ is the average frame length, M is the number of competing stations, τ is the maximum propagation delay, and $E[\text{Coll}]$ and $E[S]$ are the expected time of collision and data transmission, respectively. Based on the above equation, $\rho_{\max}$ is a function of M and p. Then, given a specific M, there exists a $p_{\min}$ to maximize $\rho_{\max}$. Simulation results show that the p-persistent model approximates the performance of the IEEE 802.11 DCF quite well.

4.2.2 Markov Chain Model

Bianchi [3] proposed one of the first Markov chain models for the saturation capacity of the IEEE 802.11 DCF. In this model, it is assumed that at each transmission, regardless of the retransmission suffered, each frame collides with constant and independent probability, which is defined as the probability that a collision seen by a frame is being transmitted on the channel and is referred to as *conditional collision probability*. Then, a bidimensional process $\{s(t), b(t)\}$ is defined. In this model, nonnull on-step transition probabilities are

$$P\{i, k|i, k+1\} = 1 \qquad k \in (0, W_i - 2) \quad i \in (0, m)$$

$$P\{0, k|i, 0\} = (1-p)/W_0 \qquad k \in (0, W_i - 1) \quad i \in (0, m)$$

$$P\{i, k|i-1, 0\} = p/W_i \qquad k \in (0, W_i - 1) \quad i \in (1, m)$$

$$P\{m, k|m, 0\} = p/W_m \qquad k \in (0, W_m - 1)$$

where m is the maximum backoff stage, i.e., maximum allowed data retransmissions, and $W_i = 2^i W$, $i \in (0, m)$, is the CW at the ith backoff stage with W being the initial/minimum CW.

The state probability $b_{i,k} = \lim_{t\to\infty} P\{s(t) = i, b(t) = k\}$, $i \in (0, m)$, $k \in (0, W_i - 1)$, is the stationary distribution of the chain and it is obvious that $b_{i,0} = p^i b_{0,0}$, $i \in (0, m)$, and $b_{m,0} = p^m b_{0,0}/(1-p)$. Then, with appropriate derivation, we have that $b_{i,k} = (W_i - k) b_{i,0}/W_i$, $i \in (0, m)$, $k \in (0, W_i - 1)$. Finally, $b_{0,0}$ can be calculated according to the normalization condition that summation $b_{i,k}$ equals to 1 and is expressed as

$$b_{0,0} = \frac{1(1-2p)(1-p)}{(1-2p)(W+1) + pW(1-(2p)^m)}$$

Define τ as the probability that a station transmits in a randomly chosen slot time. Then, according to [17], $\tau = 2/(W+1)$. On the other hand, it is obvious that $p = 1 - (1-\tau)^{n-1}$. Thus, these two equations can be solved jointly to obtain a unique solution for unknown variables τ and p.

Finally, the saturation throughput can be calculated as

$$S = \frac{P_s P_{\mathrm{tr}} E[P]}{(1 - P_{\mathrm{tr}})\sigma + P_{\mathrm{tr}} P_s T_s + P_{\mathrm{tr}}(1 - P_s)T_c}$$

where $E[P]$ is the average frame payload size and P_{tr} is the probability that at least one station is transmitting with a total of n competing stations. Therefore, $P_{\mathrm{tr}} = 1 - (1-\tau)^n$. Also, $P_s = n\tau(1-\tau)^{n-1}/P_{\mathrm{tr}}$ is the probability that a transmission is successful given that exactly one station is transmitting with conditional probability that at least one station is transmitting, T_s is the average time the channel is sensed busy because of a successful transmission, T_c is the average time the channel is sensed busy by a station during a collision, and σ is the duration of an empty slot time.

Results in [3] show that RTS/CTS overhead does decrease network capacity slightly due to the extra overhead at light traffic load. However, when the number of stations increases, the basic DCF suffers severely from throughput degradation. Also, it is clear that without RTS/CTS it is more desirable to use larger window sizes to alleviate the channel contention so that throughput does not decrease significantly. However, with RTS/CTS, the CW does not have much effect on the saturation throughput due to its effectiveness on preventing transmission failure of large-size frames. The Markov chain has been widely used for analysis of IEEE 802.11 performance under various scenarios [5, 12, 13, 15, 18, 19].

4.2.3 Optimal Throughput of IEEE 802.11 WLANs

To investigate the performance of the IEEE 802.11 DCF with different numbers of competing stations, Ma et al. [5] utilized a system model similar to Bianchi's model [3] and use different $\mathrm{CW_{min}}$ to calculate the theoretical effective system throughput when the number of competing stations varies from 1 to 100. Numerical results in [5] show that none of the fixed $\mathrm{CW_{min}}$ from 13 to 726 yields optimal performance with all numbers of competing stations. When N is small, using large $\mathrm{CW_{min}}$ leads to very low channel utilization. However, when N is big, using small $\mathrm{CW_{min}}$ causes too many frame collisions and data retransmissions. Instead, each $\mathrm{CW_{min}}$ has its own optimal operating zones, or subareas. To reach optimal throughput under dynamic network conditions, it is desirable that $\mathrm{CW_{min}}$ be dynamically changed according to the number of competing stations. We will discuss this in Section 4.3.1.

In Section 4.2.1, we introduced the p-persistent model, where the normalized throughput $\rho_{\max}$ is a function of competing station M and constant p. Then, given a specific M, there exists a $\rho_{\min}$ to maximize $\rho_{\max}$. Cali et al. [4] also found the optimal throughput of IEEE 802.11 WLANs with the appropriate $p_{\min}$ from analysis. Then, they proposed a distributed scheme to estimate M and then configure the network with the best $p_{\min}$. A performance comparison shows that this approach yields throughput very close to the theoretical limit.

4.2.4 Theoretical Throughput and Delay Limit of WLANs

Xiao and Rosdahl [2, 6] investigated the theoretical throughput and delay limits of IEEE 802.11. The intention of this research was to see potentially how high the achievable throughput can be given arbitrarily high data rates. To derive the throughput upper limit (TUL) and the delay lower limit (DLL), two performance metrics, the achievable MT and the achievable minimum delay (MD), are derived first. To derive the MT and the MD, the system must be at the best-case scenario where (i) the channel is an ideal channel without errors and (ii) at any transmission cycle there is one and only one active station which always has a frame to send and other stations can only accept frames and provide acknowledgments. In this situation, we have that

$$\mathrm{MT} = \frac{8L_{\mathrm{DATA}}}{T_{\mathrm{D\_DATA}} + T_{\mathrm{D\_ACK}} + 2\tau + T_{\mathrm{DIFS}} + T_{\mathrm{SIFS}} + \overline{\mathrm{CW}}}$$

$$\mathrm{MD} = T_{\mathrm{D\_DATA}} + \tau + T_{\mathrm{DIFS}} + \overline{\mathrm{CW}}$$

where $T_{\mathrm{D\_DATA}}$, $T_{\mathrm{D\_ACK}}$, T_{DIFS}, and T_{SIFS} are the time for data transmission, ACK transmission, DIFS, and SIFS, respectively, L_{DATA} is the payload of the data frame, and τ is the propagation delay. The average backoff time $\overline{\mathrm{CW}}$ is given by $\mathrm{CW}_{\min}T_{\mathrm{slot}}/2$. Then, further analysis shows that

$$\mathrm{TUL} = \frac{8L_{\mathrm{DATA}}}{2T_P + 2T_{\mathrm{PHY}} + 2\tau + T_{\mathrm{DIFS}} + T_{\mathrm{SIFS}} + (CW_{\min}T_{\mathrm{slot}}/2)}$$

$$\mathrm{MD} = T_P + T_{\mathrm{PHY}} + \tau + T_{\mathrm{DIFS}} + \frac{CW_{\min}T_{\mathrm{slot}}}{2}$$

where T_P is the transmission time of the preamble frame. Thus, TUL and MD correspond to the situation where channel data rate is infinity. Although this limit is not practical, it provides a theoretical upper bound and lower bound for the throughput and delay, respectively. Figure 4.1 shows the TUL upper bounds of all the stations for IEEE 802.11a. When the payload size is 1000 bytes, the MT for 54 Mbps is 24.7 Mbps and the TUL is 50.2 Mbps. The MT for 54,000 Mbps with the same set of overhead parameters almost reaches the TUL. Figure 4.2 shows the DLL lower bounds of all the MDs for IEEE 802.11a. The DLL is the same for all payload sizes, i.e., 122.5 μs. When the payload size is 1000 bytes, the MD for 54 Mbps is 278.5 μs. The MD for 54,000 Mbps with the same set of overhead parameters almost reaches the DLL. Furthermore, it is surprising to

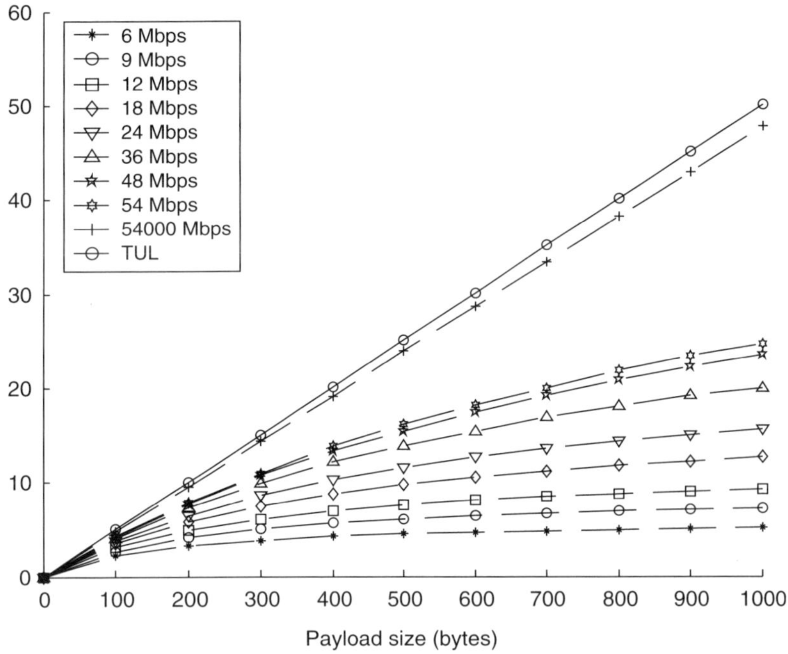

FIGURE 4.1 MT and TUL (Mbps) of IEEE 802.11a. *Source*: Y. Xiao and J. Rosdahl, *IEEE Commun. Lett.* 6(8), 2002, 355–357.

observe that when payload is very small (100 bytes), only around 5-Mbps throughput limit can be achieved. This is due to the fact that with small payload size the overhead of control frames and backoff time is much larger and most channel time is not used for useful data transmission.

In summary, the existing IEEE 802.11 MAC protocol does not achieve effective network capacity. From our discussion on the analysis of the DCF performance and its theoretical limit, it is obvious that two types of inefficiencies significantly reduce the overall throughput:

- *Operating Inefficiency*. The fixed CW setting in the 802.11 DCF does not adapt well under either low traffic load or congestion due to the large number of competing stations. Existing research results have shown that using one CW, no matter big or small, yields very low throughput in many scenarios.
- *Protocol Inefficiency*. The overhead of the RTS/CTS/ACK and backoff procedure per channel access consumes a significant portion of channel time, especially with high physical channel data rates and small frame sizes. Existing research results indicate that with the existing overhead simply increasing the data rate cannot improve the throughput beyond a certain bound.

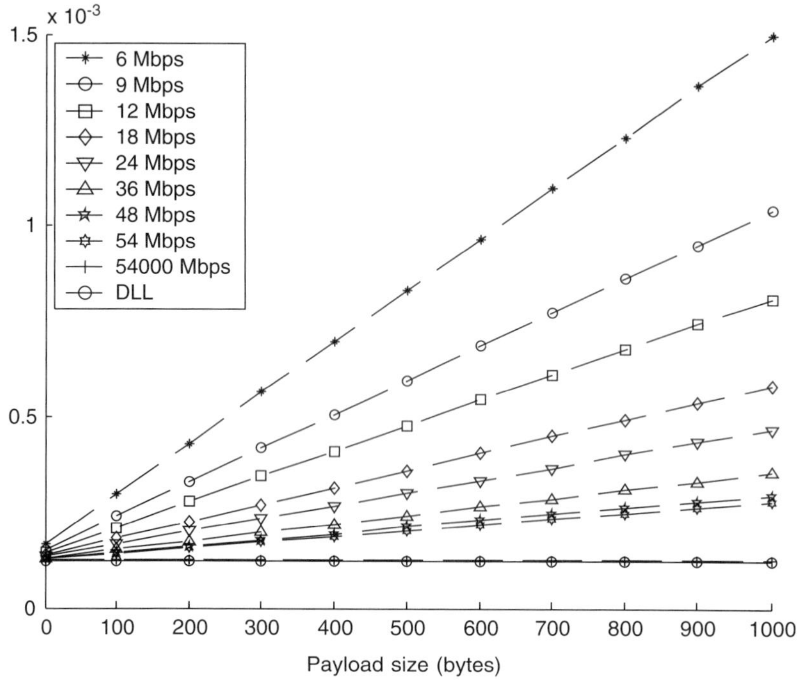

FIGURE 4.2 MD and TLL (seconds) of IEEE 802.11a. *Source*: Y. Xiao and J. Rosdahl, *IEEE Commun. Lett.* 6(8), 2002, 355–357.

On the other hand, given the various available rates in the IEEE 802.11 standard, it is desirable to use higher rates for data transmission to improve network capacity. In the next section, we will discuss various rate adaptation protocols.

4.3 RATE ADAPTATION PROTOCOLS

Basically, rate adaptation is quite complicated. Given a certain BER and maximum channel data rate, the higher the data rate used, the higher the loss ratio will be. With higher loss ratio, data retransmission and collision probability significantly increase, yielding lower throughput. Thus, finding an appropriate channel rate for data transmission to handle this trade-off and maximize network capacity is a challenging issue.

Figure 4.3 shows the taxonomy of rate adaptation schemes. According to how the data transmission rate is determined, we can classify rate adaptation approaches into three categories:

- *Loss Ratio Based*. In these approaches the channel quality is measured by sending probing frames and then the data transmission rate is increased or

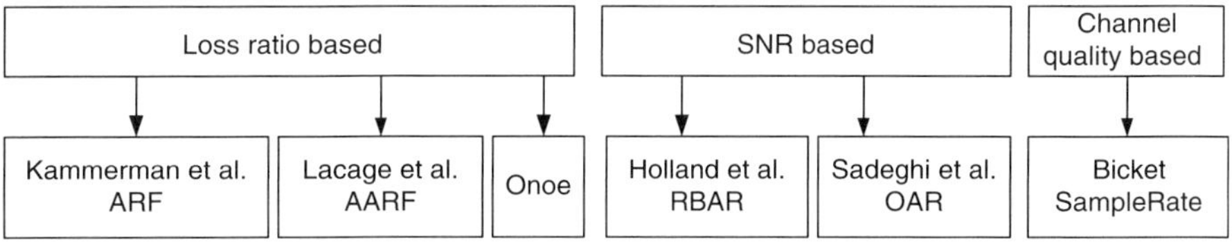

FIGURE 4.3 Taxonomy of rate adaptation protocols.

decreased based on the experienced loss ratio of the links. The lower the loss ratio, the more frequently are higher data rates selected and vice versa. Representative protocols are ARF [7], adaptive autorate fallback (AARF) [20], and Onoe [9].

- *SNR Based.* In these approaches the SNR of some frames (such as the RTS) is used to decide the data transmission rate. The higher the SNR is, the higher the data rate selected and vice versa. Representative protocols are the RBAR [10] and opportunistic autorate (OAR) [11].
- *Link Quality Based.* In these approaches an overall link quality measure, instead of only the loss ratio or SNR, is the criteria for rate selection. The higher the measured link quality, the higher the data rate selected and vice versa. The representative protocol is SampleRate [8].

Although the OAR determines the link rate in the same way the RBAR does, it also introduces a novel frame concatenation scheme to further improve the throughput of RBAR. From this aspect, the OAR is generally considered as a throughput improvement technique and will be discussed in details in Section 4.4.3.1.

4.3.1 Loss-Ratio-Based Rate Adaptation

Kamerman and Monteban [7] proposed the Lucent's WaveLan-II 802.11 solution on rate adaptation, known as ARF. In ARF, multiple data transmission rates are provided. In order to adaptively adjust the rate increase/decrease, an approach similar to the transmission control protocol (TCP) is adopted. Initially, the sender uses a higher rate after a fixed number of successful data transmissions. If 1 or 2 frames in a row are lost, i.e., not acknowledged, the rate is decreased to the next lower one and a timer is started. If 10 frames in a row are acknowledged or the timer expires, then the next higher rate is used for data transmission. In this way, ARF tries to reach the best possible rate if higher channel quality is indicated. However, this approach may generate frequent frame failure and retransmission and thus suffers from performance oscillation and low throughput. For example, let us assume that the loss rates at 5.5 and 11 Mbps are 0.1 and 0.5, respectively, and a sender is transmitting data at 5.5 Mbps. Then, the probability that 10 successive frames get acknowledged is around 35%, which is not low. So, the sender may still possibly switch to

11 Mbps. In this case, the probability that 1 or 2 frames get lost is quite high. Thus, the sender will very soon switch back to the lower rate after the frame failure. Therefore, ARF cannot stabilize to the best rate. In addition, unnecessary frame retransmissions are frequently triggered due to this type of adaptive probing, which further reduces application layer throughput. The cause of this issue comes from the fact that ARF neither has any knowledge of the loss ratio at higher rates nor keeps any record of this information after a failure occurs. In many cases, it is inappropriate to simply assume a low loss ratio at the higher rate as a result of a low loss ratio at the lower rate.

To avoid the above instability issue, Lacage et al. [20] proposed AARF to dynamically adjust the threshold on the number of consecutive successful data transmission before switching to a higher data rate. The idea is to use binary exponential backoff (BEB) [21]. When the transmission of the probing frame fails, a sender immediately switches back to the previous lower rate at the same time doubling the number of consecutive successful data transmission before switching to the higher data rate. This threshold is initialized to 10 with a maximum bound of 50. With this simple adaptation mechanism, the average time period between switching from the current rate to the next higher rate is increased. As a result, the number of failed transmissions and retransmissions is reduced for better overall throughput than ARF.

Similar to ARF, the Onoe algorithm [9] tries to find the highest bit rate based on the loss ratio. In Onoe, the initial bit rates used for 802.11a/g and 802.11b are 24 and 11 Mbps, respectively. Basically, a sender moves to the next lowest bit rate under the following two conditions: (i) no frames have succeeded and (ii) 10 or more frames have been sent and the average number of retries per frame was greater than 1. Then, if more than 10% of the frames need a retry, a number called a *credit* is decreased. Otherwise, the credit is increased. With more data frames being transmitted, the credits that accumulated for each specific bit rate at the sender change dynamically. If the current bit rate has 10 or more credits, then the data rate is increased to the next highest level. Onoe has been implemented for Atheros 802.11 cards in Linux and FreeBSD systems.

Due to the use of accumulative credits over many data transmissions, Onoe is much less sensitive to individual frame failure than ARF. Also, the credit provides a history of how well a bit rate performed in the past. Finally, since Onoe moves to the next highest bit rate when the average number of retries per frame is greater than 1, it actually tries to find the *highest* bit rate with loss ratio less than 50%. Of course, it is difficult to guarantee that this decision will always lead to better performance.

4.3.2 SNR-Based Rate Adaptation

Because there is a strong correlation between the SNR and the BER, it is natural to determine the link rate based on the SNR. Given a specific SNR, it is possible to find an appropriate link rate such that the corresponding BER is low enough. By choosing the link rate based on the SNR, the instability problem in ARF can be eliminated.

Holland et al. [10] proposed RBAR to determine the best link rate a sender should use for data transmission at the receiver side. In RBAR, let M_1, M_2, ..., M_N be the set of modulation schemes in increasing order of their rate and θ_i be the SNR thresholds at which BER(M_i) = 1×10^{-5}; the modulation scheme is chosen according to the following criteria:

$$
\begin{array}{ll}
M_i & \text{if SNR} < \theta_1 \\
M_i & \text{if } \theta_i \leq \text{SNR} < \theta_{i+1} \quad i = 1, \ldots, N-1 \\
M_N & \text{otherwise}
\end{array}
$$

RBAR can be incorporated with the IEEE 802.11 DCF, as illustrated in Fig. 4.4. When the RTS is sent from the source, a fixed *temporary* rate is used and the the RTS network allocation vector (NAV) is calculated based on the temporary rate. Upon receipt of the RTS, the destination station estimates the SNR of the RTS frame and determines the best data rate for this data transmission. Then, it calculates its CTS NAV based on the new rate and sends back the CTS to the sender. With the CTS from the destination, the sender now has a final best rate for the specific frame. Then, it includes a reservation header (RSH) in the data frame for confirmation of the new rate. Any neighboring stations such as A and B, when they overhear the RTS, CTS, or RSH, will readjust the corresponding reserved time period for the data transmission. Since the RBAR opportunistically gets the best rate for each specific frame, it does not suffer from instability and in many scenarios improves network capacity by using higher data rates than ARF.

However, the major limitation of RBAR is that in practical networks it is impossible to obtain the BER characteristics and then determine the SNR threshold. Thus, trying to set a fixed threshold may in fact lead to situations

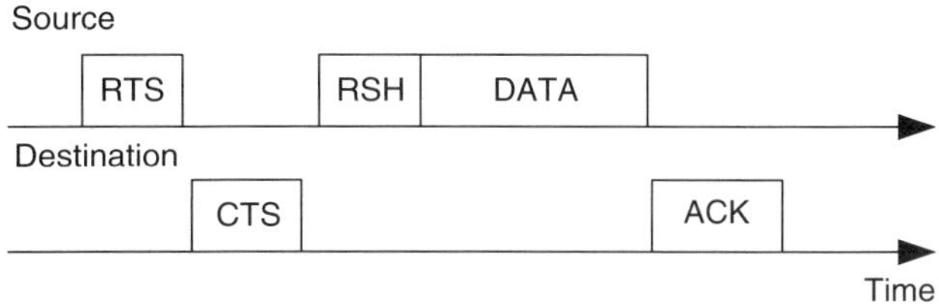

FIGURE 4.4 Access sequence of RBAR. *Source*: G. Holland, N. Vaidya, and P. Bahl, "A rate-adaptive MAC protocol for wireless networks," paper presented at the ACM MobiCom, July 2001.

where some links do not work at higher rates (due to low threshold) or some links do not operate at the optimal rate (due to higher threshold) [8].

4.3.3 Link-Quality-Based Rate Adaptation

Neither ARF nor the RBAR fully considers link quality. In ARF, the loss ratio is the only criterion for rate selection. In the RBAR, only the SNR is used for rate selection. In some scenarios, it is better to use a higher rate to improve network throughput even though the corresponding loss ratio is also high. In other scenarios, RBAR may yield low throughput due to an inaccurate SNR threshold. Thus, it is better to consider a more comprehensive and realistic metric that helps choose the best possible data rate with the highest throughput.

Bicket [8] proposed SampleRate to maximize throughput over wireless links that are capable of multiple rates. In SampleRate, when a link begins to send frames, it first uses the highest possible bit rate. A bit rate is discarded if the link experiences four successive failures. In this case, the bit rate is decreased until a bit rate that is capable of sending frames is found. Then, for every tenth frame, a random bit rate is sampled to see if it can work better than the current one. A bit rate is not eligible to be sampled if the link experiences four successive failures or its lossless transmission time is greater than the average transmission time of the current rate, i.e., that bit rate is even worse than the current one in the best scenario. To avoid high sampling overhead, the number of bit rates that are picked is restricted to a certain threshold.

Let us assume that for a frame of size S the total transmission time TX_Time of sending the frame over a link with rate r with N number of retransmission can be calculated as

$$\text{DIFS} + \text{Backoff}(N) + (N+1)(\text{DIFS} + T_{\text{ACK}} + T_{\text{HMAC}} + 8S/r)$$

where T_{HMAC} and T_{ACK} are the transmission times of the MAC header and ACK frames, respectively. Backoff(N) is obtained through experiment. Obviously, if the total transmission time per each byte of frame is minimized, the overall network throughput can be maximized. Therefore, SampleRate outperforms the RBAR and ARF in most scenarios.

In summary, due to the characteristic of the dynamic channel quality change in IEEE 802.11 WLANs, accurate and efficient rate adaptation is very difficult. On the one hand, using only the loss ratio is not enough since it is incorrect to conclude that the higher rate will yield lower throughput due to more data loss. On the other hand, the SNR threshold proposed in the RBAR is impractical in real networks. In addition, even when the loss ratio is the only criterion for rate selection, how to set an appropriate threshold that supports a better decision is still an open issue. Therefore, the design of realistic and efficient rate adaptation algorithms deserves further investigation.

4.4 TECHNIQUES FOR IMPROVING IEEE 802.11 NETWORK CAPACITY

Many techniques [4–6, 11, 22–24] have been proposed to enhance the performance of the IEEE 802.11 DCF. Figure 4.5 shows the taxonomy of throughput improvement techniques. According to the approaches taken to address throughput inefficiency in 802.11 WLANs, these techniques can be classified into three categories:

- *Adaptive Parameter Tuning*. Through dynamic tuning of parameters such as the backoff counter or the CW, the collision probability decreases and less channel time is wasted on idle waiting, thus improving the network capacity. Representative approaches are the optimization based on the p-persistent model [4], dynamic optimization on range (DOOR) [5], and slow CW decrease [22].
- *Frame Concatenation*. Instead of transmitting only one data frame per each successful channel access, multiple frames can be sent back to back from the sender with a single set of overhead of backoff time and control frames, thereby improving the channel utilization. Representative approaches are the frame concatenation mechanism (CM) [24], OAR [11], and adaptive frame concatenation (AFC) [23].
- *Frame Piggyback*. If the receiver has frames to transmit (e.g., interactive TCP traffic), one or more frames can be sent back along with the legacy ACK frame without another channel contention. Thus, the overhead of frame transmission is further reduced. A representative approach is the frame piggyback mechanism (PM) [24].

While adaptive parameter tuning techniques improve network throughput by setting the appropriate network configuration at run time to address operating inefficiency, frame concatenation and PMs try to reduce the protocol overhead by exploiting better channel utilization. Furthermore, the frame concatenation approach can be integrated with the rate adaptation approach [11, 23] to further improve network capacity.

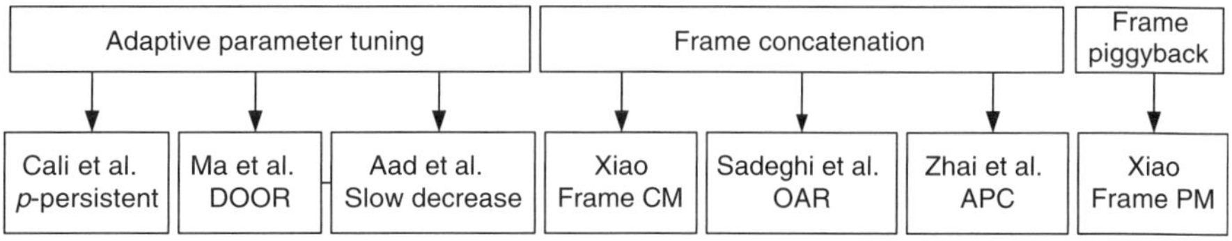

FIGURE 4.5 Taxonomy of IEEE 802.11 throughput improvement techniques.

4.4.1 Adaptive Parameter Tuning

In Section 4.2.2, we have shown that optimal throughput can be approximated when the appropriate CW_{min} is used for different N, the number of competing stations. Ma et al. [5] proposed DOOR to first divide N into multiple subareas and then use a reference CW_{min} for each specific subarea. DOOR works as follows. First, the access point (AP) measures N and selects the proper subarea. Then, the AP broadcasts the optimal CW_{min} of the subarea to all the stations with a beacon frame. Upon receipt of the message, each station refreshes its CW_{min} and adjusts its CW_{max} to $2^m CW_{min}$, where m is the maximum number of allowed retransmissions. A new station can simply consult the AP on the appropriate CW_{min} value to get started. For infrastructure mode WLANs, this approach is quite practical and efficient. For the ad hoc mode, one of the stations has to serve as the central point to coordinate CW_{min} selection.

On the other hand, one of the issues with the IEEE 802.11 DCF is the handling of channel congestion. According to the existing standard, upon successful data transmission, the CW is reset to the predefined CW_{min} (31 for IEEE 802.11b). If collision occurs when two or more stations try to send frames in the same time slot, CW is increased exponentially until the CW_{max} is reached. However, under heavy traffic load, the probability of collision increases significantly, especially with relatively small CW. Thus, with the DCF, if the channel is congested, the CW is reset to CW_{min} after each successful data transmission and it is very likely that collision occurs due to severe channel contention. In this case, CW is increased again to an appropriate size to restrict the frequency of channel access and thus reduce the channel collision probability. It is clear that due to this type of CW decrease/increase fluctuation under a congested network environment, unnecessary frame retransmissions are triggered, leading to low network throughput.

In fact, the DCF made an invalid assumption that every successful data transmission indicates a noncongestion channel status. If the congestion level does not decrease, it is better to keep the same CW for the best performance. Based on the argument that congestion usually disappears slowly, Aad et al. [22] proposed a slow CW decrease scheme to modify the DCF. Thus, instead of resetting the CW to CW_{min} after succeeding a data transmission, a station only decreases the CW slightly according to appropriate functions. In [22], two basic functions are introduced:

- *Multiplicative Decrease*. $CW_{new} = \max(CW_{min}, \delta CW_{old})$, where δ is a constant with a recommended value from 0.6 to 0.8. When CW_{old} is much larger than CW_{min} due to multiple collisions and retransmissions, δCW_{old} will give a more reasonable prediction of the best CW rather than CW_{min}.
- *Linear Decrease*. $CW_{new} = \max(CW_{min}, CW_{old} - \alpha)$, where α is a constant with a recommended value of less than 100. When α is small, a slow CW decrease is expected.

Even though a multiplicative or linear CW decrease approach works better than the DCF under congestion scenarios, it may add significant protocol overhead under low traffic load. When network traffic is light, using large CWs will increase the idle backoff time and reduce channel utilization. Thus, Aad et al. [22] and Ni et al. [18] proposed an *adaptive CW decrease scheme* where the multiplication factor is dynamically adjusted based on the channel condition. Similar to the TCP, the collision ratio can act as an important indicator of the channel congestion level. Thus, the estimated collision rate f_{curr} is calculated using the number of collisions N_{coll} and the total number of frames sent N_{data} during a constant period as N_{coll}/N_{data}. Then, to smooth out the fluctuation of f_{curr}, the average collision rate is calculated as $f^i = (1 - \varphi) \times f^i_{curr} + \varphi \times f^{i-1}$, where i refers to the ith update period and φ is the smoothing factor. Then, the new CW is obtained as $\max(\mathrm{CW}_{min}, (\min(f^i, 0.8)) \times \mathrm{CW}_{old})$. Obviously, if the congestion level is low, the collision probability is very small. In this case, a small CW_{new} comparable to CW_{min} will be calculated and the performance is similar to the DCF. Thus, the adaptive CW decrease scheme works well in various channel conditions.

4.4.2 Frame Concatenation and Piggybacking

The intuition of frame concatenation can be explained from the way the existing IEEE 802.11 standard treats long frames. In CSMA/CA DCF, if a MAC frame is longer than a threshold (*FragmentationThreshold*), it will be split to multiple short frames and concatenated. With fragmentation, each fragment is acknowledged individually but there is no necessity to go through the channel contention and backoff procedure for other fragments in the same frame. Given that the same overhead is used for a long frame, the bandwidth efficiency is much better than sending a short frame. Thus, segmentation handling in the DCF shows that it is possible to send multiple data frames with only one channel access.

4.4.2.1 Burst Transmission Acknowledgment and CM. Aware of the theoretical throughput and delay limit of IEEE 802.11 WLANs, Xiao and Rosdahl [6] proposed using burst transmission and acknowledgment (BTA) for the reduction of protocol overhead. BTA allows a burst of frames to be transmitted before any acknowledgment. After sending a burst of frames, the sender sends a burst acknowledgment request (BurstAckReq) frame, and the receiver must respond by sending the burst acknowledgment (BurstAck) frame in which the correctly received frame information is included. All the frames, including BurstAckReq and BurstAck, are separated by an SIFS period. Figure 4.6*b* shows a typical sequence of BTA. Since there is only one time overhead of the BurstAckReq and BurstAck frames for multiple frames, the original overhead of one ACK and backoff time per frame data transmission (Fig. 4.6*a*) is significantly reduced. Results show that with a limit in the number

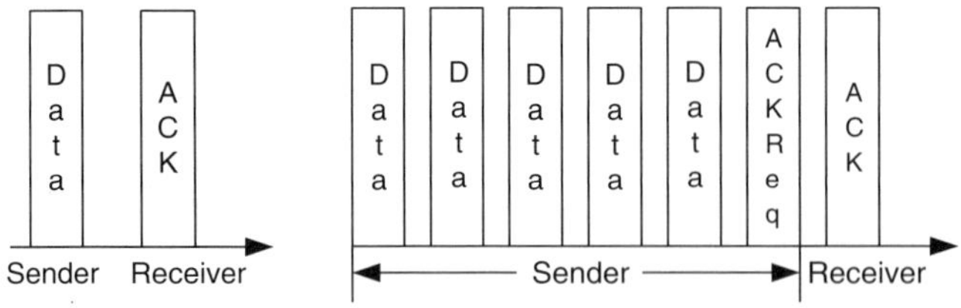

FIGURE 4.6 Comparison of overhead per channel access without RTS/CTS: (*a*) 802.11; (*b*) BTA. *Source:* Y. Xiao and J. Rosdahl, *ACM Sigmobile Mobile Comput. Commun. Rev.* 7(2), 2003, 6–19.

of bursting MAC protocol data units (MPDUs) of 16, significant throughput and delay improvement can be achieved.

BTA provides a novel idea on reducing frame overhead of the IEEE 802.11 DCF MAC protocol. However, BTA has its issues. For example, BurstAckReq adds additional overhead. Also, BurstAck has to acknowledge each and every MPDU and thus has to maintain a record of the state of each MPDU sent from the sender. To further reduce the overhead, Xiao [24] proposed the CM as an extension of BTA. In the CM, a *virtual frame* that includes a concatenation header (CH) frame and concatenated frames is transmitted per channel access. In the CH, the frame control type field indicates that it is a concatenation virtual frame, and the payload includes the count of concatenated frames (2 bytes) and a total-length field (2 bytes). After the destination station receives the CH frame, it will receive the followed concatenated frames one by one and acknowledges only the *last* concatenated frame. Figure 4.7 shows the overhead of the CM. The total length of the virtual frame is limited by a threshold that is set to 1000 bytes. Thus, a sender keeps concatenating frames back to back until the threshold is met. It is obvious that instead of several sets of overheads for

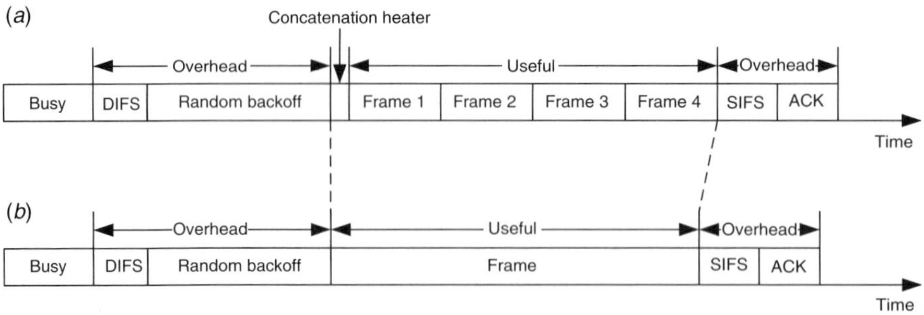

FIGURE 4.7 Concatenation of multiple frames (not to scale): (*a*) concatenating multiple short frames and (*b*) long-frame transmission. *Source*: Y. Xiao, *IEEE Trans. Wireless Commun.* 4(5), 2005, 2182–2192.

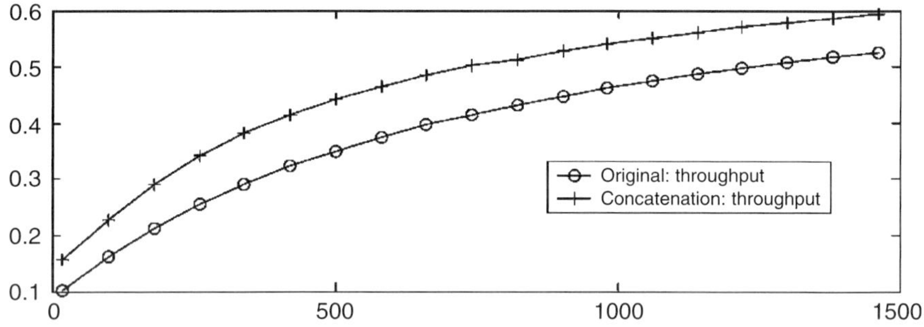

FIGURE 4.8 Throughput of CM; *x* axis represents payload size (bytes). *Source*: Y. Xiao, *IEEE Trans. Wireless Commun.* 4(5), 2005, 2182–2192.

different frames, only one set of overheads is used, yielding much higher bandwidth efficiency.

On the other hand, the CM is not the reverse of segmentation but a compliment. If a concatenated frame is longer than the fragmentation threshold, it will be segmented to multiple frames. So, the CM will not be used for very long frames. Also, the CM is for a concatenation of different frames, not for segments of a single frame. Figure 4.8 shows the normalized throughput comparison of the CM and DCF. It can be seen that the CM achieves much higher throughput than the standard 802.11, especially when payload size is small.

4.4.2.2 Frame PM. The frame CM is very effective in reducing the overhead for frame transmission from only the sender's point of view. For interactive data applications such as TCP or voice over Internet protocol (VOIP), receivers usually have frames to send back to the sender in the same communication session. With the existing IEEE 802.11 DCF, all the frames from a receiver are considered separate from frames from the corresponding sender and thus require another channel contention period for the receiver to win the channel. Thus, this incurs two sets of overheads for a sender and receiver to deliver a frame to each other. Xiao [24] proposed the PM to reduce this type of overhead by allowing the receiver to piggyback a data frame to the sender once if the receiver station has a frame to send to the sender. Figure 4.9 shows the overhead of the PM. Intuitively, the PM further improves the bandwidth efficiency. Figure 4.10 shows the normalized throughput comparison of the PM and DCF. It can be seen that the PM achieves significantly higher throughput than the standard 802.11.

4.4.3 Integration of Rate Adaptation and Frame Concatenation

In existing rate adaptation schemes such as ARF [7] and the RBAR [10], the benefit of using high data rates has not been fully explored. Since control

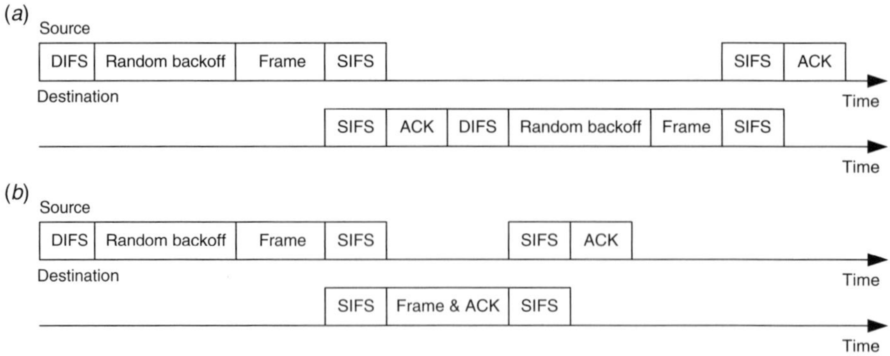

FIGURE 4.9 Overhead of PM without RTS/CTS: (*a*) without piggybacking; (*b*) with piggybacking. *Source*: Y. Xiao, *IEEE Trans. Wireless Commun.* 4(5), 2005, 2182–2192.

frames are sent with basic rates and data frames are sent with high rates, control frames consist of a higher portion of total channel time compared to single-rate (2-Mbps) channel access where more time is consumed for useful data transmission. When multiple data rates coexist in a network, since it takes much more time to send data over a low rate channel than over a high rate channel, each channel access over the low rate channels consumes more channel time than over the high rate channels. So, given approximately the same number of channel accesses for low and high rate channels, more total channel time is used for low rate channel data transmission, which is known as *temporal unfairness*. Therefore, even though the high rate capability is enabled, the existing MAC protocol does not take its full advantage and the bandwidth efficiency still suffers.

4.4.3.1 Opportunistic AutoRate. To achieve temporal fairness, i.e., grant stations approximately the same channel time for data transmission per each

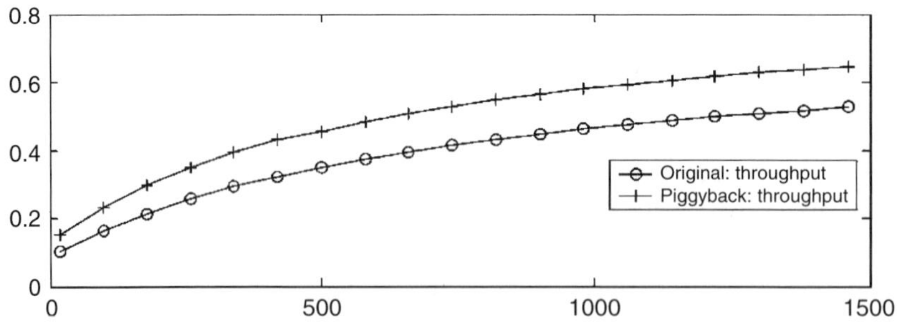

FIGURE 4.10 Throughput of PM; *x* axis represents payload size (bytes). *Source*: Y. Xiao, *IEEE Trans. Wireless Commun.* 4(5), 2005, 2182–2192.

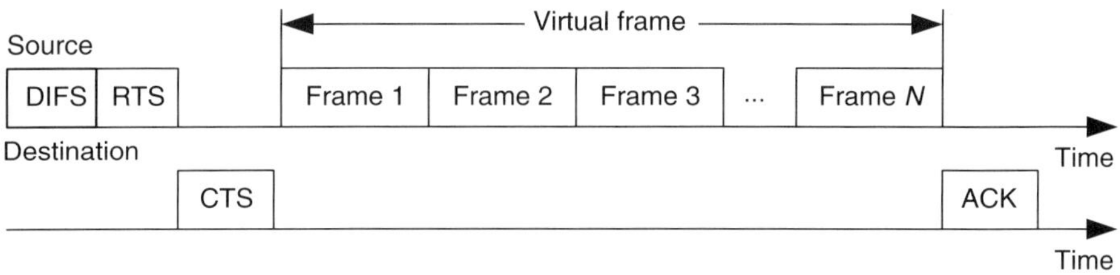

FIGURE 4.11 Access mode of OAR.

channel access, Sadeghi et al. [11] proposed the OAR, which allows frame concatenation when a high data rate is available. Based on the principle of maintaining temporal fairness, i.e., keeping the total transmission time over different rate channels the same, the OAR proposed sending one, three, and five concatenated frames per each data transmission when the best channel rate is 2, 5.5, and 11 Mbps in IEEE 802.11b, respectively. The frame format is similar to the CM. The only difference is that the OAR takes the same approach as the RBAR to get the appropriate channel data rate determined by the receiver through RTS/CTS handshaking. In this case, the OAR potentially incurs higher overhead than the CM. Figure 4.11 shows a typical frame sequence in the OAR when the higher data rate (5 or 11 Mbps) is present. Due to the frame concatenation under the OAR, more frames are transmitted over higher data rates per channel access. Thus, the OAR achieves significantly higher throughput than the RBAR and DCF. In general, the OAR gains 40%–55% more throughput than the RBAR and the performance gain increases with the number of flows.

4.4.3.2 Adaptive Frame Concatenation. Obviously, if we consider the single-rate channel, then the CM also maintains temporal fairness by sending the same amount of data whenever a station accesses the channel. On the other hand, if we consider uniform frame sizes, the OAR maintains good temporal fairness in that the number of concatenated frames is proportional to the channel data rate. Thus, both the CM and OAR maintain temporal fairness under certain conditions but only consider the situation partially.

Zhai and Fang [23] proposed AFC to improve network capacity by considering both frame sizes and link rates. Senders in the AFC scheme adaptively concatenate several short frames which are destined to be sent to the same next hop in a long frame for MAC layer transmission according to the congestion status as well as the observed channel status. First, the concatenation threshold is defined as

$$L_{\mathrm{th}} = r_{\mathrm{data}} \times (T_{\mathrm{cc}} - T_{\mathrm{Hphy}} - T_{\mathrm{HMAC}} - T_{\mathrm{RTS}} - T_{\mathrm{CTS}} - T_{\mathrm{ACK}} - 3\mathrm{SIFS})$$

where T_{Hphy} and T_{HMAC} are the transmission times of the physical and MAC headers of a DATA frame, respectively; T_{RTS}, T_{CTS}, and T_{ACK} are the transmission times of the RTS, CTS, and ACK frames, respectively; and T_{cc} is the channel coherent time at the stable state and is obtained from empirical data [11]. Then the average total length of the frame concatenation is approximately L_{th}. Under various channel rates and frame sizes, AFC always achieves much higher throughput than the basic DCF. The reason for this is that AFC uses frame concatenation to reduce the overhead and the concatenation length is appropriate for data rate and frame sizes.

4.5 CONCLUSION AND FUTURE TRENDS

In this chapter, we have provided a survey of the literature on the capacity analysis and rate adaptation in IEEE 802.11 WLANs. We have shown that the existing 802.11 DCF standard does not function efficiently and sufficiently to support multimedia applications that usually require high bandwidth availability. To improve the performance of WLANs, many protocols have been proposed to optimize network throughput by either providing better network configuration or reducing protocol overhead. With higher data rates being exploited, rate adaptation provides further improvement over single-rate channels.

Despite the effort being discussed, there are still several critical issues to be investigated in the future:

- *Very High Capacity WLANs*. Continuous effort will provide considerable bandwidth increase for future WLANs and will enable higher network capacity, as much as several hundred megabytes per second, in a cost-effective fashion. At the same time, how to further reduce the overhead of IEEE 802.11 WLANs and overcome its *operating inefficiency* and *protocol inefficiency* to achieve an overall throughput close to the physical bandwidth is critical.
- *Stable and Effective Rate Adaptation*. With complexity such as obstacles and channel fading in real networks, using the highest possible rate may not be feasible or does not yield the optimal performance due to the high loss ratio. As discussed, it is hard to obtain the accurate threshold of the SNR for different BERs. On the other hand, using probing frames to test channel quality may lead to incorrect channel rate decisions since it is still possible to have multiple consecutively successful data transmissions under high loss ratio, though not frequently. Thus, how to design stable and effective rate adaptation algorithms to maximize network throughput is still an open issue.
- *QoS Support*. Even though overall throughput is significantly improved, multimedia applications may not be sufficiently supported until QoS

mechanisms such as service differentiation and resource management are enforced. For example, when real-time flows and data traffic coexis, the performance of real-time flows can be easily degraded when data traffic is heavy. Many existing QoS strategies [25–29] focus on the IEEE 802.11 DCF, EDCF, and HCF and have not fully considered new throughput improvement techniques, rate adaptation, and data loss ratio. Design of efficient QoS protocols that take full advantages of emerging high capacity MAC protocols such as APC [23] is thus desirable.

REFERENCES

1. IEEE 802.11, "Wireless LAN medium access control (MAC) and physical layer (PHY) specification," IEEE, New York, Aug. 1999.
2. Y. Xiao and J. Rosdahl, "Throughput and delay limits of IEEE 802.11," *IEEE Commun. Lett.* 6(8), 355–357 (2002).
3. G. Bianchi, "Performance analysis of the IEEE 802.11 distributed coordination function," *IEEE J. Sel. Areas Commun.* 18(3), 535–547 (2000).
4. F. Cali, M. Conti, and E. Gregori (2000) "IEEE 802.11 protocol: Design and performance evaluation of an adaptive backoff mechanism," *IEEE J. Sel. Areas Commun.* 18(9), 1774–1786.
5. H. Ma, X. Li, H. Li, P. Zhang, S. Luo, and C. Yuan, "Dynamic optimization of IEEE 802.11 CSMA/CA based on the number of competing stations," paper presented at the IEEE ICC, 2004.
6. Y. Xiao and J. Rosdahl, "Enhancement for the current and future IEEE 802.11 MAC protocols," *ACM Sigmobile Mobile Comput. Commun. Rev.* 7(2), 6–19, (2003).
7. A. Kamerman and L. Monteban, "WaveLAN II: A high performance wireless LAN for the unlicensed band," *Bell Labs Tech J.* 2(3), 118–133, 1997.
8. J. Bicket, "Bit-rate selection in wireless networks," Master's Thesis, Massachusetts Institute of Technology, Cambridge, MA, 2005.
9. Onoe Rate Control, available: http://madwifi-project.org.
10. G. Holland, N. Vaidya, and P. Bahl, "A rate-adaptive MAC protocol for wireless networks," paper presented at The Annual International Conference on Mobile Computing and Networking (ACM MOBICOM), 2001 July.
11. B. Sadeghi, V. Kanodia, A. Sabharwal, and E. Knightly, "Opportunistic media access for multirate ad hoc networks," paper presented at The Annual International Conference on Mobile Computing and Networking (ACM MOBICOM), 2002 Sep.
12. E. Ziouva and T. Antonakopoulos, "CSMA/CA performance under high traffic conditions: Throughput and delay analysis," *Comput. Commun.* 25(3), 313–321 (2002).
13. H. Wu, Y. Peng, K. Long, S. Cheng, and J. Ma, "Performance of reliable transport protocol over IEEE 802.11 wireless LANs: Analysis and enhancement," paper presented at the IEEE International Conference on Computer Communications (INFOCOM), New York, 2002, pp. 599–607.

14. Y. C. Tay and K. C. Chua, "A capacity analysis for the IEEE 802.11 MAC protocol," *Wireless Networks* 7(2), 159–171 (2001).
15. H. Zhai and Y. Fang, "Performance of wireless LANs based on IEEE 802.11 MAC protocols," paper presented at The Annual IEEE International Symposium on Personal, Indoor, and Mobile Radio Communications (IEEE PIMRC, January 2003), Beijing, China, Sept. 2003.
16. D. Qiao, S. Choi, and K. Shin, "Goodput analysis and link adaptation for IEEE 802.11a wireless LANs," IEEE Trans Mob Comput, 1(4), 278–292, 2002.
17. G. Bianchi, L. Fratta, and M. Oliveri, "Performance evaluation and enhancement of the CSMA/CA MAC protocol for 802.11 wireless LANs," paper presented at the IEEE International Symposium on Personnel, Indoor and Mobile Radio Communications (PIMRC), Taipei, Taiwan, Oct. 1996, pp. 392–396.
18. Q. Ni, I. Aad, C. Barakat, and T. Turletti, "Modelling and analysis of slow CW decrease for IEEE 802.11 WLAN," paper presented at the IEEE International Symposium on Personnel, Indoor and Mobile Radio Communications (PIMRC 2003), Beijing, China, Sept. 2003.
19. Y. Xiao, "Performance analysis of priority schemes for IEEE 802.11 and IEEE 802.11e wireless LANs," *IEEE Trans. Wireless Commun.* 4(4), 1506–1515 (2005).
20. M. Lacage, M. H. Manshaei, and T. Turletti, "IEEE 802.11 rate adaptation: A practical approach," paper presented at The ACM International Conference on Modeling, Analysis and Simulation of Wireless and Mobile Systems (ACM MSWiM), Venice, Italy, October 4–6, 2004.
21. R. M. Metcalfe and D. R. Boggs. "Ethernet: distributed frame switching for local computer networks," *ACM Commun.* 19(5), 395–404 (1976).
22. I. Aad, Q. Ni, C. Castelluccia, and T. Turletti. "Enhancing IEEE 802.11 performance with slow CW decrease," IEEE 802.11e working group document 802.11-02/674r0, IEEE, New York, Nov. 2002.
23. H. Zhai and Y. Fang, "'A distributed adaptive frame concatenation scheme for sensor and ad hoc networks," paper presented at the IEEE Military Communications Conference (Milcom'05), Atlantic City, NJ, Oct. 17–20, 2005.
24. Y. Xiao, "IEEE 802.11 performance enhancement via concatenation and piggyback mechanisms," *IEEE Trans. Wireless Commun.* 4(5), 2182–2192 (2005).
25. Y. Xiao, F. H. Li, and S. Choi, "Two-level protection and guarantee for multimedia traffic in IEEE 802.11e distributed WLANs," Wireless Networks, accepted for publication.
26. Y. Xiao and F. H. Li, "Local data control and admission control for QoS support in wireless ad hoc networks," *IEEE Trans. Vehic. Technol. (TVT)* 53(5), 1558–1572 (2004).
27. Y. Xiao and F. H. Li, "Voice and video transmissions with global data parameter control for the IEEE 802.11e enhanced distributed channel access," IEEE Trans. Parallel Distrib. Syst. (TPDS) 15(11), 1041–1053 (2004).
28. M. Li and B. Prabhakaran, "MAC layer admission control and priority re-allocation for handling QoS guarantees in non-cooperative wireless LANs," *ACM/Springer Mobile Networks Appl.* (MONET) 10(6), 947–959 (2005).

29. H. Zhu, M. Li, I. Chlamtac, and B. Prabhakaran, "Survey of quality of service in IEEE 802.11 networks," *IEEE Wireless Commun.*, Special issue on Mobility and Resource Management,11(4), 6–14 (2004).
30. D. Qiao and S. Choi, "Fast-responsive link adaptation for IEEE 802.11 WLANs," paper presented at the IEEE International Conference on Communications (IEEE ICC), Seoul, Korea, May 16–20, 2005.
31. L. Romdhani, Q. Ni, and T. Turletti, "Adaptive EDCF: Enhanced service differentiation for IEEE 802.11 wireless ad hoc networks," paper presented at the IEEE WCNC'03, New Orleans, LA, Mar. 16–20, 2003.

PART II

IEEE 802.15.1 BLUETOOTH AND IEEE 802.15.2

CHAPTER 5

OVERVIEW OF IEEE 802.15.1 MEDIUM ACCESS CONTROL AND PHYSICAL LAYERS

KAVEH GHABOOSI, YANG XIAO, and JEFF J. ROBERTSON

5.1 INTRODUCTION

Hand-held devices have already become an integral part of our daily lives. New radio frequency (RF) technologies enable devices to connect effortlessly and with little, if any, intervention from their user through standardized air interfaces. Wireless personal area network (WPAN) communication technologies differ from other conventional wireless network technologies. WPANs primarily target the consumer market and are used for ease of connectivity of personal wearable or hand-held devices. WPANs are thus designed to be inexpensive, small in size, easy to use, and power efficient. However, independent of the specific technology used, WPANs bring a new concept in communications, that of the personal operating space (POS). Contrary to infrastructure networks that are installed, a POS is a space of small coverage around an individual where communications occur in an ad hoc manner. The POS is tethered to an individual, in particular to his or her personal devices, and moves as the individual moves. The POS empowers the individual with communication capabilities, allowing him or her to communicate with other devices that enter the individual's POS.

Bluetooth is a standard for short-range, low power, low cost wireless communication that uses radio technology. Although originally envisioned as a cable replacement technology by Ericsson in 1994, embedded Bluetooth capability is becoming widespread in numerous types of devices. They include intelligent devices, e.g., personal digital assistants (PDAs), cell phones, personal computers (PCs), data peripherals such as mice, keyboards, cameras, digital

Emerging Wireless LANs, Wireless PANs, and Wireless MANs. Edited by Y. Xiao and Y. Pan

pens, printers, local area network (LAN) access points, audio peripherals including headsets, speakers, stereo receivers, and finally embedded applications comprising automobile power locks, grocery store updates, industrial systems, and musical instrument digital interface (MIDI) capability. Ericsson joined forces with Intel, International Business Machines (IBM), Nokia, and Toshiba to form the Bluetooth special interest group (SIG) in early 1998. 3Com, Lucent/Agere Technologies, Microsoft, and Motorola joined the group in late 1999. Joint work by the SIG members allowed the Bluetooth vision to evolve into open standards to ensure rapid acceptance and compatibility in the marketplace. The resulting Bluetooth specification, developed by the Bluetooth SIG, is open and freely available. Bluetooth technology is already supported by over 2100 companies around the world.

The WPAN technology, based on the Bluetooth specification, is now an IEEE standard under the denomination of 802.15 WPANs [1]. The IEEE 802.15.1 standard, designated for compatibility with Bluetooth specifications (version 1.1), covers both medium access control (MAC) and physical (PHY) layer specifications for short-rang wireless networking. This chapter covers both MAC and PHY layers of the IEEE 802.15.1 standard. The IEEE 802.15.1 MAC protocol is divided into three layers: the logical link control and adaptation protocol (L2CAP) layer, the link manager protocol (LMP) layer, and the *baseband* layer. This standard also supports two forms of communication in the WPAN, asynchronous connectionless (ACL) communication and synchronous connection-oriented (SCO) communication. Consequently, we explore all specifications regarding the MAC and PHY layers of IEEE 802.15.1 and all aforementioned technical issues are covered subsequently. We start our discussion with an introduction of the WPAN architecture and then the structures of different layers are elucidated in detail.

5.2 WPAN ARCHITECTURE OVERVIEW

Intelligent personal devices such as PDAs, cellular phones, and tablet PCs have become much more popular than before and their usage has become integrated with each other. Consequently, the need to synchronize the information contained in such devices has become much greater. For instance, the calendar in one's smart phone and PC should be synchronized so that appointments, phone numbers, etc., stored in one can be easily transferred to and accessed in the other. Current consumers require devices that allow the integration of information among their personal smart devices. However, this integration should not come at a loss to the primary function or a cost to their personal devices. Therefore, a smart cellular phone should allow for wireless connectivity to an existing WPAN without requiring expensive hardware that would drive up the cost of the phone or have high energy requirements that would significantly reduce battery life. Furthermore, such devices may need to interact with other devices that come within their POS.

The WPAN is sometimes confused with the wireless LAN (WLAN). However, the WPAN differs from the WLAN in four key areas. First, WPAN devices need to operate using as little power as possible since many of them will be merely battery operated. Conversely, power consumption is normally not a critical issue for WLAN devices. Second, the coverage area for WPANs is much smaller than that of WLANs, as the POS of the network is typically within $10\,m^3$. Third, the WPAN is not required to maintain a management information base (MIB) which is required of WLANs. Fourth, the lifespan of the WPAN is specified, unlike that of the WLAN.

The layout of the IEEE 802.15.1 standard for WPANs is shown in Fig. 5.1. This standard uses the master–slave paradigm with the basic unit called a piconet. A piconet consists of a device serving as the unique master and at least one device acting as a slave. Moreover, piconets may be interconnected via devices common to multiple adjacent piconets. These interconnected piconets are termed scatternets.

The relationship between the International Organization for Standardization (ISO) seven-layer model and the IEEE 802.15.1 standard is illustrated in Fig. 5.2. Note that the application, presentation, session, transport, and network layers are not within the scope of the IEEE 802 standards. The data link layer of the open systems interconnection (OSI) seven-layer model is mapped directly to the logical link control (LLC) and MAC layers.

An IEEE 802.15.1 WPAN is created when a device requests to send information and ends when all such requests have been satisfied. These kinds of networks exploit two types of communication channels: *synchronous* channels used for audio communications as required by cellular phones and *asynchronous* channels deployed for data communications as utilized by all equipment for file transfer. The IEEE 802.15.1 WPAN operates in the 2.4-GHz industrial, scientific, and medical (ISM) band using a so-called fast frequency-hopping (FFH) transceiver. The 2.4-GHz ISM band is one of the few unlicensed bands available; however, this is also the band used by WLAN

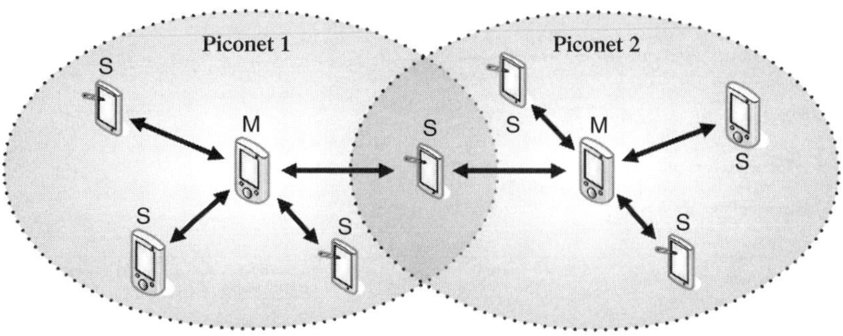

FIGURE 5.1 Scatternet formed by two interconnected piconets. A piconet consists of a master (M) and at least one slave (S).

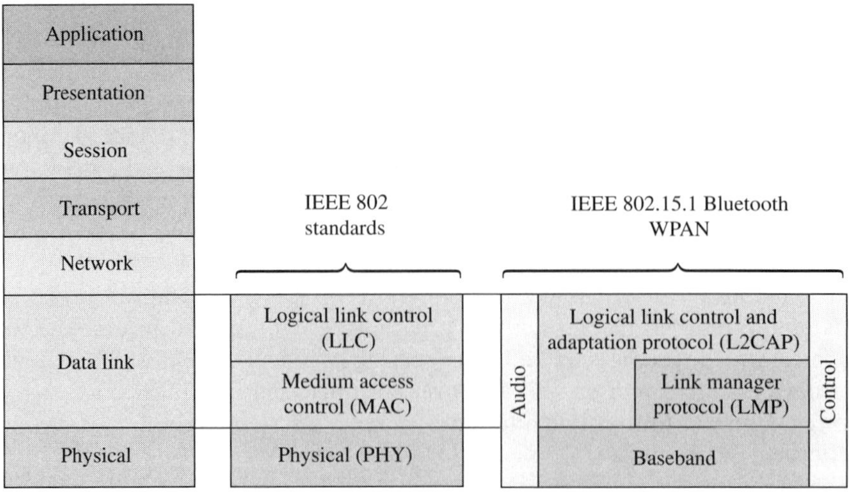

FIGURE 5.2 Mapping of ISO seven-layer open systems interconnection (OSI) standard model to IEEE 802 standard and then to IEEE 802.15.1 Bluetooth WPAN standard.

equipment (such as 802.11b/g). So, radio interference can easily result from either WLAN or WPAN equipment within the networks' POS. For that reason, FFH was chosen in order to minimize the effect of such interferences. In addition, communication between devices occurs using a slotted channel and with designated packets transmitted on the frequency bands which are changed based upon hopping sequence. Although by default data packets are only one slot in length, they can consist of one, three, or even five slots. The layout of each slot/packet is illustrated in Fig. 5.3.

The duration of defined slots in IEEE 802.15.1 is chosen based on the transceiver speed, which is 1600 hops/s, and is 625 μs. Typically, each slot

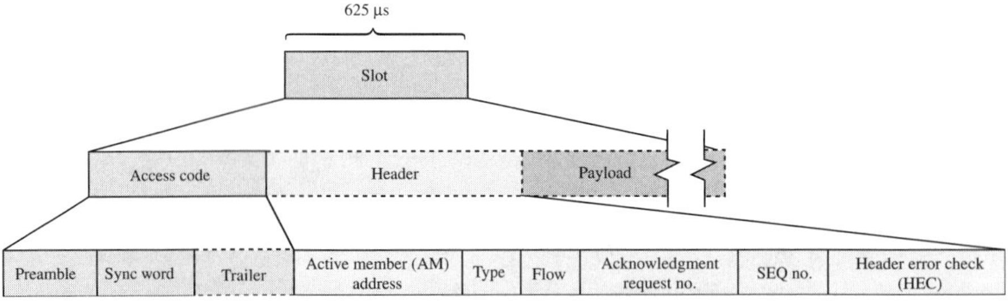

FIGURE 5.3 Standard format of IEEE 802.15.1 packet. The dotted sections/fields are optional. The header data are encoded, which results in the header becoming 54 bits in length with only 18 bits of actual field information.

consists of an *access code*, *header*, and *payload*. The access code is used to identify particular WPANs from each other, the header part is used for packet transmission management, and the payload carries the information transmitted.

As stated earlier, a WPAN using either Bluetooth or the IEEE 802.15 standard is called a piconet and consists of at least one master and one slave node. The master can communicate with slave nodes using either point-to-point or point-to-many communication. For the latter case, the master node is able to communicate with more than one slave node at a given time instance. On the other hand, slave nodes are only allowed to communicate directly with the master node by point-to-point communication. In addition, as mentioned before, a device can be a slave node in multiple piconets but can be a master node in only one piconet. Also, a device might be a slave node in one piconet at the same time it is a master node in another piconet. Moreover, it should be noted that a piconet can have access to another network, e.g., LAN, with the help of an attachment gateway (AG).

5.3 BASEBAND SPECIFICATION

5.3.1 General Characteristics

The IEEE 802.15.1 standard operates in the 2.4-GHz ISM band using a short-range radio link. A time division duplex (TDD) paradigm is used to simulate full-duplex communication. This standard is able to support one ACL channel, a maximum of three SCO channels, or a combination of asynchronous and synchronous communication channels. In addition, a piconet consists of only one master and up to seven active slave devices. More slaves can be part of the piconet but only in the inactive state. Despite not being in the active state, inactive slave nodes are still synchronized to the master's clock.

5.3.2 Physical Channels Structure

In the physical layer, a so-called pseudorandom hopping scheme is exploited. The hopping sequence is determined with the help of the master node's address and, apparently, for each hop a totally different frequency band is deployed. The aforementioned frequency hopping is performed in order to minimize the negative effect of radio interferences elucidated formerly. Besides, short slot durations are chosen in order to effectively make full-duplex communication possible. Furthermore, each device in a given piconet should be perfectly synchronized to the master node with regard to both *time* and *hop sequence*.

As pointed out earlier, an intended channel is divided into time slots each of which has a duration of 625 μs. These slots are numbered from 0 to $2^{27} - 1$ (mod 2^{27}); thus, slot numbers are repeated upon reaching $2^{27} - 1$. The master and slave devices transmit on alternating numbered slots; the master node transmits only on even-numbered slots while slave nodes deliver their packets only on odd-numbered slots, as shown in Fig. 5.4.

The rule of transmission on either even or odd slots is followed even when a packet extends to three or five slots in length. In such cases, depending on whether the delivered packet extends to three or five slots, regular transmission slot(s) will simply be skipped.

Each SCO channel is a one-to-one link between a master and a slave node. This type of communication link is usually reserved for only audio communication. On the other hand, an ACL channel is a one-to-many link between the master and all slaves. The SCO link supports time-sensitive communication, e.g., voice traffic. The time interval between consecutive transmissions, T_{SCO}, is counted in number of elapsed slots. This type of transmission imitates a so-called circuit-switched connection. In contrast, the ACL supports transmission of time-insensitive communication. As such, data packet retransmission, which is not available in communications through SCO, is utilized to avoid packet corruption during information exchange. ACL packets can be addressed either to a single slave node or to all slave devices with the absence of a particular address. In addition to the aforementioned SCO and ACL channels, the so-called extended SCO (eSCO) logical channel is a symmetric/asymmetric, point-to-point link between the master and a specific slave device. The eSCO also reserves the slots on the physical channel similar to SCO. Hence this is also considered as a circuit-switched connection between the master and the slave. The eSCO links offer a number of extensions and advantages over the standard SCO links. They support a more flexible and selectable slot periods, allowing a range of synchronous bit rates to be supported. The eSCO also can offer limited retransmission of packets, which is a cutting advantage over SCO links in which there is no retransmission. If these retransmissions are required, they take place in the slots that follow the reserved slots; otherwise the slots may be

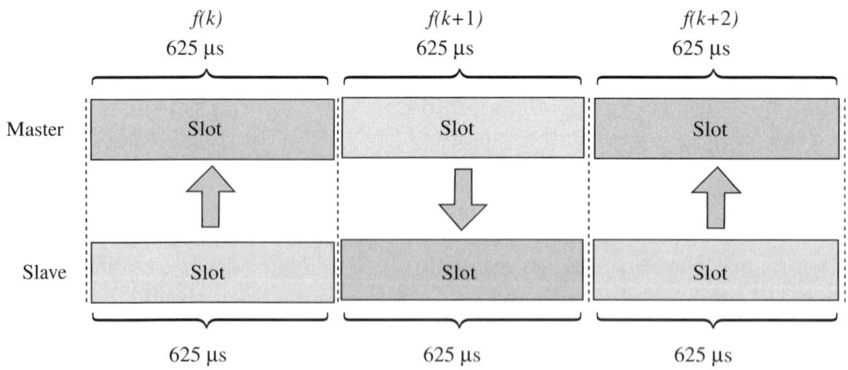

FIGURE 5.4 Time division duplex scheme utilized in IEEE 802.15.1 standard. The direction of each arrow shows whether the master node is transmitting a packet to one of its associated slaves or an active slave device is transmitting a packet to its master.

used for other traffic. Apart from these transport channels, there are two extra logical channels that are defined in Bluetooth core 1.2 and ratified in late 2003.

5.3.3 Packet Structure

The information sent across a piconet is in the form of data packets. Typically, each packet consists of an access code, a header, and a payload. However, a packet might have only an access code if it is used for signaling purposes. Packets are sent with the least significant bit (LSB) being transmitted first.

The access code comes in one of three types: channel access code (CAC), which is used for piconet identification; device access code (DAC), which is exploited for paging; and finally inquiry access code (IAC), which is utilized for discovering compatible devices that are within the piconet's POS. These access codes consist of a *preamble*, *sync word*, and a *trailer* if there is a header. The preamble will be 0101 if the LSB of the sync word is 0; otherwise, it will be set to 1010. The trailer is 1010 if the most significant bit (MSB) of the sync word is 0; else, it will be 1010. The sync word is a 64-bit word that is based upon a 24-bit address. This address is called the lower address part (LAP) and is used by the master for CAC, the slave for DAC, or a dedicated LAP for IAC.

The packet header has six fields: The active member address (AM_ADDR) field is used to identify active members of a piconet. The type field is used to indicate the packet type, such as SCO versus ACL. The flow field is used to determine overflow. The acknowledgment request notification (ARQN) field is exploited to acknowledge successful transmissions. The sequence (SEQN) field is used to allow for correct merging of large data files that were segmented for transmission over the air interface. The header error check (HEC) field ensures data integrity.

There are numerous predefined packet types in the IEEE 802.15.1 standard. The ID packet is used for identification purposes. The null packet is used to provide acknowledgment information. The poll packet is exploited to elicit a response from an intended slave device. The frequency hop synchronization (FHS) packet is a control packet that is mainly used for synchronization purposes. Besides, there are SCO packets for synchronous communication and ACL packets for asynchronous communication. SCO packets are used for synchronous communication such as voice transmission. These packets come in several types. For example, HV1, HV2, and HV3 are *high-quality voice* packets. HV1 has the highest quality but requires transmission once every two slots. HV2 is in the middle requiring packets every four time slots. HV3 is the lowest quality requiring transmission once every six time slots. Thus, their T_{SCO}, which is the number of time slots between successive transmissions, can be two, four, or six. There are also data–voice (DV) packets which combine both data and voice communication into a single packet. ACL packets are used for asynchronous communication, i.e., data transmission. These packets also come in several types. The data– medium rate (DM) packet comes in three formats: DM1 is one slot long carrying 18 bytes of information, DM3 is three slots long and carries 123 bytes of information, and DM5 is extended to five slots and carries 226 information bytes. There are also data– high rate (DH) packets which do not use packet payload encoding in order to increase the achieved data rate. DH1 carries a maximum of 28 information bytes in one slot, DH3 carries a maximum of 185 information bytes in three slots, and DH5 is five slots long and carries a maximum of 341 bytes. As can be seen, data transmission becomes much more efficient when the packet is extended to three or five slots.

5.3.4 Error Control

The IEEE 802.15.1 standard uses three error correction strategies: the 1/3 and 2/3 rate forward error corrections (FECs) and the acknowledgment request (ARQ) scheme. The 1/3 FEC scheme simply repeats each bit three times. Errors are detected by having any of three bits not being in agreement. The 2/3 FEC scheme uses a (15/10) shortened Hamming code to create five error correction and detection bits that are appended to each 10 bits of information. This scheme allows all single-bit errors to be corrected. Double-bit errors, while not able to be corrected, are detected. The ARQ scheme works by requiring acknowledgment for successful data transmission. Thus, data packets are repeatedly transmitted until an acknowledgment or a timeout signal is received. Successful receipt of a data packet is verified with a lack of cyclic redundancy check (CRC) errors, access code errors, or HEC errors whichever is applicable. The ARQ scheme is implemented independently per slave functionality when the device has membership to multiple overlapping piconets. To check the received packet, first, the CAC is verified to ensure that the packet has been received from the correct piconet. Second, the HEC is checked to find errors in the packet header. Third, the CRC is checked to find errors in the packet payload. Typically, the HEC and CRC codes are generated using 8 and 16-bit linear feedback shift register (LFSR) circuits, respectively.

5.3.5 Logical Channels

In the IEEE 802.15.1 standard, five logical channels are defined: link control (LC), link manager (LM), user asynchronous (UA), user isochronous (UI), and user synchronous (US) channels. The LC channel is used for low level control information such as ARQ. The LM channel is used to control information shared between the master and slaves. The UA channel is used for L2CAP asynchronous data. The UI channel is used for timing data during a packet transmission. Finally, the US channel is used for synchronous user data.

5.3.6 Data Whitening

In IEEE 802.15.1, both the header and payload of a packet are encrypted using a *polyalphabetic* cipher technique. This technique encrypts a given symbol/letter in plaintext format to a set of varying symbols/letters in *ciphertext*. By applying the aforementioned technique, further randomization of appended information which is being transmitted is achieved, leading to removal of redundant patterns seen in many nonrandomized data streams. Typically, this procedure is performed prior to FEC encoding. In addition, it should be noted that LFSR is initialized using a portion of the master's clock.

5.3.7 Transmission and Reception Routines

In IEEE 802.15.1, there are separate transmission/reception buffers for ACL/SCO packets. The master device should have a separate ACL transmission buffer for each slave node. In addition, there should be at least one SCO transmission buffer for each SCO slave; however, one ACL reception buffer and possibly one SCO reception buffer are sufficient. Each buffer consists of two first in, first out (FIFO) registers. For packet transmission, only one register is exploited while the LM puts the new information into the other buffer for the upcoming traffic transmission. A switch controlled by the link controller (LC) is then used to alternate between the two registers. For reception of asynchronous data, the LM reads one register while the LC loads data into the other register. For synchronous communication, the voice-processing unit can read one register while the LC loads voice information into the other register.

Bit stream processes are performed for each packet before its transmission. For the header, an HEC code is added to increase reliability followed by a whitening word and FEC encoding for security purposes. Upon reception of a packet, the reverse bit stream processes are accomplished to decode and check the information for any possible error. A similar process is performed for the payload portion of each packet; however, a CRC code is used instead of an HEC code. The CRC and encoding/decoding processes are optional and their requirements depend on the packets type.

5.3.8 Master–Slave Timing Management

Each piconet is synchronized to the master's system clock. Slave devices adjust their system clocks to the master's system clock by using a timing offset. Clock alignment is done by comparing the actual-versus-predicted reception times. The master and all slave nodes have a small ± 10-μs window to allow for timing mismatches when receiving information. A slave device in the hold/park or sniff mode may use larger window sizes when it wakes up to resynchronize with the master node. Also, if a master node has addressed a slave device in a preceding time slot or if it has already established an SCO communication link, it would have used a FHS packet during both "connection setup" and the "master–slave switch" procedure. The slave will then respond by sending an ID packet during the next time slot. This response will be followed by a protocol-specific process to establish timing and frequency synchronization. The slave will then use an offset to match its transmission/reception to or from the master device.

5.3.9 Channel Control

In an 802.15.1 piconet the master device initiates the communication and is responsible for setting piconet system parameters. Both piconet timing and phase are synchronized to the master's system clock, which uses a polling scheme to control the flow of traffic in the network. The FH sequence and CAC are determined by the master's Bluetooth device address (BD_ADDR). Every Bluetooth unit has a free-running internal clock which is used for timing and frequency regulation of each transceiver. In this case, the local clock is signified as CLKN, for *native clock*. The slave clocks use an *offset* to synchronize with the master clock. This offset, when added to the native clock, is labeled CLKE, for *estimated clock*. This process is illustrated in Fig. 5.5.

In Fig. 5.5, the master clock, which has an offset value of zero, is shown as CLK. The clock runs in one of two possible modes: In its higher accuracy mode, the clock accuracy is ± 20 ppm, while in the lower accuracy mode its accuracy is ± 250 ppm. The clock not only allows synchronization between master and slave devices but also tracks critical time periods which are used to trigger special events. These time intervals are 312.5 μs, 625 μs, 1.25 ms, and 1.28 s and are related to bits 0, 1, 2, and 12, respectively, of the clock, as shown in Fig. 5.6.

Two main states are used in the link controller: *standby* and *connection*. There are also substates that are used to add new devices to the network. These substates are *page*, *page scan*, *inquiry*, *inquiry scan*, *slave response*, *master response*, and *inquiry response*. The standby state is the default state, while the connection state is used when the device successfully completes a page attempt. These states and substates are shown in Fig. 5.7.

5.3.9.1 Access Procedures. With the help of DAC and IAC, "paging" and "inquiry" are used to set up a connection required for access establishment. Normally, inquiries are used to establish the device's addresses and clocks while paging is used to establish the actual connection.

A device is allowed to enter the page scan substate to check for any new connection with other devices. The slave device may be in the standby state when entering the page scan state. In this case, it is required to periodically enter the page scan state to see if another device is attempting to establish a connection with it. If, however, this device is in the connection state, then it will devote as much time as possible to scanning upon transition into the page scan

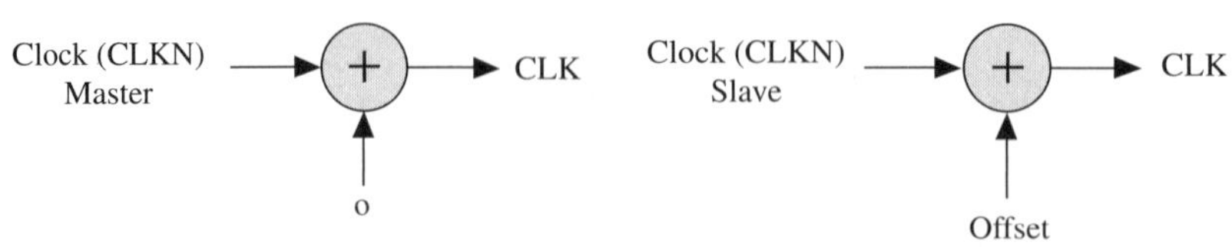

FIGURE 5.5 Derivation of CLK in (*a*) master and (*b*) slave device.

CLK 3.2 kHz						
27	26-13	12	11-3	2	1	0
Most significant bit (MSB)	Noncritical	1.28 s	Noncritical	1.25 ms	625 μs	312.5 μs

FIGURE 5.6 Bluetooth clock. While all bits are used in the *free-running* clock, only those bits that have significant meaning with regard to the operation of the standard are shown, namely CLK_0, CLK_1, CLK_2, and CLK_{12}.

state. This state is achieved by placing ACL connections into either the hold or park mode. SCO connections, which have a higher priority, can be compensated for by the use of lower quality SCO modes and by increasing the number of slots reserved for page scanning. In a page scan, while the unit monitors a particular channel in order to establish a connection, it is also required to alternate the deployed channels using a unique FH sequence. The page substate is used by either a master device or a potential master to establish a connection

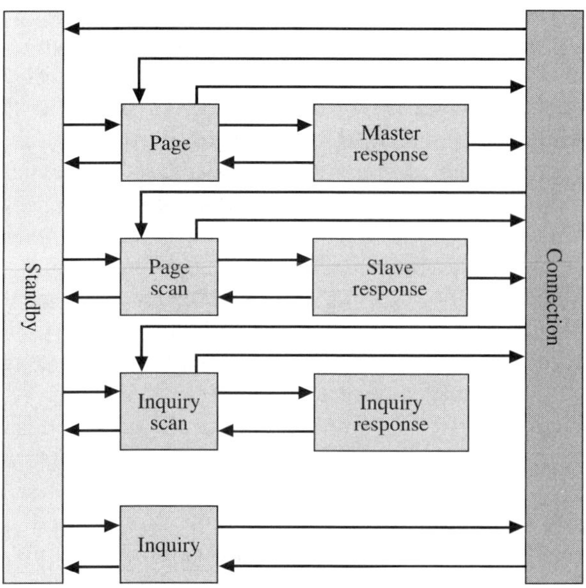

FIGURE 5.7 Bluetooth link controller state diagram.

with either a slave node or a potential slave device. The master tries to connect with the intended slave by sending a series of packets, at different hopped channels, tagged by the slave device's DAC.

Since the master device does not know the slave's true CLK time, it may use an estimate based upon the most recent previous experience. If the estimation is either inaccurate or unavailable, then the master attempts to contact the intended slave device during both hopped frequencies before and after the predicted one as well as during different wake-up frequencies. During each transmission slot, the master will transmit on two different, sequential frequency hops. In addition, it will listen to both corresponding frequency hops for a response during the following reception slot. The difference in CLK times between master and slave devices is taken into account by alternating page trains, named A and B, which is simply a division of page hopping sequence. Train A is designated to detect differences in CLK times between -8×1.28 and $+7 \times 1.28$ s. This covers hops from $f(k-8), \ldots, f(k+7)$, where $f(k)$ is determined by $CLKE_{16-12}$. Train B is designated for time differences greater than 7×1.28 s or less than -8×1.28 s, which covers hops from $f(k-16), \ldots, f(k-9), f(k+8), \ldots, f(k+15)$. Both train A and train B are repeated at least 128 or 256 times (depending on the hop system used) and are alternated until a connection is established or until the timeout point (i.e., pageTO) is reached.

Like the page scan substate, the page substate can be entered from either the standby or connection state. Also, like the page scan substate, it will devote as much time as possible to this endeavor. To achieve this substate while departing from the connection state, it will place ACL connections into the hold or park state. Likewise, SCO connections should be relegated to the lower quality modes when applicable. Once the slave device successfully receives the page message, it enters a response protocol with the master to exchange information required to establish the connection. The master then responds with a FHS packet. This packet contains the master device's BD_ADDR, clock and many other parameters. The master's BD_ADDR is used to derive the channel-hopping sequence. In addition, the master's clock is also used for synchronization of the master and slaves. Table 5.1 shows the sequence of exchanged messages for the paging procedure. Once the first four steps have been completed, the master can then send messages to the slave using its access code and clock. Message delivery is conducted with a poll packet. In order to complete this connection protocol, the master node should send the aforementioned poll packet to the receiver. Upon reception of the poll message, the receiver should issue a response to the master, which must then be received within the newconnectionTO number of slots. The newconnectionTO time starts once the master device receives the slave's response to its FHS message, i.e., after step 4 in Table 5.1.

5.3.9.2 Inquiry Procedures. The inquiry process is used to discover all Bluetooth devices that are within listening range of the inquiring device. The responding units will then give the inquiring unit the information necessary to establish connections with them. A general inquiry access code (GIAC) is used

TABLE 5.1 Message Sequence during Paging Process

Step	Message	Direction	Hopping Sequence	Access Code and Clock
1	Slave ID	Master to slave	Page	Slave
2	Slave ID	Slave to master	Page response	Slave
3	FHS	Master to slave	Page	Slave
4	Slave ID	Slave to master	Page response	Slave
5	First MP	Master to slave	Page	Master
6	First SP	Slave to master	Page response	Master

to poll for any Bluetooth device, while the dedicated inquiry access code (DIAC) is used to poll for specific types of devices.

The inquiry scan substate is similar to the previously mentioned page scan substate. However, a device in this substate is listening for the IAC instead of its personal DAC. The device may scan for one or more DIACs and/or the GIAC. The device should dedicate all of its time to the inquiry scan substate if it is coming from the standby state. However, if it is coming from the connection state, then it should reserve as many time slots as possible to the inquiry scan substate by placing ACL connections in the hold or park mode and by using the lowest capacity SCO connection mode allowable. Furthermore, the inquiry scan window should be increased to compensate for any ongoing SCO connections. The inquiry substate is used to establish a connection with those devices currently within the radio range of inquiring device. This substate is functionally similar to the page substate from the timing sequence point of view. The response packet sent by a device answering the inquiry is actually a FHS packet. Thus, all of the information required to establish a connection with this device is transmitted to the master device. This process will be continued; otherwise, it is aborted early by the Bluetooth link manger, if the inquiry TO timer times out. Similar to the inquiry scan substate, a device entering the inquiry substate from the standby state may dedicate all of its capacity to the inquiry process. If it is entering this substate from the connection state, then it can place current ACL connections into either the hold or park modes. SCO connections should be placed in the lowest capacity SCO mode allowed for the given connection.

The message sequence for the inquiry process is shown in Table 5.2. As seen in this table, the master node does not respond to the slave. Instead of

TABLE 5.2 Message Sequence during Inquiry Process

Step	Message	Direction	Hopping Sequence	Access Code and Clock
1	ID	Master to slave	Inquiry	Inquiry
2	FHS	Slave to master	Inquiry response	Inquiry

responding to the slave, it simply records the information sent by the slave. This information is an FHS packet and as such contains all of the information that the master requires to establish a connection with the slave, such as the slave's device address and clock.

5.3.9.3 Connection State. Once a connection is established, the involved devices are in the connection state and data packets may be exchanged. Then, the involving devices use both the CAC and clock of the master node to transmit packets using the channel-hopping sequence.

There are four modes associated with the connection state: the *active*, *sniff*, *hold*, and *park* modes. In the active mode, the device can actively communicate with other devices as well as maintain synchronization with the master. In the sniff mode, the device only has to listen for ACL transmissions carrying its AM_ADDR during *select* slots, which are spaced at intervals of T_{SNIFF} slots. In the hold mode, ACL links can be suspended, which will allows the device to free up capacity for other functions, such as inquiry, or to enter a low power sleep mode. After an agreed-upon time interval, i.e., holdTO, the device will wake up. In the park mode, the slave device gives up its AM_ADDR for a parked member address (PM_ADDR) and an access request address (AR_ADDR) and then enters a low power sleep mode. The parked device is activated at regular intervals to resynchronize its clock with the master and to listen for broadcast messages.

The master uses a beacon channel to manage parked slaves. This channel is used by the parked slaves to maintain synchronization with the master, to alert parked slaves to changes in the beacon channel, and to allow for unparking of slaves. The master can initiate a slave's unpark process by sending the slave's PM_ADDR or BD_ADDR in the beacon broadcast message as well as the slave's new AM_ADDR. The slaves also have a beacon access window in which they can request to be unparked. The slave waits for its AR_ADDR slot within the beacon access window to request unparking. Once it receives the unpark message from the master, it will enter the active mode and wait for the poll message containing its new AM_ADDR. Finally, after responding to the message, it will enter the mode dictated by the master.

5.3.9.4 Polling Scheme. In order to follow the TDD scheme, slaves are only allowed to communicate with the master, and only in slots following a slot in which their AM_ADDR was in the header sent by the master. For an SCO slot the slave is allowed to transmit in its designated slot as long as another slave device's AM_ADDR is not addressed in the preceding packet header sent by master node. Slaves in the parked mode are allowed to send messages only in the access window following a beacon broadcast by the master.

5.3.9.5 Scatternet. As stated earlier, a scatternet is an interconnected group of piconets. A device can only be the master of one piconet; however, a device can serve as a slave node in multiple piconets. These piconets can operate over the same area since they have different masters and thus different CACs and operate on different frequency hops and phases. A device in multiple piconets uses time multiplexing to alternate between them. The device must maintain synchronization between two different masters, each with their own drift, so the device must update its offset frequently. The device may also go to a lower capacity mode, such as sniff, hold, or park, in one piconet to increase its capacity in another piconet.

A master and a slave of a piconet may switch roles by using the master–slave (MS) switch. The role of MS switching is needed in a few different situations, e.g., when a device pages the master of a piconet that it wants to join. This MS switch consists of a TDD switch followed by a piconet switch between the master and slave. This process is followed by a piconet switch between the new master and each old slave that wishes to be a part of the new piconet.

5.3.9.6 Power Management. To ensure low power consumption of Bluetooth devices, the standard ensures that packet handling and active slot monitoring are minimized and has built in energy-efficient modes of operation. With regard to packet handling, the time a device is actively transmitting or receiving data is minimized by only passing the information required for a particular purpose. For packets that span multiple slots, all nonaddressed devices may remain inactive during that span. Moreover, for devices in the connection mode, they may use the more power efficient modes of sniff, hold, and park when applicable.

5.3.9.7 Hop Selection. The Bluetooth standard supports both 79- and 23-hop systems. There are five hopping sequences defined for each of these systems, which include a page-hopping sequence, a page response sequence, an inquiry sequence, and an inquiry response sequence. These channels all have 32 (for 79-hop) or 16 (for 23-hop) unique frequencies. There is also a long period channel-hopping sequence used to distribute the hop frequencies equally over the available bandwidth.

The hopping sequence is generated from a selection box, as shown in Fig. 5.8. The selection box uses the 27 MSBs of the master's CLK in the connection state, while all 28 bits of the CLK are used in the page and inquiry substates. The address input consists of 28 bits, i.e., the four LSBs of the upper address part (UAP) and all 24 bits of the LAP. In the connection state, this is the master node's address, in the page substate it is the address of the paged device, and in the inquiry substate it is the GIAC. The output constitutes a pseudorandom sequence covering either 79 or 23 hops depending on the state.

The frequency hop selection is based upon selecting one of the existing sets of 32 frequencies, which is called a segment. These segments and their overlap are shown in Fig. 5.9. In the connection state, the selection is pseudorandom, while the order set is based upon the device address for the page substates. The segments are listed in order with all even hop frequencies listed together

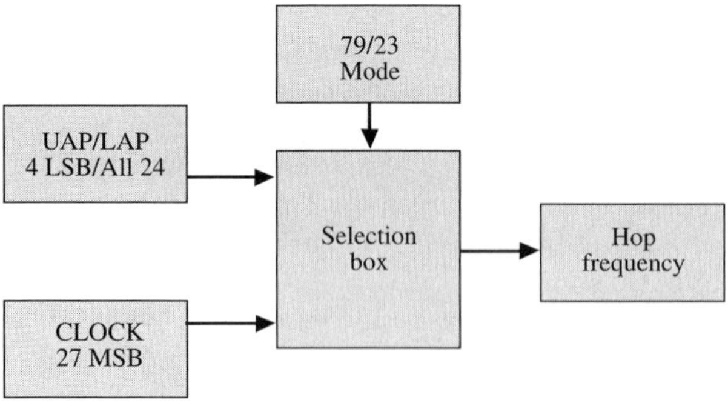

FIGURE 5.8 General block diagram of hop selection scheme.

followed by all odd hop frequencies. This process allows for better distribution of hop frequencies within a given segment.

The hop selection kernel for the 79-hop system is chosen based upon the scheme shown in Fig. 5.10. The 23-hop system is nearly identical except for the modulus numbers and the deployed permutation algorithm. The inputs to this scheme are denoted by letters in Fig. 5.10. The X input determines the phase of the hop segment. The Y_1 and Y_2 inputs are used for selecting either master-to-slave or slave-to-master transmission. The A, B, C, and D inputs are used to determine the ordering in the segment. The E and F inputs determine the hop frequencies.

The first ADD box is a modulus addition. This is modulo 32 for a 79-hop system or modulo 16 for a 23-hop system. For the XOR operation the four LSBs of the first addition are modulo 2 added to the AM_ADDR bits A_{22-19}. The permutation operation involves switching either four or five inputs to get either four or five outputs for the 23- or 79-hop system, respectively. The second addition is also modulus addition. However, this addition is modulo 79 for a 79-hop system or modulo 23 for the 23-hop system. The output of the

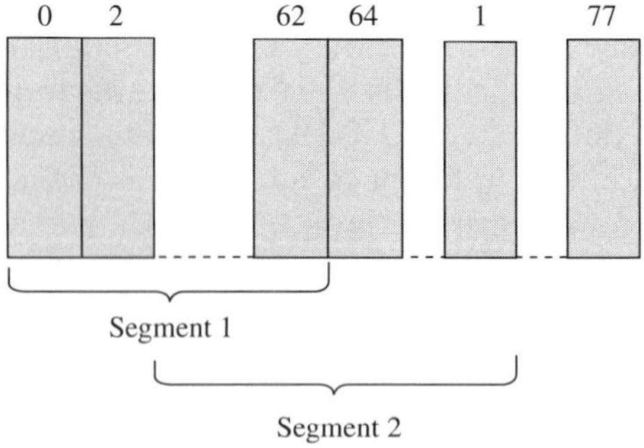

FIGURE 5.9 Hop selection scheme in connection state.

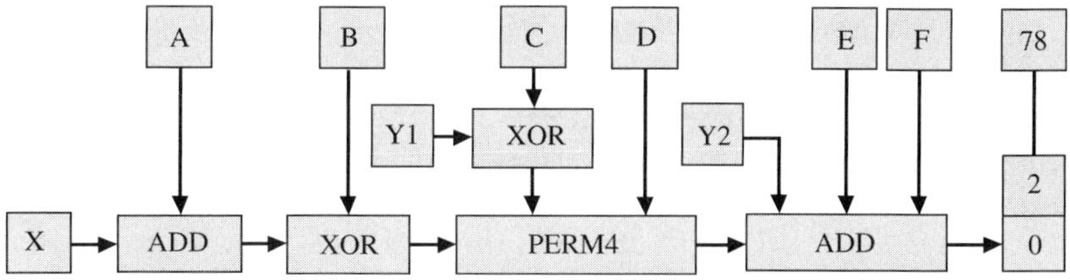

FIGURE 5.10 Hop selection kernel block diagram for 79-hop system.

second addition indexes a set of 79 or 23 registers depending upon the hop frequency.

5.3.9.8 Audio. The voice channels support a 64-kbps audio stream. This audio stream comes in either a log pulse-coded modulation (PCM) compression or continuous variable slope delta (CVSD) modulation format. In addition, errors are handled by an FEC scheme in HV1 and HV2 packets.

5.3.9.9 Addressing. A unique 48-bit BD_ADDR is assigned to each Bluetooth device. The 24 LSBs of the BD_ADDR constitute the LAP. The next eight LSBs represent the UAP. The 16 MSBs represent the nonsignificant address part (NAP). There are three access codes used in the IEEE 802.15.1 standard. These are CAC, DAC, and IAC. All of these codes are derived from the the BD_ADDR's LAP. The CAC is generated from the master's BD_ADDR LAP and is used in the preamble of every packet exchanged in the piconet. The DAC is used in the paging substates. The GIAC is used for inquiry of all devices, while a DIAC limits the inquiry to a particular class of devices. Each slave is assigned a 3-bit AM_ADDR, which is used to uniquely identify all active members. The all-zero AM_ADDR is used for broadcast messages, which are accepted by all devices. When parked, a device gives up its AM_ADDR and receives a PM_ADDR and an AR_ADDR. These are used to allow the device to stay synchronized to the system and for later reentry as an active participant.

5.3.9.10 Security. The Bluetooth standard provides security with use of the following four entities: a 48-bit BD_ADDR that is unique to each device, an 128-bit authentication key (referred to as the *link key*), a variable 8–128-bit (1–16-octet) *encryption key*, and a 128-bit random number (RAND). The link key is an 128-bit randomly generated number which is used for authentication as well as encryption key generation. There are four types of link keys: *combination key* (K_{AB}), *unit key* (K_A), *temporary key* (K_{master}), and *initialization key* (K_{init}). There is also an *encryption key* (K_c). The security standards will only be discussed further when directly related to MAC as it is out of the scope of this chapter.

5.4 LINK MANAGEMENT PROTOCOL

This protocol is used to for establishment, maintenance, security, and control of communication links between devices inside piconets. Upon receiving a packet with an LM message, the LM message is filtered out before the packet is sent to the higher layer. This procedure is shown in Fig. 5.11. Also note that the LC is only required to communicate with each slave device once per T_{POLL}; thus, this is the maximum time required to get an LMP message to a slave.

The LM protocol data units (PDUs) can be of two types: *mandatory* or *optional*. The LM should respond to all mandatory PDUs and optional PDUs that require a response. Furthermore, the source or destination of the delivered PDU is determined by the AM_ADDR included in the packet header. In addition, the "Op Code" and "transaction ID" are located in the payload body.

5.4.1 Procedure Rules

Each PDU procedure is described using a sequence diagram as shown in Fig. 5.12. A solid line indicates communication in the direction of the arrow. A double arrow indicates communication in both directions. A dashed line indicates optional communication.

Authentication is based upon the challenge–response scheme. The verifier sends a PDU packet with a randomly generated authentication number, LMP_au_rand. The claimant calculates a response, LMP_sres, and sends it back to the verifier. Authentication is granted only if the response is correct; otherwise, authentication is denied. The response is a function of the challenge, the claimant's BD_ADDR, and their shared secret key. An exponential backoff scheme is followed for failed authorization attempts. Figure 5.13 illustrates a successful authentication sequence.

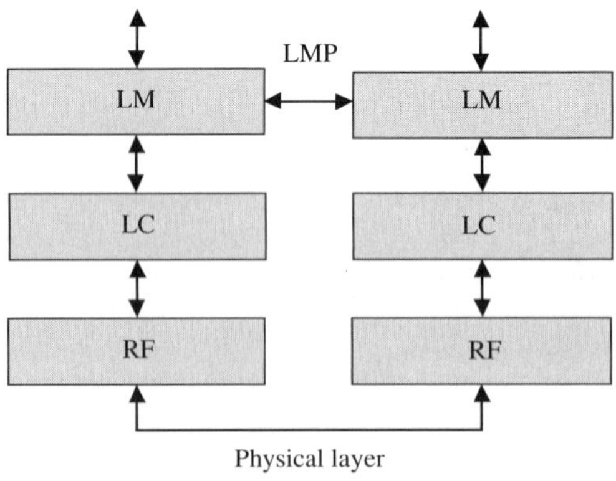

FIGURE 5.11 LM's communication path in layered architecture.

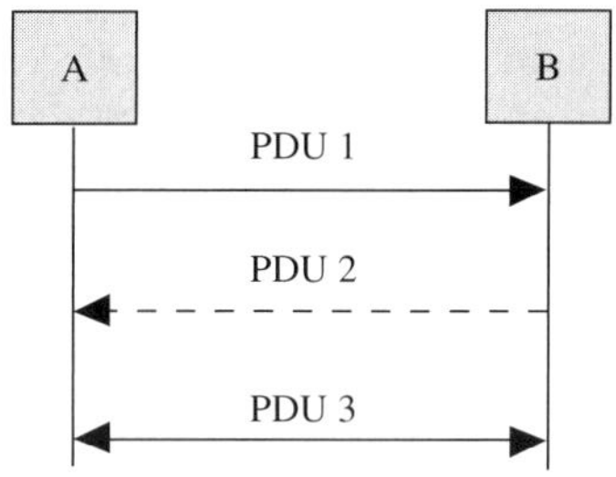

FIGURE 5.12 Sequence diagram showing types and meaning of LM PDU messages. PDU1 is a mandatory PDU sent from A to B. PDU2 is an optional PDU sent from B to A. PDU3 is a mandatory PDU that may be sent from either A or B.

When units communicate using encryption, they must share a link key. This link key can be semipermanent or temporary. A semipermanent link key lasts for the duration of the session. The master can issue a temporary link key, K_{master}, if it wishes to send an encrypted broadcast. The semipermanent link key will then replace the temporary link key whenever the encrypted broadcast is completed. Encrypted broadcast completion is always signaled by the master device. The encryption mode must be negotiated and agreed upon by the master and slave devices. The negotiation phase includes whether encryption will be used. If encryption is to be used, then it must be determined if this also includes point-to-multipoint as well as point-to-point transmissions. Finally, the encryption key size, from 1 to 16 octets, must be determined.

The LM allows devices to switch modes, roles, and power control mode as well as detach from the piconet. The modes supported, as stated previously in the discussion of baseband layer access procedures, are sniff, hold, and park. Either the master or the slave can initiate these modes. The transmission power might be adjusted by reception of a request based upon the receiver's signal strength indicator (RSSI). The LM also controls the communication quality of service (QoS). The poll interval T_{poll} is the maximum time between ACL

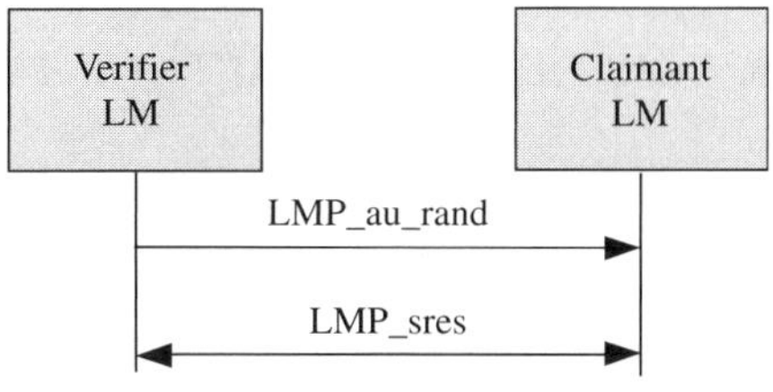

FIGURE 5.13 Authentication sequence. The verifier sends an 128-bit RAND and with the use of this number the claimant's BD_ADDR, and the shared link key generates the correct response. This is compared to the claimant's response. If correct, the claimant has been authenticated.

transmissions from the master to a slave. The master and slave can also renegotiate the QoS upon request. The master and slave are also able to decide upon the number of repetitions for broadcast messages N_{BC}. In addition, either the master or the slave can initiate a SCO link. The SCO link is uniquely identified by an SCO handle. Each SCO link has an SCO interval, T_{SCO}, which is in fact the number of time slots between two successive transmissions. Bluetooth supports three different voice-coding formats: μ-law log PCM, A-law log PCM, and CVSD. The connection can be relinquished by a request from either the master or the slave.

5.4.2 Connection Establishment

To create a connection between devices using a layer above the LM layer, an LMP_host_connection_req request is sent. If this request is accepted, then security procedures must be agreed upon as well. These include pairing, authentication, and encryption considerations. Finally, the procedure is concluded by sending an LMP_setup_complete message.

5.4.3 Test Modes

The test mode is used to ensure a device complies with Bluetooth standards. The device's transmission and reception frequencies, hopping mode, and poll period are among the parameters that can be tested. This mode is activated by sending a packet, LMP_test_activate to a device under test (DUT) and exited by either an LMP_test_control or an LMP_detach packet.

5.4.4 Error Handling

The LM will reject errors due to either an invalid Op Code or parameters with an LMP_not_accepted message. This message also includes the reason for the rejection. If the maximum response time is exceeded, the procedure will also result in a failure. Finally, if a link loss is detected, the procedure will be terminated by the waiting device.

5.5 LOGICAL LINK CONTROL AND ADAPTATION PROTOCOL

The L2CAP supports both connection-oriented and connectionless data services with packet segmentation and reassembly, protocol multiplexing, and abstraction. All possible communication transactions for this protocol are shown in Fig. 5.14.

The reason L2CAP should support protocol multiplexing is due to lack of a dedicated-type field in the baseband layer. This protocol must also be able to segment and reassemble packets. Due to the small baseband payload size required for this standard, packets at the L2CAP layer level should often be

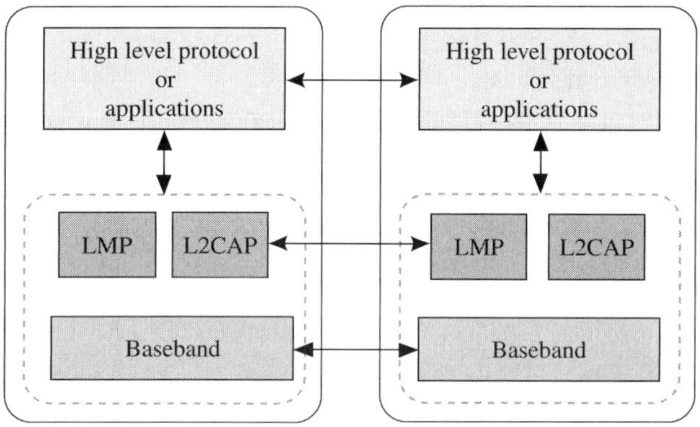

FIGURE 5.14 L2CAP within protocol layers. The left and right solid boxes are two different Bluetooth devices. The left and right dashed boxes indicate the MAC layers of each device. The solid arrows indicate actual communication, while the dashed arrows indicate a communication abstraction.

broken down into segments small enough to be sent to the baseband layer and later reassembled at the receiving L2CAP layer. This layer also monitors the QoS status and transmits measurements of the level of QoS to ensure constraints are being met. Finally, this layer uses abstraction to allow higher layer protocols and applications to correctly address other devices within the network without dealing with the actual addressing at the baseband layer.

The L2CAP is only defined for ACL communication links. L2CAP communication requires CRC for data integrity checks. AUX1 packets, which do not use CRC, cannot be used for L2CAP. The ACL packet header uses the 2-bit L_CH field to distinguish between L2CAP and LMP packets. The L2CAP in the Bluetooth architecture is shown in Fig. 5.15. The L2CAP interfaces with higher level communication protocols such as Bluetooth service discovery

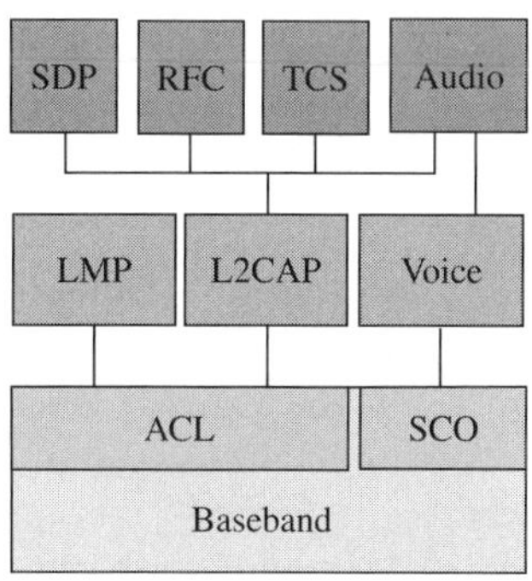

FIGURE 5.15 Architecture of Bluetooth L2CAP. The L2CAP and LMP layers manage ACL packets, while voice is handled with SCO packets in the baseband layer.

(BSD) and telephony control system (TCS). The L2CAP should strive to be power efficient and bandwidth efficient but of low complexity to run on all Bluetooth-supported devices.

The L2CAP is built with a number of assumptions taken into account: First, at most one ACL link can be established by the LMP between several communicating devices. Second, the baseband layer provides a reliable means of delivering packets and, whenever applicable, it is functionally delivered in a full-duplex manner. Third, the protocol provides a reliable channel by utilizing the baseband layer to resend data units until it passes a data integrity check or has timed out. The L2CAP layer does not perform the retransmission or checksums but instead uses the baseband layer for these functions. The L2CAP does not support real-time audio, which requires SCO links. Furthermore, this layer does not support global group names.

5.5.1 General Operation

The L2CAP layer is based on the concept of logical channels for which a user's local name, called a channel identifier (CID), is employed to represent the logical endpoint of each channel. The CIDs ranging from 0×0001 to $0 \times 003F$ are reserved for the L2CAP layer. The 0×0001 CID corresponds to a particular logical channel which is required to meet Bluetooth standards and is reserved for signaling purposes (i.e., a signaling channel). The signaling channel is used for creating connection-oriented channels. The 0×0002 CID is reserved for connectionless data traffic. In particular, it is deployed for connectionless communication.

The CIDs are used in directing communications between L2CAP devices. These communications can come in two forms, connection-oriented channels and connectionless channels. Connection-oriented channels are two way and connectionless channels are one way. Connectionless channels support "group" or multipoint communication. These communication channels are illustrated in Fig. 5.16.

Connection-oriented channels are formed through the use of signaling channels. Each connection-oriented channel has a CID at each end. Each device maintains these CIDs locally, so they are unique for a given device. These communications can pass through a common device, as shown in Fig. 5.16, with connection-oriented data channels between devices 1 and 4. The connectionless data traffic channel, 0×0002, is used for group or multipoint data communications. This process is shown in Fig. 5.16 as connectionless channels going to devices 2 and 3. The L2CAP layer must allow for communication between not only the L2CAP layers of different devices but also the higher layer protocols and lower layer protocols. This procedure is illustrated in Fig. 5.14, where the higher layer protocols and applications communicate via the L2CAP, LMP, and baseband layers.

The L2CAP layer is responsible for packet segmentation and reassembling. This layer allows for improved efficiency by reducing the overhead required to

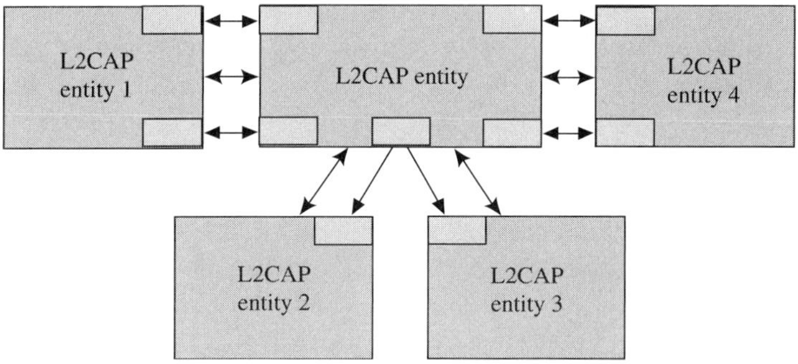

FIGURE 5.16 Diagram of connectionless and connection-oriented data channels between Bluetooth devices. The rectangles represent CIDs. Arrows with two solid ends represent connection-oriented channels while the arrows with one solid end represent the connectionless channels. The arrows without solid ends represent the signaling channels. ♯*n* represents the L2CAP for device *n*, where $n = 1, 2, 3, 4$.

send information. This improvement in efficiency is achieved by supporting a maximum transmission unit (MTU) size that is much larger than that of the baseband layer. The larger transmission size results in a higher percentage of actual payload to overheads. For sending information, the L2CAP breaks higher layer packets into segments using the receiving MTUs. These segments are then passed to the LM via the host controller interface (HCI). The L_CH bits are set to 10 for the first segment and 01 for all subsequent segments. For receiving information, the L2CAP reassembles these segments using information passed to it via the sender's HCI. This process is shown in Fig. 5.17.

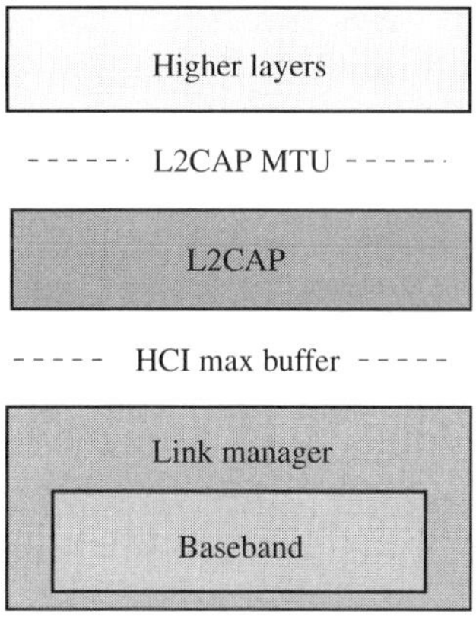

FIGURE 5.17 L2CAP segmentation and reassembly variables. Both the L2CAP MTU and the HCI maximum buffer size must be transferred between the designated layers for proper segmentation and reassembly of data packets passed through the L2CAP layer from higher and lower layers.

Besides, all L2CAP segments associated with a single higher level data packet will be transmitted before any other L2CAP packet that might be transmitted to that particular device.

5.5.2 State Machine

The L2CAP state machine handles all events that cause state transitions. The state machine only pertains to connection-oriented CIDs and is not representative of either connectionless or signaling channels. The terminology used for packets coming from a higher protocol layer are called requests and the replies to these requests are called confirms. Packets coming from a lower protocol layer are called *indications* and the replies to these indications are called *responses*. The L2CAP events and state transitions are illustrated in Fig. 5.18.

The actual message-passing sequence is shown in Fig. 5.19. The initiator is the entity that initiates or starts the communication and the acceptor is the entity that is replying. The L2CAP of the initiating device has received a request from higher protocol layers. In order to deliver this request to the acceptor device, it propagates this message to the lower protocol layers via a request. The acceptor device receives this request in the form of an indication from its lower protocol layers. It propagates that indication to the higher levels and awaits a response, which will then travel back the way it came in the form of responses in the acceptor device and confirms in the intended initiating device. A series of confirm packets propagate back to the higher layers from which the request has originated.

Events are particular designated messages that are directed to the L2CAP layer. Events may be initiated from higher protocol layers as requests and

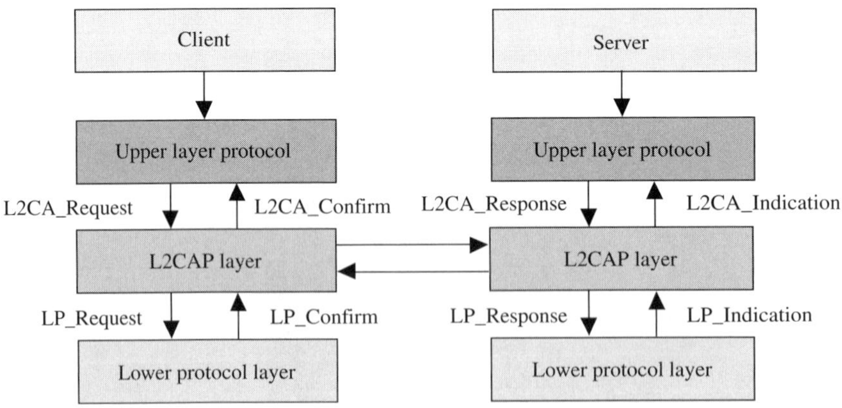

FIGURE 5.18 L2CAP interactions. Communications from higher to lower layers or vice versa are indicated without the letter P in the layer receiving the request for service. Communications from the same level include the letter P in the name of the receiving layer. *Client* is the initiating entity and *server* is the responding entity.

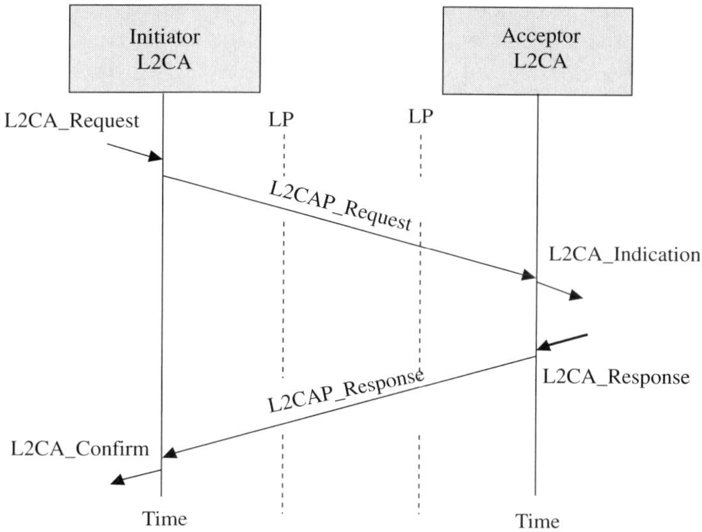

FIGURE 5.19 Message sequence chart (MSC) for interacting L2CAP layers. This chart illustrates the standard sequence of events between entities via their respective L2CAP layers. LP indicates the lower protocols for both devices.

confirms, initiated from lower protocol layers as indications and confirms, in the form signals to or from peers, or as timeouts. These events include connection, detachment messages, and QoS messages, among many others. On the other hand, channels operate in different states. The open state indicates that a connection has been established and configured and the intended data communication may commence. The closed state indicates that there is no connection. In the configuration state, a connection exists but has not been configured yet or is being reconfigured. The connect and disconnect states are used to create or terminate a connection, respectively. Actions are mapped to events according to the current state and allow for state transitions, creating and terminating connections, and tracking timeouts, among others.

5.5.3 Data Packet Format

The L2CAP packet structure consists of an L2CAP *header* and the *payload*. The L2CAP header is composed of a *length* field followed by a *channel ID* field. The length field is 16 bits and represents the payload size in bytes, which has a limit of 2^{16}, or 65,536, bytes. This length field is used for a simple data integrity check in the packet-reassembling procedure. The channel ID is also 16 bits and is used to identify the destination device for a particular packet. The payload carries the actual data being transmitted.

Unlike connection-oriented data, connectionless data are not guaranteed to be received by the destination devices as no acknowledgment scheme is used. In connectionless data transactions, the data packet is similar to that of connection-oriented transmissions with the channel ID being set to 0×0002 and an additional field, i.e., the protocol/service multiplexer (PSM). The PSM field requires that the LSB of the least significant octet be a 1 and the LSB of the most significant octet be a 0. This notation allows this field to be extended beyond 16 bits.

5.5.4 Signaling

The signaling channel is used to establish connections between different L2CAP entities. This kind of channel uses the reserved 0×0001 CID. L2CAP signal commands are sent as requests and responses while multiple commands should be sent in a single data packet. These signals must be accepted if the packet does not exceed the device's MTU, which is required to be at least 48 bytes in size. These signals come in many forms: Connection requests are used to establish a connection while a connection response is used to return the response to the device requesting a connection. A configuration request is used to negotiate or renegotiate the connection parameters. A configuration response is used to return the response to the configuration request. Disconnect requests and responses are used to terminate a connection. There are also echo requests, information requests, and their corresponding responses. These responses are used to elicit a response or information from a device, respectively.

5.5.5 Configuration Parameter Options

Configuration parameter options make negotiation for connection requirements possible. These requirements include MTU size, flush time, and QoS constraints. Each Bluetooth device is required to support a minimum MTU size in order to allow for channel connections to be established. However, if both devices are able to support larger MTU sizes and QoS constraints can still be met, then a larger MTU size can be agreed upon. These values are agreed to by each receiver and can be different between two involving devices in a given channel.

The flush time is used to denote the number of attempts a transmitting device will continue resending a certain packet. The source device keeps on resending the packet until it either is received successfully or has reached the flush time limit. The flush time is based upon time units and not on the number of retries to send a particular packet. If all bits for the time variable are set to 1, then the flush time is considered to be infinite.

The QoS parameters are completely optional. If no parameter is specified, the last used parameters agreed upon by the two communicating devices should be used. These parameters specify certain constraints, such as peak bandwidth (bytes per second), latency (microseconds), and delay variation (i.e., delay jitter). If any of these constraints cannot be met, a QoS violation is raised.

5.6 IEEE 802.15.1 RADIO SPECIFICATIONS

The Bluetooth transceiver is operating in the 2.4-GHz ISM band. This section briefly explains the requirements for a Bluetooth transceiver operating in this unlicensed band.

5.6.1 Frequency Bands and Channel Arrangement

In a vast majority of countries around the world the range of the ISM frequency band is 2400–2483.5 MHz. Some countries have national limitations in the frequency range. To comply with these national limitations, special frequency-hopping algorithms have been specified for these countries. It should be noted that products implementing the reduced frequency band will not work with products implementing the full band. The products implementing the reduced frequency band must therefore be considered as local versions for a single market. The Bluetooth SIG has launched a campaign to overcome these difficulties and reach total harmonization of the frequency band. Table 5.3 shows all available operating frequency bands.

5.6.2 Transmitter Characteristics

The requirements stated in this section are given as power levels at the antenna connector of the equipment. If the equipment does not have a connector, a reference antenna with 0 dBi (isotropic decibel) gain is assumed. Table 5.4 summarizes the power classes defined for IEEE 802.15.1.

Power control is required for power class 1 equipment. Power control is used for limiting the transmitted power over 0 dBm (power ratio in decibels relative to 1 mW). The power control capability under 0 dBm is optional and could be used for optimizing the power consumption and overall interference level. The power steps should form a monotonic sequence, with a maximum step size of 8 dB and a minimum step size of 2 dB. Class 1 equipment with a maximum transmit power of +20 must be able to control its transmit power down to 4 dBm or less. Equipment with power control capability optimize the output power in a link with LMP commands. It is done by measuring the RSSI and reporting back if the power should be increased or decreased.

TABLE 5.3 Operating Frequency Bands

Geography	Regulatory Range (GHz)	RF Channels
United States, Europe, and most other countries	2.4000–2.4835	$f = 2402 + k$ MHz ($k = 0, \ldots, 78$)
Spain	2.4450–2.4750	$f = 2449 + k$ MHz ($k = 0, \ldots, 22$)
France	2.4465–2.4835	$f = 2454 + k$ MHz ($k = 0, \ldots, 22$)

TABLE 5.4 IEEE 802.15.1 Power Classes

Power Class	Maximum Output Power (P_{max})	Nominal Output Power	Minimum Output Power	Power Control
1	100 mW (20 dBm)	N/A	1 mW (0 dBm)	$P_{min} < 4$ dBm to P_{max} Optional: P_{min} to P_{max}
2	2.5 mW (4 dBm)	1 mW (0 dBm)	0.25 mW (−6 dBm)	Optional: P_{min} to P_{max}
3	1 mW (0 dBm)	N/A	N/A	Optional: P_{min} to P_{max}

In IEEE 802.15.1, the modulation is Gaussian frequency shift keying (GFSK) with BT = 0.5. The modulation index (MI) must be between 0.28 and 0.35. A binary 1 is represented by a positive frequency deviation, and a binary 0 is represented by a negative frequency deviation. The symbol timing should be better than ±20 ppm.

5.6.3 Receiver Characteristics

The actual sensitivity level is defined as the input level for which a raw bit error rate (BER) of 0.1% is met. The requirement for a Bluetooth receiver is an actual sensitivity level of −70 dBm or better. The receiver must achieve the −70 dBm sensitivity level with any Bluetooth transmitter compliant to the transmitter specification.

5.7 SUMMARY AND CONCLUDING REMARKS

In this chapter, a brief overview of IEEE 802.15.1 MAC and PHY layers for WPANs has been provided. Starting from the MAC layer, we reviewed the well-known IEEE 802.15.1 MAC architecture and then pursued our discussion with a short overview of the IEEE 802.15.1 PHY layer.

ACKNOWLEDGMENTS

This work was supported partially by Nokia Foundation and Elisa Foundation.

REFERENCE

1. IEEE Standard 802.15.1, 2002. Part 15.1: "Wireless medium access control (MAC) and physical layer (PHY) specifications for wireless personal area networks (WPANs)," IEEE, New York, June 2005.

CHAPTER 6

OVERVIEW OF IEEE 802.15.2: COEXISTENCE OF WIRELESS PERSONAL AREA NETWORKS WITH OTHER UNLICENSED FREQUENCY BANDS OPERATING WIRELESS DEVICES

KAVEH GHABOOSI, YANG XIAO, MATTI LATVA-AHO, and BABAK H. KHALAJ

6.1 INTRODUCTION

Nowadays, concurrent deployment of IEEE 802.15 devices with other wireless equipment operating in the same unlicensed frequency bands is becoming popular. The core purpose of the IEEE 802.15.2 standard is to facilitate coexistence of IEEE 802.15 wireless personal area network (WPAN) devices with other wireless appliances operating in unlicensed frequency bands. The intended users of this standard include IEEE 802.11 wireless local area network (WLAN) developers as well as designers and consumers of wireless products being developed to operate in unlicensed frequency bands [1].

IEEE 802.15.2 defines several coexistence mechanisms that can be deployed to make the coexistence of WLAN and WPAN networks possible. These mechanisms are categorized into two distinct classes: *collaborative* and *noncollaborative*. A collaborative coexistence mechanism can be used when there is a communication link between the WLAN and WPAN networks. This is best implemented when both WLAN and WPAN devices are embedded into the same piece of equipment (e.g., an IEEE 802.11b card and an IEEE 802.15.1 module embedded in the same laptop computer). The so-called noncollaborative coexistence mechanism does not require any communication link between the WLAN and WPAN [1].

Emerging Wireless LANs, Wireless PANs, and Wireless MANs. Edited by Y. Xiao and Y. Pan

In this chapter, after a detailed explanation of interference problems, we discuss the defined coexistence mechanisms in the standard followed by the alternalting wireless medium access (AWMA) rules and its related technical issues. Next, we deal with 802.15.2 packet traffic arbitration (PTA). We will also look at packet selection approaches in IEEE 802.15.2 and conclude with a brief overview of the packet-scheduling mechanism proposed in 802.15.2.

6.2 INTERFERENCE PROBLEM STATEMENT

Since both IEEE 802.11b and IEEE 802.15.1 specify operation in the same 2.4-GHz unlicensed frequency band, there is mutual interference between the two wireless systems that might result in severe performance degradation [1–4]. In fact, there are numerous factors that affect the level of interference, specifically separation between WLAN and WPAN equipment, the amount of data traffic generated by each wireless network, i.e., WLAN and WPAN, power levels of different deployed devices, and the data rate of WLANs. Furthermore, different types of information being sent over the wireless networks have different levels of sensitivity to the interference. For instance, a voice link is usually more sensitive to radio interference in comparison to a data link being used to transfer a data file.

There are several versions of the IEEE 802.11 physical layer (PHY). However, all versions use a common medium access control (MAC) sublayer. When implementing the distributed coordination function (DCF), 802.11 MAC uses carrier sense multiple access with collision avoidance (CSMA/CA) for MAC. Basically, the scope of 802.15.2 is limited to DCF implementations of IEEE 802.11 and does not include the point coordination function (PCF) access scheme [2]. Initially, 802.11 included both a 1- and 2-Mbps frequency-hopping spread spectrum (FHSS) PHY layer as well as a 1- and 2-Mbps direct-sequence spread spectrum (DSSS) PHY layer. The FHSS PHY layer uses 1-MHz channel separation and hops pseudorandomly over 79 channels. The DSSS PHY layer uses a 22-MHz channel and may support up to three nonoverlapping channels in the unlicensed band. Subsequently, the IEEE 802.11 DSSS PHY layer was extended to include both 5.5- and 11-Mbps data rates using complementary code keying (CCK). This high rate PHY layer is standardized to be named IEEE 802.11b. This high rate version includes four data rates: 1, 2, 5.5, and 11 Mbps. The channel bandwidth of the IEEE 802.11b PHY layer is 22 MHz [2, 3].

On the other hand, the considered WPAN covered in IEEE 802.15.2 is IEEE 802.15.1-2002, which is a 1-Mbps FHSS system. The IEEE 802.15.1 PHY layer uses the same seventy-nine 1-MHz-wide channels that are utilized by the FHSS version of IEEE 802.11. IEEE 802.15.1 hops pseudorandomly at a nominal rate of 1600 hops/s. The IEEE 802.15.1 MAC sublayer supports a master–slave topology referred to as a piconet. The master device controls medium access by polling the slaves for data and using scheduled periodic transmission for voice packets [4].

As stated above, the IEEE 802.11 frequency-hopping (FH) WLAN has the same hopping channels as the IEEE 802.15.1 WPAN. However, these two systems operate at different hopping rates. IEEE 802.11 FH specifies a hopping rate of greater than 2.5 hops/s, with typical systems operating at 10 hops/s [2]. IEEE 802.15.1 specifies a maximum hopping rate of 1600 hops/s for data transfer [4]. So, while IEEE 802.11 FH dwells on a given frequency for approximately 100 ms, IEEE 802.15.1 will have hopped 160 times. So the odds are that IEEE 802.15.1 will hop into the frequency used by IEEE 802.11 FH several times while IEEE 802.11 FH is dwelling on a given channel. IEEE 802.11 FH frames will be corrupted by the IEEE 802.15.1 interference whenever IEEE 802.15.1 hops into the channel used by IEEE 802.11 FH, assuming the IEEE 802.15.1 power level is high enough to corrupt the IEEE 802.11 FH frames at the IEEE 802.11 FH receiver. It is also possible for the IEEE 802.11 FH WLAN frames to be corrupted by the IEEE 802.15.1 interference if the IEEE 802.15.1 frame is sent in an adjacent channel to the IEEE 802.11 FH data [1].

For the case of IEEE 802.11b, there is a potential frame collision between IEEE 802.11b and IEEE 802.15.1 frames when the WPAN hops into the WLAN passband [1, 3, 4]. Due to the fact that IEEE 802.11b WLAN bandwidth is 22 MHz, as the IEEE 802.15.1 WPAN hops around the unlicensed band, 22 of the 79 IEEE 802.15.1 channels fall within the WLAN passband [3, 4]. An important issue that affects the level of observed interference is the WLAN automatic data rate scaling. If it is enabled, it is possible for the WPAN interference to cause the WLAN to scale to a lower data rate. At a lower data rate the temporal duration of the WLAN frames is increased. The increase of frame duration may lead to an increase in frame collisions with the interfering WPAN frames [1].

IEEE 802.15.1 uses two different types of links between the piconet master and the piconet slave. Basically, for data transfer IEEE 802.15.1 uses the asynchronous connectionless (ACL) link. The ACL link incorporates automatic repeat request (ARQ) to guarantee reliable transmission of information. IEEE 802.15.1 voice communications utilize the synchronous connection-oriented (SCO) link. Since SCO links do not support ARQ, perceivable degradation in the voice quality is observable for the duration of the IEEE 802.11 FH interference [1].

Finally, for the case of IEEE 802.15.1 in the presence of an IEEE 802.11b interferer, as we know, IEEE 802.15.1 uses FHSS while IEEE 802.11b uses DSSS and CCK. The bandwidth of IEEE 802.11b is 22 MHz. Thus, 22 of the 79 hopping channels available to the IEEE 802.15.1 hops are subject to interference. An FH system is susceptible to interference from the adjacent channels as well. This increases the total number of interference channels from 22 to 24 [1].

6.3 IEEE 802.15.2 COEXISTENCE MECHANISMS

There are two categories of coexistence mechanisms: collaborative and non-collaborative. Collaborative coexistence mechanisms exchange information between two types of wireless networks. In this case, the collaborative coexistence mechanism requires communication between the IEEE 802.11 WLAN and the IEEE 802.15 WPAN. Noncollaborative mechanisms do not exchange information between two wireless networks. These coexistence mechanisms are only applicable after a WLAN or WPAN is established and user data are to be sent [1, 5, 6]. Both types of coexistence mechanisms are designed to mitigate interference resulting from the operation of IEEE 802.15.1 devices in the presence of frequency static or slow-hopping WLAN devices. Note that the interference due to multiple IEEE 802.15.1 devices is mitigated by FH. All collaborative coexistence mechanisms, described in the standard (and this chapter), are intended to be deployed when at least one WLAN station and WPAN device are collocated within the same physical unit. When the aforementioned WLAN and WPAN equipment is collocated in the same place, there should be a dedicated communication link between them that can be either a wired connection or an integrated solution.

6.3.1 Collaborative Coexistence Mechanisms

Three collaborative coexistence mechanisms are defined in IEEE 802.15.2. Two of them are MAC sublayer–based techniques and one is based on a PHY layer method. Both MAC sublayer techniques involve coordinated scheduling of frame transmission between two wireless networks (i.e., WLAN and WPAN). The PHY layer scheme is a programmable notch filter in the IEEE 802.11b receiver to notch out the narrowband IEEE 802.15.1 interferer [1, 5, 7]. The above-mentioned collaborative mechanisms might be utilized separately or combined with others to provide a better coexistence mechanism.

The collaborative coexistence mechanism provides coexistence of a WLAN (in particular IEEE 802.11b) and a WPAN (in particular IEEE 802.15.1) by sharing information between the collocated IEEE 802.11b and IEEE 802.15.1 radios while locally controlling transmissions to avoid interference [6, 7]. These mechanisms are interoperable with existing legacy devices that do not include such features. Two modes of operation are chosen depending upon the network topology and supported traffic. In the first mode, both IEEE 802.15.1 SCO and ACL traffic is supported where SCO traffic is given higher priority than the ACL traffic in scheduling. The second mode is based on time division multiple access (TDMA) and is utilized when there is ACL traffic in high piconet density areas. In the TDMA mode, the IEEE 802.11b beacon-to-beacon interval, i.e., target beacon transmission time (TBTT), is subdivided into two subintervals: one subinterval dedicated to IEEE 802.11b and the other for IEEE 802.15.1. Since each radio has its own subinterval, both radios will operate properly. This technique needs an additional feature to restrict when the IEEE

802.15.1 master transmits. The mode to be used is chosen under the command of the access point (AP) management software. Frequency nulling might be deployed in conjunction with these modes to further reduce the observed interference [1, 6, 7].

Both AWMA and PTA can be combined to produce a smarter coexistence mechanism. In Fig. 6.1, the overall structure of the combined collaborative coexistence mechanisms is illustrated [1].

Essentially, deployment of either AWMA or PTA collaborative coexistence mechanisms is strongly recommended by the standard [1]. If the PTA mechanism is used, it is also recommended that the deterministic interference suppression mechanism be deployed in concert with the PTA mechanism. While PTA can be employed without deterministic interference suppression, combination of the two mechanisms leads to increased WLAN/WPAN coexistence and better system performance. If there is a high density of physical units incorporating both WLAN and WPAN devices in a common area (greater than or equal to three units in a circle of radius 10 m as the standard defines [1]) and the WPAN SCO link (voice link) is not being utilized, then it is highly recommended that the AWMA mechanism be used. If the density of units

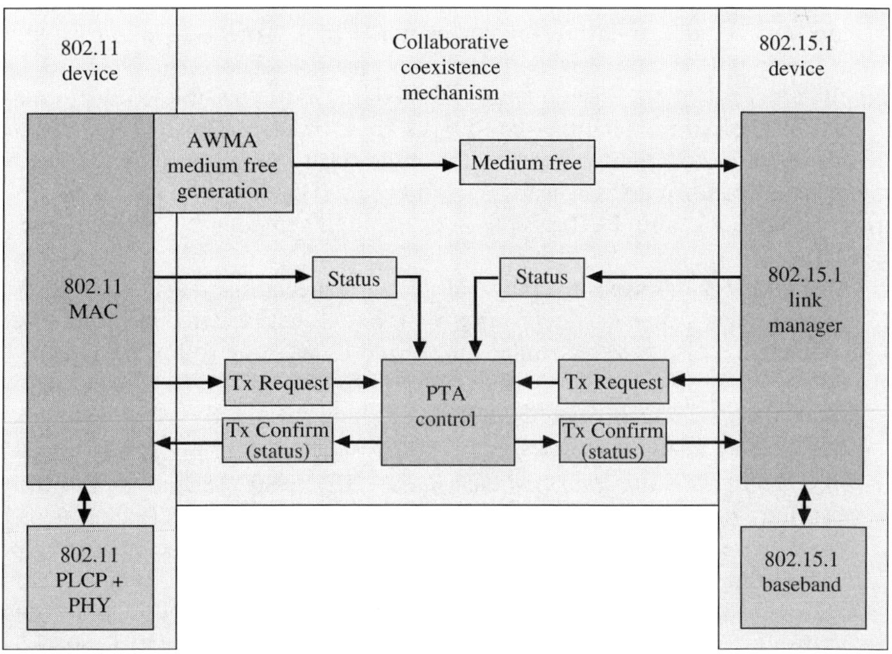

FIGURE 6.1 Overall structure of 802.11b/802.15.1 combined AWMA and PTA collaborative coexistence mechanism. *Source*: IEEE 802.15.2 WG, Part 15: "Coexistence of wireless personal area networks with other wireless devices operating in unlicensed frequency bands," IEEE, New York, Aug. 2003.

incorporating both the WLAN and WPAN devices is low (less than three units in a circle with a radius of 10 m) or the WPAN SCO link is used, then it is suggested that the PTA mechanism be used in concert with the deterministic interference suppression mechanism [1, 5, 6].

6.3.2 Noncollaborative Coexistence Mechanisms

IEEE 802.15.2 describes several methods to improve the performance of both IEEE 802.15.1 and IEEE 802.11 networks through one of the following methods: adaptive interference suppression of IEEE 802.11b devices, adaptive packet selection, and packet scheduling for ACL links. These methods do not require collaboration between IEEE 802.11 devices and IEEE 802.15.1 devices. Therefore, they belong to the general category of noncollaborative coexistence mechanisms. Two other methods, i.e., packet scheduling for SCO links and adaptive frequency hopping (AFH) for IEEE 802.15.1 devices, are provided as well [1, 5–7].

The key concept of adaptive packet selection and scheduling methods is to adapt the transmission according to current channel conditions. If the channel is dominated by interference due to an IEEE 802.11b network, the packet error rate (PER) will be mainly caused by collisions between IEEE 802.15.1 and IEEE 802.11 systems, instead of bit errors resulting from noise. Packet types that do not include forward error correction (FEC) protection could provide better throughput if combined with intelligent packet scheduling. The basis for the effectiveness of the aforementioned methods is to be able to accurately figure out the current channel conditions in a timely fashion. In addition, channel estimation might be performed in a variety of ways: received signal strength indication (RSSI), header error check (HEC) decoding profile, bit error rate (BER), and PER profile and an intelligent combination of these techniques [1, 5, 6].

IEEE 802.15.2 describes five noncollaborative mechanisms [1]. At least two of them share a *channel classification* common function. In addition, three mechanisms are also covered under the second item in the following [1]:

Adaptive Interference Suppression. A mechanism based solely on signal processing in the physical layer of the WLAN.

Adaptive Packet Selection and Scheduling. IEEE 802.15.1 systems utilize various packet types with varying configurations such as packet length and degree of employed error protection. By selecting the best packet type according to the channel condition of the upcoming frequency hop, better data throughput and network performance can be achieved. Additionally, by carefully scheduling packet transmission so that IEEE 802.15.1 devices transmit during hops that are outside the WLAN frequencies and desist from transmitting while in band, interference to the WLAN systems can be minimized and at the same time it increases the obtained throughput of IEEE 802.15.1 systems.

Adaptive Frequency Hopping. IEEE 802.15.1 systems hop among 79 channels at a nominal rate of 1600 hops/s in the connection state and 3200 hops/s in the inquiry and page states. By identifying the channels with interference, it is possible to change the sequence of hops such that those channels with interference are avoided. From the traffic type and channel condition, a partition sequence is generated as input to the frequency remapper, which modifies hopping frequencies to avoid or minimize interference effects.

6.4 ALTERNATING WIRELESS MEDIUM ACCESS

AWMA utilizes a portion of the IEEE 802.11 beacon interval for the operation of IEEE 802.15. From a timing point of view, each type of wireless system is restricted to the appropriate time segment, which prevents interference between coexisting technologies. In AWMA, the WLAN radio and the WPAN equipment are collocated in the same physical unit. The AWMA mechanism uses the shared clock within all the WLAN-enable devices and thus all WLAN devices connected to the same WLAN AP share common WLAN and WPAN time intervals. As a result, all devices connected to the same AP restrict their WLAN traffic and WPAN traffic to nonoverlapping time intervals. As such, there will be no WLAN/WPAN interference for any devices connected to the same WLAN AP. In the case of multiple APs, typically the APs are not synchronized. In that case there will be some residual interference between WPAN devices synchronized with one WLAN AP and WLAN devices synchronized with another AP. If APs are synchronized, then the residual interference can also be easily eliminated [1].

The IEEE 802.11 WLAN AP periodically generates a beacon frame. The beacon period is T_B. AWMA subdivides this interval into two subintervals: one for WLAN traffic and one for WPAN information exchange. Figure 6.2 illustrates the separation of the WLAN beacon interval into two subintervals [1].

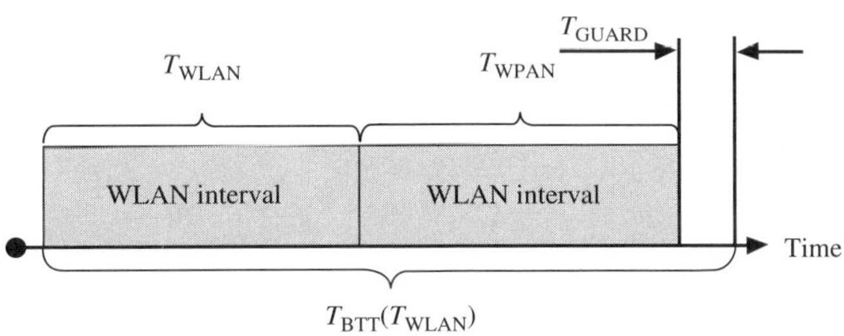

FIGURE 6.2 Timing of WLAN and WPAN subintervals. *Source*: IEEE 802.15.2 WG, Part 15: "Coexistence of wireless personal area networks with other wireless devices operating in unlicensed frequency bands," IEEE, New York, Aug. 2003.

The WLAN interval commences at the WLAN TBTT. The length of the WLAN subinterval is T_{WLAN}, which is specified in the offset field of the medium-sharing element (MSE) residing in the delivered beacon frame. The WPAN subinterval begins at the end of the WLAN interval. The length of the WPAN subinterval is T_{WPAN}, which is specified in the duration field of the MSE. The combined length of these two subintervals should not be greater than the beacon period [1, 6].

In addition to the WLAN and WPAN subinterval duration, the MSE may also specify a guard band (T_{GUARD}) by setting a nonzero value in the guard field [1]. The purpose of the guard band is to specify an interval immediately preceding the next expected beacon. Basically, each guard period should be totally free of either WLAN or WPAN traffic. On the other hand, this guard band may be compulsory to ensure that all WPAN traffic has completed by the WLAN beacon time. If the offset field in the MSE is greater than the beacon interval, then no WPAN subinterval should exist [6]. If the total value of the offset field and the duration field is greater than the beacon time, T_{WPAN} should end at the next TBTT. If the guard field is set to nonzero, and the beacon period minus the total value of offset and duration fields is less than the current value of the guard field, then T_{WPAN} should finish prior to the subsequent TBTT. If the value in the offset field is less than the beacon interval but the value of the offset field plus the guard field is equal to or greater than the beacon interval, then there should be no WPAN subinterval [1, 6, 7].

As explained previously, AWMA necessitates that WLAN equipment and the WPAN master node are collocated in the same physical unit. AWMA requires the WLAN entity to control the timing of both WLAN and WPAN subintervals. All WLAN stations connected to the same AP are synchronized and hence have the same TBTT. As a result, all units that implement AWMA

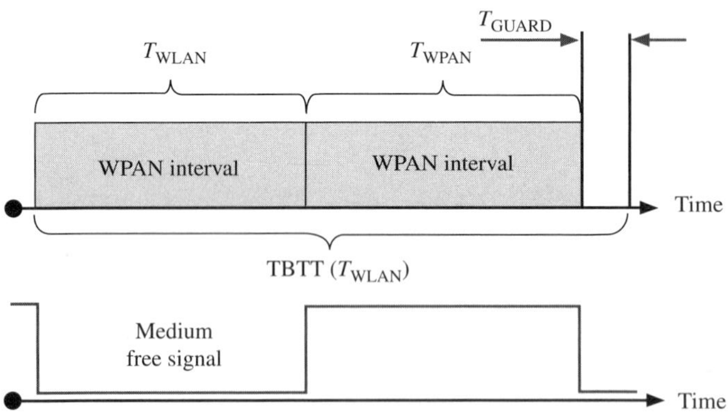

FIGURE 6.3 Medium free signal. *Source*: IEEE 802.15.2 WG, Part 15: "Coexistence of wireless personal area networks with other wireless devices operating in unlicensed frequency bands," IEEE, New York, Aug. 2003.

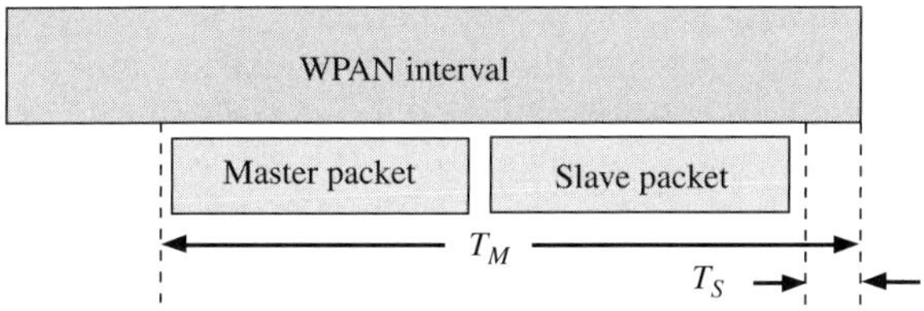

FIGURE 6.4 Timing of WPAN packets. *Source*: IEEE 802.15.2 WG, Part 15: "Coexistence of wireless personal area networks with other wireless devices operating in unlicensed frequency bands," IEEE, New York, Aug. 2003.

have synchronized WLAN/WPAN subintervals. The WLAN entity is mandated to send out a physical synchronization signal to the WPAN master, which is in the same physical unit as the WLAN equipment. That synchronization signal specifies both the WLAN interval and the WPAN interval. This synchronization signal is called the *medium free signal*. Therefore, the medium is free of WLAN traffic when the medium free signal is true. Figure 6.3 illustrates the medium free signal [1].

The WPAN device collocated with the WLAN entity should be a WPAN master device. In particular, if the WPAN device conforms to IEEE 802.15.1 [4], then all ACL data transmissions are controlled by the WPAN master entity. Especially, WPAN slaves may only transmit ACL packets if in the previous time slot the WPAN slave received an ACL packet. For that reason, the WPAN master should end the transmission long enough before the end of the WPAN subinterval so that the longest allowed slave packet (e.g., a five-slot IEEE 802.15.1 packet) will complete its transmission prior to the end of the WPAN interval. Figure 6.4 illustrates the aforementioned timing requirement. The value of T_M should be large enough so as to guarantee that the value of T_S is greater than zero [1].

IEEE 802.15.1 supports SCO packets for voice traffic [4]. These packets are generated on a regular basis with a fixed period. There are numerous SCO packet types, depending on the level of FEC. As an example, an HV3 (where

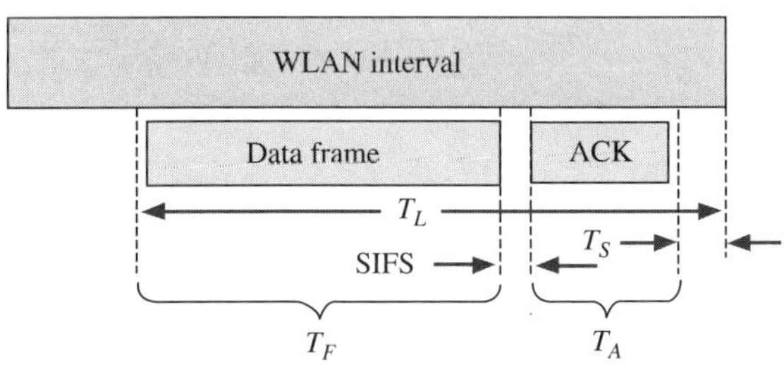

FIGURE 6.5 Timing of WLAN frames. *Source*: IEEE 802.15.2 WG, Part 15: "Coexistence of wireless personal area networks with other wireless devices operating in unlicensed frequency bands," IEEE, New York, Aug. 2003.

HV refers to *high quality voice*) link repeats every six slots. The first two slots are utilized for SCO packets and the last four packets might be used for ACL packets. In IEEE 802.15.1 a time slot is 0.625 μs and the SCO HV3 period is 3.75 μs. This is a small fraction of the typical WLAN beacon period [2–4]. As a result, if the WLAN beacon period is subdivided into two subintervals, the WPAN SCO packets may not be restricted to the WPAN interval. Consequently, the AWMA coexistence mechanism does not support IEEE 802.15.1 SCO links [1, 5, 6].

The WLAN entity should also restrict all WLAN transmissions to the WLAN subinterval. Figure 6.5 illustrates the timing of WLAN traffic [1]. Note that before a WLAN device may transmit a frame it should ensure that the value of T_S is greater than zero [1, 5, 7].

6.5 PACKET TRAFFIC ARBITRATION

The PTA control entity provides per-packet authorization [1]. In PTA, both IEEE 802.11b and IEEE 802.15.1 nodes are supposed to be collocated in the same physical entity. An attempt for transmission by either IEEE 802.11b or

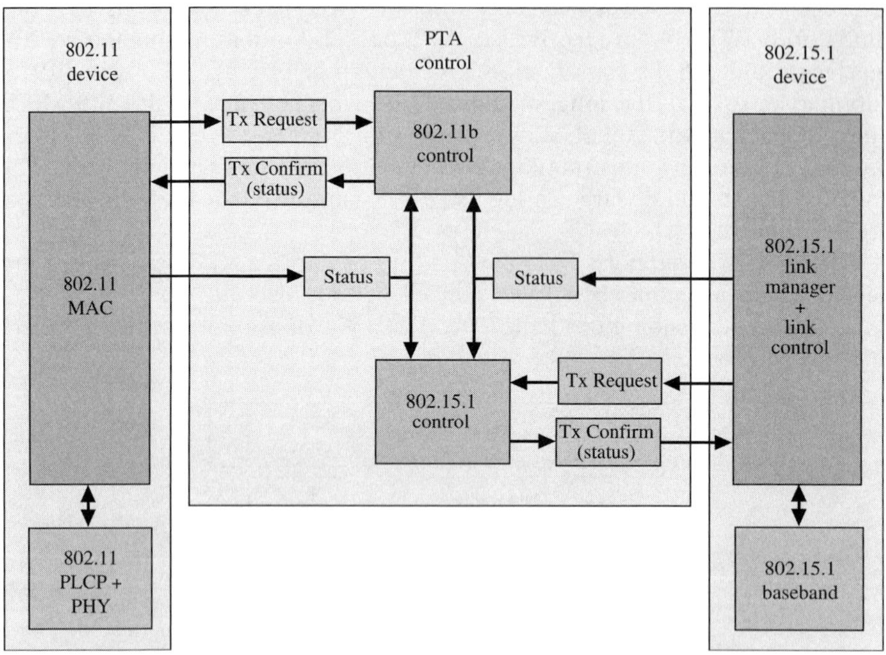

FIGURE 6.6 Structure of PTA entity. *Source*: IEEE 802.15.2 WG, Part 15: "Coexistence of wireless personal area networks with other wireless devices operating in unlicensed frequency bands," IEEE, New York, Aug. 2003.

IEEE 802.15.1 is referred to PTA for approval. PTA may deny a transmit request that would result in collision. The PTA mechanism can also sustain IEEE 802.15.1 SCO links. Based on the existing traffic load, PTA dynamically coordinates sharing of radio resources between two coexisting wireless systems. PTA uses its knowledge of IEEE 802.11b and IEEE 802.15.1 future activities to predict all probable collisions. Whenever a collision is expected to take place, PTA prioritizes different transmissions based on a set of simple rules that depend on the priorities of the various packets [1, 5–7]. Figure 6.6 illustrates the general structure of a PTA control entity [1]. Each device has a corresponding control entity to which it forwards all its intended transmission requests. This control entity either accepts or rejects the received requests based upon known states of both radios [1].

The goal of the IEEE 802.11b control entity is to permit or refuse the received transmission requests from the IEEE 802.11b MAC. Upon reception of a TX Request signal, the IEEE 802.11b control entity instantly generates a so-called TX Confirm signal comprising a status value that is either allowed or denied. Figure 6.7 shows how the status value is chosen [1].

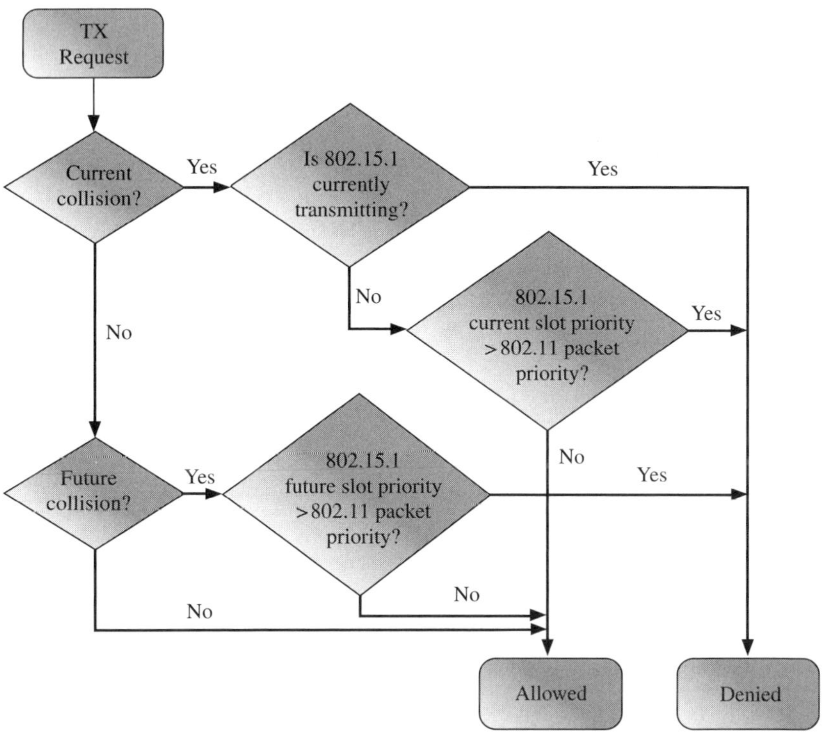

FIGURE 6.7 Decision algorithm for 802.11b TX Request. *Source*: IEEE 802.15.2 WG, Part 15: "Coexistence of wireless personal area networks with other wireless devices operating in unlicensed frequency bands," IEEE, New York, Aug. 2003.

For the IEEE 802.15.1 control entity and in response to the received TX Request signal, the control entity immediately generates a TX Confirm signal containing a status value that is either allowed or denied. Figure 6.8 shows how the status value is determined [1].

The decision-making algorithm that accepts or rejects a received packet transmission request uses a priority comparison between the state of the requested packet transmission and the known state of the other protocol stack. Basically there are two different priority comparison schemes: fixed and randomized. In a fixed priority assignment, an IEEE 802.15.1 SCO packet

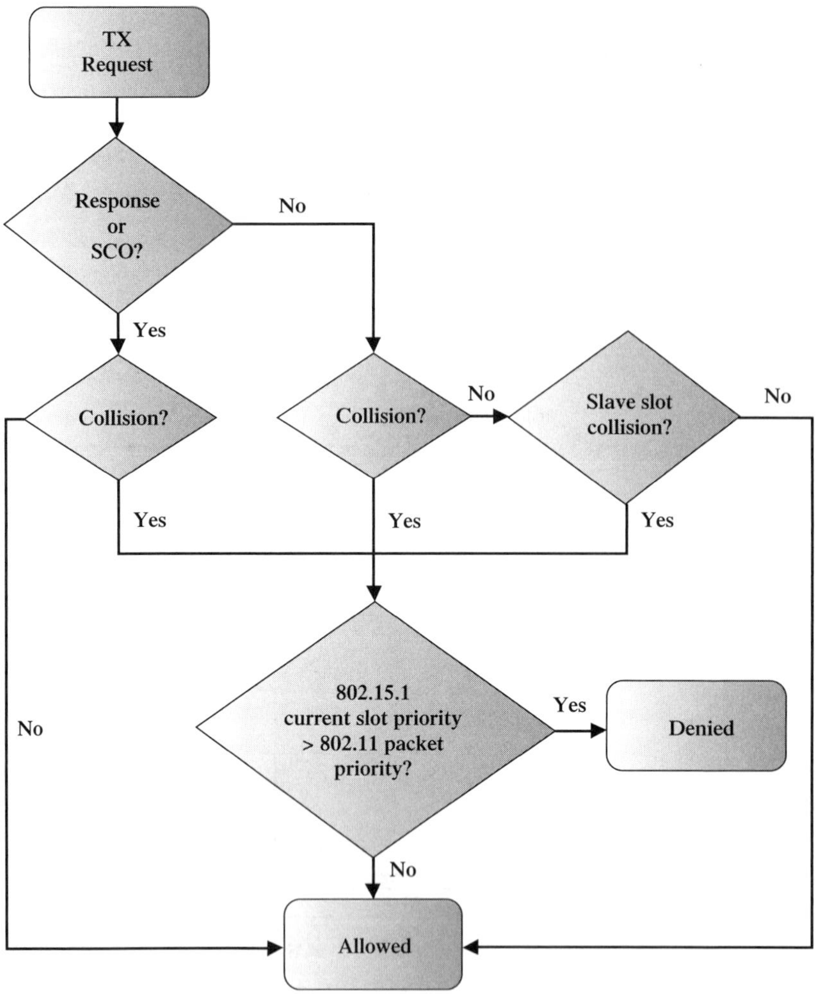

FIGURE 6.8 Decision algorithm for 802.15.1 TX Request. *Source*: IEEE 802.15.2 WG, Part 15: "Coexistence of wireless personal area networks with other wireless devices operating in unlicensed frequency bands," IEEE, New York, Aug. 2003.

should have a higher priority than IEEE 802.11b data type MAC protocol data units (MPDUs) while an IEEE 802.11b acknowledgment (ACK) MPDU should have a higher priority than all IEEE 802.15.1 packets. On the other hand, in the randomized scheme, the priority of packets might be assigned in a randomized fashion. For this purpose, a random variable r uniformly distributed between [0, 1] and a threshold $T\,(0 \leq T < 1)$ are used. If the incoming packet is from an IEEE 802.11b device, a priority of 2 is assigned to it if the random number r is smaller than T. Otherwise, a priority of 0 is assigned. If the incoming packet is from an IEEE 802.15.1 device, a priority of 1 is assigned [1].

6.6 ADAPTIVE PACKET SELECTION

IEEE 802.15.1 specifies a variety of packet types with different combinations of payload length, slots occupied, FEC codes, and ARQ options [1]. The motivation is to supply the required flexibility for the implementers and applications so that the packets can be chosen in an optimized fashion based on the existing traffic and channel conditions. In this section, a mechanism is described to take advantage of different packet types in order to enhance the achieved network capacity for coexistence scenarios.

6.6.1 IEEE 802.15.1 Packet Types for SCO and ACL

IEEE 802.15.1 provides four types of packets [i.e., HV1, HV2, HV3, and data–voice (DV)] that can be delivered over an SCO link [4]. These packets differ mostly in the FEC code employed and the amount of channel occupied by the SCO link [1, 4]. The choice of different packet types provides intriguing trade-offs of error protection at the bit level and the amount of incurred interference. The ACL link, in addition to the use of different FEC protections, incorporates the choice of multislot packets [1]. Apparently, the different ACL packet types allow the applications to make trade-offs among different considerations of traffic flow, channel conditions of the current hop, duty cycles, and interference generated to neighboring networks [1, 5, 6, 8, 9].

6.6.2 Methods of Adaptive Packet Selection

The fundamental scheme is to dynamically choose packet types, given either an ACL or SCO link, such that maximal total network capacity is achieved. This implies not only optimizing throughput for the IEEE 802.15.1 piconet but also reducing interference to the coexisting IEEE 802.11b network, which will increase the throughput of the IEEE 802.11b network [1, 6, 7].

For SCO links, when the network performance is range limited, meaning that the associated stations are separated by a distance such that only a small noise margin is maintained [1], random bit errors are the main problem. Choosing a packet type that utilizes more error protection will consequently increase the performance of the SCO link. Hence, for range-limited applications, the HV1 packet is preferred over the HV2 packet and the HV2 packet is preferred over the HV3 packet. By monitoring the RSSI and the signal-to-noise ratio (SNR) of the IEEE 802.15.1 radio, IEEE 802.15.1 may determine if the choice of more error protection is beneficial [4]. For SCO links in the coexistence scenarios, usually the dominant reason for packet drop is not noise or range but rather the strong interference produced by the collocated network such as an IEEE 802.11b network. In this case, increasing FEC protection will cause IEEE 802.15.1 devices to generate more packets (HV1 packets occupy the channel three times more often than HV3 packets) and thus more interference to the IEEE 802.11b network [1, 5–7]. For similar reasons, the same guidelines apply to the selection of ACL packets. When IEEE 802.15.1 network performance is range limited, ACL packets with FEC protections, which include DM1, DM3, and DM5, should be employed (where DM stands for *data–medium*). On the other hand, when the system is interference limited, the 802.15.1 device should reduce the number of bits transmitted by choosing a more bandwidth-efficient packet format such as DH1, DH3, or DH5 (where DH stands for *data–high rate*) [1].

6.7 PACKET SCHEDULING

6.7.1 Packet Scheduling for ACL Links

In this section, a scheduling mechanism for IEEE 802.15.1 to alleviate the effect of interference with the IEEE 802.11 DSSS is introduced. This scheduling mechanism comprises two distinct components: *channel classification* and *master delay policy* [1].

Channel classification is accomplished on every IEEE 802.15.1 receiver and is based upon measurements conducted per frequency or channel in order to determine the presence of interference [1]. A frequency is determined to be *good* if a device can correctly decode a packet received on it. Otherwise it is marked as *bad*. A number of criteria can be utilized in determining whether a frequency (or channel) is good or bad, such as RSSI, PER measurements, or negative ACKs. A channel classification table capturing the frequency status (good or bad) for each device in the piconet is kept at the master device. Depending on the classification method used, an explicit message exchange between the master and the slave device may be required. Implicit methods such as negative ACKs do not require the slave to send any communication messages to the master concerning its channel classification [1, 5].

The master delay policy makes use of the information available in the channel classification table in order to avoid packet transmission in a bad channel. Bearing in mind that the IEEE 802.15.1 master device controls all transmissions in the piconet, the delay rule should only be implemented in the master device. Moreover, following each master transmission there is a slave data communication. Thus, the master checks both the slave's receiving frequency and its own receiving frequency before choosing to transmit a packet in a given frequency hop [1].

6.7.2 Packet Scheduling for SCO Links

Voice applications are among the most sought-after applications for IEEE 802.15.1 devices, and they are most susceptible to interference [1, 4]. An in-band adjacent WLAN network will almost certainly make the voice quality of the IEEE 802.15.1 SCO link unacceptable for users. In this section we briefly discuss existing approaches for improving the quality of service (QoS) of SCO links.

The key idea is to allow the SCO link the flexibility of choosing hops that are out of band with the collocating IEEE 802.11b network spectrum for transmission [1]. Basically, the duty cycle or channel utilization of the SCO link does not change. The only proposed change is to allow the piconet master the flexibility of choosing when to initiate the transmission.

Given that only the original HV3 packet allows for sufficient flexibility in moving the transmission slots around (two additional choices), the focus is on modifying the structure of the HV3 packet. A new SCO packet type, the EV3 packet, is defined which has the following features [1]: (a) no FEC coding, (b) 240 bits payload, (c) one EV3 packet for every six slots (delay < 3.75 ms), and (d) a slave that will only transmit when addressed by the master. For HV3 packets, data transmission of master and slave devices should occur at the fixed slots whether the hops are good or bad. An EV3 packet is not transmitted during the two bad hops but waits for the next pair of slots, which happens to be a good channel. The throughput for IEEE 802.15.1 will be higher while interference is reduced. Finally, a score of 0–3 is assigned to each pair and the pair with the highest score is selected [1, 5, 6].

6.8 SUMMARY AND CONCLUDING REMARKS

In this chapter, a brief overview of the IEEE 802.15.2 standard was presented. After an explanation of the interference problem, we presented the defined coexistence mechanisms in the standard. AWMA rules and related technical issues were covered subsequently. We also dealt with 802.15.2 PTA and gave a cursory look at packet selection approaches in IEEE 802.15.2. Finally a brief overview of the packet-scheduling mechanism proposed in 802.15.2 was presented.

ACKNOWLEDGMENTS

This work was supported partially by Nokia, Elektrobit, and Elisa Corporation Foundations.

REFERENCES

1. IEEE 802.15.2 WG, Part 15: "Coexistence of wireless personal area networks with other wireless devices operating in unlicensed frequency bands," IEEE, New York, Aug. 2003.
2. IEEE 802.11 WG, Part 11: "Wireless LAN medium access control (MAC) and physical layer (PHY) specification," IEEE, New York, Aug. 1999.
3. IEEE 802.11b WG, Part 11: "Wireless LAN medium access control (MAC) and physical layer (PHY) specifications: High-speed physical layer extension in the 2.4 GHz band," supplement to IEEE 802.11, IEEE, New York, Sept. 1999.
4. IEEE 802.15.1 WG, Part 15.1: "Wireless medium access control (MAC) and physical layer (PHY) specifications for wireless personal area networks (WPANs)," IEEE, New York, June 2005.
5. J. Lansford, A. Stephens, and R. Nevo, "Wi-Fi (802.11b) and Bluetooth: Enabling coexistence," *IEEE Network* 15(5), 20–27 (2001).
6. C. F. Chiasserini and R. R. Rao, "Coexistence mechanisms for interference mitigation in the 2.4-GHz ISM band," *IEEE Trans. Wireless Commun.* 2(5), 964–975 (2003).
7. I. Howitt, "WLAN and WPAN coexistence in UL band," *IEEE Trans. Vehic. Technol.* 50(4), 1114–1124 (2001).
8. I. Howitt, "Bluetooth performance in the presence of 802.11b WLAN," *IEEE Trans. Vehic. Technol.* 51(6), 1640–1651 (2002).
9. A. Conti, D. Dardari, G. Pasolini, and O. Andrisano, "Bluetooth and IEEE 802.11b coexistence: Analytical performance evaluation in fading channels," *IEEE J. Sel. Areas Commun.* 21(2), 259–269 (2003).

CHAPTER 7

COEXISTENCE OF BLUETOOTH PICONETS AND WIRELESS LAN

JINGLI LI and XIANGQIAN LIU

7.1 INTRODUCTION

With the upcoming pervasive deployment of wireless networks and devices on the unlicensed industrial, scientific, and medical (ISM) band, multiple (homogeneous and/or heterogeneous) networks using the same frequency band are likely to coexist in a physical environment. For example, it is common for users to be connected to the network from their IEEE 802.11b–enabled notebooks in the presence of Bluetooth devices or to form multiple Bluetooth piconets in a conference room. Without coordination among colocated networks, cochannel interference due to frequency collision has become a major performance-limiting factor. When collisions occur, the received packets are discarded without recovering any data, and retransmissions must follow, possibly inducing new collisions, hence throughput decreases and delay can become excessive.

In recent years, the coexistence of wireless personal area network (WPAN), represented by Bluetooth, and wireless local area network (WLAN), represented by IEEE 802.11a/b/g technologies, has been studied extensively due to the wide spread of those two types of networks. This chapter provides an overview of various coexistence techniques developed for the simultaneous operation of Bluetooth piconets and WLAN. Most coexistence mechanisms are implemented on Bluetooth devices because Bluetooth is more vulnerable to cochannel interferences due to its low power. The rest of this chapter is organized as follows. Section 7.2 is an overview of the coexistence issue, including some background on Bluetooth and WLAN. Section 7.3 reviews collaborative mechanisms for coexistence of Bluetooth and WLAN. In the collaborative case, it is assumed that Bluetooth and WLAN can communicate with each other, typically when they are colocated on the same device.

Emerging Wireless LANs, Wireless PANs, and Wireless MANs. Edited by Y. Xiao and Y. Pan

Noncollaborative mechanisms that do not require direct communication between networks are reviewed in Section 7.4. Section 7.5 presents a frequency diversity method that can deal with cochannel interferences from both Bluetooth piconets and WLAN. Conclusions are drawn in Section 7.6.

7.2 THE COEXISTENCE PROBLEM

The radio transmission technology adopted in Bluetooth is frequency-hopping spread spectrum (FHSS). Gaussian frequency shift keying (GFSK) is the modulation scheme used in Bluetooth. The Bluetooth transceiver operates in the unlicensed 2.4-GHz ISM band, where signals can hop among 79 frequency channels between 2.4 and 2.480 GHz with 1-MHz channel spacing. The nominal hop rate is 1600 times per second. The typical transmission range for Bluetooth devices is up to 10 m. A Bluetooth transmitter has three levels of radio transmission power: 100 mW (20 dBm), 2.5 mW (4 dBm), and 1 mW (0 dBm).

A Bluetooth piconet is formed when one master device and up to seven slave devices are connected via Bluetooth technology. Each piconet has a unique hop sequence that is determined by the clock and address of the master device. A time division duplex (TDD) scheme is used for the master and slave devices to transmit alternatively. In Bluetooth, different piconets have different frequency hop sequences, but they share the same 79 frequency channels on the 2.4-GHz band. Access technique in one piconet is time division multiple access (TDMA). As shown in Fig. 7.1, three packet sizes are available for Bluetooth: one slot, three slots, and five slots. Each slot has a duration of 625 μs. Because an idle time is allocated at the end of each packet for transient time settling, the actual transmission time for each packet is less than the duration of the slots occupied. For a multislot packet, its frequency is determined by the first slot and remains unchanged throughout the packet.

Bluetooth transmissions can be either symmetric or asymmetric. A symmetric link occurs when both the master node and slave node in a piconet transmit packets of the same size. An asymmetric link occurs when the master sends a packet of one size and receives a packet of a different size as a response from the slave. Real-time information such as voice is transmitted using a

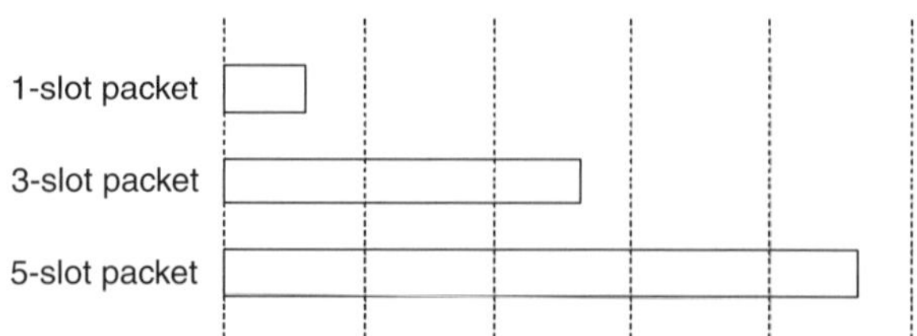

FIGURE 7.1 Types of Bluetooth packets.

synchronous connection-oriented (SCO) link. The SCO link is a symmetric point-to-point link between a master and a single slave in the piconet. The master maintains the SCO link by using reserved slots at regular intervals. SCO packets are never retransmitted. Non-time-critical application such as data information is transmitted using an asynchronous connectionless (ACL) link. The ACL link is a point-to-multipoint link between the master and all the slaves participating in the piconet. Retransmission can be applied for most ACL packets.

The same 2.4-GHz band is also used by IEEE 802.11 WLAN, which usually adopts direct-sequence spread spectrum (DSSS). WLAN has a data rate up to 11 Mb/s and a transmission range up to 100 m. Its radio transmission power is typically between 30 and 100 mW. According to Federal Communications Commission (FCC) regulation, there are 11 channels crossing the acceptable 83.5 MHz of the 2.4-GHz frequency band; each occupies a fixed frequency band of 22 MHz. These channels are centered at 2412, 2417, ... , 2462 MHz, respectively. A WLAN system can utilize any of these channels. A maximum of three WLAN networks can coexist without interfering with one another, since only three 22-MHz channels can fit within the allocated band without overlapping. Several wireless stations can build up a basic service set by sharing the same spreading sequence and using the same medium access control (MAC) function. Using the carrier sense multiple access with collision avoidance protocol, several wireless stations form an ad hoc network where they can communicate with each other directly or communicate with a wired network through a centralized access point (AP).

With the increasing popularity of wireless networks sharing the unlicensed ISM band, multiple homogenous and/or heterogenous networks would be collocated in a physical environment. For example, an office scenario is shown in Fig. 7.2. where two Bluetooth piconets are colocated with a WLAN. A Bluetooth packet may be destroyed if the transmission is overlapped by other transmissions from Bluetooth and/or WLAN both in time and frequency. Cochannel interference (CCI) caused by neighboring networks degrades performance significantly. A number of results have been reported on the collision and throughput analysis of the coexistence of multiple piconets [1–4]. An upper bound of the packet error rate (PER) of single-slot packets is obtained in [1]. A more general analysis is provided in [2], where multislot packets are considered. Packet interference under different traffic conditions is analyzed in [3]. An analytical model based on Bluetooth interference and radio propagation is derived in [4] and validated by empirical tests. Performance analyses for coexistence of Bluetooth and WLAN are shown in [5–10]. In [5], the performance of Bluetooth when operating in close proximity to a WLAN system is quantified. Interference measurements are performed in [6, 7] for coexistence of Bluetooth and WLAN. A closed-form expression for the probability of collision in terms of the network and radio propagation parameters is derived in [8]. A method to analytically evaluate the impact of an 802.11b on a Bluetooth piconet is developed and validated by empirical

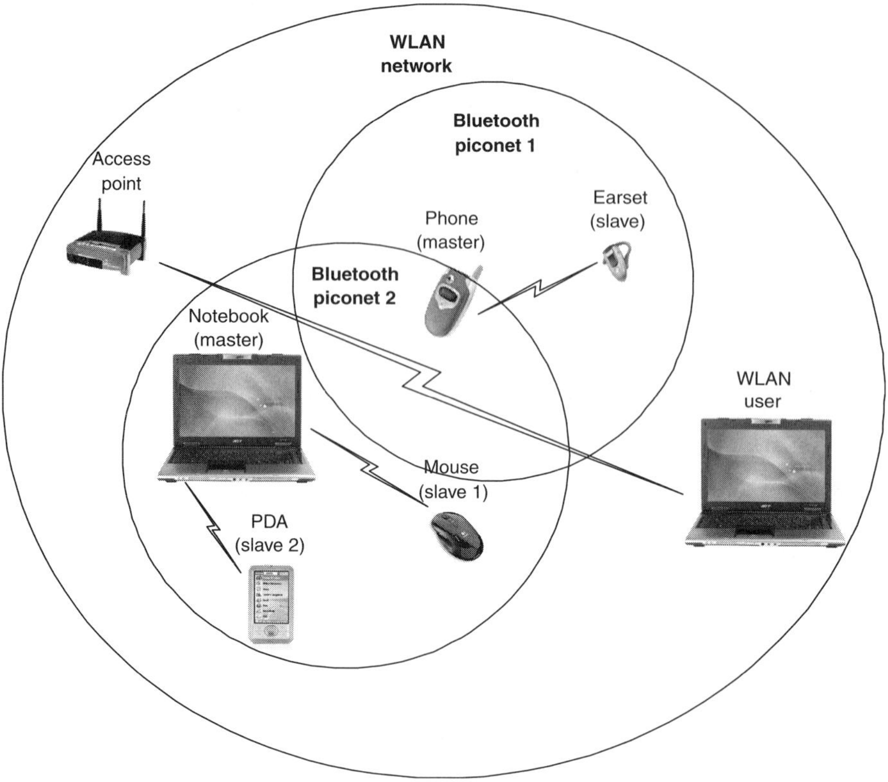

FIGURE 7.2 Coexistence of Bluetooth piconets and a WLAN.

results in [9]. Performance evaluation for Bluetooth and WLAN coexistence in fading channels are provided in [10].

Based on the interference analysis for Bluetooth piconets in [1], Fig. 7.3 shows the PER of one-slot packets when multiple synchronized Bluetooth piconets coexist. It can be seen that the PER is up to 5% if five Bluetooth piconets coexist. It is also shown in [5] that a Bluetooth receiver may experience up to 27% packet loss for data traffic and 25% packet loss for voice applications in the presence of IEEE 802.11–based WLAN interference. Theoretically, the performance degradation for a Bluetooth piconet in the presence of a WLAN is about $22/79 \approx 28\%$, which is validated by experimental results in [6]. According to [6], when the distance between Bluetooth the transmitter and receiver is 5 m, the throughput for Bluetooth versus the distance between Bluetooth and a WLAN station (STA) is shown in Table 7.1. If the Bluetooth and WLAN station are located in the same device such as a notebook, the mutual interference is worse. From the experimental results in [7], the throughput for Bluetooth versus the distance between WLAN STA and AP is shown in Table 7.2. If multiple Bluetooth piconets coexist with a WLAN, the performance degradation will be even worse.

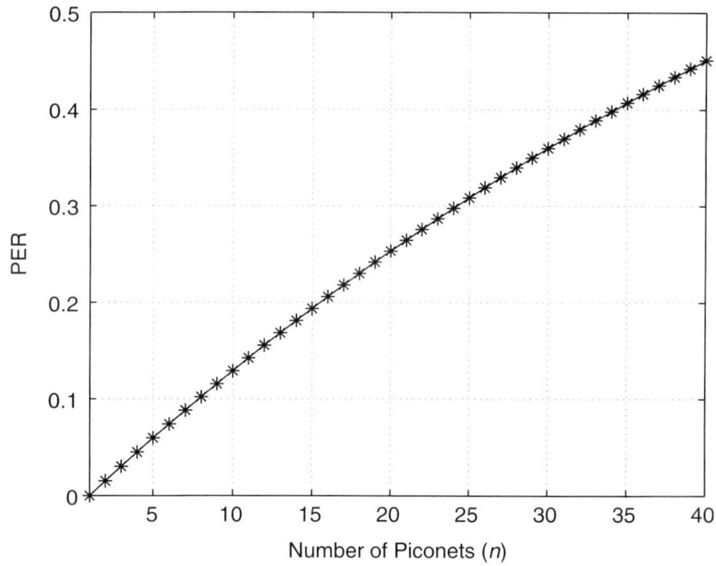

FIGURE 7.3 PER for multiple synchronized Bluetooth piconets.

To address the cochannel interference, many coexistence mechanisms have emerged. These coexistence mechanisms can be classified as collaborative and noncollaborative methods. For the collaborative case, attractive data transmission rates and throughputs can be achieved by using a communication link between the Bluetooth and WLAN when they are embedded on the same device [11–13] or by coordinating the hopping frequencies of the colocated Bluetooth devices [14], but the centralized control mechanism confines their applications to certain situations. Noncollaborative methods do not require direct communication between coexisting networks, and they usually rely on monitoring the channel to detect interference and estimate traffic. For example, to avoid hopping onto preoccupied frequency channels, adaptive frequency hopping (AFH) [15, 16] modifies the Bluetooth frequency hopping sequence, and the Bluetooth interference-aware scheduling (BIAS) [15, 17] strategy postpones the transmission, both detecting preoccupied frequency bands by monitoring PERs on all channels. The overlap avoidance (OLA) schemes proposed in [18] are based on packet scheduling and traffic control. Power control is proposed based on PER [19] or the received signal strength [20] to sustain the quality of service for a Bluetooth link. There are also hybrid methods that combine AFH and

TABLE 7.1 Throughput for Bluetooth versus Bluetooth–WLAN Distance

Distance (m)	0	1	3	5	7	9	∞
Throughput (kb/s)	300	300	325	340	345	385	420

TABLE 7.2 Throughput for Bluetooth versus WLAN STA–AP Distance

Distance (m)	2	6	10	20	30	40	50
Throughput (kb/s)	225	250	270	350	380	390	390

Bluetooth carrier sense (BCS) [21] or combine power control, listen-before-talk (LBT) and AFH [22] to achieve better performance or to deal with more complicate scenarios when there are both dynamic interference and static interference. A collision resolution method [23] is also proposed to this problem.

Most noncollaborative interference detection and collision avoidance schemes are developed for coexistence of Bluetooth and WLAN, and they are not applicable to the coexistence of multiple Bluetooth piconets because the frequency channels are constantly changing and the hop sequence of one piconet is not known to another. A frequency diversity technique using dual-channel transmission (DCT) is proposed in [24] for multiple coexisting Bluetooth piconets to combat CCI. Unlike existing approaches, this method does not require channel sensing, PER monitoring, or extra delay in transmission, and it is effective when multiple Bluetooth piconets coexist or when Bluetooth piconets coexist with a WLAN. The idea of DCT is to transmit the same packet on two distinct frequency-hopped channels simultaneously and the power used in each channel is half of what would be used in single-channel transmission (SCT). The two channels are intentionally separated by at least 22 MHz so that DCT is also robust to WLAN interference. A packet is successfully received if at least one channel survives. Theoretic analysis shows that with DCT, the PER can be reduced significantly compared to SCT when a small number of piconets coexist.

7.3 COLLABORATIVE MECHANISMS

Based on a central control scheme, collaborative mechanisms achieve attractive data transmission rates and throughput. But the central control mechanism needed for collaborative schemes confines their applications to certain situations. Four collaborative techniques: the MAC-enhanced temporal algorithm (META) [11], the alternating wireless medium access (AWMA) [12], the deterministic frequency-nulling scheme (DFNS) [13], and the coordinated colocated access point (CCAP) scheme [14] are described in the following.

7.3.1 MAC-Enhanced Temporal Algorithm

META [11] is an intelligent scheduling algorithm with queuing aimed to facilitate collaborative coexistence between Bluetooth and WLAN, as well as among Bluetooth piconets/scatternets. The META [11] assumes that the Bluetooth and WLAN are collocated in the same device and they can communicate with each other. A centralized MAC layer controller monitors the Bluetooth and WLAN traffic and predicts collisions. The MAC layer coordination allows precise timing of packet traffic. Each attempt to transmit by either the Bluetooth or the WLAN is submitted to META for approval. If it foresees that a collision will happen, it schedules proper transmission activities for both Bluetooth and WLAN to execute. META can deny a transmit request that would result in collision.

Specifically, the META control entity receives a per-transmission transmit request and issues a per-transmission transmit confirm to each stack to indicate whether the transmission can proceed. The transmit confirm signal carries a status value that is either allowed or denied. The transmit request and confirm signals are exchanged for every packet transmission attempt. By using META, Bluetooth and WLAN transmit their packets sequentially according to the schedule, and collisions are avoided. The schedule that one transmits after another is made based on the packet types. For example, WLAN acknowledgment packets have the highest priority, and the Bluetooth SCO traffic has higher priority than WLAN data packets. That is, if a WLAN acknowledgment packet is to collide with a Bluetooth packet, Bluetooth should delay its transmission. If a Bluetooth SCO packet is about to collide with a WLAN data packet, WLAN should delay its transmission.

Simulation results in [11] show that during Bluetooth ACL operation, META optimizes WLAN throughput, and during Bluetooth SCO operation, it attempts to improve SCO performance, even if it reduces WLAN throughput. META meets required Bluetooth and WLAN timing constraints, for example, the acknowledgment (ACK). No modification in the physical layer is required with META. It supports both ACL and SCO links in Bluetooth. But it introduces some latency due to the delay of transmission.

7.3.2 Alternating Wireless Medium Access

AWMA [12] is a MAC layer mechanism based on TDMA. It assumes that the Bluetooth radio and the WLAN radio are colocated in the same physical unit. In order to avoid overlap in time between their transmissions, the Bluetooth and WLAN devices transmit alternately according to their assigned time intervals.

A WLAN sends out beacons periodically. Let us denote the beacon period as T_B, which is split into three subintervals in AWMA: the WLAN subinterval T_{WLAN}, the Bluetooth subinterval T_{BT}, and the guard interval T_G. The guard interval is optional, it may be used to guarantee that all Bluetooth traffic has completed before the next beacon is sent. Figure 7.4 illustrates the time segmentation of the WLAN and Bluetooth intervals. T_{WLAN}, T_{BT}, and T_G are

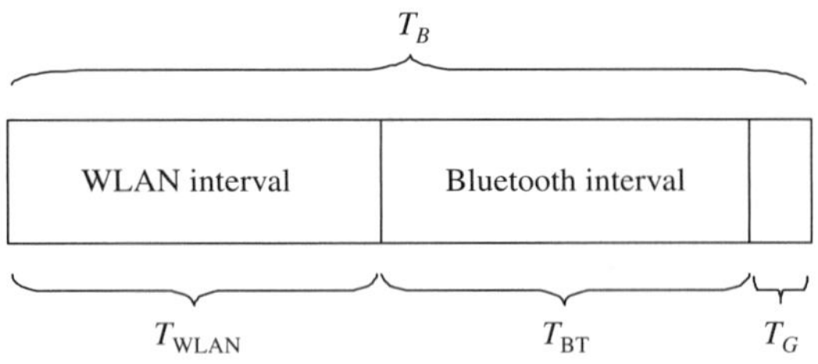

FIGURE 7.4 Time segmentation of WLAN and Bluetooth intervals in AWMA.

specified by the medium sharing element (MSE) in the beacon. The time allocation for Bluetooth and WLAN intervals obeys the following rules:

1. $T_{\text{WLAN}} + T_{\text{BT}} \leq T_B$.
2. If $T_{\text{WLAN}} > T_B$, $T_{\text{BT}} = 0$.
3. If $T_{\text{WLAN}} + T_{\text{BT}} > T_B$, $T_{\text{BT}} = T_B - T_{\text{WLAN}}$.
4. If $T_G \neq 0$ and $T_B - T_{\text{WLAN}} - T_{\text{BT}} < T_G$, $T_{\text{BT}} = T_B - T_{\text{WLAN}} - T_G$.
5. If $T_{\text{WLAN}} < T_B$ and $T_{\text{WLAN}} + T_G \geq T_B$, $T_{\text{BT}} = 0$.

Because the Bluetooth master and the WLAN node are colocated in the same physical unit, the WLAN node can control the timing of the Bluetooth and WLAN. AWMA requires the WLAN node to send a synchronization signal to the Bluetooth master. This signal contains the specification for the Bluetooth interval and the WLAN interval. Management of the AWMA coexistence mechanism is handled over the WLAN by utilizing the MSE in the beacon.

The performance analysis of WPAN and WLAN utilizing AWMA is also provided in [12]. The results are highlighted here. The Bluetooth throughput with AWMA enabled is the throughput of the Bluetooth with no WLAN present multiplied by T_{BT}/T_B. Similarly, the WLAN throughput with AWMA enabled is the throughput of the WLAN with no WPAN present multiplied by T_{WLAN}/T_B. The AWMA coexistence mechanism also increases the latency of each packet sent over the WPAN and WLAN networks. The extra latency introduced by AWMA over the Bluetooth transmission is $T_{\text{WLAN}}^2/2$, while the extra latency introduced by AWMA over the WLAN transmission is $T_{\text{BT}}^2/2$.

By scheduling the WLAN and the Bluetooth radio transmissions, the AWMA coexistence mechanism prevents CCI between WLAN and Bluetooth. It is recommended to use AWMA when the density of devices with collocated Bluetooth and WLAN is high, or when the Bluetooth and/or WLAN bandwidth allocation needs to be deterministically controlled irrespective of its traffic load. Note that the AWMA mechanism cannot be applied to the case when SCO links is utilized in Bluetooth.

7.3.3 Deterministic Frequency-Nulling Scheme

The DFNS [13] is a collaborative method for colocated Bluetooth and WLAN devices. It aims to mitigate the effect of Bluetooth on WLAN. It is primarily a physical layer solution. By employing a Bluetooth receiver as part of the WLAN receiver, the WLAN receiver can obtain the hop frequencies, hop timing, and hop pattern of the Bluetooth transmitter. Because WLAN occupies approximately 22 MHz bandwidth and Bluetooth occupies approximately 1 MHz bandwidth at each hop, the Bluetooth signal can be assumed as a narrowband interferer for WLAN. Then WLAN can put a null in its receiver at the frequency of the Bluetooth signal in order to suppress the interference from Bluetooth.

The implementation of DFNS is described as follows. Between the chip-matched filter and the pseudorandom noise (PN) correlator in the WLAN device, there is an adjustable transversal filter. The optimal coefficients of this filter are estimated and then used to update the filter. By assuming that the interferer is a pure tone and that the PN sequence is sufficiently long, the PN signal samples at the different taps are considered to be uncorrelated and the solutions for the optimal tap weights are simply related to the signal power, the interferer power, the noise power, and the frequency of the interferer. Because the interfering frequency is assumed known a priori and the signal-to-noise ratio (SNR) is often quite high for WLAN systems, only an estimation of the carrier-to-inference (CIR) ratio is necessary to determine the optimal tap coefficients.

The bit error rate (BER) performance analysis in [13] for a 1-Mb/s WLAN system in an additive white Gaussian noise (AWGN) channel with Bluetooth interference shows that the performance of WLAN is greatly improved by using DFNS. As the Bluetooth frequency is known, the offset between the WLAN and the interferer can be calculated. As shown in the experiments in [13], without using DFNS, the BER is 5% for a 5-MHz offset at approximately −11 dB CIR. With DFNS, even for the worst case of a 1-MHz offset, a CIR of −20 dB can achieve a BER less than 0.1%.

Unlike META and AWMA, DFNS causes no delay for the transmission. Because the frequencies of Bluetooth may not always fall into the WLAN band, in META and AWMA, the bandwidth is not fully utilized as in DFNS. By using DFNS, the total data throughput for WLAN increases because there is no packet loss or retransmission due to collision.

7.3.4 Coordinated Colocated Access Point

The CCAP [14] scheme reduces CCI in colocated Bluetooth devices by coordinating their hopping frequencies in a scatternet scenario. A group of piconets in which connections exist between different piconets is called a scatternet [16]. Figure 7.5 is an example of the scatternet consisting of three piconets, with master node denoted by a square and slave node denoted by a

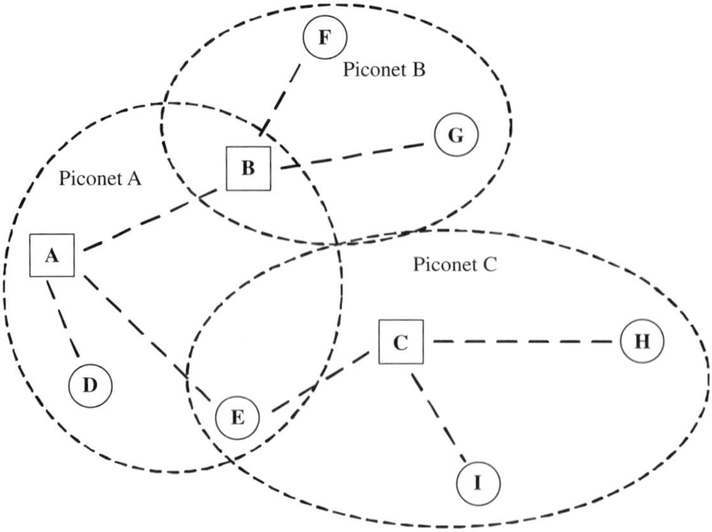

FIGURE 7.5 Bluetooth scatternet.

circle in each piconet. In a scatternet, slaves can participate in different piconets on a time division multiplex basis. Slave E is an example for this case. In addition, a master in one piconet can be a slave in other piconets. For example, device B is the master of piconet B and also a slave in piconet A. In the Bluetooth specification [16], piconets are not frequency synchronized and each piconet has its own hopping sequence. As shown in [14], by breaking this rule and using CCAP, there can be a significant gain in capacity and throughput.

It is defined in [14] that the colocated devices forming a Bluetooth AP are essentially the masters of the piconets that they form. The CCAP technique operates by coordinating hop frequency selection between colocated master nodes. First, the hop timings of the master nodes are synchronized; then the same hopping sequence with different frequency offset is applied to each master node. Therefore no two devices use the same frequency at the same time, and CCI is eliminated between Bluetooth piconets. Because the frequency-hopping sequence for Bluetooth is uniquely determined by the Bluetooth device address and the clock counter, the coordination of the hop frequency selection can be done in two different methods. The first method is to let every piconet use the same device address and clock counter to generate the same hopping sequence and add multiples of a fixed frequency offset to each hopping sequence. The second method is to add offsets to the clock counter, so that there will be frequency offset in the output of each hopping sequence. Note that the offsets for the clock counter should be predetermined. A universally available device address of 0×000000 is suggested in [14].

In Bluetooth, a control protocol for the baseband and physical layers is carried over logical links in addition to user data. This is the link manager protocol (LMP) [16]. Devices that are active in a piconet have a default asynchronous connection-oriented logical transport that is used to transport the LMP signaling. It is pointed out in [14] that some interference can occur when the LMP signaling takes place. This problem can be alleviated by using clock offsets that are not closely spaced. Interference is also possible when the nodes of a CCAP system are in an inquiry or paging state. This interference can be reduced if the least loaded node of the AP is required to enter the inquiry or paging scan state frequently in order to respond to potential clients in the shortest possible time [14]. By using CCAP in the connection state, no CCI occurs when all master nodes use packets of the same length. The capacity and throughput of the Bluetooth networks is greatly increased when compared to conventional Bluetooth. In addition, because of the known address used, the handover time can be decreased considerably by using CCAP.

7.4 NONCOLLABORATIVE MECHANISMS

Noncollaborative mechanisms do not require direct communication between two coexisting networks so that they provide more flexibility for implementation, but most of them may rely on monitoring the channel to detect interference and estimate traffic. Since a variety of mechanisms are embraced in this category, in the following, we only describe several typical mechanisms in detail.

7.4.1 Adaptive Frequency Hopping

We begin with a description of AFH [15, 16] one of the most widely adopted coexistence mechanisms. The idea of AFH is that Bluetooth devices can maintain a performance measurement for each channel visited and periodically classify "good" and "bad" frequency channels, then modify their hop patterns to avoid frequency bands occupied by a WLAN. Because a WLAN usually occupies a fixed frequency band for a relatively long time, it is possible to monitor/detect frequencies occupied in this band by performance measurement.

Interference estimation can be performed by measuring the signal-to-interference ratio (SIR), BER, packet loss, or frame error rate in a Bluetooth receiver. Here PER measurement is adopted. If the PER on certain carrier frequency is greater than the packet loss threshold of 0.5, this frequency is considered to be a bad frequency, otherwise, it is considered to be a good one. Because the master in a Bluetooth piconet is in charge of all packet transmissions, the channel information collected by the slave must be made available to the master. One way to achieve this is that the master and the slave exchange channel information via management messages periodically. Another way is that the master makes use of the acknowledgment information in each slave's response packets to determine the channel information. The second way can speed up the estimation time. It is important to remark that there is a trade-off between the classification update interval and the performance improvement. A higher update rate can capture rapid environment changes, thus guaranteeing channel information accuracy. But it incurs a higher communication overhead if the information is distributed via management messages.

Frequency hopping in traditional Bluetooth is executed as follows: 79 carrier frequencies are sorted into a list of even and odd frequencies in the 2.402- to 2.480-GHz range, with all the even frequencies followed by all the odd frequencies. A segment consisting of the first 32 frequencies in the sorted list is chosen. After all 32 frequencies in that segment are visited once in a random order, a new segment is set including 16 frequencies of the previous segment and 16 new frequencies in the sorted list. In AFH, bad frequencies are eliminated in the sequence. Given a segment of 32 good and bad frequencies, each good frequency is visited exactly once. Each bad frequency in the segment is replaced with a good frequency selected from outside the original segment of 32. Due to the FCC regulation, at least 15 different frequencies should be kept in the hop sequence. If there are less than 15 good frequencies left, some of the bad frequencies will be used.

AFH requires modification of the original Bluetooth specification. This has already been turned into reality in the updated version of Bluetooth specification [16]. The adaptive hop selection mechanism in [16] is shown in Fig. 7.6. When AFH is enabled in a Bluetooth device, the basic hop selection procedure for conventional Bluetooth is initially used to determine a hop frequency. Parameters generated by the basic hop selection algorithm from the clock and the address of the master are sent to the basic hop selection kernel, where a frequency is selected from the 79 frequencies in the basic mapping table. If this frequency is a good frequency, no adjustment is made, and it is used as the frequency for the next packet. If it is a bad frequency according to the channel classification information, the frequency is replaced with a good frequency by executing the remapping function. A good frequency is selected from the AFH mapping table, which consists of U good frequencies; and this new frequency is used as the frequency for the next packet. Since only bad frequencies are

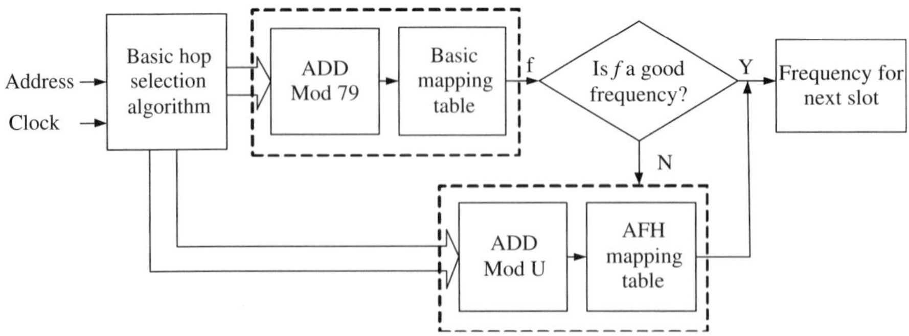

FIGURE 7.6 Adaptive hop selection mechanism.

replaced by new frequencies, good frequencies remain unchanged in the hop sequence, and non-AFH slaves remain synchronized while other slaves in the piconet are using the adapted hopping sequence [16].

In order to keep all devices in the piconet updated with the new hopping pattern, advertisement of the new hopping sequence is typically done by using LMP messages exchanged between the master and slaves in the piconet. How often a new hopping pattern should be advertised could be dynamically adjusted so that it tracks changes in the channel.

Performance evaluation results of AFH are provided in [15] for file transfer protocol (FTP) and voice applications. Maximizing the throughput is the goal for FTP application and minimizing the delay is the goal for voice application. AFH improves the throughput by 25% for FTP application. The improvement is more obvious than that for voice application. When Bluetooth and WLAN coexist, the PER for WLAN drops if AFH is adopted. AFH allows additional scheduling techniques to be used simultaneously if there is a need to control the transmission of packets on the medium. AFH has been demonstrated as effective in dealing with static interference that comes from WLAN, but it is not applicable for multiple colocated Bluetooth piconets because in Bluetooth frequency channels are constantly changing, the hop pattern of one piconet is not known to another one, and multiple piconets are not necessarily synchronized in a typical noncollaborative scenario. Besides, the performance of AFH is also dependent on the update rate of the frequency classification to track the channel dynamics [15].

7.4.2 Bluetooth Interference-Aware Scheduling

By taking advantage of the fact that devices in the same piconet are not subject to the same levels of interference on all channels of the band, the BIAS [17] algorithm dynamically distributes channels to devices in order to maximize their throughput while maintaining fairness of access among users.

Interference estimation is used to identify whether a frequency is good or bad based on SIR or PER measurement. The estimation and classification is done continuously. The master collects frequency classification information and schedules the transmission about sending packet to which slave and using which frequency. Before each transmission, the master chooses a slave device if no retransmission is required at that moment. Then the master verifies whether the two frequencies that will be used by itself and the slave device are good frequencies. The new transmission will begin only if both frequencies are good. Otherwise, the master postpones the transmission of the packet until a good pair of frequencies becomes available. When the master schedules transmission, retransmission has the highest priority and is transmitted first; then data packets and finally the acknowledgment packets are transmitted. In all cases, the frequency pair must consist of good frequencies. In Bluetooth, a slave device must respond to the master device even if it does not have any data to send (in this case, a NULL packet will be sent). Therefore a slave transmission always follows a master transmission. When using BIAS, the master avoids receiving data on a bad frequency, by avoiding a transmission on a frequency preceding a bad one in the hopping pattern.

BIAS needs to be implemented in the master device only. It adopts a backoff strategy for Bluetooth to avoid collision with WLAN. BIAS is a neighbor-friendly strategy for WLAN because it avoids transmission on the WLAN band. BIAS may outperform AFH for delay jitter and packet loss constrained applications such as voice and video [15]. Performance results obtained in [17] show that BIAS eliminates packet loss of Bluetooth even in the worst interference case when more than three-fourths of the spectrum are occupied by other networks. The increased delay compared to the case that no interference is present varies between 1 and 5 ms on average. Furthermore, BIAS is adaptable to rapid changes of the channel. But it is not very effective to mitigate interference that comes from other Bluetooth piconets.

7.4.3 Overlap Avoidance

The OLA scheme [18] assumes that Bluetooth and WLAN devices can acquire the interference information from each other by channel sensing, PER calculation, and received signal power monitoring in a noncollaborative scenario. It can also be used in a collaborative scenario by assuming that the traffic information is exchanged between WLAN and Bluetooth. The basic idea of this scheme is traffic scheduling at the MAC layer. It consists of voice OLA (V-OLA) to deal with Bluetooth voice traffic and data OLA (D-OLA) to deal with Bluetooth data traffic. The two algorithms are jointly applied when both voice and data are transmitted in Bluetooth devices.

For Bluetooth voice traffic, SCO link is adopted. Packets are transmitted in a predetermined pattern. The channel idle time also obeys a deterministic pattern. By taking advantage of the fixed pattern, WLAN packets are scheduled to be transmitted with an adjusted length and during the Bluetooth channel idle time. This is how V-OLA works. V-OLA is still applicable when there are both SCO and ACL links by assuming that the CCI caused by ACL traffic is negligible compared to the CCI caused by SCO traffic. When the channel is occupied, the WLAN can choose to transmit a shortened packet [called the shortened-transmission (ST) mode] or delay the transmission [called the postponed-transmission (PT) mode]. When dealing with the detection of the ending time of SCO, it is pointed out in [18] that the WLAN considers the Bluetooth SCO transmission ended if it does not detect any interference for a certain time period. Note that the timing of the Bluetooth packet transmission may drift so that imperfect information may be obtained.

For Bluetooth data traffic, the ACL link is adopted. D-OLA is employed by assuming that the Bluetooth master has the information about the frequency bands occupied by WLAN. If the next Bluetooth carrier frequency falls into the WLAN band, the Bluetooth master will schedule to transmit a long packet using the current carrier frequency. Therefore, the next carrier frequency is

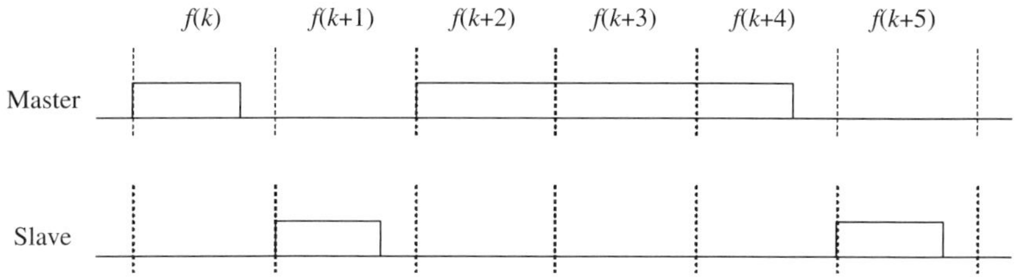

FIGURE 7.7 Example for D-OLA transmission.

skipped and CCI is eliminated. Figure 7.7 gives an example. Suppose that $f(k + 3)$ falls into the WLAN band, the master knows this and chooses to transmit a three-slot packet instead of a single-slot packet, so that transmission on frequency $f(k + 3)$ is avoided. This method works on the condition that enough data are buffered for transmission. If there is not enough data, the transmission that causes CCI can be postponed. By taking advantage of the variety of packet lengths and scheduling a packet with proper duration, collision is avoided. According to the FCC regulation, the average occupation time of any frequency should not be greater than a threshold. Sometimes, an intelligent schedule is required to use some of the bad frequencies while maximizing the usage of all the good frequencies.

Performance results in [18] show that OLA greatly improves the system throughput. V-OLA PT causes more delay for WLAN than V-OLA ST does, while D-OLA does not cause much delay for Bluetooth. When operating in the case that interfering devices are collocated in the same physical unit, collaborative methods such as META outperforms OLA. With noncolocated interfering devices, OLA outperforms META. The OLA scheme only requires a minor change on the Bluetooth standard and the 802.11 specification. It not only avoids overlap between Bluetooth and WLAN but also copes with interference from microwave ovens, which also operates at the 2.4-GHz band.

7.4.4 Power Control Scheme Based on SIR

Provided that Bluetooth devices can dynamically change their transmission power, a power control scheme based on the SIR is proposed in [20]. Since no information about other systems is available in the Bluetooth receiver, it can only measure the interference to obtain the SIR. Based on each measurement, the current power is updated by multiplying itself with a ratio. The ratio is the target SIR divided by the measured SIR. With the target SIR, the signal power level is adjusted to no more than what is needed. The updated power is bounded by the minimum and maximum transmission power range. If there is no change in the interfering signal, the transmitted power can converge to its final value in one step.

Bluetooth transmitter has three levels of radio transmission power: 100 mW (20 dBm), 2.5 mW (4 dBm), and 1 mW (0 dBm). Power control method only works well for the first two cases since the maximum power of 1 mW limits the power changing range. The Bluetooth specification [16] suggests that the transmitted power should be adjusted based on the received signal strength indicator (RSSI) measurements at the receiver. By assuming that the noise can be neglected, RSSI corresponds to the SIR. LMP messages are used to transmit the measured SIR from the receiver to the transmitter so that it can perform the power updating. It is pointed out in [20] that there is a trade-off between the value of the update interval and the signaling traffic required. A small value of the update interval makes the system more adaptive but requires more signaling information to be exchanged.

Experimental results are reported in [20]. The maximum power for Bluetooth is set to 100 mW.The WLAN power is set to 25 mW. By using power control, the PER for Bluetooth decreases from 18 to 4% when the distance between Bluetooth and WLAN is 0.5 m. Bluetooth with power control effectively reduces PER for distance greater than 0.5 m. But when the distance is less than 0.5 m, since the transmitted power is bounded by the maximum power, it cannot achieve further performance improvement.

The power control scheme does not change the Bluetooth frequency-hopping pattern. It requires no change in the Bluetooth specification. It can be easily implemented by using a scheduling rule on current Bluetooth chip set. It is compatible to devices without power control. The power control scheme is effective in the scenario that the interference power is not too large because the maximum achievable SIR is limited by the maximum transmission power. What's more, increasing the Bluetooth transmission power will inevitably incur more interference to neighboring devices.

7.4.5 A Hybrid Scheme

A hybrid scheme that combines power control, LBT and AFH to achieve low PER and improve system throughput is proposed in [22]. AFH and power control methods are discussed above. LBT is a carrier sense technique. Before each transmission, the transmitter senses the channel in the turn-around time of the current slot. If it detects that the channel is occupied, it will postpone its current transmission until another chance. LBT combats dynamic frequency interference by withdrawing packet transmission that may become potential interference. As pointed out in [22], even with the ideal carrier sense of Bluetooth, LBT cannot totally avoid all packet collisions between Bluetooth and WLAN. LBT can effectively avoid the collision when the Bluetooth packet is to be transmitted during the WLAN transmission duration. If a WLAN begins its transmission in the middle of a Bluetooth transmission and causes collision, LBT cannot avoid it. By avoiding hopping into WLAN band, AFH effectively deals with this static interference. Since there are more Bluetooth transmissions outside the WLAN band, if several Bluetooth piconets coexist with a WLAN, AFH may introduce more dynamic frequency interferences to neighboring Bluetooth piconets. Performance results in [22] show that by combining LBT and AFH together, the Bluetooth throughput is higher than if using these methods separately, when there are fewer than 70 piconets coexisting with a WLAN.

By using power control and LBT, unnecessary power usage is avoided. LBT is used to deal with dynamic interference. AFH is used to deal with static interference. The hybrid method is implemented by adding a few extensions to the MAC layer, and it is compatible with the current Bluetooth specification.

To summarize this section, we note that most noncollaborative methods are applied in Bluetooth to avoid collision with WLAN because Bluetooth is more vulnerable to CCI if Bluetooth and WLAN coexist. AFH [15] and interference-source-oriented AFH (ISOAFH) [25] are effective in dealing with WLAN interference but not applicable for multiple colocated Bluetooth piconets. The performance of AFH is also dependent on the update rate of the frequency classification to track the channel dynamics [15]. Approaches based on scheduling such as BIAS [15, 17], OLA [18], and master delay MAC scheduling (MDMS) [25] cause delay in the transmission, hence they may not be bandwidth efficient. Power control methods [19, 20] depend on the accuracy of channel sensing and cannot provide much improvement if the Bluetooth device is very close to the interfering device. Carrier-sensing-based schemes inevitably suffer from the hidden terminal problem [21, 26]. A hybrid method of power control, LBT and AFH proposed in [22] can achieve better performance with added complexity.

7.5 DUAL-CHANNEL TRANSMISSION

In this section, we describe a frequency diversity technique using DCT [24] for Bluetooth piconets to combat CCI. In DCT, the same packet is transmitted on two distinct frequency-hopped channels simultaneously, and the power used in each channel is half of what would be used in SCT. A packet is successfully received if at least one channel survives. In order to make DCT robust to the 22 MHz WLAN bandwidth, the two channels of DCT are separated by at least 22 MHz.

The hopping sequences for DCT can be generated as shown in Fig. 7.8. At the kth hop, one frequency $f_{1,k}$ is generated according to the conventional Bluetooth specification [16], where an index is obtained based on the master device's address and clock (see [16] for details of the basic index generator); then the index modulo 79 is used to select a frequency from mapping table 1. Mapping table 1 contains 79 frequencies with all of the even frequencies in ascending order followed by all of the odd frequencies in ascending order. The other frequency, $f_{2,k}$, is generated as follows. We first translate the master's address to another address (e.g., by taking its complement as shown in Fig. 7.8). This address and the clock are used to obtain an index from the

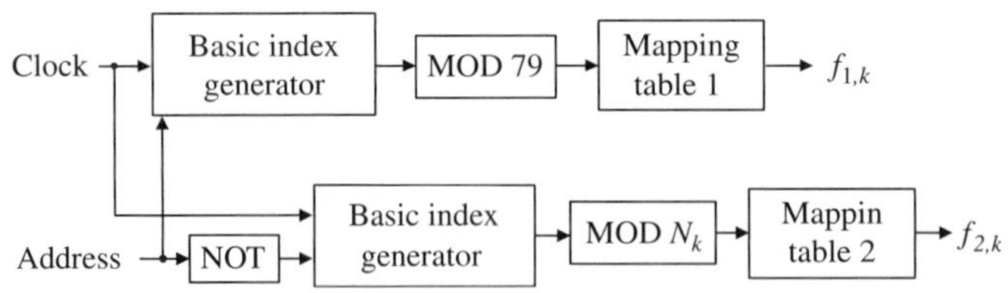

FIGURE 7.8 Diagram for dual-hop sequence generation.

basic index generator. We obtain mapping table 2 by eliminating $f_{1,k}$ and all frequencies with a distance less than 22 MHz to $f_{1,k}$ from mapping table 1. Let us denote the number of frequencies in mapping table 2 as N_k. The index modulo N_k is subsequently used to select the corresponding frequency from mapping table 2 as $f_{2,k}$.

For example, suppose that the frequency $f_{1,k} = 12$ MHz is obtained using the master's address 101 and the clock (for simplicity, we use a 3-bit device address here). We take the complement of 101 to obtain 010. Since we use 28 bits of the Bluetooth address, it is unlikely its complement will result in another address that is already in the system (the probability for two addresses to be identical is $\frac{1}{2}^{28} = 3.7 \times 10^{-9}$). An index is generated from 010 and the clock. Since $f_{1,k} = 12$ MHz, mapping table 2 is obtained by eliminating frequencies $1, 2, \ldots, 33$ MHz from mapping table 1. The number of frequencies in mapping table 2 is $N_k = 79-33 = 46$. Mapping table 2 has the even frequencies in ascending order followed by the odd frequencies in ascending order as follows: $34, 36, \ldots, 78, 35, 37, \ldots, 79$. Finally $f_{2,k}$ is generated by selecting one frequency from this table according to the index modulo 46.

DCT does come with increased implementation complexity, but compared with other approaches mitigating CCI, it has the following merits:

1. It works without the need to detect/estimate the interference pattern or measure the PER, signal strength, and signal-to-interference ratio so that it benefits from reduced processing complexity.
2. It is robust to both static (WLAN) and dynamic (Bluetooth) interferences. It works independently without communicating with other networks, and no control information is transmitted among networks. Hence, it saves on overhead, control channel requirements, and potential instability due to errors over such channels.
3. It is delay efficient, that is, it does not incur extra delay compared to conventional Bluetooth. While some approaches need an initialization period, others delay the transmission when encountering interference.
4. It does not change the MAC layer protocol. DCT-enabled Bluetooth devices can use the same association mechanism to establish a link within

a piconet as SCT. DCT is also backward compatible to SCT by using only one channel.

5. Although DCT uses two channels, its effect on adjacent channel interference (ACI) to other networks is the same as SCT, since with DCT, the probability of ACI is doubled but the power for each channel is only half of that of SCT. For the same reason, the interference of DCT enabled Bluetooth to WLAN is also the same as that of SCT.

It is clear that for the same transmitter–receiver link, the received signal power at each channel in DCT is 3 dB less than that of SCT. We will show in Section 7.5.1.2 that at low SNR, when PER is dominated by channel noise, DCT underperforms SCT; when PER is dominated by frequency collisions, DCT outperforms SCT when the number of piconets is less than 20. In the following, we evaluate the performance of DCT. We begin with the performance analysis for the coexistence of multiple piconets, and then we use the results to evaluate the case that multiple piconets coexist with a WLAN.

7.5.1 Analysis for Coexistence of Multiple Piconets

Here, we analyze the performance of SCT and DCT when multiple Bluetooth piconets coexist. Similar to the analysis for Bluetooth in [1, 2], for simplicity we do not consider adjacent channel interference, propagation characteristic, and error correction. We note that the extension of our results to the case where error correction coding is taken into account can be obtained similar to [10]. Suppose n piconets coexist in sufficiently close vicinity so that a frequency collision in two or more packets for the time duration of at least one bit will destroy all the packets involved. Since there is no coordination between these piconets, each piconet has $n - 1$ potential competitors. Different from [1, 2], here we also consider channel noise. The channel noise is assumed to be AWGN.

First, we consider the effect of noise on SCT and DCT in AWGN channels. The demodulation of GFSK can be approximated as a coherent continuous phase index $h = 0.32$ FSK demodulation. Without CCI, the BER due to noise is given in [27] and shown to be approximately achievable with a zero-IF Bluetooth receiver in [28]:

$$P_b(\alpha) = Q\{\sqrt{\alpha[1 - \text{sinc}(h)\cos(\pi h)]}\} \tag{7.1}$$

where $Q(x) = (1/\sqrt{2\pi})\int_x^\infty e^{-t^2/2}dt$, $\alpha = E_b/N_0$ is the SNR, E_b is the bit energy, and N_0 is the noise power spectral density. The success probability for a k-slot SCT packet with m_k bits is

$$\beta_k(\alpha) = [1 - P_b(\alpha)]^{m_k} \tag{7.2}$$

where $k = 1, 3, 5$. The success probability of one channel for a k-slot DCT packet with m_k bits is

$$\beta_k\left(\tfrac{1}{2}\alpha\right) = \left[1 - P_b\left(\tfrac{1}{2}\alpha\right)\right]^{m_k} \tag{7.3}$$

where $\frac{1}{2}\alpha$ reflects that the received SNR in each of the two channels in DCT is half of what would be in SCT. We note that in practice, the BER can be inferior to that achieved by the optimal coherent scheme depending on the actual detectors used [29].

7.5.1.1 Analysis for Piconets with the Same Packet Type. Suppose that all piconets use the same type of packets. For one-slot packet, although 625 μs is allocated for each packet, the transmission time of a packet is only 366 μs, and the remaining 259-μs idle time is used for transient time settling. The ratio of the packet to slot duration is $r_1 = 366/625$. For three-slot packets, the ratio is $r_3 = (625 \times 2 + 372)/(625 \times 3)$. And for five-slot packets, the ratio is $r_5 = (625 \times 4 + 370)/(625 \times 5)$.

We consider two extreme cases: One is that all piconets are synchronized; the other is that all piconets are fully unsynchronized, that is, no two piconets are synchronized. If some piconets happen to be synchronized, the PER and throughput will fall within the range of the aforementioned two extreme cases. If piconets are synchronized, a packet of interest may be affected by one packet from each colocated piconet. If piconets are unsynchronized, the packet of interest may be affected by one or two packets from each colocated piconet. The probabilities that there are one or two dangerous packets from a colocated piconet in the unsynchronized case are $2 - 2r_k$ and $2r_k - 1$, respectively [1], where $k = 1, 3, 5$. Assuming a transmission rate of 1 Mbps in each piconet, the average data throughput for each piconet is defined as $(1-\text{PER})r_k$ Mbps. The throughput includes both original and retransmitted packets.

7.5.1.1.1 Packet Error Rate of SCT. When multiple piconets coexist, with SCT a packet is successfully received only when no bit error occurs due to frequency collision or channel noise effect. The PERs for n synchronized and fully unsynchronized piconets with SCT in AWGN channels are

$$P^s_{\text{sct}}(k, n) = 1 - s_0^{n-1}\beta_k(\alpha) \tag{7.4}$$

$$P^u_{\text{sct}}(k, n) = 1 - \left[(2 - 2r_k)s_0 + (2r_k - 1)s_0^2\right]^{n-1}\beta_k(\alpha) \tag{7.5}$$

respectively, where $s_0 = (M-1)/M$ is the probability that one interfering piconet chooses another frequency instead of the one chosen by the piconet of interest. $k = 1, 3, 5$ denotes the packet type. The superscript s in $P^s_{\text{sct}}(k, n)$ indicates the synchronized case, and the superscript u in $P^u_{\text{sct}}(k, n)$ indicates the fully unsynchronized case.

7.5.1.1.2 Packet Error Rate of DCT. With DCT, a packet is successfully received if at least one of the two channels is not destroyed by frequency collisions or channel noise. Let us first consider synchronized piconets. We use an example of three colocated piconets to illustrate the collision analysis. Suppose A is the piconet of interest, and B and C are the interfering piconets. Let a_1, a_2 be the two frequencies used in piconet A for a particular packet transmitted. Similarly, let b_1, b_2 and c_1, c_2 be the corresponding frequencies in piconets B and C, respectively. When three piconets coexist, there are three cases that a packet in piconet A can be successfully received: the first case is that b_1, b_2, c_1, and c_2 are chosen from frequencies other than a_1 and a_2; the second case is that only one of a_1 and a_2 is overlapped by one of the frequencies from b_1, b_2, c_1, and c_2; and the third case is that only one of a_1 and a_2 is overlapped by the same frequency from both B and C: b_1 or b_2 and c_1 or c_2. Therefore with DCT, the PER for three synchronized piconets in AWGN is

$$P_{\text{dct}}^{s}(k,3) = 1 - d_0^2\left\{1 - \left[1 - \beta_k\left(\tfrac{1}{2}\alpha\right)\right]^2\right\} - 4d_1 d_0 \beta_k\left(\tfrac{1}{2}\alpha\right) - 4d_1 d_2 \beta_k\left(\tfrac{1}{2}\alpha\right) \quad (7.6)$$

In Eq. (7.6), $k = 1, 3, 5$, d_0 is the probability that frequencies of two piconets do not overlap, d_1 is the probability that one frequency of piconet A is overlapped by a frequency of another piconet, and d_2 is the probability that a frequency of piconet C is overlapped with a frequency of piconet A, which is already overlapped by a frequency of piconet B. The functions for d_0, d_1, d_2 are derived in the following.

First, we know that a_1 and a_2 are two frequencies selected from the $M = 79$ frequency bins and $|a_1 - a_2| \geq D$ MHz, where $D = 22$. Specifically, a_1 is uniformly selected from the M frequency bins, a_2 is chosen from mapping table 2, which is based on a_1. Let us use index $1, 2, \ldots, m, \ldots, M$ to denote the M frequency bins. For example, 1 is a possible choice for a_2 only if $a_1 = D + 1$, $D + 2, \ldots, M$. So if a_1 is selected as $1, 2, \ldots$, or M, then 1 is a possible choice for a_2 for $M-D$ times. Let v_m denote the times that a particular frequency m is a possible choice for a_2 when a_1 is chosen as 1,2, … , or M. It can be verified that

$$v_m = \begin{cases} M - D - m + 1 & \text{for } m = 1, \ldots, D \\ M - 2D + 1 & \text{for } m = D + 1, \ldots, M - D \\ m - D & \text{for } m = M - D + 1, \ldots, M \end{cases} \quad (7.7)$$

Consequently, the probability for $a_2 = i$ is obtained as

$$t_i = \frac{v_m}{\sum_{m=1}^{M} v_m} \quad (7.8)$$

for $i = 1, \ldots, M$, which is a nonuniform distribution.

Based on the distribution of the two frequencies for a piconet, we obtain

$$d_0 = \sum_{i=1}^{M}\sum_{j=1}^{M} \frac{t_j}{M}\left[\frac{M-2}{M}(1-t_i-t_j)\right] \tag{7.9}$$

where t_j/M is the average probability for $a_1 = i$ and $a_2 = j$. Also in Eq. (7.9), $(M-2)(1-t_i-t_j)/M$ is the probability for piconet B to choose a frequency pair such that neither b_1 nor b_2 is equal to i and j. Therefore, d_0 can be considered as the probability that four frequencies of piconet A and B are all distinct. Similarly, we obtain the probability that either a_1 or a_2 is overlapped by a frequency of piconet B as

$$d_1 = \frac{1}{2}\sum_{i=1}^{M}\sum_{j=1}^{M} \frac{t_j}{M}\left[\frac{2}{M}(1-t_i-t_j) + \frac{M-2}{M}(t_i+t_j)\right] \tag{7.10}$$

Since frequencies of different piconets are generated independently, the probabilities d_0 and d_1 can also be applied for coexistence of piconets A and C. When three piconets coexist, there is a specific case that one frequency of piconet A is overlapped by a frequency of piconet B and also by a frequency of piconet C, which has a probability of $d_1 d_2$, where $d_2 = d_1/2$.

Summarizing the above cases in which a packet of piconet A can be successfully received, we conclude that the PER of DCT for three synchronized piconets in AWGN is given by Eq. (7.6). We generalize Eq. (7.6) to obtain the PER of DCT for n synchronized piconets in AWGN as

$$\begin{aligned} P^s_{\mathrm{dct}}(k,n) = 1 - \sum_{i=0}^{n-1} 2^i \binom{n-1}{i} d_0^{n-1-i} d_1^{\langle i>0\rangle} \left(d_2^{i-1}\right)^{\langle i>1\rangle} \\ \times \left\{1 - \left[1 - \beta_k\left(\frac{\alpha}{2}\right)\right]^2\right\}^{\langle i=0\rangle} \beta_k\left(\frac{\alpha}{2}\right)^{\langle i>0\rangle} \end{aligned} \tag{7.11}$$

where $E^{\langle F\rangle}$ is equal to E if F is true, otherwise $E^{\langle F\rangle}$ is equal to 1. In Eq. (7.11), $\binom{n-1}{i}$ is the number of ways of selecting i unordered piconets from $n-1$ piconets, where each selected piconet has one interfering frequency, and 2^i is the number of possible permutations of the interfering frequencies from the i piconets.

Next we consider the case where the piconets are fully unsynchronized. Suppose i interfering piconets have one dangerous packet, each with probability $2-2r_k$, for $i = 0, \ldots, n-1$, and the other $n-1-i$ interfering piconets have two dangerous packets, each with a probability of $2r_k-1$. This is equivalent to that with probability $(2-2r_k)^i(2r_k-1)^{n-1-i}$, there are a total of $1+i+2(n-1-i)$ synchronized piconets, which give a PER of $P^s_{\mathrm{dct}}(k, 1+i+2(n-1-i))$. Therefore the PER of DCT for the fully unsynchronized case is

$$P^u_{\text{dct}}(k,n) = 1 - \sum_{i=0}^{n-1} \binom{n-1}{i} (2-2r_k)^i (2r_k - 1)^{n-1-i} \times \left[1 - P^s_{\text{dct}}(k, 1+i+2(n-1-i))\right] \tag{7.12}$$

7.5.1.1.3 Simulations and Discussion. We use Monte Carlo simulations to validate the theoretical analysis. We evaluate the performance for DCT with two channels separated by at least $D = 22\,\text{MHz}$. Each channel uses half the power that would be used in SCT. AWGN channel is simulated. Figure 7.9 depicts the PER for synchronized piconets versus SNR when the number of piconets is 5, 10, and 20. The results for piconets with one-slot and five-slot packets are shown. The results for piconets with three-slot packets will fall within the range of the aforementioned two cases and are omitted here. In Fig. 7.9, "DCT (10)" means that 10 synchronized piconets with DCT coexist. Since the signal power for each channel of DCT is only half of what would be used in SCT, DCT outperforms SCT only when PER is dominated by collisions, which is the case when SNR is greater than 18 dB for SCT (equivalently, 15 dB in each channel of DCT). The error floors in Fig. 7.9 indicate that when SNR is sufficiently high, the noise effect on PER can be neglected.

A Bluetooth transmitter has three levels of radio transmission power: 20, 4, and 0 dBm. As shown in [30], considering a 70-dB loss for a 10-m link, a noise level of −114 dBm at the receiver input, and a receiver noise figure of 23 dB, a typical transmit power of 0 dBm in Bluetooth will result in a receive SNR of 21 dB. Therefore, Bluetooth typically operates in an SNR range where DCT outperforms SCT.

In Fig. 7.10, the PER and throughput per piconet are plotted for SCT and DCT as a function of the number of piconets. Synchronized and fully unsynchronized cases with one-slot packet are simulated. The results for three-slot and five-slot packets are similar to these results. The SNR is 18 dB, under which the effect of noise can be ignored compared to that of collisions. It can be seen that the simulation results match well with the theoretic analysis. DCT outperforms SCT when the number of piconets is less than 30. For example, when the number of piconets is 10, for the synchronized case, the PERs for SCT and DCT are 0.11 and 0.043, respectively, the throughput per piconet for SCT and DCT are 0.5213 and 0.5605 Mbps, respectively. Therefore, with DCT, the PER reduces by about 60% and the throughput increases by 8% compared to SCT. We note that in a practical scenario, it is rare to have more than a dozen colocated piconets due to the short transmission distance of Bluetooth.

Figures 7.9 and 7.10 show that DCT outperforms SCT under certain conditions. Naturally, one may think to use triple or even more channels simultaneously to transmit a packet in order to further reduce PER. However, because more frequency channels are used, there are also more collisions. Assuming channels are distinct, we have obtained the PER for multiple-channel

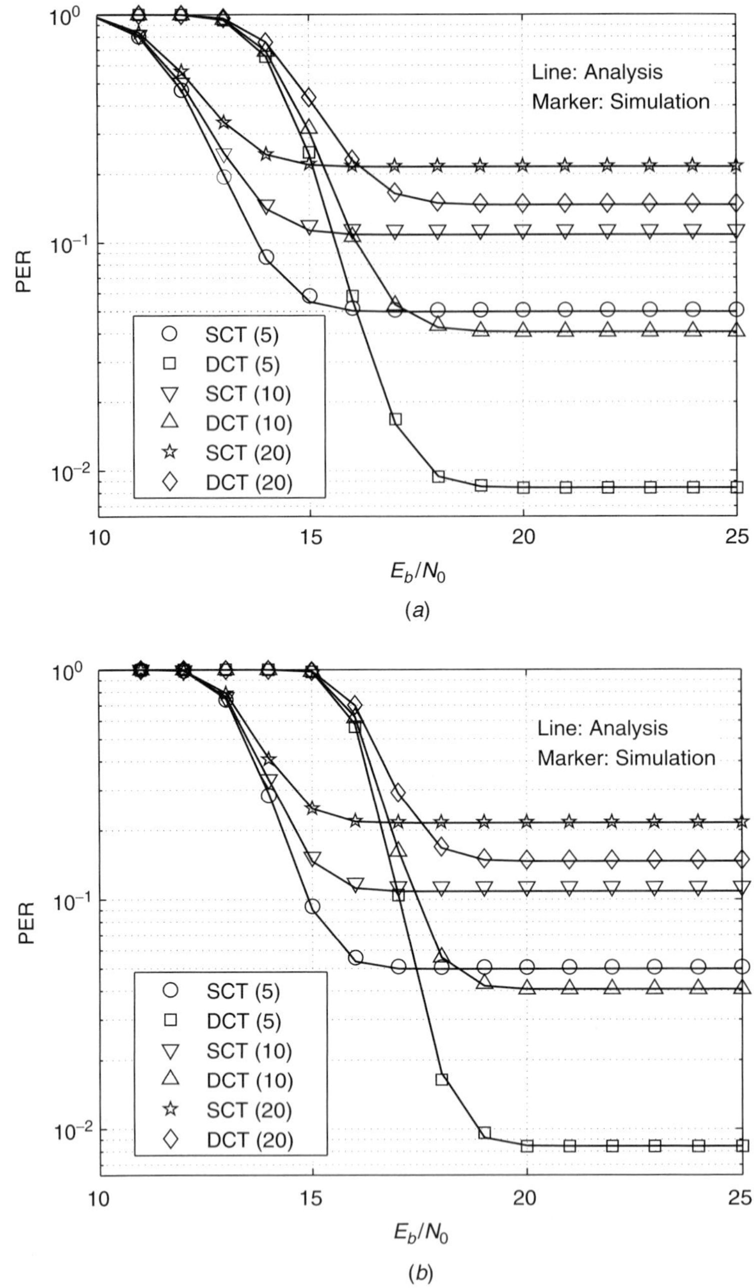

FIGURE 7.9 PER for synchronoized piconets versus E_b/N_0: (a) one-slot packet; (b) five-slot packet.

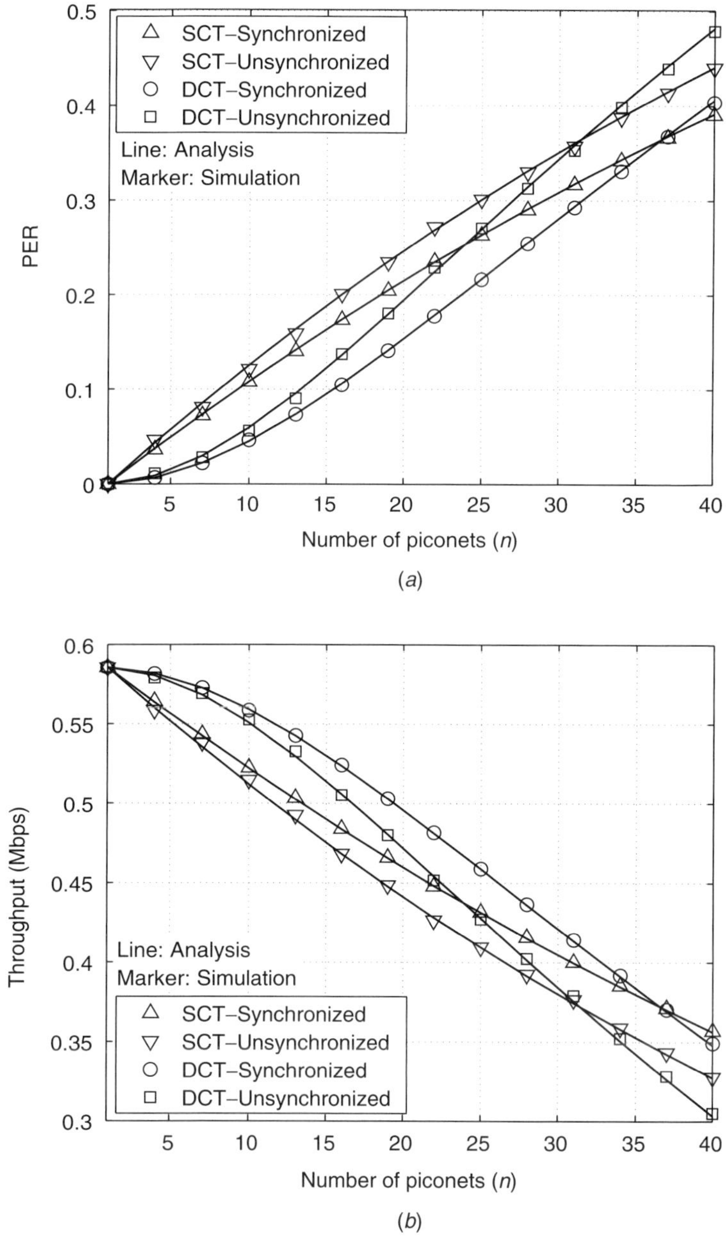

FIGURE 7.10 Performance for multiple colocated piconets with one-slot packets: (a) PER; (b) throughput per piconet.

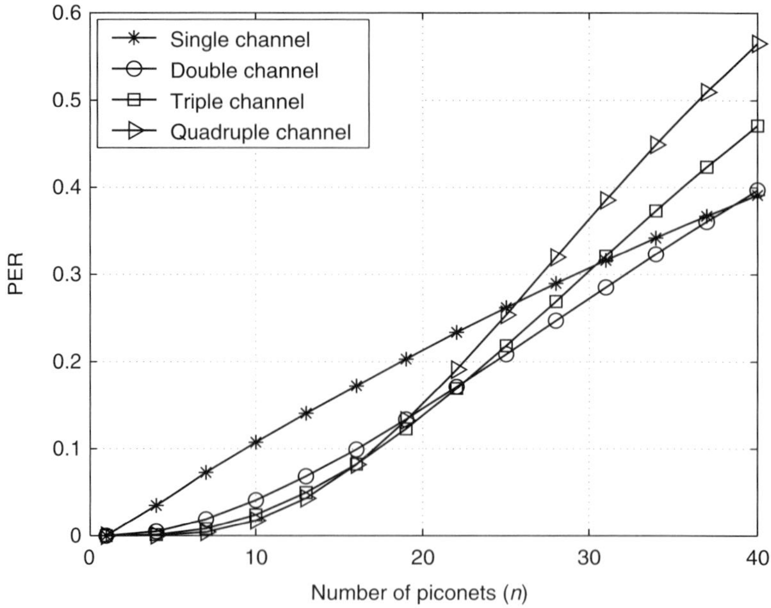

FIGURE 7.11 PER for frequency diversity methods with different number of channels.

transmissions using simulations. The results for SCT, DCT, triple-channel transmission (TCT), and quadruple-channel transmission (QCT) are shown in Fig. 7.11, where we assume that all piconets transmit synchronized one-slot packets and the SNR is infinity. It can be seen that further PER reduction over DCT by using TCT or QCT is marginal, compared to the improvement provided by DCT over SCT. Considering the balance of complexity and performance, DCT is the preferred choice.

7.5.1.2 Analysis for Piconets with Mixed Packet Type. In this section we generalize the collision analysis to the case of mixed packet type (one slot, three slot, and five slot), and the model is similar to [2] except that we also consider an AWGN channel. The system traffic load may vary as we add idle slots (single slot with no traffic load) into our model.

We first derive the PER for SCT based on the result in [2]. Let $F_{\text{sct}}^{m}(k,n)$ denote the probability of success for the k-slot packet in the piconet of interest when n unsynchronized piconets with SCT coexist without noise, where $k = 1, 3, 5$. The superscript m denotes that mixed packet types coexist. Then, the PER for a k-slot packet when n piconets with SCT coexist in AWGN channels is

$$P_{\text{sct}}^{m}(k,n) = 1 - F_{\text{sct}}^{m}(k,n)\beta_k(\alpha) = 1 - \left[F_{\text{sct}}^{m}(k,2)\right]^{n-1}\beta_k(\alpha) \tag{7.13}$$

Between transmissions, there are idle times used for transient time settling. Let the data occupancy ratio at the last slot for a k-slot packet be $\hat{r}_1 = 366/625$, $\hat{r}_3 = 372/625$, $\hat{r}_5 = 370/625$. For simplicity, we approximate them by a single value $\hat{r}$, that is, $\hat{r} = r_3$. Poisson traffic is assumed in the piconets, and λ_1, λ_3, and λ_5 are the arrival rates of one-, three-, and five-slot packets, respectively; λ_{10} is the arrival rate of the idle (one-slot) packets. As given in [2], let B_j, $j = 1, \ldots, 10$, be the delimiter of time slots, where B_1, B_2, and B_5 are the beginnings of one-, three-, and five-slot packets, respectively; B_3 and B_4 are the beginnings of the second and third slots of a three-slot packet, respectively; B_6, B_7, B_8, and B_9 are the beginnings of the second, third, fourth, and fifth slots of a five-slot packet, respectively; and B_{10} is the beginning of an empty slot. The arrival rate of B_j is λ_j, and $\lambda_2 = \lambda_3 = \lambda_4$, $\lambda_5 = \lambda_6 = \lambda_7 = \lambda_8 = \lambda_9$. Given any B_j, $g(j)$ is defined as the number of slots that follow delimiter B_j and belong to the same packet. Therefore $g(1) = 1, g(2) = 3, g(3) = 2, g(4) = 1, g(5) = 5, g(6) = 4,$ $g(7) = 3$, $g(8) = 2$, $g(9) = 1$, and $g(10) = 1$. It is shown in [2] that

$$F^{m}_{\text{sct}}(k,2) = \sum_{j=1}^{10} ((1-\hat{r})\lambda_j\{\tilde{f}(j)L[k-g(j)] + f(k,j)\tilde{L}[k-g(j)]\}. \\ +(2\hat{r}-1)\lambda_j f(j)L(k-g(j))) \tag{7.14}$$

where the following definitions are duplicated from [2]:

$$\tilde{f}(j) = \begin{cases} 1 & \text{if } j = 10 \\ s_0 & \text{otherwise} \end{cases} \tag{7.15}$$

$$f(j) = \begin{cases} \dfrac{(\lambda_1+\lambda_3+\lambda_5)s_0^2+\lambda_{10}s_0}{\lambda_1+\lambda_3+\lambda_5+\lambda_{10}} & j = 1, 2, 5 \\[2ex] \dfrac{(\lambda_1+\lambda_3+\lambda_5)s_0+\lambda_{10}}{\lambda_1+\lambda_3+\lambda_5+\lambda_{10}} & j = 10 \\[2ex] s_0 & \text{otherwise} \end{cases} \tag{7.16}$$

$$f(k,j) = \begin{cases} \dfrac{(\lambda_1+\lambda_3+\lambda_5)s_0+\lambda_{10}}{\lambda_1+\lambda_3+\lambda_5+\lambda_{10}} & k = 1, j = 1, 2, 5 \\[2ex] \dfrac{(\lambda_1+\lambda_3+\lambda_5)s_0^2+\lambda_{10}s_0}{\lambda_1+\lambda_3+\lambda_5+\lambda_{10}} & k = 3, 5, j = 1, 2, 5 \\[2ex] \dfrac{(\lambda_1+\lambda_3+\lambda_5)s_0+\lambda_{10}}{\lambda_1+\lambda_3+\lambda_5+\lambda_{10}} & j = 10 \\[2ex] s_0 & \text{otherwise} \end{cases} \tag{7.17}$$

$$\begin{aligned} L(i) = & \frac{\lambda_{10} L[i-g(10)]}{\lambda_1+\lambda_3+\lambda_5+\lambda_{10}} + \frac{\lambda_1 s_0 L[i-g(1)]}{\lambda_1+\lambda_3+\lambda_5+\lambda_{10}} \\ & + \frac{\lambda_3 s_0 L[i-g(2)]}{\lambda_1+\lambda_3+\lambda_5+\lambda_{10}} + \frac{\lambda_5 s_0 L[i-g(5)]}{\lambda_1+\lambda_3+\lambda_5+\lambda_{10}} \end{aligned} \tag{7.18}$$

for $i>0$, and $L(i) = 1$ for $i \leq 0$, and

$$\begin{aligned} \tilde{L}(i) = & \frac{\lambda_{10} \tilde{L}[i-g(10)]}{\lambda_1+\lambda_3+\lambda_5+\lambda_{10}} + \frac{\lambda_1 s_0 \tilde{L}[i-g(1)]}{\lambda_1+\lambda_3+\lambda_5+\lambda_{10}} \\ & + \frac{\lambda_3 s_0 \tilde{L}[i-g(2)]}{\lambda_1+\lambda_3+\lambda_5+\lambda_{10}} + \frac{\lambda_5 s_0 \tilde{L}[i-g(5)]}{\lambda_1+\lambda_3+\lambda_5+\lambda_{10}} \end{aligned} \tag{7.19}$$

for $i>1$, and $\tilde{L}(i) = 1$ for $i \leq 1$; and s_0 is defined in Eq. (7.4). We note that compared to [2], here Eqs. (7.16) and (7.17) slightly modified to reflect the probabilities for different types of slots.

Notice that by expanding Eq. (7.14), we can write $F^m_{\text{sct}}(1,2) = a_2 s_0^2 + a_1 s_0 + a_0$; s_0^0, s_0, and s_0^2 are the probabilities of packet success when there are one, two, and three synchronized piconets with the same type packets coexisting in ideal channels without noise, and a_0, a_1, a_2 are the corresponding parameters accounting for the effect of mixed types of multi-slot packets. Generalizing this result to n piconets, we obtain

$$\begin{aligned} P^m_{\text{sct}}(k,n) &= 1 - \sum_{j=0}^{\gamma_k(n-1)} \beta_k(\alpha) a_j^{\langle n>1 \rangle} s_0^j \\ &= 1 - \sum_{j=0}^{\gamma_k(n-1)} a_j^{\langle n>1 \rangle} \left[1 - P^s_{\text{sct}}(k, j+1)\right] \end{aligned} \tag{7.20}$$

where k = 1,3,5, $\gamma_1 = 2$, $\gamma_3 = 3$, $\gamma_5 = 5$, and $P^s_{\text{sct}}(k, j+1)$ is defined in Eq. (7.4)

Similarly, for DCT, the PER for k-slot packets when n piconets with mixed packet types coexist in AWGN channels is

$$P^m_{\text{dct}}(k,n) = 1 - \sum_{j=0}^{\gamma_k(n-1)} a_j^{\langle n>1 \rangle} \left[1 - P^s_{\text{dct}}(k, j+1)\right] \tag{7.21}$$

where k, γ_k, a_j are defined in Eq. (7.20), and $P^s_{\text{dct}}(k, j+1)$ is defined in Eq. (7.11).

The throughputs per piconet when n piconets coexist for SCT and DCT are

$$R_{\text{sct}}(n) = \lambda_1\left[1 - P^m_{\text{sct}}(1,n)\right]r_1 + 3\lambda_3\left[1 - P^m_{\text{sct}}(3,n)\right]r_3 + 5\lambda_5\left[1 - P^m_{\text{sct}}(5,n)\right]r_5 \qquad \text{Mbps} \tag{7.22}$$

$$R_{\text{dct}}(n) = \lambda_1\left[1 - P^m_{\text{dct}}(1,n)\right]r_1 + 3\lambda_3\left[1 - P^m_{\text{dct}}(3,n)\right]r_3 + 5\lambda_5\left[1 - P^m_{\text{dct}}(5,n)\right]r_5 \qquad \text{Mbps} \tag{7.23}$$

The PERs for different data packets (DH$_i$ means i-slot packet) and throughput per piconet are plotted in Fig. 7.12 for SCT and DCT. The SNR is 18 dB. The arrival rates $\lambda_1 = \lambda_3 = \lambda_5$, and the traffic load is 70%. Compared to SCT, DCT reduces PER by as much as 50% and increases throughput by up to 6%, when the number of piconets is small (less than 20).

7.5.2 Coexistence of Multiple Piconets and a WLAN

In this section we evaluate the performance of DCT when there are multiple Bluetooth piconets colocated with a WLAN. A hop frequency falling into the 22-MHz WLAN band results in a collision in that channel. We compare the performance of the proposed DCT to those of SCT and AFH. For simplicity, we only consider the PER for synchronized Bluetooth piconets with the same type packets in this section. The other cases can be analyzed using a similar method.

A packet from a piconet with SCT is successfully received only when no bit error occurs due to noise or collision, where the collision can happen between Bluetooth piconets or between the Bluetooth piconet and WLAN. The PER for n synchronized piconets with SCT, when colocated with a WLAN in AWGN channels, is given by

$$P^w_{\text{sct}}(k,n) = 1 - s_0^{n-1}\beta_k(\alpha)\gamma \tag{7.24}$$

where the superscript w in $P^w_{\text{sct}}(k,n)$ stands for the case that multiple Bluetooth piconets coexist with a WLAN; $\gamma = (M-D)/M$ denotes the probability that a Bluetooth frequency falls outside the WLAN band.

We then analyze the PER performance of AFH. Suppose that after an initial time period, the AFH scheme successfully detects[1] the D MHz band occupied by a WLAN and subsequently avoids hopping onto this band. The probability that one interfering piconet chooses another frequency instead of the one

[1] A false alarm (the detector says that WLAN is active when it is not) leads to wasted opportunities but not to collisions and errors. A missed detection leads to collisions and hence bit errors. For simplicity, we assume perfect detection.

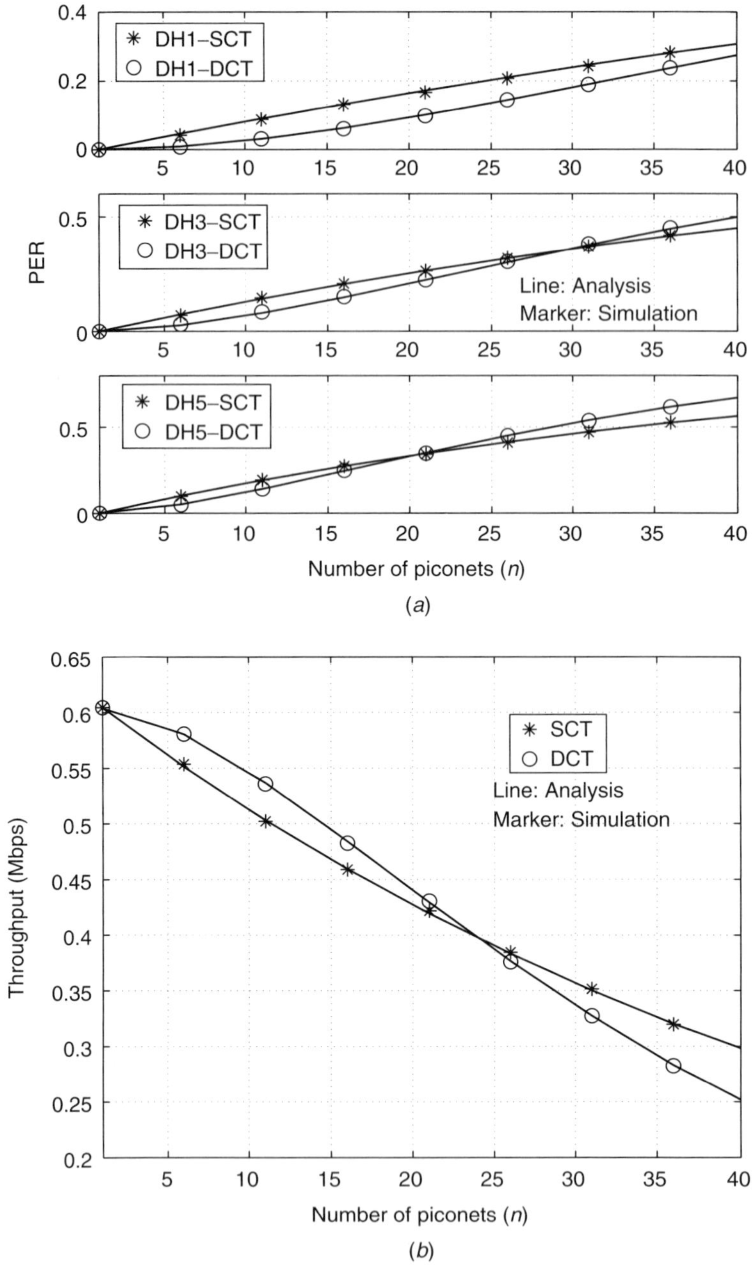

FIGURE 7.12 Performance for multiple colocated piconets with mixed types of packets: (a) PER; (b) throughput per piconet.

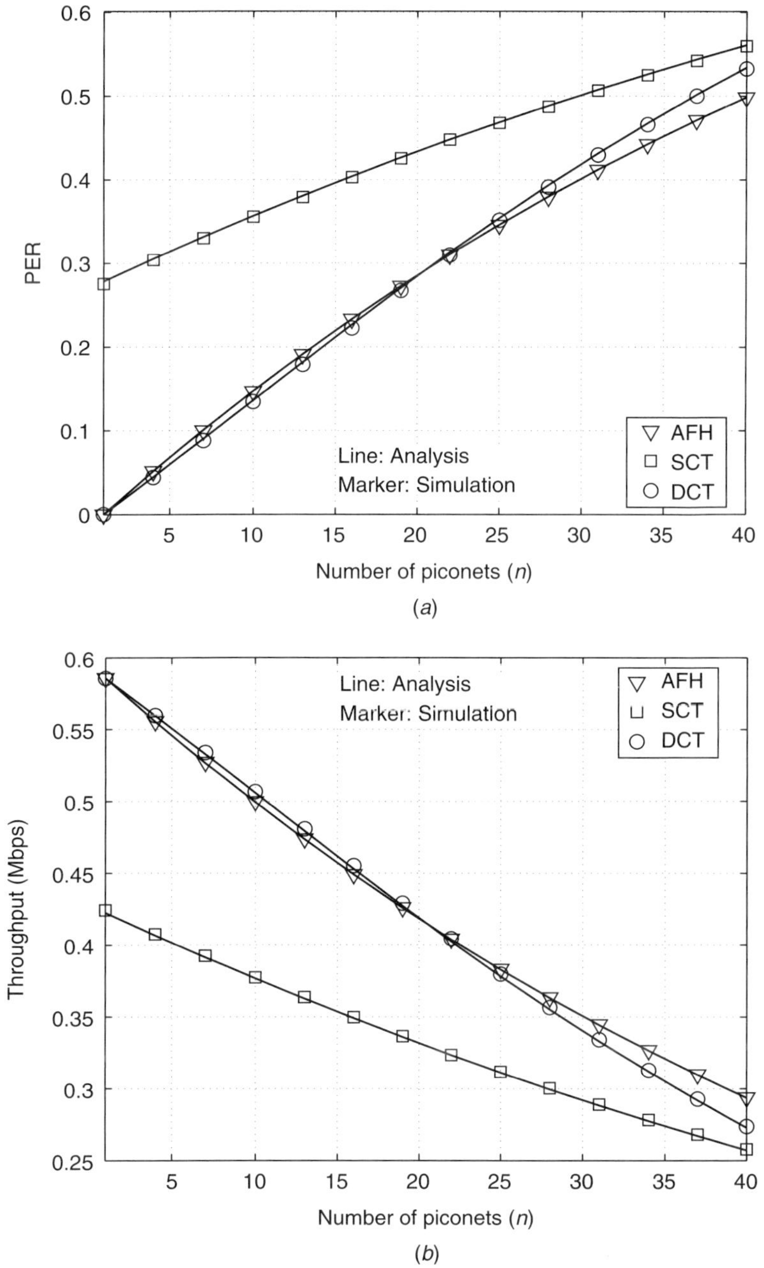

FIGURE 7.13 Performance for synchronized Bluetooth piconets when they coexist with a WLAN: (a) PER; (b) throughput per piconet.

chosen by the piconet of interest is $\tilde{s}_0 = (M-1-D)/(M-D)$. With AFH, a packet is successfully received only when no bit error occurs due to noise or collision between Bluetooth piconets. The PER for n synchronized piconets with AFH, when colocated with a WLAN in AWGN channels is

$$P^{w}_{\mathrm{afh}}(k,n) = 1 - \tilde{s}_0^{n-1}\beta_k(\alpha) \tag{7.25}$$

Similar to the SCT case, by revising Eq. (7.11), the PER for n synchronized Bluetooth piconets with DCT, when colocated with a WLAN in AWGN channels, can be obtained as

$$\begin{aligned} P^{w}_{\mathrm{dct}}(k,n) = 1 - \sum_{i=0}^{n-1} 2^i \binom{n-1}{i} d_0^{n-1-i} d_1^{\langle i>0\rangle} \left(d_2^{i-1}\right)^{\langle i>1\rangle} \\ \times \left\{1 - \left[1 - \beta_k\left(\frac{\alpha}{2}\right)\right]^2\right\}^{\langle i=0\rangle} \beta_k\left(\frac{\alpha}{2}\right)^{\langle i>0\rangle} \gamma^{\langle i>0\rangle} \end{aligned} \tag{7.26}$$

Figure 7.13 shows the PERs for SCT, AFH, and DCT as functions of the number of piconets when multiple Bluetooth piconets coexist with a WLAN. Here we consider the case that all piconets are synchronized and they transmit single-slot packets. In each realization of the Monte Carlo simulation, the center frequency of the WLAN band is randomly generated among the 11 channels for 802.11b WLAN as listed in Table I of [25]. In addition, we assume that for Bluetooth with AFH, the band occupied by the WLAN is already known, therefore AFH is able to avoid hopping onto the WLAN band. From Fig. 7.13, we observe that DCT has performance comparable to that of AFH when the total number of Bluetooth piconets is small, and both outperform SCT significantly. In practice, if the channel status changes, the master device in a Bluetooth piconet with AFH has to communicate with each slave device to update the frequency classification information, which reduces the total useful throughput. In contrast, DCT works independently and does not require an initial period to detect the WLAN band, and it is robust to both dynamic and static interference.

7.6 CONCLUSIONS

Recently, the coexistence of Bluetooth and WLAN has drawn a lot of attention. In this chapter, we reviewed several typical collaborative and non-collaborative methods for coexistence of Bluetooth and WLAN. We discussed them on aspects such as the main idea, implementation issues, performance results, comparison with other methods, advantages and disadvantages, and the like.

Based on direct communication between Bluetooth and WLAN, central control mechanisms such as META, AWMA, DFNS, and CCAP play an important role for coexistence of these networks in collaborative scenarios. Relying on monitoring the channel to detect interference and estimate traffic, noncollaborative methods mitigate the CCI between Bluetooth and WLAN, which embrace a variety of techniques such as power control, AFH, BIAS, OLA, BCS, and LBT. There are also the hybrid method and the collision resolution method.

We investigated a frequency diversity technique for Bluetooth in detail. This method uses dual-channel transmission to deal with both dynamic and static CCI. To evaluate its performance, several metrics are analyzed, including PER and throughput. Theoretic analysis and numerical simulations demonstrated that when the number of colocated piconets is less than about 20 and SNR is higher than 18 dB, the DCT design offers significant performance improvement over SCT, which makes it attractive in most practical scenarios where only a small number of piconets can possibly coexist in a physical environment due to the short transmission range of Bluetooth. With channels separated by at least 22 MHz, the DCT design is also robust to WLAN interference. This method is characterized by its independence, efficiency, and robustness, requiring no interference detection, transmission delay, or traffic control. Its key limitation is the requirement for additional channel and reduced transmission range.

REFERENCES

1. A. El-Hoiydi, "Interference between Bluetooth networks—Upper bound on the packet error rate," *IEEE Commun. Lett.* 5, 245–247 (2001).
2. T.-Y. Lin, Y.-K. Liu, and Y.-C. Tseng, "An improved packet collision analysis for multi-Bluetooth piconets considering frequency-hopping guard time effect," *IEEE J. Sel. Areas Commun.* 22, 2087–2094 (2004).
3. K. Naik, D. S. L. Wei, Y. T. Su, and N. Shiratori, "Analysis of packet interference and aggregated throughput in a cluster of Bluetooth piconets under different traffic conditions," *IEEE J. Sel. Areas Commun.* 23, 1205–1218 (2005).
4. I. Howitt, "Mutual interference between independent Bluetooth piconets," *IEEE Trans. Vehic. Technol.* 52(3), 708–718 (2003).
5. N. Golmie and F. Mouveaux, "Interference in the 2.4 GHz ISM band: impact on the Bluetooth access control performance," in *Proc. IEEE Int. Conf. on Communications*, Vol. 8, Helsinki, Finland, June 2001, pp. 2540–2545.
6. O. Karjalainen, S. Rantala, and M. Kivikoski, "The performance of Bluetooth system in the presence of WLAN interference in an office environment," in *Proc. 8th IEEE Int. Conf. on Communication Systems*, Vol. 2, Singapore, Nov. 2002, pp. 628–631.
7. J. Lansford, A. Stephens, and R. Nevo, "Wi-Fi (802.11b) and Bluetooth: Enabling coexistence," *IEEE Network* 15(5), 20–27 (2001).
8. I. Howitt, "WLAN and WPAN coexistence in UL band," *IEEE Trans. Vehic. Technol.* 50(4), 1114–1124 (2001).

9. I. Howitt, "Bluetooth performance in the presence of 802.11b WLAN," *IEEE Trans. Vehic. Technol.* 51(6), 1640–1651 (2002).
10. A. Conti, D. Dardari, G. Pasolini, and O. Andrisano, "Bluetooth and IEEE 802.11b coexistence: Analytical performance evaluation in fading channels," *IEEE J. Sel. Areas Commun.* 21(2), 259–269 (2003).
11. J. Lansford, "MEHTA: A method for coexistence between co-located 802.11b and Bluetooth systems," IEEE 802.15-00/360r0, available: http://www.ieee802.org/15/pub/TG2.html, Nov. 2000.
12. IEEE 802.15.2, "IEEE recommended practice for information technology—Part 15.2: Coexistence of wireless personal area networks with other wireless devices operating in the unlicensed frequency bands, available http://standards.ieee.org/getieee802/802.15.html, 2003.
13. R. E. Van Dyck and A. Soltanian, "IEEE 802.15.2, Clause 14.1—Collaborative co-located coexistence mechanism," IEEE P802.15 Working Group for WPANs, IEEE 802.15-01/364r0, available http://www.ieee802.org/15/pub/TG2-Draft.html, July 2001.
14. J. Dunlop and N. Amanquah, "High capacity hotspots based on Bluetooth technology," *IEE Proc. Commun.* 152(5), 521–527 (2005).
15. N. Golmie, N. Chevrollier, and O. Rebala, Bluetooth and WLAN coexistence: Challenges and solutions," *IEEE Wireless Commun.* 10(6), 22–29 (2003).
16. "Specification of the Bluetooth system," Version 2.0 + EDR, Bluetooth Special Interest Group, Nov. 2004.
17. N. Golmie, "Bluetooth dynamic scheduling and interference mitigation," *ACM MONET*, 90(1) (2004), pp. 21–31.
18. C. F. Chiasserini and R. R. Rao, "Coexistence mechanisms for interference mitigation in the 2.4-GHz ISM band," *IEEE Trans. Wireless Commun.* 2(5), 964–975 (2003).
19. E. C. Arvelo, "Open-loop power control based on estimations of packet error rate in a Bluetooth radio," *Wireless Commun. Networking* 3(5), 1465–1469 (2003).
20. N. Golmie and N. Chevrollier, "Techniques to improve Bluetooth performance in interference environments," in *Proc. Military Communications Conference*, Vol. 1, Vienna, VA, Oct. 2001, pp. 581–585.
21. C. D. M. Cordeiro, S. Abhyankar, R. Toshiwal, and D. P. Agrawal, "A novel architecture and coexistence method to provide global access to/from Bluetooth WPANs by IEEE 802.11 WLANs," in *Proc. IEEE Int. Conf. on Performance, Computing, and Commun.*, Phoenix, AZ, 2003, pp. 23–30.
22. B. Zhen, Y. Kim, and K. Jang, "The analysis of coexistence mechanisms of Bluetooth," in *Proc. IEEE 55th Vehicular Technology Conference*, Vol. 1, Birmingham, AL, 2002, pp. 419–423.
23. J. Li and X. Liu, "A collision resolution technique for robust coexistence of multiple Bluetooth piconets," in *Proc. 64th IEEE Vehicular Technology Conference*, Montreal, Canada, Sept. 2006, pp. 25–28.
24. J. Li and X. Liu, "Interference mitigation using frequency diversity for coexistence of Bluetooth piconets and WLAN," in *Proc. IEEE International Conference on Communications*, Glasgow, Scotland, June 2007.
25. K. Yu-Kwong and M. C.-H. Chek, "Design and evaluation of coexistence mechanisms for Bluetooth and IEEE 802.11b systems," in *Proc. IEEE Int.*

Symposium on Personal, Indoor and Mobile Radio Communications, Vol. 3, Barcelona, Spain, Sept. 2004, pp. 1767–1771.

26. A. Willig, K. Matheus, and A. Wolisz, "Wireless technology in industrial networks," *Proc. IEEE* 93(6), 1130–1151 (2005).
27. W. P. Osborne and M. B. Luntz, "Coherent and noncoherent detection of CPFSK," *IEEE Trans. Commmun.* 22, 1023–1036 (1974).
28. S. Samadian, R. Hayashi, and A. A. Abidi, "Demodulators for a zero-IF Bluetooth receiver," *IEEE J. Solid-State Circuits* 38(8), 1393–1396 (2003).
29. L. Lampe, M. Jain, and R. Schober, "Improved decoding for Bluetooth systems," *IEEE Trans. Commun.* 53(1), 1–4 (2005).
30. J. C. Haartsen and S. Mattisson, "Bluetooth—A new low-power radio interface providing short-range connectivity," *Proc. IEEE* 88(10), 1651–1661 (2000).

PART III

IEEE 802.15.3 WIRELESS PANs

CHAPTER 8

FRAME FORMAT, CHANNEL ACCESS, AND PICONET OPERATION OF IEEE 802.15.3 WIRELESS PANs

YANG XIAO, MICHAEL J. PLYLER, BO SUN, and YI PAN

8.1 INTRODUCTION

IEEE 802.15.3 [1] is a standard for wireless personal area networks (WPANs), which consume a lower amount of power and have the capability of using multiple formats. These capabilities also allow support for scalable data rates. IEEE 802.15.3 WPAN is indented for relatively short distances, about 10 m. This standard also allows for a dynamic environment because mobile devices may enter and exit a piconet often. With some of these features also comes the capability for the user to control devices in the WPAN. In this chapter, we focus on frame format, channel access, and piconet operation.

8.1.1 Piconet

The IEEE 802.15.3 WPAN [1–4] works through a dynamic topology known as a piconet. Peer-to-peer connections comprise the main organization of this ad hoc network. A piconet is a collection of one or more associated devices that share a single identifier with a common coordinator. This common coordinator is called a piconet controller (PNC).

Figure 8.1 gives an example of what we start with for our WPAN. First, each of the wireless devices communicate with each other to determine which is best suited to be the PNC, and they all concur with it. In this example, the point coordinator (PC) has been selected as the PNC. Next, each device requests to join the piconet and performs an authentication protocol with the PNC. Once they have established a connection with the PNC and are given access to the piconet, the controller establishes the time slots for each of the devices and

Emerging Wireless LANs, Wireless PANs, and Wireless MANs. Edited by Y. Xiao and Y. Pan

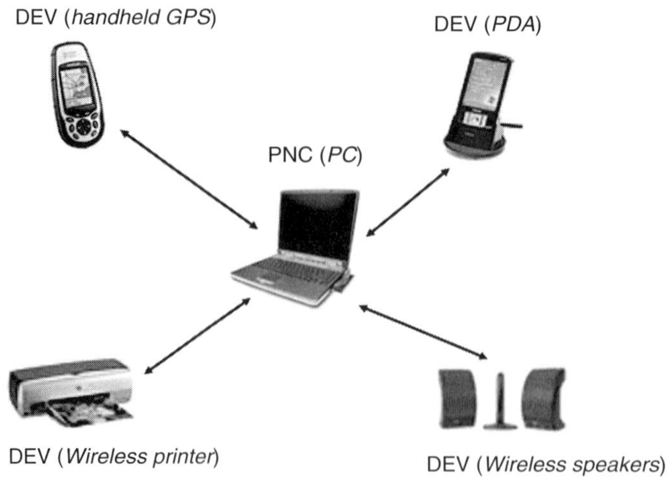

FIGURE 8.1 Representation of a piconet.

distributes payload protection keys. Figure 8.1 represents a fairly small piconet with only five devices. However, depending on the capabilities of the PNC, there can be as many as 255 devices in the piconet.

Once the PNC has established control of the piconet, devices can transmit protected data to the other devices in the piconet during the arranged time slots. This is depicted in Fig. 8.2 with the letter A. Each device in the piconet has connectivity with every other device in the piconet [1]. In addition to sending and receiving data among all devices (DEVs), two devices may

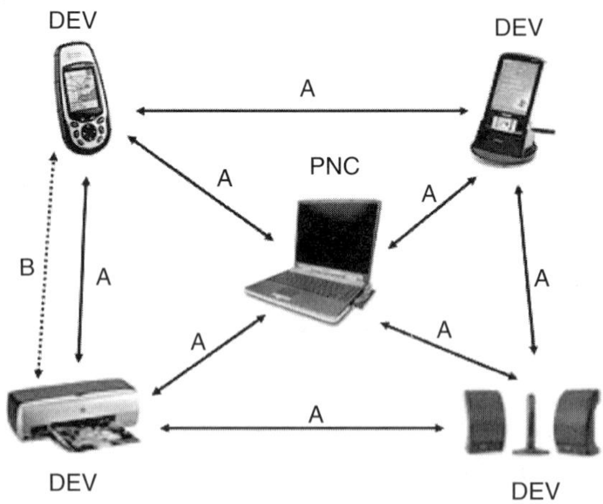

FIGURE 8.2 Piconet after all connections have been set up by PNC—a functional piconet.

optionally establish their own secure connection sub-network within the established piconet. This is depicted as the letter B in Fig. 8.2.

One thing to keep in mind is that the PNC is not a special type of device. It is merely the device that has the most capabilities and is able to manage the other devices. With that said, each device in the piconet can be a parent to another piconet. For instance, the global positioning system (GPS) from Fig. 8.1 is connected to another three devices but these devices do not have access to our original piconet. The new piconet with the GPS as the PNC would then be a child piconet of the original. The reason is that the PC has greater capabilities than the GPS, but the GPS has already established different mutual key authentications than the main piconet and is the controller. The child piconet is operating at a different frequency than the original.

8.1.2 Transmission/Frequency Information

The transmission information of the IEEE 802.15.3 protocol is similar to but at a higher rate than the Bluetooth standard. One of the general characteristics of this WPAN is that it operates in the 2.4-GHz frequency band. Some differences and similarities between these protocols are given in Table 8.1. They both use 2.4–2.4835 GHz frequency. One thing to note about these frequencies is that they have unlicensed use in much of the world.

The data rate of the IEEE 802.15.3 standard is 55 Mbps compared to Bluetooth's <1 Mbps. The range of transmission is ~ 10 m. The cost of implementing this network is medium compared to wireless locl area networks (WLANs) and the Bluetooth WPAN. In addition to data applications, IEEE 802.15.3 also supports voice and multimedia applications. This type of network is supported worldwide, as are most other wireless networks. The interesting thing about IEEE 802.15.3 compared to Bluetooth is that it has five video channels. Its power consumption is low, although not as low as that of Bluetooth. As currently set up, the PNC controls the power usage of the other devices within the piconet. It does this by allowing them to go into a state similar to standby, so that power consumption is minimal.

TABLE 8.1 The Similarities and Differences between 802.15.1 (Bluetooth) and 802.15.3

	Bluetooth 1.1	802.15.3
Frequency band, GHz	2.4	2.4
Data rate, Mbps	<1	55
Range, m	10	10
Cost	Low	Medium
Target applications	Voice and data	Voice, data, and multimedia
Number of video channels	0	5
Power	Very low	Low

8.1.3 PNC/DEV Connections

This section will give a general idea of how a PNC gives a handover to another PNC if it chooses and how the PNC is chosen in the first place.

Alternate coordinators (ACs) broadcast the capabilities of the devices to other devices that want to participate in the piconet. Based on certain criteria, the *best* AC is chosen and it then becomes the PNC. Once the PNC is established, it issues a beacon to all of the other devices, establishing protocols, security, time slots, and other communication information. If, along the way, another *more capable* AC joins the piconet, the current PNC must hand over the task of being the controller. This only happens if the second, more capable, AC also passes the security protocol of the piconet that has already been established.

The DEVs join with association commands. The PNC allows and checks each device based upon each DEV resource. These resources could include how many channels the device can operate on, which protocols it has for security, and which frequency it can/will operate on. Once the DEV is authenticated, it can send data in the contention access period (CAP) if the PNC has allowed it, and the DEV can request guaranteed time slots (GTSs) for specific connections to other devices within the piconet. If the DEV is given a GTS by the PNC, it then can stream data, with quality-of-service (QoS) requirements, or it can send nonstreamed data, which has no QoS requirements.

The two main types of GTS, which have different persistence, are dynamic GTS and pseudostatic GTS. The dynamic GTS may change its position within the superframe. This means that it does not have to send/receive information in the same frame slot. With pseudostatic GTS, the PNC has the capability of changing, but before it can change, it must communicate and confirm with any devices that are transmitting within a GTS.

8.1.4 Media Access Control Support

The primary contributions of media access control (MAC) to high rate WPAN (HR-WPAN) are peer discovery, multirate support, repeater service, dynamic channel service and selection, transmitter control, and better power management. Once again, keep in mind that we are talking about high rate, 55-Mbps wireless connections—much faster than that of Bluetooth 1.1. We are also talking about ultra wideband (UWB) frequencies.

8.2 MAC FRAME FORMATS

The IEEE 802.15.3 MAC layer has a coordinator–device topology. This means that the PNC controls the DEVs in the piconet. The PNC assigns times and access for connections, and all the commands go to and come from the PNC. If you think about it in this manner, the PNC works like a master and the DEVs in the piconet act as the slaves. The communication in the piconet is peer to peer. This means that all devices can "talk" to all other devices in the piconet with permission from the PNC. QoS is achieved by a time division multiple access (TDMA) architecture with GTSs. The MAC layer also contains the information for implementation of the security and authentication sequences.

The MAC layer is responsible for implementing authentication protocols and symmetric-key management. The MAC layer also implements command and data payload protection. In addition, the MAC layer has its own public- and private-key pairs.

The MAC in all DEVs should be used to validate the error-free reception of all the frames from the physical (PHY) layer using the frame check sequence (FCS). Every device in the piconet should be able to create and maintain a subset of the command frames for transmission and should also be able to decode possibly different subsets of the command frames once the frame has been received.

8.2.1 Basic MAC Superframe

To satisfy QoS requirements, the basic MAC superframe is divided into three main segments, shown in Fig. 8.3*a*: the beacon, access slots, and GTSs. The time slots are allocated to different devices for the transfer of data. Once again, we must remember that this is dictated by the PNC.

The parts of the superframe are shown in Fig. 8.3*b*. These more detailed sections of the superframe are the beacon, the CAP, and the contention-free period (CFP). The beacon section of the frame is TDMA and can only be sent by the PNC.

The CAP is carrier sense multiple access with collision avoidance (CSMA/CA); the data types that can be sent through this period of the frame can be

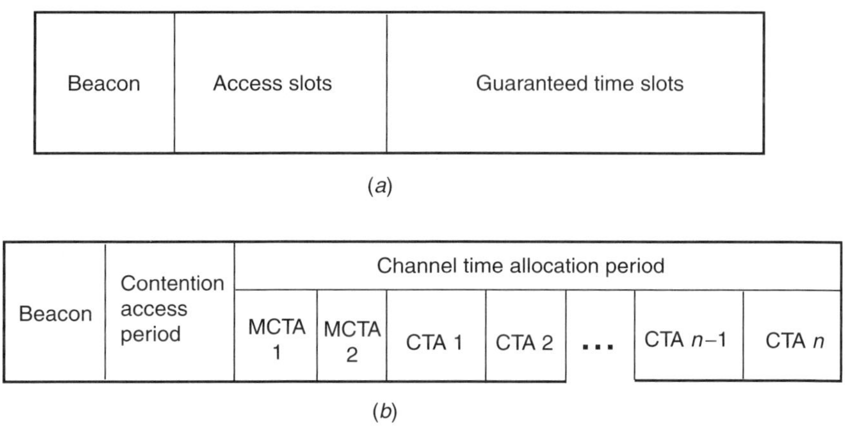

FIGURE 8.3 (*a*) Basic superframe; (*b*) superframe.

restricted by the PNC. The PNC also replaces the CAP with a managed time slot (MTS) using slotted-aloha access during this section of the frame. The CFP is TDMA, as is the beacon, because it is assigned by the PNC. The CFP can be either unidirectional GTS or MTS as the PNC chooses.

The general MAC frame format consists of a set of fields that occur in a fixed order in all frames. Each of these frames has a MAC header and a MAC frame body. The MAC frame body consists of a variable-length frame payload and an FCS.

Figure 8.4 shows the frame control field format; we can observe that the first bits of information sent in the frame indicate the protocol version followed by the frame type, security mode (SEC), acknowledgment (ACK) policy, retry, more data, and reserved sections.

Protocol Version Field. The protocol version field of this frame control format is a constant size. For all practical purposes, the default value of this bit sequence is 0b000. Since there aren't any fundamental incompatibilities with revisions, all other bit sequences are reserved for the time being.

Frame Type Field. The frame type field b5–b3 indicates the types of frames to be sent. The format for those types is as follows: 000 (beacon), 001 (Imm-ACK), 010 (Dly ACK), 011 (command), 100 (Data), and 101–111 (reserved).

SEC Field. The bit b6 is for the SEC. This bit is set to 1 when the frame body is protected by using the SECID. Otherwise, this bit is set to 0. In order to send a secure frame, for example, between DEVs, the bit must be set to 1, consistent with proper security protocol. Secure frames enable capabilities that keep unauthorized DEVs from accessing information. It is also important to note that secure DEVs will only read frames that are secure. DEVs that are not secure will only read nonsecure frames, shown with a 0 in the SEC bit field.

ACK Policy Field. This field is used to help indicate the type of acknowledgment procedures to be used in the frame. The bits in b7 and b8 can be set to 00 for the No-ACK setting. In this case, the recipient does not acknowledge the transmission. It is also interesting that the sender resumes transmitting without waiting for an acknowledgment. It is possible for a sender to transmit 25 unreceived frames without even

bits: b15–b11	b10	b9	b8–b7	b6	b5–b3	b2–b0
Reserved	More data	Retry	ACK policy	SEC	Frame type	Protocol version

FIGURE 8.4 Frame control field format (from right to left).

knowing it. The immediate acknowledgment (Imm-ACK), type bit 01, requires that a user return an Imm-ACK to the sender after a successful reception. With a bit value of 10, the ACK policy type is set to a delayed acknowledgement (Dly ACK). What happens with this policy is that the receiver keeps accepting frames until requested to send a Dly ACK frame back to the sender. The last setting for the ACK policy field is 11, a Dly ACK request. Under this policy, the receiver returns to the sender an ACK frame, either an Imm-ACK or Dly ACK, depending upon the procedures established by the PNC.

Retry Field. This bit, b9, is set to 1 only when the current frame is a retransmission of an earlier frame. Otherwise, the value is always 0. This lets the DEV know whether the frame that it is receiving is an extension of previous frames or not. This way it does not mistake old information for new information.

More Data Field. The more data field, b10, is set to 0 if there is no more information that the DEV wishes to send during the rest of the channel time allocation (CTA) that was allotted. Otherwise, the bit is set to 1 to show that it is part of the extended beacon. This allows data that extend beyond the frame size to be sent in subsequent frames.

Reserved Field. These bits, b11–b15, are reserved for each vendor. Different type values and frame controls can be added as needed.

8.2.2 Individual Frames

The IEEE 802.15.3 MAC layer uses four main types of frames: beacon, ACK, command, and data frames.

Beacon. As stated before, the PNC uses the beacon to synchronize with the other devices in the piconet. The beacon is only sent from the PNC. The beacon is a signal that is sent out often in order to ensure piconet handling and security.

ACK Frames. The second type of frame is the ACK frame [1]. This frame explains what type of ACK should be expected.

Command Frames. The third type of frame is the command frame. Multiple commands can be sent in one frame. Examples of supported commands are PNC selection and hardware selection, association and disassociation between devices in the piconet, information request commands, repeater services, power management commands, device information, retransmission requests, and requests from devices to the PNC for GTS allocations.

Data Frames. The fourth type of frame is the data frame. These frames may contain encryption information for transmission of data between DEVs. For encrypted data the SEC bit is set to 1. The data can only be read by a DEV within that piconet that has the same security keys. Figure 8.3 shows how this superframe structure works.

8.2.3 Information Element

Figure 8.5*a* shows the information element (IE) format used in the data frame. The first octet is the ElementID. The second octet is the length of the payload of the IE. The following octets are for the actual payload. In this situation, the payload is the actual bits of data sent through the frames during the time allocated by the PNC.

Figure 8.5*b* shows how the CTA IE is formatted. Keep in mind that the length parameter of the IE only supports 255 octets of IE payload. If more information is sent, the PNC can split the CTA information into more than one CTA IE block. These CTA blocks are formatted according to Fig. 8.5*c*. In this frame, the first octet has the destination ID (DestID) of the device where the frames are sent.

The next octet indicates the source ID (SrcID) of the DEV to which the channel time is allocated. The third octet is the stream index, which indicates to the receiver the stream that corresponds to the appropriate CTA.

The next two octets contain the CTA location, which indicates the start time of the allocation. This is an offset value from the start of the beacon. The last octets of the block contain the CTA duration. The value stored here is the end time of the allocation. This can be calculated by adding the start time to the CTA duration.

Some of the other elements that the IE defines are the beacon source identifier (BSID), which is used to identify the piconet. The parent piconet IE

Octets: L_n	1	1
IE payload	Length (= L_n)	Element ID

(*a*)

Octets: 7	···	7	7	1	1
CTA block *n*	···	CTA block 2	CTA block 1	Length = (7 *n*)	Element ID

(*b*)

Octets: 2	2	1	1	1
CTA duration	CTA location	Stream index	SrcID	DestID

(*c*)

FIGURE 8.5 (*a*) IE format; (*b*) CTA IE frame format; (*c*) CTA block.

Octets: 3	1	1	1
DEV capabilities	DEV status	DEVID	DEV address

FIGURE 8.6 DEV association frame format.

contains the DEV address and the BSID in the parent piconet's beacon. The DEV association fields are formatted as shown in Fig. 8.6.

The DEV address contains the address of the DEV that corresponds to the DEVID. The DEVID is the identifier assigned to the DEV by the PNC. The DEV status declares whether or not the DEV is disassociated, indicated with a 0, or associated, indicated with a 1. The remaining octets, the DEV capabilities, announce the physical capabilities of the DEV.

PNC shutdown messages are also transmitted in the IE section of MAC frame formatting. These frames are used to show that the PNC is shutting down. Any parameters of the piconet that change are also indicated by the IE.

Figure 8.7 shows the format of the frame for a PNC handover. The first octet is for the ElementID. The next is the length of the transmission frame. The next eight octets give the address of the new PNC. These eight octets are followed by the PNC DEVID. The last octet of this frame is the handover beacon number. The handover beacon number is the beacon number of the first beacon to be sent by the new PNC.

The capabilities of a DEV are communicated in a frame similar to that of Fig. 8.8. The first section of the frame specifies the maximum number of devices that can be associated with the DEV. The next octet is the maximum number of channel time request blocks (CTRqBs) that can be handled. If the DEV is a non-PNC DEV, the value is zero. The following octets signify the transmission power of the DEV. The last octet, the PNC rating, gives the DEV capabilities, modes, security protocols, and so on. Supposedly, the most capable DEV will become the PNC using this frame format.

8.2.4 MAC Command Types

This section discusses some of the MAC command types and their capabilities. These can be seen in Table 50 in [1]. Before these are mentioned, it is important

Octets: 1	1	8	1	1
Handover beacon number	New PNC DEVID	New PNC address	Length (=10)	Element ID

FIGURE 8.7 PNC handover IE format.

Octets: 1	1	1	1
PNC rating	Max TX power	Max CTRqBs	Max associated DEVs

FIGURE 8.8 PNC capabilities field format (what information goes into PNC selection).

to remember that all DEV-to-DEV communications are peer to peer. In the table [1], these are shown with an "X" in the last two columns. If there are not any "X"s in either column, it is safe to assume that the command is PNC to DEV, which does not require a peer-to-peer connection.

Some of the commands described in Table 50 in [1] can be broken down into general categories. There are association and disassociation commands which are used to join a DEV to a piconet or separate a DEV from a piconet; security commands that establish security and privacy functions between the PNC and all DEVs of the piconet; PNC handover commands as discussed above; information request commands (commands to gain information about a DEV within the piconet); information announcement commands that carry information to the piconet and specify protocols to be followed to handle commands; commands used to request, modify, and terminate channel time within the channel time allocation period (CTAP); channel status commands used to provide and request remote DEV information to change transmission power based on channel conditions; power management commands used to manage DEV power consumption; and other special commands that include security messages and vendor specifications.

8.3 802.15.3 CHANNEL ACCESS

Channel time for the IEEE 802.15.3 proposed standard is divided into three major parts within each superframe: the beacon, a CAP, and a CTAP. The purpose of the beacon is to synchronize the piconet and allow the PNC to set ACK policies and security policies, among other management procedures. The CAP section of the superframe is used for commands and nonstream data. The third part, the CTAP, is used for sending asynchronous data streams (data streams in which there is no timing requirement for the transmission) and isochronous data streams (data streams that have set, uniform durations). This is shown in Fig. 8.9. Before any more details are discussed about the CAP and the CTAP and their roles in the channel access, it is important to look at the interframe spacing (IFS) of the superframe and its role in the superframe structure.

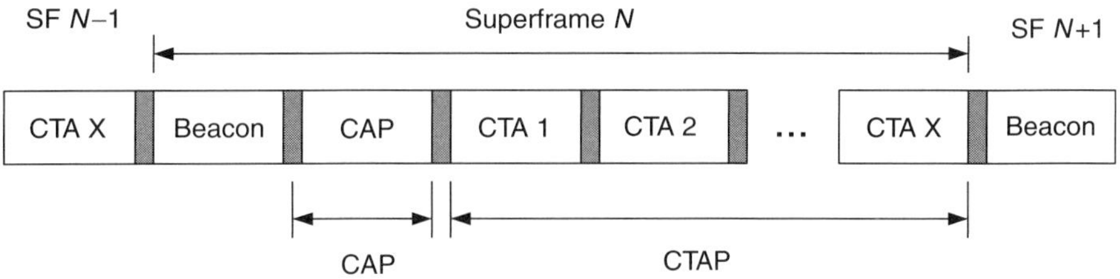

FIGURE 8.9 Superframe *N* representing basic superframe structure to be used for WPAN channel access and transmission.

8.3.1 Interframe Spacing

There are four different IFSs for the 802.15.3 WPAN MAC: backoff IFS (BIFS), short IFS (SIFS), minimum IFS (MIFS), and retransmission IFS (RIFS). All of these IFSs are device dependent. It is also important to know that the IFSs for the beacon are established during the CTA and assigned by the PNC.

BIFS. This type of IFS is used along with the backoff algorithm for channel access during the CAP. It is based on the wireless interface and is basically the measure of time needed before a DEV can begin transmitting data. Before transmission occurs, a BIFS must be sensed, meaning that the previous transmission has ceased or the PNC has acknowledged the request (REQ) by the DEV to begin transmission. At this point, the backoff algorithm begins. The BIFS is basically an idle time that elapses before a DEV can begin its transmissions.

SIFS. When an ACK for transmission is set to an Imm-ACK or Dly-ACK, the frames are preceded by an SIFS. This occurs if the ACK is requested. It is then followed by an SIFS and then the ACK will be given.

MIFS. This type of IFS is used in the CTA between the first frame and any sequential frames if the ACK policy is set to an Imm-ACK or Dly-ACK in the first frame.

RIFS. Any DEV that uses this type of IFS is sending transmissions in the CTAP. These IFSs are used for spacings between retransmissions. The PNC dictates and ensures the ACK and retransmission of the frames. The transmitting DEV must wait for an IFS in order for it to know when it can begin retransmission, according to the rules that have been established by the PNC.

8.3.2 Contention-Based Period

Channel access during the CAP occurs using different distribution styles. It uses CSMA/CA and a backoff algorithm. Before continuing to describe CAP access, it is important to remember that this is an optional period that is used for command and nonstream data. Any transmissions during this period are regulated by the PNC. The PNC controls the data type that can be sent as well as the commands that are transmitted in the CAP. The PNC does this by modifying the CAP control bits in the piconet mode field of the synchronization parameters field in the beacon frame of the superframe. The frames that are sent can only be of the type specified by the beacon for the current superframe. With the permission of the PNC, it is possible to change these bits from superframe to superframe. However, such changes are not permitted from frame to frame.

A CAP transmission cannot be extended from the CAP into the CTAP. If there is a policy of Imm-ACK in effect, there needs to be enough time to have the current frames, two SIFS times, and the Imm-ACK before the transmission time ceases. If there is not enough time, the transmission will never begin.

During the CAP, it is important to remember that the DEVs are all competing for channel access. As mentioned earlier, there are collision minimization techniques used in order to optimize channel access. One of these techniques requires a DEV that wants to transmit to wait an idle period of time before transmitting. The length of this time is dependent on the capabilities of the DEV PHY attributes. The PHY of each DEV also allows it to check to see if a channel is idle or not. Clear-channel assessment (CCA) is the part of the PHY that checks for this. After CCA checks the channel and finds it to be idle, the transmission may commence. This process of waiting for idle time is referred to as a backoff. Backoff is not applied to the beacon frame at the beginning of each superframe but can be applied to every frame in between. It is important to note that the backoff procedure does not apply to the Imm-ACK frame that is sent in accordance with an Imm-ACK policy set in the superframe. At this point there are several things the PNC can do. It can send an SIFS after the Imm-ACK of a frame or it can send an SIFS to a command following a frame that does not have an Imm-ACK policy. Whenever this happens, the PNC is not required to perform a backoff procedure.

8.3.2.1 Backoff Algorithm. The backoff procedure for channel access in the CAP uses four different variables: the retry counter, which is never set to a number greater than 3 or less than 0; a backoff window table, which has values used for determining transmission times; a PHY parameter called the pBackoffSlot which is dependent on the device's capability to sense the channel; and a backoff window random integer, which can be drawn from a distribution in the backoff window table. It is imperative that any random number based on a DEV PHY attribute hold no correlation with any other device. If for some reason there is no PHY attribute to derive a random number, a pseudo-random-number generator (PRNG) is used.

During an attempt to transmit data, the DEV waits for a BIFS duration while ascertaining that the medium is idle. This happens before any backoff procedures start. At the beginning of the CAP, the DEV can begin the backoff algorithm an SIFS time after the end of the beacon transmission. If for some reason the beacon is extended, the DEV must wait an SIFS time after the announce command is given. Then it may apply the backoff algorithm to the subsequent frames. The DEV will choose a random number by means explained earlier and keep a backoff counter. This counter is only decremented when the medium is idle for the pBackoffSlot time. The DEV retry counter is set to zero for the first attempt at frame transmission. If for some reason a channel is deemed to be busy or if the DEV is transmitting outside the CAP, the backoff procedures are suspended.

Once the channel is determined to be idle for a BIFS period, the countdown can be resumed. When a DEV backoff counter has reached zero after an idle BIFS period, the DEV is given permission to transmit. If for some reason there is not enough time for the transmission to be completed within the transmission duration given by the PNC, the counter is suspended because the transmission has halted. Other than these few exceptions, the counter is not reset from superframe to superframe. When a frame is able to transmit but does not receive the correct type of ACK, the retry counter for the DEV is set to no more than 3 and will allow this many retransmission attempts for the frames of data.

8.3.3 Contention-Free Period

The contention-free period, or CTAP, does not require that DEVs compete for channel access. Rather than this, the CTAP uses the TDMA method of giving CTA to DEVs. It gives the DEV a GTS in which the DEV can send its data. This enables the DEV to be concerned only with QoS aspects and also power saving because a fixed start time and duration are given. All of the CTAs for the current frames are broadcast in the beacon of each superframe. Although a DEV is given a set amount of time for the channel, it does not have to use all of the access that it has been given. During the CTAP, a collection of streaming data, commands, and asynchronous data types determined locally by the DEV can be transmitted. These types are usually prioritized and sent in a queue-type fashion. Depending upon the priority, a frame may be sent or may have to wait for other frames to be sent first.

There are two primary types of CTAs: dynamic and pseudostatic. A third type, the private CTA, will also be discussed below. All of these are indicated in the channel time request (CTRq) frame.

8.3.3.1 Dynamic CTA. With a dynamic CTA, the PNC gains the capability to move the CTAs around from superframe to superframe. This allows the PNC to optimize channel usage and not waste excessive amounts of channel time. This is done through the PNC beacon. The CTA parameters of the beacon change in order to let the DEVs that have membership in the piconet know what is taking place. Dynamic CTA is used for both asynchronous and isochronous data streams.

If a DEV were to request multiple CTAs, the PNC would have the capability to spread the CTAs over the superframe evenly using the dynamic aspects of the CTA. Besides CTA management, the PNC also has the job of allocating all synchronous power save (SPS) DEV CTAs at the beginning of the superframe. It is important for these to be at the beginning so power-saving capabilities are not implemented at inopportune times.

There are several exceptions to the rules of dynamic CTA. First, QoS streams needing multiple CTAs are located following the beacon. This is only the case if a CAP is used. If no CAP is used, then the CTA that follows the beacon is a management channel time allocation (MCTA) with the PNCID set as the SrcID. The other exception is that the CTA following a beacon must be a pseudostatic CTA.

8.3.3.2 Pseudostatic CTA. A pseudostatic CTA is used for isochronous streams of data. These are used when the PNC needs to change the duration of channel access. It can do this by changing the CTA blocks in the beacon. It can change these parameters from superframe to superframe.

In this case the PNC can overlay new and old time slots of the same pseudostatic CTA within a superframe. This ensures that frames do not collide. The PNC cannot create, but can move around, new stream indexes that overlap with old or changed time intervals for mMaxLostBeacon number of superframes. When the source DEV of a pseudostatic CTA receives a beacon with a new CTA, it automatically stops using the old CTA. However, a destination DEV that receives the new CTA can receive transmissions from the old and the new CTA.

While the PNC is changing the time interval for a pseudostatic CTA, it is possible for a destination DEV to miss transmissions for up to mMaxLostBeacon superframes. In order to avoid this loss of data reception, the destination DEV is able to go into the listen mode for the entire superframe transmission once it misses a beacon. This enables it to wait until the next beacon to synchronize reception of the transmission with the source DEV.

8.3.3.3 Private CTA. A private CTA is the same as a regular CTA, except for the fact that the DEVID for the requesting DEV is used for the source and destination for its transmission and reception. Private CTAs are not used for communication over the piconet. However, they do reserve channel time for other things. For example, the DEV could be reserving the time for a dependent piconet to which it belongs. Remember that any of the devices in neighboring piconets or child piconets still rely on the parent PNC for channel access.

8.3.3.4 More Data Bit. The more data bit is used in the frame control field. This data bit helps the PNC determine whether or not a DEV can switch into one of its power save (PS) modes. Setting it to 1 indicates to the destination DEV that the source DEV could be sending more frames in the CTA. If the bit is set to 0 and an Imm-ACK or Dly-ACK is required, this ensures that the source DEV does not retransmit a frame just because it did not receive an ACK when it was expecting one. The source DEV can also send a zero length frame with the more data bit set to 0 when it has no more frames in the CTA. If for some reason the more data bit is set during the CAP, it is ignored. It is sometimes used by the PNC in an extended beacon for announce commands.

8.3.3.5 Management Channel Time Allocations. MCTAs are the same as any other CTA, except for the fact that the PNCID is either the SrcID or DestID during the time allocation. The MCTA uses the TDMA method of channel access.

It is the prerogative of the PNC to send commands during these MCTAs instead of during the CAP. The PNC can do this only when the PHY attributes of the DEVs do not prohibit it. There may be one, many, or no MCTAs in a superframe. It is up to the PNC to determine how many MCTAs are appropriate in each superframe. If a CAP is not used, there must be at least one MCTA per every mMCTAAssocPeriod within the superframe.

A DEV that wants to request an MCTA must send a CTRq command to the PNC. Then the stream index for the superframe is set to the MCTA stream index. The CTA rate function is also used to set the DEVs desired interval for uplink MCTAs. The target ID is also set to the PNCID.

Open MCTA. An open MCTA occurs when the SrcID is the DestID. When a DEV is associated with a piconet, it can attempt to send a command frame to the PNC in the open MCTA, instead of the CAP. These types of MCTAs are very useful because the PNC is able to control large numbers of DEVs by using a small number of MCTAs. The reason for this is that all associated DEVs in the piconet can send commands.

Association MCTA. An MCTA with the UnassocID set as the SrcID is called an association MCTA. In one of these MCTAs, a DEV that is not a member of a piconet cannot send information REQs during this time allocation. Association REQs are only sent during this type of MCTA. If there are few DEVs in a piconet, it is sometimes more efficient to use association MCTAs to manage DEVs within the piconet.

Both open and association MCTAs use slotted-aloha channel access. This access is controlled by a contention window CW_a, which is maintained by each DEV. This contention window is derived from the number of retransmission attempts made by each DEV. The size of this contention window is shown in the following equation:

$$CW_a = \begin{cases} 256 & \text{for } 2^{a+1} \geq 256 \\ 2^{a+1} & \text{for } 2^{a+1} < 256 \end{cases} \tag{8.1}$$

The MCTA used for the ath retransmission is chosen by a uniformly distributed random integer value. This random integer value r_a should be different for every device. If, for some reason, the DEV does not possess a unique ID number, a PRNG should be used to obtain r_a. To gain channel access, the DEVs start counting r_a beginning with either MCTA from superframe to superframe. If there is not an Imm-ACK, then this indicates that there has been a failure with the previous attempts to gain access.

Once the first MCTA begins its access process, it is specified with $r=1$. If $r=r_a$, then the DEV is given access. Once this access has been acquired and an ACK has been set, a is set back to 0.

The MCTA allocation rate field is used to indicate three different things: the rate at which the PNC will be allocating MCTAs, whether it is using any special type of MCTA, and whether the rate will be guaranteed or not. This field helps the DEVs in the piconet to determine the length of time that it has to send a command to the PNC.

8.3.4 Channel Time Usage

In a given piconet, members, or DEVs, use CTRqs when they want to make a change to a CTA. Once a DEV sends one of these requests, the PNC remembers the settings of this request throughout the duration of superframes. If it is only one, it should be "a number of superframes" until the transmission has timed out or the DEV has issued a new CTRq. The CTAs in the CTAP are based on two things: the pending requests from DEVs and the currently available channel time. The beginning time of a CTA is referenced by the starting time of the beacon.

When a source DEV has any type of frame for a destination DEV, the source can send it directly during the CTA or it can use the CAP, according to the PNC policy, to send the frame. A source DEV can send to any destination DEV in any CTA that has been assigned to the source DEV. This can occur as long as the destination DEV receives the data in that CTA.

It is possible for a destination DEV to refuse source DEV data if the destination DEV is in the PS mode.

On the topic of transmission errors, it is also possible for a DEV in a given piconet not to receive a beacon from the PNC. If for some reason this happens, the DEV cannot transmit in the CAP or MCTA or dynamic CTA unless it is sending an ACK frame. DEVs with pseudostatic CTAs are allowed to transmit as long as the number of beacons that have been missed is not greater than or equal to mMaxLostBeacon. If for some reason this number has been exceeded, the DEV will stop transmitting and listen for the next beacon. The reason for this action is that the CTA may have been moved by the PNC at anytime.

We note a few additional details about channel time usage and CTAs. First, a DEV cannot extend its CTA transmission from one CTA to another. If there is not enough time to transmit, the DEV must stop transmitting right away. It is also true that the PNC can compute several superframes at a time. It repeats these until a change is made. At any point in time a PNC can make a change to a CTA. However, if a DEV needs to make a change to a CTA, it uses a stream modification procedure.

8.3.5 Guard Times

In a TDMA access architecture, guard times are used to keep adjacent CTA transmissions from colliding with each other. These guard times are basically the time between the end of one CTA and the start of the next sequential CTA. Figure 8.10 shows where guard times are allocated and how they are also separated by at least a SIFS time.

The guard time depends on the maximum drift between the DEV local time (time from their internal clock) and the ideal time (time that the PNC deploys throughout the piconet). The maximum drift can be calculated with the following equation:

$$\text{MaxDrift} = \frac{\text{clock accuracy (ppm)}}{1 \times 10^6} \times \text{interval} \tag{8.2}$$

The PNC calculates MaxDrift by using the superframe duration and clock accuracy. The accuracy of its clock falls under the PHY attributes of each DEV. Propagation delay has been omitted from this equation, but since the DEVs in the piconet are limited to a 33–66-ns propagation delay, this factor is ignored in the guard time calculation since it is minimal.

Within the piconet, it is the responsibility of the PNC to calculate the single worst case guard times for all CTAs in a superframe. Usually this is based on the type of CTA, dynamic or pseudostatic, and its position in the superframe. It is important to note that pseudostatic CTAs require more guard time because, as was mentioned earlier, pseudostatic CTAs allow transmissions even up to the mMaxLostBeacon. The PNC makes sure that there is enough guard time

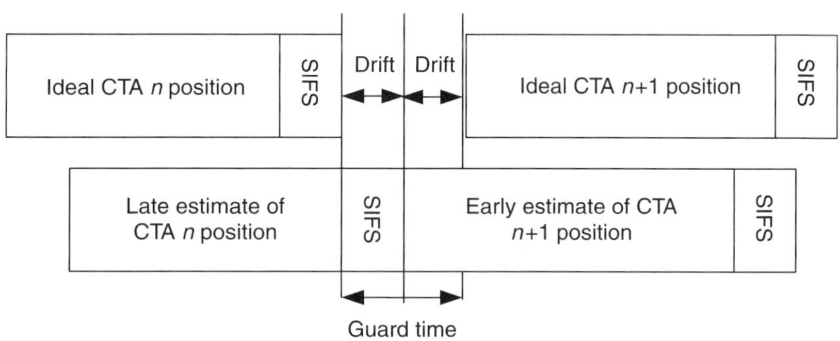

FIGURE 8.10 Reason for guard times (to avoid colliding of ideal and estimated time frames).

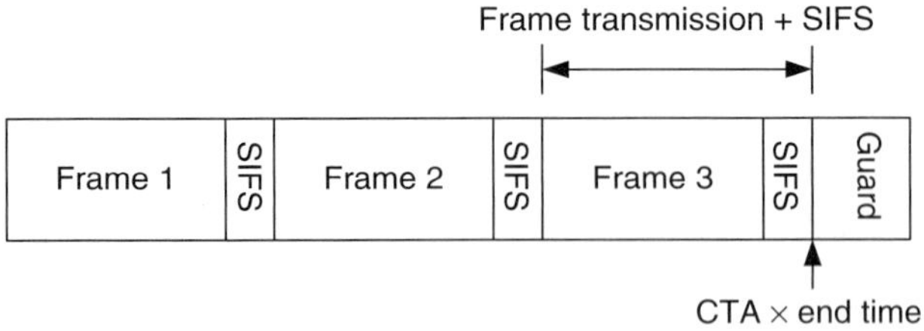

FIGURE 8.11 CTA and guard time at end with No-ACK.

between a given CTA and the following one. This is shown in Fig. 8.11. A DEV that is transmitting in a CTA actually starts transmitting the preamble at the calculated point. The CTA will start according to its own local clock. If there is a No-ACK or Dly-ACK, the transmission DEV ensures that there is enough time to transmit and for an SIFS. Likewise, for an Imm-ACK policy, the DEV needs to make sure that there is enough time for another ACK and an SIFS. This is seen in Fig. 8.12.

It is also the job of the PNC to make sure that there is enough guard time between the last CTA in the superframe and the beacon of the next superframe. This is seen in Fig. 8.13.

Once the PNC starts its preamble, the other DEVs in the piconet resynchronize their clocks. Because clocks in different DEVs may be faster or slower than others, each DEV must account for drift when receiving transmissions. This is done at the start of the beacon, CAP, or CTAP.

8.3.6 Channel Time Requests

It is up to each DEV in the piconet to send a request to the PNC for channel time. If the amount of time is known, the DEV should also include the frame transmission time and an ACK transmission frame, if an ACK policy is used, and one MIFS or SIFS per frame and/or ACK. This is seen in Fig. 8.14. Figure 8.15 shows a CTRq when a No-ACK policy is being used. It is important to remember that it is also possible for a CTRq time in the CTA

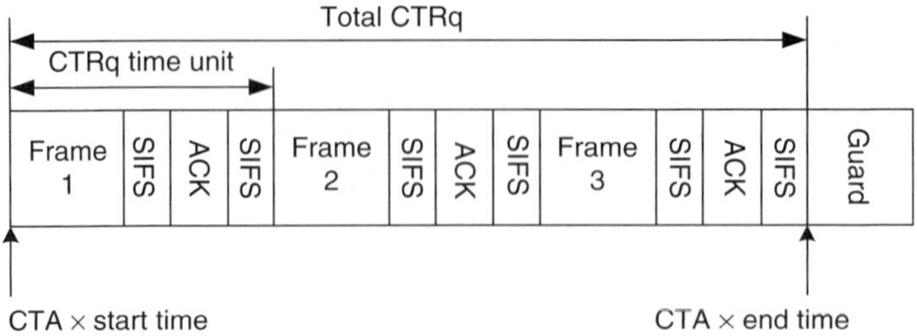

FIGURE 8.12 SIFS, ACK, and guard time for Imm-ACK.

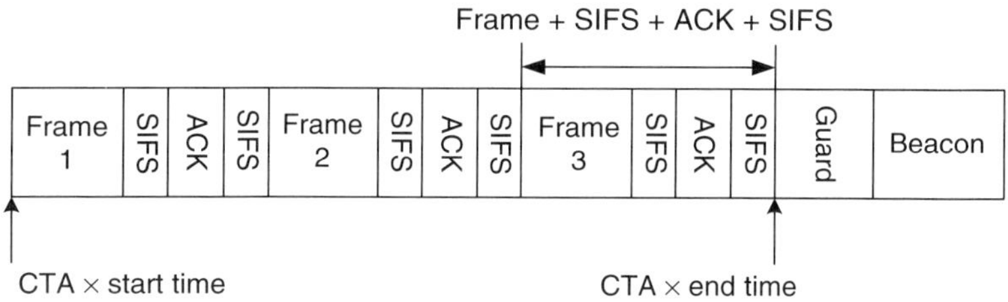

FIGURE 8.13 Guard frames between superframes.

to cover more than one frame. If for some reason the frame size is not known, the DEV is responsible for the channel time amount for the CTA REQ.

8.3.7 Channel Time Management

For channel time management (CTM) within a HR-WPAN, there must be creation, modification, and ending of isochronous data streams between two or more DEVs. There must also be time reservation and termination of asynchronous CT for the exchange of asynchronous data between two or more DEVs.

Isochronous. Isochronous streams are accomplished in a piconet through negotiation between the source DEV and the PNC using a CTRq and a response command. Only the source DEV or the PNC can modify an isochronous stream. For isochronous stream termination, the PNC, source DEV, or destination DEV can terminate the established stream. For multicasting or broadcasting, only the source DEV or PNC can terminate streams.

Asynchronous. There are only two methods for obtaining asynchronous channel time. The first is to request a single CTA for multiple target DEVs. The second is to request individual CTAs for each of the target DEVs. To modify asynchronous channel time, the target and the source

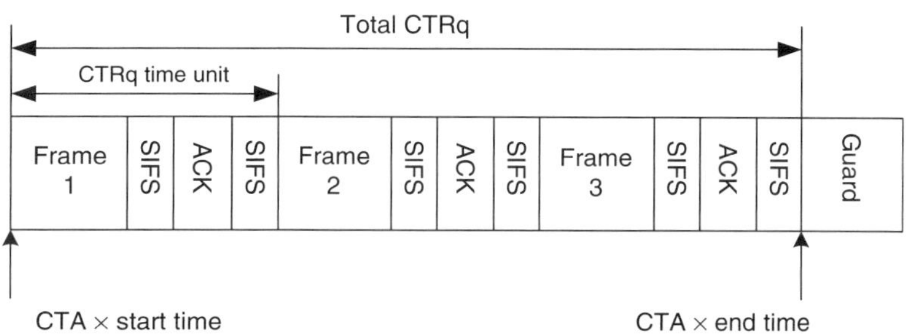

FIGURE 8.14 CTRqs for frames with Imm-ACK.

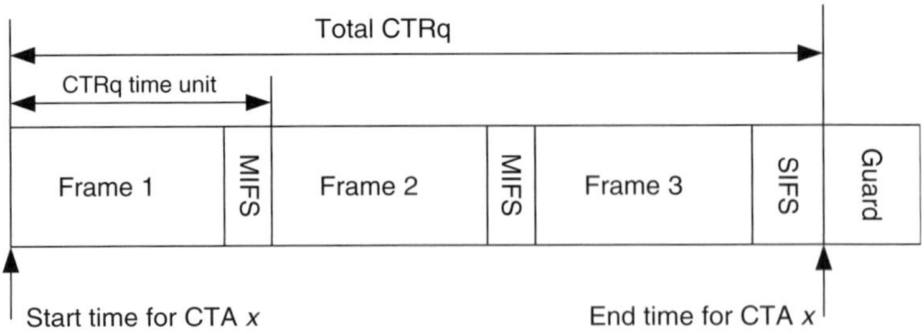

FIGURE 8.15 CTRqs between superframes with No-ACK.

DEVs can send a CTRq command. To terminate an asynchronous CTA, the source DEV must send a request to the PNC. The PNC then sends an ACK that the CTA has expired.

8.3.8 Synchronization

Synchronization concerns an entire piconet. All of the DEVs within the piconet must be synchronized with that piconet's PNC clock. Any child or neighbor piconet must also be synchronized to the parent PNC clock. The information necessary to do this is sent in the beacon, which is at the very beginning of every superframe. Looking at Fig. 8.16, we can see that each DEV sets its clock to zero in the preamble of the beacon frame. The synchronization of the superframe helps to synchronize the DEVs of the piconet. If for some reason the DEV does not hear the beacon, its clock is reset to zero until it receives the beginning of another beacon preamble.

As we can see in Fig. 8.16, there is sometimes an extended beacon. This occurs when the PNC decides that the beacon frame is too large and splits its information.

The PNC sends announce commands with the SrcID and the DestID set to the PNCID and the BestID, respectively. All DEVs in the piconet use the

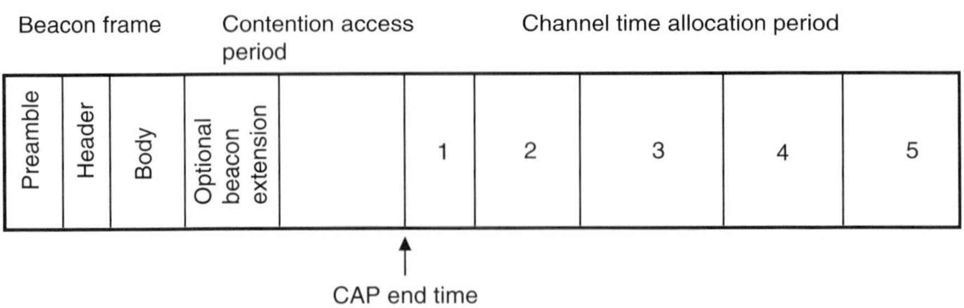

FIGURE 8.16 Piconet timing in comparison to beacon. The frame is divided into three main parts: beacon, CAP, and CTAP.

beacon time to start their transmissions. It is also important to realize that all synchronizations through the beacon come from the PNC.

8.3.9 Others

ACK and Retransmission. As stated above, there are three kinds of ACKs: Imm-ACK, Dly-ACK, and No-ACK. In addition, the functionality of the ACKs is dependent upon which bit sequence has been used in the ACK policy field of the superframe. If, for some reason, during CTAs within CTAPs an ACK is expected but not received, the source DEV will start a retransmission of the frame. This only happens if there is enough CTA for the entire frame to be resent. If there is not enough time, another CTA can be requested from the PNC.

DEV Discovery. Using the PNC information request, channel status request, channel status response, and probe request commands gives DEVs within a piconet many capabilities. They can request information about another or all DEVs in the piconet. They can obtain frequency and channel information. With this information, they can request peer-to-peer communication with a device from the PNC. It is also possible for a DEV to request its own or another DEV's link information in order to reevaluate its connection. These commands also help the DEV to determine which power management mode to select or which another device has chosen. It is also possible to obtain information about what channel a DEV is using. This information enables a DEV to decide which commands or requests can be sent to the other DEVs.

Dynamics of the Piconet. Details of the dynamic aspects of a piconet are beyond the scope of this chapter but deserve mentioning. A PNC can change its beacon position within the superframe but retain the superframe duration. However, the PNC can also change the duration. These things may seem somewhat arbitrary but can result in minimizing wasted frame sequences, downtime between DEV requests, and ACK between the DEV and the PNC.

8.4 PICONET OPERATIONS

In the 802.15.3 standard [1], piconet operations begin when a controller-capable DEV undertakes the role of the PNC for the network. There are only three different types of piconets under this standard: the independent piconet, parent piconet, and dependent piconet. Within the dependent piconet, there are two types, the child and neighbor.

Independent piconets are usually autonomous piconets that do not have parents or dependent piconets attached to them. Parent piconets are piconets

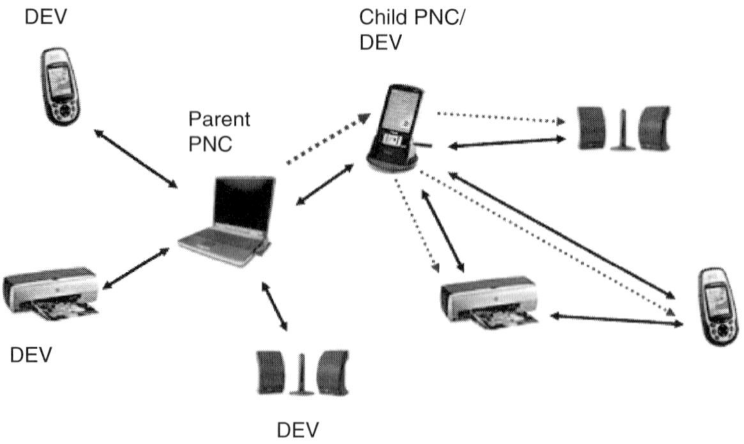

FIGURE 8.17 Child piconet and its relation to parent piconet.

on which dependents rely. The child piconet is dependent when the child PNC is a member of the parent piconet. This relationship can be seen in Fig. 8.17. In a neighbor piconet, the neighbor PNC is not a member of the parent piconet but requires private CTA from the parent for its operations. Figure 8.18 gives an example of a neighbor piconet.

DEVs use passive scanning to search for other DEVs for a period of time determined by the MAC layer management entity (MLME)-SCAN. Request. During this scanning, DEVs look for beacons from a PNC. This takes place to determine whether a piconet has been established and whether the DEV can join the piconet. While this scanning is taking place, information is being gathered, for example, channel information. This information is stored in a

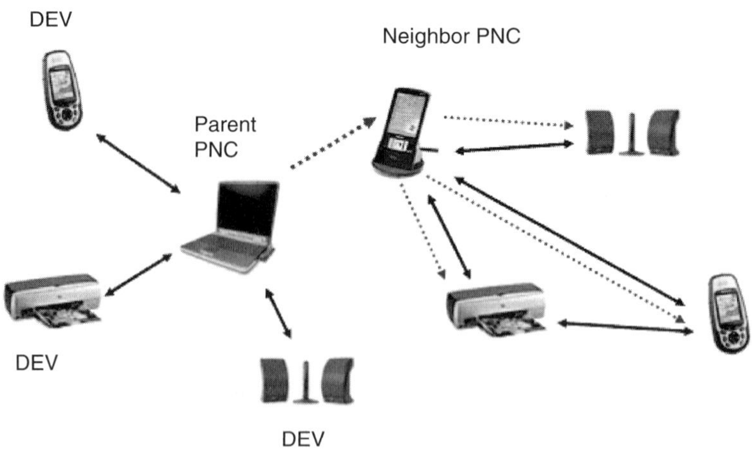

FIGURE 8.18 Neighbor piconet and communication paths.

TABLE 8.2 An Example of Inventory

	Information	Preferred value
1	PNC Des-Mode bit in PNC capabilities field	PNC Des-Mode = 1
2	SEC bit in PNC capabilities field	SEC = 1
3	Power source (PSRC) bit in PNC capabilities field	PSRC = 1
4	Maximum associated DEVs	Higher value preferred
5	Maximum CTRqBs	Higher value preferred
6	Transmitter power level (PHY dependent)	Higher value preferred
7	Maximum PHY rate (PHY dependent)	Higher value preferred
8	DEV address	Higher value preferred

channel rating list to be used at a later time to decide which channels are open or to make a piconet association. The results of these searches also provide information about any parent, child, neighbor, or 802.15.3 piconets and provide a complete inventory of pertinent items. Table 8.2 shows an example of this table and the information stored within.

If a frame is found while this searching and scanning are taking place, the DEV will remain on that channel for a period of time to see if there is a beacon on that particular channel. If there is a beacon, association procedures take place.

8.4.1 Starting a Piconet

It is possible for DEVs to be implemented without providing PNC support. In this case, these DEVs would not be able to establish a piconet themselves but only work in conjunction with a PNC-capable DEV.

Once a DEV has been instructed to start a piconet, part of the MLME-START.Request command, it tries to start its own piconet instead of trying to associate with an existing one. Before this process begins, though, a device management entity (DME) should have recently completed a scan so that the DEV can choose its channel for operation. Once again, this selection is based on a table similar to Table 8.2.

There are several steps in starting a piconet. After the PNC is chosen, the DEV chooses the channel with the least amount of interference. This is done using the information collected in the scan. The channels are ranked from clearest to the ones with the most noise. After a channel has been chosen, a beacon is broadcast to the other DEVs associated with the piconet.

Once a piconet has been established, the PNC listens periodically and scans the channel for other piconets. The rationale for this is to optimize piconet performance. This also allows for improvements on two coexisting piconets. The PNC can also improve the original piconet by changing transmissions to another channel, reducing transmission power, or by becoming a child or neighbor of the other piconet. This process is much like combining two binary trees.

The PNC will remain the controller of the piconet until is chooses to leave or a more capable DEV initiates the PNC handover procedure.

8.4.1.1 PNC Handover. When the controller's responsibilities move from the PNC to another DEV, it is the PNC's job to choose the most capable DEV. The PNC capability bit is used to determine the rank of a DEV's capabilities. This information, derived from the initial scanning, is what the PNC uses to determine the most qualified DEV. The new PNC must also be a member of the piconet.

To initiate the handover sequence, the PNC sends a PNC handover request command to the DEV that it has chosen.

When the piconet is a dependent piconet and the DEV is a current PNC of a child or neighbor, it can refuse to be the new PNC. This can be done by sending a reason code such as "Handover refused, unable to act as PNC for more than one piconet" [1].

If the PNC and DEV are members of the same dependent piconet, the DEV will agree to be the PNC. If it is unable to join the parent piconet, it will send a reason stating why. If the DEV is a member of an independent piconet, the DEV accepts and prepares to be updated with the piconet information. Once the handover procedure begins, it will not stop unless the timer expires or a PNC shutdown procedure is taking place.

The PNC allocates channel time to the new PNC so that it can transfer information about the piconet to the new PNC. The PNC sends a PNC information request command followed by a PNC handover information command. After the PNC handover information is successfully sent, a PS set information response command is sent. However, if the PNC does not have any CTRqBs to transfer, it is not necessary to send PNC handover information. In one of these information transfers, all information except asynchronous CTRqBs and CTA locations will be sent.

Once the new PNC has all of the necessary information, it signals that it is ready to take over control of the piconet. The old PNC acknowledges this by placing a PNC handover IE in the last beacon that it sends so that the rest of the piconet recognizes this transfer of management.

After the new PNC receives an ACK to the PNC handover response command, it broadcasts its first beacon to the piconet. The beacon should be sent at the same time as the previous PNC would have sent it but may be a little off due to inconsistencies with clocks between DEVs. This will be corrected, however, once the DEVs in the piconet synchronize with the PNC.

It is the responsibility of the new PNC to make sure that the handover announcement complies with all of the beacon rules for 802.15.3 beacons.

One thing to note is that a PNC handover does not cancel an isochronous data connection. However, since the new PNC may require considerable CTRqB data, traffic flow could be interrupted. This depends on the flow and channel clarity of the piconet at that given time.

When a DEV joins a piconet, it is checked to see if it is more PNC capable than the current PNC. If it is, handover procedures are initiated.

Table 8.2 shows the fields and their priority in the PNC selection process. Des-Mode has the highest priority. A DEV will be considered as the PNC only if this bit is set. Bit two of the same table shows the SEC bit. If this bit is set, then during the handover process the old and new PNCs must follow all of the security protocols laid out by the piconet.

8.4.1.2 Dependent PNC Handover. The PNC handover process for dependent DEVs is similar to a regular handover. If the newly designated PNC is not a member of the parent piconet, the DEV can join the piconet as a neighbor or member.

While the DEV is associating itself with the piconet, the current PNC can send the DEV information about the piconet. The target DEV can also request piconet security information. Since both of these transmissions can occur at the same time, it is more important for the DEV to be associated. Because of this priority, there are no conflicts between these transmissions: One occurs in the time reserved for the dependent piconet and the other occurs in the time set aside for the parent piconet.

After all information has been transferred and the new DEV has associated with the parent piconet, the DEV sends a PNC handover response command to the PNC with the result code value set to the DEVID. At this time the DEV takes over the management of the piconet and then sends its first beacon, starting the new superframe sequence.

Although this seems like a foolproof method, there are several points at which the handover process can fail. The current PNC can cancel the process at any time. If the DEV fails to join the piconet as a member or a neighbor, the handover can also fail. If the process is unsuccessful or is canceled, the DEV may disassociate itself from the piconet. This might happen, for example, to free up resources.

8.4.1.2.1 Child. If a DEV in a piconet wants to form a child piconet, it can use a CTRq command to ask for a private CTA In this case, the SrcID and the DestID are the same. The PNC recognizes a request to form a child piconet when the SrcID and TrgtID fields are set to the DEVID of the source DEV, the stream index field is set to zero, and the PM CTRq type field is set to active. In this case, depending on the available resources and security of the piconet, the PNC may allocate a private CTA for the child.

A message of "request denied" is returned to any DEV that is denied unless there was insufficient CTA or the DEV was unable to acquire private CTA. However, once the DEV is a new PNC, it begins sending its beacon in the allocated CTA. In order to distinguish itself from the parent PNC, the child uses a PNID that is different from the parent's PNID.

It is important to note that this standard does not provide for direct connections between members of the parent and child piconets. However,

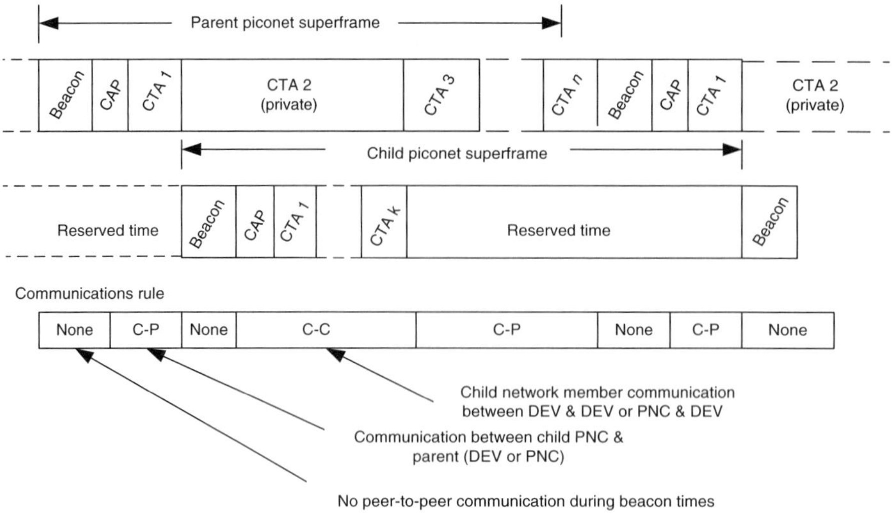

FIGURE 8.19 Parent and child piconet superframe relationship.

communications can be sent indirectly through the child PNC since it is a member of the parent piconet.

Figure 8.19 shows the correlation between the parent and child beacons. At the bottom of this figure is a set of communication rules by which it can be seen how communication between the parent and child piconets is possible.

Except for a private CTA, which must come from the parent PNC, the child is its own entity. It has its own rules for security, communication, etc.

8.4.1.2.2 Neighbor. If a DEV has scanned and found no open channels, it can then try to start a neighbor piconet. A neighbor piconet can exist on the same channel as the original piconet.

When the process is initiated, a PNC-capable DEV tries to be associated with the parent piconet. However, a neighbor PNC does not have to become associated with the parent PNC.

If the request to start a neighbor piconet is accepted, the correct codes are set and the neighbor PNC transmits its beacon. If the request is rejected, a reason code is given. Based on the reason, it is possible for the DEV to try the request at a different time in the future. If neighbor piconets are not supported by the parent PNC, then this process cannot take place until another piconet is found.

Once the request is accepted and the beacon is sent, the new neighbor PNC sends a CTRq to the parent PNC. If the parent PNC determines that there is enough channel time available, it will allocate private CTA to the neighbor PNC. At this time, the neighbor PNC can begin sending its beacon.

If this occurs in an 802.15.3 piconet, private CTA should be included in the neighbor PNC beacon, not so that the neighbor piconet can send information

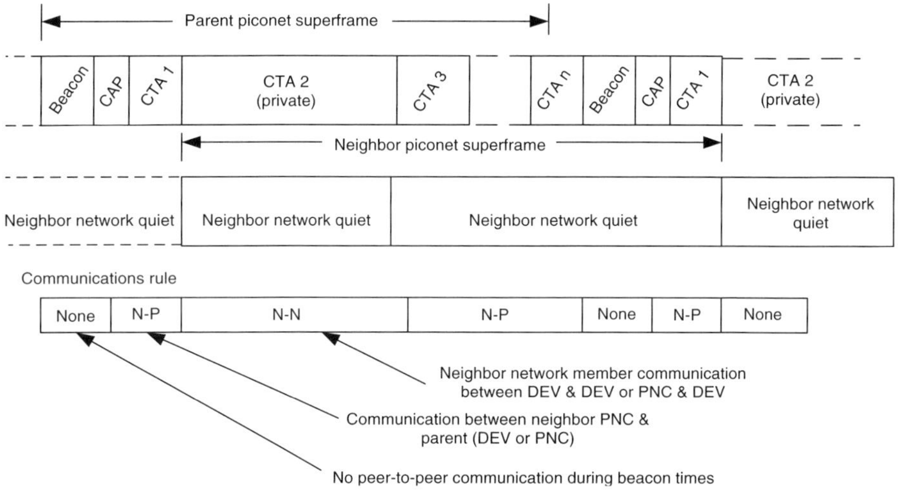

FIGURE 8.20 Relationship between parent and neighbor piconet superframe.

to the parent PNC, but for the opposite reason. If the neighbor is not an 802.15.3 piconet, there need not be any type of private CTA for the parent PNC. In any case, the neighbor PNC is not allowed any transmission outside of its CTA. Figure 8.20 shows the relationship between the parent and neighbor superframes. If the neighbor reaches the maximum number of lost beacons, it will stop its transmissions and wait and listen for a beacon before continuing.

Since the neighbor PNC is not a member of the parent piconet, it can only send association/disassociation request commands, CTRqs, vendor-specific commands, security commands, probes, and Imm-ACK frames.

8.4.2 Stopping Piconets

When a PNC needs to leave a piconet and there are no successors to take its place, it must shut down all piconet operations.

8.4.2.1 Independent. Before shutting down a parent piconet, a PNC makes sure that all of its shutdown announcements are in line with beacon announcement policies. If for some reason there is not enough time, an exception is made.

If the parent PNC that is shutting down has a dependent PNC attached to it, it may request that the dependent PNC continue PNC operations. In this case any of the other dependent piconets must join that child's piconet, stop operating, or change channels.

Once this happens, the dependent PNC is no longer dependent and will change this bit in its beacon frame to show its independence.

8.4.2.2 Parent Stopping Child/Neighbor. When a parent PNC wants to terminate a child piconet, it can simply cancel the data stream to the child. You can see this in Fig. 8.17 (the stream from the parent PNC to the child PNC). If the parent PNC wants to stop a neighbor, it must go through a disassociation process from the neighbor PNC.

In each case, the PNCs must initiate shutdown procedures, change their channels, or join other piconets. All of these processes have a time limit and must be done within the allocated channel time. If the shutdown does not complete successfully, the parent PNC can remove the CTA and the DEVs then time out, ceasing to function.

8.4.2.3 Child/Neighbor Stopping Child/Neighbor. PNCs of both types of dependent piconets can cease operations of their piconets and then inform the parent PNC that they do not require any more CTA. At this time, they would either be disassociated from the piconet or terminate the piconet.

8.5 CONCLUSION

This survey of the IEEE 802.15.3 standard has provided a detailed look at piconet topology and MAC layer structure. Information about piconets, transmissions, and MAC layer frames and functionality has been given. Channel access for the HR-WPAN has been subdivided into two major parts: the CAP and the CTAP. The DEV sets many device-dependent variables in its access assignments. The main access mechanisms used in this standard are CSMA, TDMA, and other backoff and distributed random sequences for minimizing REQ and transmission collisions. All of these mechanisms are used to optimize wireless access for this HR-WPAN. Piconet operations for the IEEE 802.15.3 standard seem to work fine for this network structure. It is not clear how this type of network would perform on a larger scale, but since this network was designed for a 10-m area, it really does not matter.

REFERENCES

1. LAN/MAN Standards Committee of the IEEE Computer Society. Part 15.3: "Wireless medium access control (MAC) and physical layer (PHY) specifications for high rate wireless personal area networks (WPAN)," IEEE, New York, Feb. 2003.
2. Y. Xiao, "MAC layer issues and throughput analysis for the IEEE 802.15.3a UWB," *Dynamics of Continuous, Discrete and Impulsive Systems—An International Journal for Theory and Applications (Series B) Special Issue: Ultra-Wideband (UWB) Wireless Communications* 12(3), 443–462 (2005).
3. Y. Xiao and X. Shen, "Adaptive ACK schemes of the IEEE 802.15.3 MAC for the ultra-wideband system," *Proc. of IEEE Consumer Communications and Networking Conference 2006*.
4. Y. Xiao, X. Shen, and H. Jiang, "Optimal ACK schemes of the IEEE 802.15.3 MAC for the ultra-wideband system," *IEEE J. Sel. Areas Commun.* 24(4), 836–842 (2006).

CHAPTER 9

POWER MANAGEMENT AND SECURITY OF IEEE 802.15.3 WIRELESS PANs

YANG XIAO, MICHAEL J. PLYLER, BO SUN, and YI PAN

9.1 INTRODUCTION

Before we look at the power management (PM) and security aspects of the MAC layer in IEEE 802.15.3 wireless personal area networks (WPANs) [1], we need to understand what type of applications will be used with this standard. For example, using a high rate wireless connection, we would be able to stream audio and video content; establish interactive audio and video; and output data to personal digital assistants (PDAs), personal computers (PCs), printers, projectors, and so on. At the same time, we can still have the still image and video capabilities of existing wireless networks. The main focus of the IEEE 802.15.3 standard [1–6] is device management, quality of service, and security. This chapter surveys the PM and security aspects of IEEE 802.15.3 WPANs.

It is obvious that wireless networks face security challenges. A piconet is a WPAN, so it also faces similar problems. In the context of security, the high rate WPAN (HR-WPAN) supports two different modes: no security and strong cryptography. It uses a 128-bit advanced encryption standard (AES) security suite.

9.1.1 Mechanisms

The security mechanisms provided with this standard include security memberships and key establishments that the piconet controller (PNC) assigns to the piconet as well as key transports. Keys that are transmitted from one device (DEV) to another are encrypted. Data encryption helps the information within the piconet to not be read by any outside parties or devices.

Emerging Wireless LANs, Wireless PANs, and Wireless MANs. Edited by Y. Xiao and Y. Pan

Encrypted data can only be decrypted with cipher keys. Data can also only be modified by DEVs with the cryptographic key. The beacon transmitted to the DEVs of the piconet provides evidence that the beacon itself, from the PNC, is protected. To also help with retransmission of old frames, there is a strict time token included in the bits that hold the beacon.

9.1.2 Modes

There are basically two modes of security for the HR-WPAN.

Mode 0. If a DEV is operating in this mode, it cannot perform cryptographic operations on any of the MAC frames. It discards any frames with the secure mode (SEC) bit set to 1. It can only read frames with the SEC bit set to 0.

Mode 1. If a DEV is operating in this mode, it can perform any operation, cryptographic or otherwise, to the MAC frame it may receive. Note that a DEV in mode 1 will only read frames from other DEVs that have the frame with cryptography. The DEV receives pertinent information about encoding and decoding from the PNC.

9.1.3 Support

The security mechanisms support that a HR-WPAN receives flows from all aspects of the piconet. During PNC handovers the integrity and security are upheld for the piconet. The support also extends to the joining of a DEV to a piconet. If a DEV does not meet the security protocol set by the PNC, the DEV will not be given membership to the PNC. If by chance a DEV is given membership to the piconet, the membership list for the piconet is updated with the DEVID and security keys.

9.1.4 Protocol/Specifications

Details of the protocol for security include cryptographic components and headers for the frames. Keep in mind that in security mode 1 all frames are transmitted with immediate acknowledgment (Imm-ACK) policies unless otherwise stated. When a DEV is establishing a security relationship, it can send requests and its security information to other DEVs.

If for some reason there is a change of data key within the piconet, the PNC broadcasts the new data key to all members of the piconet.

In security mode 1, there is the capability of using symmetric-key security operations. Symmetric cryptographic operations include secure beacon integrity code generation, secure command integrity code generation, data integrity code generation, key encryption operations, and data encryption operations. These operations have been mentioned in order to show that security is a priority in the HR-WPAN.

9.2 POWER MANAGEMENT

Four PM modes are defined in tIEEE 802.15.3 [1]: active, asynchronous power save (APS), piconet synchronized power save (PSPS), and device synchronized power save (DSPS). To establish membership into the piconet, a DEV must be in the active mode. The last three modes will be referred to as power save (PS) modes. In any PM mode, there are two states in which each DEV can exist. The DEV can be in an awake state, which signifies that the DEV is either transmitting or receiving data. The DEV can also be in a sleep state. In the sleep state, the DEV is neither transmitting nor receiving data of any kind. Table 9.1 shows the different PM modes and the states that are possible for each one.

The PM mode of a DEV does not matter for it to be able to enter either of the two states. It can enter a sleep state during channel time allocation (CTA) when it is not the source or destination of any data frames that are being or will be sent. It can also enter the awake state whenever it is in a PS mode.

The wake beacon for a DEV is defined the same as the PNC—a system wake beacon for all DEVs in the PSPS mode and a wake beacon for the DSPS set for any DEV in the DSPS mode. A DEV that is in the DSPS mode is able to have multiple DSPS sets. In that case, it is possible for all of these sets to have their own wake beacon.

A DEV in the APS mode works slightly different than the other two PS modes. A wake beacon for an APS DEV occurs at points in time that the individual DEV may determine. An APS DEV wake beacon is unknown to the PNC and any other DEVs that have membership in the piconet. Something else to note is that the wake beacon of a DEV in the APS mode is not periodic. This is different than DSPS and PSPS modes. The wake beacon is only guaranteed to happen once for every association timeout period (ATP) for that particular DEV.

Before considering the PS modes, it is important to note that a PNC can support one APS set and one PSPS set. The PNC can also support one DSPS set.

TABLE 9.1 Four Power Settings for DEVs in IEEE 802.15.3 and Two States that They Can Have

Superframe Portion	Active	APS	PSPS	DSPS
Beacon	Awake	Awake	Awake	Awake
CAP	Awake	May sleep	Awake	Awake
CTA with BcstID as DestID	Awake	May sleep	Awake	Awake
CTA with McstID as DestID	May sleep	May sleep	May sleep	May sleep
CTA with DEV as SrcID or DestID	Awake	May sleep	Awake	Awake
Other CTAs and unallocated time	May sleep	May sleep	May sleep	May sleep

9.2.1 APS Mode

The APS mode cannot be used in combination with any other PS modes. The APS mode allows a DEV to conserve energy or battery life. It does this by allowing the DEV to be in a sleep state for long periods of time. Since this mode does not rely on the PNC to dictate its parameters, it is the responsibility of the APS DEV to communicate with the PNC for the purposes of keeping its membership within the piconet before the end of one of the ATPs.

In the APS mode, a DEV is not required to listen to any beacons or data traffic over the piconet, or from the PNC, until the DEV changes to an active state. It can also achieve this by changing to another PS mode. It can change modes with the PM mode change command. You can see this format in Fig. 9.1. Unlike the other two PS modes, APS cannot use the synchronous power save (SPS) configuration request command to set the APS index. However, in order for the other DEVs in the piconet to know that a DEV is in the APS mode, the PS status information element (IE) in the beacon can be set with the DEVID bitmap set to the DEVID. The APS PS index is set to 0 to signify the DEV is an APS DEV.

As was mentioned earlier, a DEV must send at least one acknowledgment (ACK) frame during the ATP in order to maintain membership within the piconet. It is important that the PNC consider this fact when assigning management channel time allocation (MCTA) if there is not enough channel time in the contention access period (CAP). Remember that the CAP is available for sending commands.

Before a DEV can enter the APS mode, it has to send a PM mode change command to the PNC. The mode field must be set to APS. The DEV must receive an ACK before assuming it is in the APS mode. Once the PNC has received the command of a mode change, it sets the DEVID bitmap field in the PS status IE to the appropriate setting. See Fig. 9.2, which shows the continued wake beacon (CWB) IE format. The PNC will then cancel all data streams and asynchronous data allocations where the DEV was a source or destination.

Although a PS set index is set for a DEV in the APS mode, each DEV in this mode acts independently from one another. All APS DEVs are still members of the piconet; they are just not members of PS sets. This works differently for DEVs that have memberships in DSPS and PSPS PS sets.

A DEV can leave the APS mode by sending a PM mode change command to the PNC with the field set to active. Once the APS DEV sends this command, it

Octets: 1	2	2
PM mode	Length (=1)	Command type

FIGURE 9.1 PM mode change command format for superframe.

Octets: 1–32	1	1	1
DEVID bitmap	Start DEVID	Length (=2–32)	Element ID

FIGURE 9.2 CWB IE format for superframe.

is considered to be an active DEV even without receiving an ACK from the PNC regarding the mode change. However, when the next wake beacon is sent and the DEV mode has not been changed, the DEV will send another PM mode change command until the mode for that DEV is changed by the PNC.

9.2.2 PSPS Mode

Any DEV in this mode must listen to all of the system wake beacons. These wake beacons are announced by the PNC. Any DEV in the PSPS mode must have the PS set index set to 1. DEVs in the PSPS mode must also be in the awake state during any system wake superframes. The actual wake beacon for the PSPS DEV is determined by the PNC. If for some reason the system wake beacon is not correctly received by the PSPS DEV, the DEV must stay in the awake state until all of the frames are correctly received.

If for some reason there are no DEVs in the PSPS mode in a given piconet, the PNC can omit the PS status IE for the PS set index. When something like this occurs within the piconet, every beacon is a system wake beacon. This happens for information announcements that the PNC might send to the DEVs regarding the policies, operations, and security of the piconet.

If a DEV decides to use PSPS, it must synchronize with the system wake beacon before entering into a sleep state. At some point in time the DEV will need to send commands to the PNC. Therefore, the PNC needs to take this into consideration when making MCTAs, especially when there is no CAP time available.

A DEV that will be using the PSPS mode will need to send the SPS configuration request command to the PNC. The DEV will do this with the operation type field set to the join state. To use the PSPS mode, the DEV must also set the SPS set index to 1 and the wake beacon interval to the desired time. Once the PNC receives the command to change the PS configuration, it will acknowledge by setting the SPS configuration command in the next wake beacon.

Before entering the PSPS mode, a DEV needs to send a PM mode change command to the PNC. Once the PNC receives this command from the PSPS DEV, all superrate streams are canceled where the DEV is the destination. The DEVID bitmap and PS structure IE fields are also changed to the appropriate settings.

The PNC uses the information that it has in the wake beacon interval field from all of the participating PSPS DEVs in the piconet to determine the system wake beacon interval. Although this information is used, the system wake beacon itself does not have to correspond to any of the PSPS DEV wake beacon intervals. If for some reason the wake beacon interval changes, the DEV can resend the SPS configuration request command as many times as needed.

If a DEV does not want to continue using the PSPS mode, it can send an SPS configuration request command with the operation type set to the leave state. The PS set index is also changed to 1. At this time, the DEV has left the PSPS set. Once a DEV chooses to change to the active mode, it will send a change mode setting to the PNC. Once the DEV has sent the command to change the status, it is considered to be in the new active mode. If the PNC does not change the PS status IE field the way it should, the DEV will send another command to change it. It will continue to do this until the mode status is changed.

9.2.3 DSPS Mode

The DSPS mode allows DEVs that are able to utilize this PS mode to synchronize its awake state with all of the other DEVs that have membership in the piconet. This is not done arbitrarily; rather DEVs with similar physical capabilities are grouped together in DSPS sets. The sets are managed by the PNC. The parameters are decided by the individual DEVs.

To use the DSPS mode, a DEV must first join a DSPS set. Each DSPS set has two associated parameters: the wake beacon interval and the next wake beacon. The DSPS PS set index can be set from 2 to 253. The wake beacon is the number of superframes that occur between wake beacons of a specific DSPS set. This value is set by the DEV, rather than the PNC, at the time the set is created. The next wake beacon is the beacon number that corresponds with the next beacon of the DSPS set. Although the DEVs set the other parameters, this one is set by the PNC as the PNC will be sending the beacons. Both of these parameters are maintained by the PNC once they are set.

The advantage to the DSPS mode is that any member of a piconet may request information about any existing member of the piconet. It may also request information about any DEV that is a member of a DSPS set. It does this with a PS set information request command. This is shown in Fig. 9.3*a*. Once it has processed this request, the PNC will return the parameters of all the PS sets currently in use by way of the PS set information response command, as shown in Fig. 9.3*b*.

A DEV is also able to select the DSPS set that it wishes to join. The DEV will choose the set based on the requirements and capabilities of the individual DEV. If there are no DSPS sets that suit the DEV, then the DEV can request the PNC to create a new set. This is done by setting the SPS set index field to the "unallocated DSPS set" value, still in a range of 2–253, with the maximum number of sets being 252, and setting the operation type field to "join." The wake beacon interval field is also set by the requesting DEV to the value that is needed. An important note is that this wake beacon interval cannot be changed

2	2
Length (=0)	Command type

(*a*)

Octets: 8–39	...	8–39	1	1	2	2
PS set structure *n*	...	PS set structure 1	Number of current PS sets	Max supported PS sets	Length (=1+sum of length of *n* PS set structure)	Command type

(*b*)

FIGURE 9.3 (*a*) PS set information request command format. (*b*) PS set information response command format.

if there are members that already exist in the set. This is the case even if there is only one member. Once all of these commands have been sent, the PNC will respond to the DEV by transmitting the SPS configuration response command. This information indicates to the DEV that there has been either a success in the change or a failure. If there is a failure, the reason that the attempt failed will be provided in the same bit sequence of that part of the frame. If the attempt is successful, the PNC will assign a DSPS set index.

It is the choice of the PNC to require that all of the PS sets have a unique wake beacon interval. If for some reason the DEV is requesting a set that is already in existence, the PNC will deny the request of the DEV. The DEV can then try to request another set or join a set that is already in use. In some cases, where DEVs have to use certain wake beacon intervals, this is the only choice that the DEV might have. If the DEV decides to join a certain set, it can send the request to the PNC with the SPS set index field set to the desired index. The operation type must also be set to "join." Once this request is sent, it is up to the PNC to confirm or deny the request. It is possible for a DEV to register more than one DSPS set at a time.

Once a DEV has decided to leave the DSPS set membership, it can send a request and change the operation type to "leave." Although the PNC does not send an ACK of this, the DEV is no longer considered to be a member of the set. Once all of the members of a particular set have left, the PNC can cancel the set and the index that is currently being used.

It is possible for DSPS DEVs to alternate between the DSPS mode and the active mode. This depends on the type and amount of traffic on the network. This can be done without leaving any of the DSPS sets that the DEV could have joined. Although the PNC does not set any of these changes, the PNC must be informed of these changes with the PM mode change command.

If a DEV chooses to change from the active to the DSPS mode, the DEV will send a mode change request to the PNC. If the DEV is the source or destination of any data streams from the other DEVs in the piconet, the PNC will cancel all of these streams once the DEV has changed to the DSPS mode. Although the transmissions are canceled automatically when the DEV goes from active to DSPS, when it changes from DSPS to active, the streams are not automatically canceled.

Once a DEV has joined the PS set index 1, indicating a PSPS mode, and any other DSPS sets before it sends the mode change command, the DEV will be in a combined mode of PSPS and DSPS. In any case, the DEV cannot consider itself in any of the two modes until it receives an Imm-ACK from the PNC. If the PS status IE bit in the beacon is set with the DEVID, then it indicates that the DEV is set in the PSPS mode.

If a DEV is going to change from the DSPS to the active mode, the same procedures are followed as before except that the PM mode field is set to active.

It is the responsibility of the PNC to create a PS status IE in the beacon for each DSPS set. Once a set is no longer in use, the PNC will stop inserting the PS status IE for that particular set, which does not exist. Along with this, it is important to remember that the PNC is also in charge of making sure that the number of sets that are allowed does not exceed the size of the extended beacon. If this number is surpassed, data and management of sets will be lost.

For a DEV in the piconet to be able to transmit to a DSPS DEV, it can use the information provided by the PNC in the PS status IE. If the DEV that is inquiring about the DSPS DEV is compatible, the PS status IE also gives the DEV the details that it needs to synchronize with the DSPS set to which the DSPS DEV belongs.

In the DSPS mode it is possible for DEVs not to be able to receive broadcast data. If the DSPS DEV is expecting broadcast data to be received, it should not use the DSPS mode but rather change to the active mode. If another DEV in the piconet needs to send data to the DSPS DEV, it can request the wake beacon interval from the PNC and send the frames during the interval in which the DSPS DEV is working.

It is also possible for the PNC to give an active DEV channel time request (CTRq) to create or modify a stream with the PM CTRq type field set to active. The DSPS DEV must also be set as the TargetID in order for the request to be processed properly.

The DSPS DEV will do one of three things if the CTA rate type field of a new allocation is set to a subrate with the DSPS as the destination but not coordinated with a DSPS set. It will stay in the DSPS mode while listening for beacons to inform it of the allocation, change to the active mode, or cancel all data streams.

A DSPS CTRq asks the PNC to allocate channel access time during the wake superframes of the specific DSPS set that the DSPS DEV requests. The total value of the CTA rate factor field cannot be less than the number of superframes between the wake beacons. This distance between wake beacons is referred to as the wake beacon interval. The CTA rate factor field and the wake beacon interval are both powers of 2. Because of this, the DSPS CTAs are also a power of 2.

In Fig. 9.4 we can see the two cases that are provided showing the DSPS sets and subrate CTAs. With the example of case 2, the CTA occurs every fourth wake beacon. This number is obtained by dividing the CTA rate factor by the wake beacon interval. The wake beacon interval should never be less than the CTA rate factor. If this happens, it indicates that the CTA would occur more often than the DEV was actually listening.

When the PNC grants the DSPS CTRq, it will allocate CTAs in the wake beacons of the DSPS sets. The only case that this is not true is when there is not enough channel time left for the allocation. If there is not enough channel time for the PNC to assign the times, it will continue trying to allocate the channel time, cancel the data stream, or set the CWB bits so the source and destination DEVIDs are both in the awake so the PNC can see if the CTA can be assigned. This method works for getting the channel time assigned, although it wastes energy and battery life.

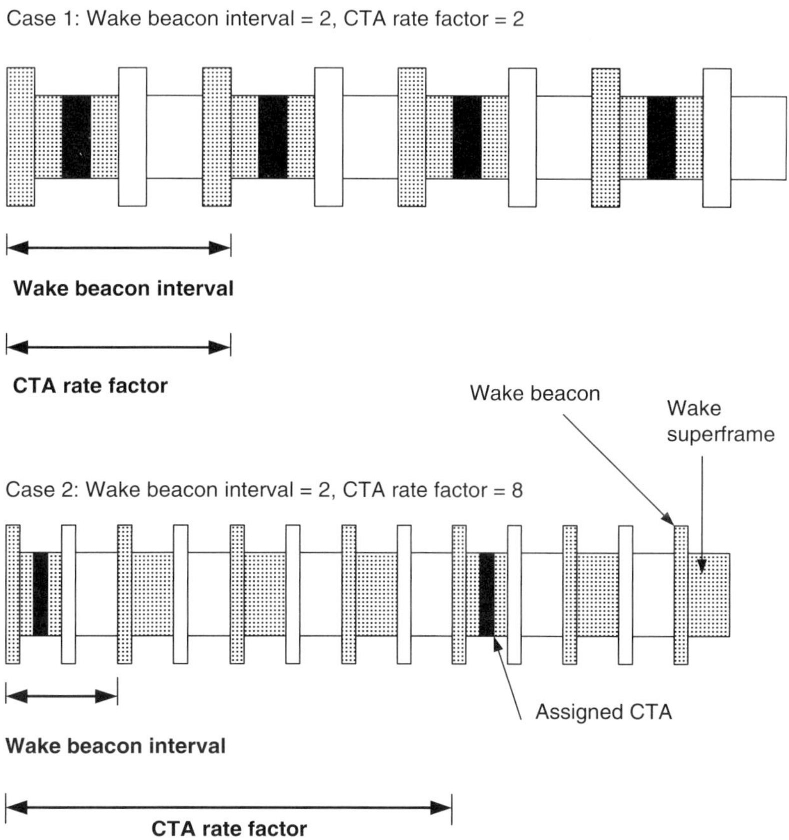

FIGURE 9.4 How the DSPS set and the CTA rate factors work together.

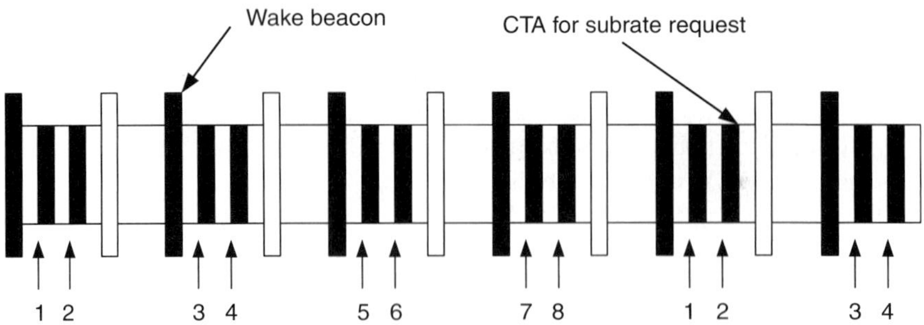

FIGURE 9.5 Minimum superframe loading with eight DEVs. There is a wake beacon interval of 2 and a CTA rate factor of 8 in this diagram.

If a DSPS DEV is the destination DEV of any CTA in a given wake beacon frame, it will listen during the assigned channel time. Figure 9.5 shows how the superframe loading and the power savings are sometime traded off. This example uses eight DEVs. If the PS features were maximized in this example, there would be a wake beacon every eight superframes. This would lead to maximum superframe loading and minimum power saving.

9.3 SECURITY OVERVIEW

Any wireless network faces security issues. With the dynamic aspects of the IEEE 802.15.3 piconet come two different security modes. This standard uses either no security or strong cryptography. There are a few assumptions made of the PHY attributes of the DEVs.

9.3.1 Mechanisms

There are several security mechanisms that allow certain security services to be implemented. These services control the transmissions between DEVs and PNCs. In the case of peer-to-peer communication, these transmissions must also be secure. The 802.15.3 standard also contains a symmetric cryptography mechanism to help provide these security services. Some of these services include security membership and key establishment, key transport, data encryption, data integrity, beacon integrity protection, command integrity protection, and freshness protection.

The method of obtaining secure membership and key establishment is not described in IEEE 802.15.3. There have been special implementation commands included in the service to help the application implementer. One such command is the security message command. Some of the security issues are handled by upper layers of the network model. For this standard, changes in security are given to the MAC layer management entity (MLME).

The key transport service is used when a DEV needs a copy of the piconet group data key. These keys are encrypted. When one of these keys is transmitted with encryption, the key protocols dictate that these transmissions be encrypted.

There are several different ways that data are encrypted in the 802.15.3 piconet. It can be done using a key shared by either all of the piconet DEVs or two peer-to-peer DEVs. The primary concern here is to protect the data from DEVs or third parties without the cryptographic key.

One way that the IEEE 802.15.3 piconet maintains data integrity is to send all secure data frames that fail integrity checks to the DME without passing these frames to the upper layers of the network model. This method uses integrity codes to protect the data from being modified by parties without the cryptographic keys.

A fifth security service, applied to beacons, is the beacon integrity protection service. This ensures that the PNC actually sent the beacon to the DEVs that are receiving it. It is a means of authenticating the beacon. This service also has a second use: to make sure that the beacon is operating properly. If the integrity check fails, it could be because the DEV is no longer synchronized with the PNC.

Command integrity is just as important as any other type of transmission. This is protected using a PNC-DEV management key. Any commands that fail the integrity checks are passed to the device management entity (DME). If this happens, the MLME does not do anything with the command frame.

In the case of cloning transmissions and channel scanning to steal frequencies, freshness protection becomes an important security service. The main use of freshness protection is to keep old messages from being replayed or resent. This is implemented by using an increasing time token. The token is found in the beacon. If a transmission does not have the current time token, then the DEV can reject the transmission. The two variables the DEV uses in a secure piconet are CurrentTimeToken and LastValidTimeToken. CurrentTimeToken is the value in the beacon for the current superframe. Any messages sent or received during this period are protected. LastValidTimeToken is used by the DEV to make sure that the beacons have not been changed or hacked.

9.3.2 Security Modes

There are two security modes in the IEEE 802.15.3 HR-WPAN piconet: modes 0 and 1. These indicate whether or not security is used in a current piconet.

If a DEV is operating in mode 0, this means that the MAC frames do not contain any type of cryptographic security. If a DEV is operating in mode 0 and receives a frame with the mode field set to 1, then the frame is discarded.

When the SEC field for a superframe is set to 1, a mechanism is used that provides use of symmetric-key cryptography. This security service protects frames by using encryption and frame integrity checks. If a PNC or DEV receiving a frame with an SEC bit set to 0 or something else, the MLME should disregard the frame and give a ReasonCode.

9.3.3 Security Policies

The policies that are adopted by the IEEE 802.15.3 piconet standard determine what actions are taken in order to provide the best security possible for the piconet. There are several different methods and procedures that support specific services. Some of them deal with PNC handovers, changes in the piconet group data key, joining a secure piconet, membership updates, secure frame generation, secure frame reception, selecting SECID for a new key, and key selection.

9.3.3.1 PNC Handover. With PNC handover security any security settings and relationships that existed with an old PNC do not apply to the new PNC. When a handover takes place, only the list of associated DEVs is passed to the new PNC. Because of this, none of the group keys need to be redone. Group memberships are not affected by a PNC handover. Although this is true, if a DEV has payload protection, the secure memberships have to be reestablished. During the course of the procedures, the new PNC has to reestablish or create CTAs for the membership DEVs. It is optional that the old PNC give the new PNC's security information to the DEVs in the piconet. This is so the new PNC will know which DEVs use secure members. This is accomplished by sending a security information command to the new PNC. This command contains the information of the piconet members. In turn, the same command is sent directly or broadcast to inform the member DEVs of the new PNC's information.

9.3.3.2 Piconet Group Data Key Changes. For DEVs that are in an active mode, the PNC has the ability to change the piconet group data key. This is done using the DistributeKey request command. Once this has been done, the PNC can also change the SECID. Since some DEVs are in a sleep state, all of the active DEVs have the ability to accept data frames with the old group data key. The active DEVs, however, have already begun transmitting with the new key. Once a DEV that was in the sleep mode receives the new group data key from the PNC, it can send a key request command to get the new key from the PNC. The next concern is how secure the procedure is if the key can be changed and the old key is still recognized by the DEVs. The fact that the old group data key is valid for a mMaxKeyChangeDuration reinforces the argument that this procedure is secure.

9.3.3.3 Joining a Secure Piconet. In order to join a secure piconet, a DEV must first send an association command to the PNC. Once the DEV has been given a local DEVID and has sent successful associations with the other DEV and exchanged information, the DEV is considered to be a secure member.

How all of this is done, specifically, is left up to the programmer or application implementer.

9.3.3.4 Membership Updates. Once it has been determined that a piconet has had membership status change with one of its DEVs, the DME issues a membership update request to the MLME. This change could be the result of termination of the security relationship, the key update process, or establishment of a security relationship. It really does not matter how this change will occur.

After the MLME has received an update request, the first thing that is checked is the TrgrtID. This is to determine which DEVs are affected. If the TrgrtID happens to be the PNC, then the data key corresponds to the group key and the management key corresponds to the PNC. The membership status contains the information telling whether or not a DEV is a secure member of the network. If the TrgrtID is not the PNC, then a peer-to-peer relationship is indicated between DEVs.

If the TrgtID is the PNCID, then there can be two different membership statuses: a member or nonmember. The membership status field is used by the MLME to determine whether or not a DEV has a secure membership to a piconet. When the MLME determines that a DEV is not a secure member, it deletes all of the DEV group keys associating it with the piconet.

If a DEV does have membership to the piconet, then all of the group keys are checked. The MLME checks a variable called KeyInfoLength. If this field is 0, then a key is being deleted. If the field is 1, then a key is being added. In the case of the previous setting, the MLME deletes the keys for the DEV. Once this happens, a DEV cannot securely transmit or receive frames; however, this is not true of peer-to-peer secure relationships. When KeyInfoLength is set to 1, the keys are updated respectively according to the current keys for the piconet.

9.3.3.5 Secure Frame Generation. Once a DEV is ready to transmit a secure frame, it will use the management key and data key to make sure that what is being sent is secure. The frames or commands that are not protected are the Imm-ACK, Dly-ACK, data frame, association request, association response, disassociation request, probe request, probe response, piconet services, announce, CTRq, channel time response, and security message. A protected frame is not sent without a correct SECID. If the key is unavailable to the DEV, then the frames are not sent.

For a PNC, the beacons are protected with the piconet group data key. This information is kept in the MLME. As stated before, the PNC uses a time token in the beacon that is incremented from superframe to superframe.

For a DEV to use the correct keys and security services, it must be completely synchronized with a beacon, or rather the PNC, or in the case of a peer-to-peer network it must be synchronized completely to the other DEV. In this way integrity, security, and cryptography all work together to make the network dependable.

9.3.3.6 Secure Frame Reception. Once a DEV receives a frame, it must first check the frame check sequence (FCS). If a frame is received with an incorrect SECID or no security, the DEV will ignore these in the secure mode.

For associated DEVs, if they have not received the group data key yet, they can accept all secure beacons for a limited amount of time until they receive the group key. Once this happens, the DEV will synchronize with the PNC beacon.

Before a DEV accepts a beacon, it must first check the LastValidTimeToken with the aMaxTimeTokenChange. If the time token that the DEV is operating with is not the same as the PNC beacon time token, then an error is indicated and the DEV will not transmit any more data during the current beacon time. The DEV can then reassociate with the piconet or choose another one.

When the security field is set, a DEV will check many IDs and group keys before a data reception is processed. The same is true of nonbeacon frames.

Once a secure frame has been received without any errors, then the cryptographic key operations are performed on the frame to modify it appropriately. If no errors occur during this procedure, then the frame is processed by the DEV. If there is no error, the operation is ceased and an error reason code is given.

9.3.3.7 New Key SECID Selection. All data and management keys that are used in the piconet contain a DEVID for the first of two octets of the SECID. If the SECID does not contain the DEVID, the frame is ignored.

When a piconet is in the initial stages of development, one of the first jobs the PNC-to-be does is to select a SECID to use for the beacon. This initial key is not distributed to any of the other DEVs until they first join the piconet, and the SECID is updated by the PNC.

9.3.3.8 Key Selection. They keys that are selected to protect frames depend upon the purpose of the frames as well as the DEV. Certain keys are used for certain types of frames. Any secure beacon frame uses a piconet group data key. Data frames using security use a piconet group data key and a peer-to-peer data key for peer-to-peer connections. Disassociation requests use the PNC-DEV management key. Request key frames, request key responses, distribute key requests, and distribute key response frames use the PNC-DEV management key and the peer-to-peer management key. The PNC handover, PNC handover response, PNC handover information, PNC information request, and PNC information commands use the PNC-DEV management key. The PNC information command also uses the piconet group data key. The probe request, probe response, and announce commands use the PNC-DEV management keys, piconet group data keys, and peer-to-peer management keys. The CTRq, channel time response, remote scan request, remote scan response, PM mode change, SPS configuration response, PS set information request, and PS set information response commands all use the PNC-DEV management key. Four other types of commands—the channel status request, channel status response, and transmit power change—use the PNC-DEV management key, piconet group data key, and peer-to-peer management key.

9.4 SECURE 802.15.3 PICONETS

This section will focus on the specifications for the types of security for the IEEE 802.15.3 HR-WPAN. This is only for the situations where security is implemented. It is possible for there to be different types of security for normal piconet operations and for peer-to-peer connections. All of the symmetric-key security operations take place in mode 1 for DEV operations.

9.4.1 Symmetric Cryptography

When frames are transmitted for this piconet standard, the first byte of the superframe that is transmitted is for the security settings and other security information. The most significant bit is first followed by the other bits all the way to the least significant bit. This order is used whether transmitting or interpreting the security material.

The data authentication mechanisms and symmetric encryption together are used in the symmetric-key security operations. These usually consist of an integrity code being generated as well as encryption of plaintext data and the integrity code. The end result is an encrypted data set and an encrypted integrity code.

The generation of an integrity code using a block cipher in the cipher block-chaining (CBC) mode computed on a nonce followed by an optional padded authentication data and plaintext data makes up the symmetric authentication operation. To verify an operation, the received integrity code is compared to the computed integrity code. If they are a match, then the data are authenticated.

The encryption operation includes the key stream generation. This generation is done using a block cipher in the counter mode with a key and nonce. It is then conducted with an XOR of the key stream using integrity and plaintext code. The decryption process uses an XOR of the key stream with the ciphertext. The end result is the plaintext and integrity code.

These counter (CTR) mode encryption operations contain the following parameters: an AES encryption algorithm, 2 octets of an *L*-length field, an 8-octet M authentication field, and a 13-octet nonce field.

The nonce value is used in CTR mode encryption and authentication. It is made up of an 8-bit SrcID and 8-bit DestID. This is followed by a 6-octet time token. The next 2 octets consist of the secure frame counter. The next 3 octets contain the fragmentation control field. This is part of the MAC header. The nonce value must be unique. The reason for this is to keep the algorithms used secure. Because of this, a DEV is not allowed to reuse any sequence numbers within a superframe for a certain DEVID. If a sequence number is repeated, it

3	2	6	1	Octets: 1
Fragmentation control field	Secure frame counter	Time token	DestID	SrcID

FIGURE 9.6 CCM nonce frame format for CCM security cryptographic algorithm.

could cause a nonce value to be repeated. The SrcID is also used to guarantee that DEVs sharing a key will use a different nonce. The time token in this case would also be different for every superframe. This uniqueness is guaranteed as long as a DEV does not send more than 65,536 frames to a DEV within the superframe. To make sure a new nonce is used for every frame, even in the case of a retransmission, the nonce value is incremented by 1. Figure 9.6 shows the CTR mode nonce format. For a protected frame, there is a fragmentation control field, secure frame counter, time token, DestID, and SrcID. The time token in this case is carried over from the beacon time token.

Not much is mentioned in the IEEE 802.15.3 standard about the AES algorithm except that its parameters consist of 128-bit keys and block size.

9.4.2 Security Implementation

The data elements related to symmetric cryptography are the encryption key, integrity code, and encrypted data. The encryption key is 16 octets in length. This key contains the result of encryption using the CTR mode encryption process. The integrity code is 8 octets in length. It is made up of the result of the CTR mode encryption computation. This is used, along with the encryption seed, to give the integrity code. The last object format is the encrypted data. The length of this variable object is dynamic. It is made up of the result of CTR mode encryption on the data. This computation is made without using the integrity code.

Cryptographic operations include secure beacon integrity code generation, data integrity code generation, key encryption operation, and data encryption generation. Secure beacon integrity code generation is done by computing the encrypted integrity code with the piconet group data key. This is done using CTR mode encryption. During this process, the entire beacon is authenticated. There are two variables, a and m, which contain the authentication data input and the plaintext input.

Secure command integrity code generation is done by calculating the encrypted integrity code with the payload protection key. Once again, this is done using CTR mode encryption. Data integrity code generation is done by computing the encrypted integrity code with the payload protection key. CTR mode encryption and data authentication are used in this case.

For the key encryption operation, CTR mode encryption and data authentication are used with the management payload protection key to encrypt the

2	2
Enc data length	Authen. data length

L_{n-1}	..	L	13	2	2	10
Information element $n-1$	..	Information element 1	Piconet synch. parameters	Secure frame counter	SECID	Frame header

FIGURE 9.7 Frame format for CCM security input.

fields. Data encryption generation is done with the CTR mode encryption algorithm and the data payload protection key. Figure 9.7 shows the input information for CTR mode encryption. This is used to obtain secure beacons. The CTR mode encryption input consists of an encryption data length, authentication data length, IEs, piconet synchronization parameters, a secure frame counter, the SECID, and the frame header.

The input for secure commands consists of a frame header, SECID, secure frame counter, command type, length, authentication data, encryption data, and authentication and encryption data lengths. The request key response and distribute key request commands are set to the length of the protected data size minus the length of the encryption key.

The secure data frame format for CTR mode encryption input is similar to the beacon and command CTR mode encryption input. It contains a frame header, SECID, secure frame counter, preencryption data, and authentication and encryption data lengths.

9.4.2.1 CTR Mode Encryption. CTR mode encryption for the IEEE 802.15.3 HR-WPAN is a general authenticate-and-encrypt process. CTR mode encryption is for use with block ciphers, such as AES. These are typically done with an 128-bit block size.

There are two parameters for the CCM mode: M, the number of octets in the authentication field, and L, the number of octets in the length field. The authentication field is 3 bits in size. Figuring out the value of M usually involves considering the trade-off between message expansion and undetected hacker penetration of the network. The encoding of the field is done using $(M-2)/2$ values.

The size of the length field is 3 bits. The trade-off for this parameter is between the message size and the nonce size. The encoding of this field takes an $L-1$ value, typically two to eight octets in length.

9.4.2.2 Input. Four inputs are needed to use CTR code encryption on a message: K, N, m, and a. Input K is the block cipher key. The size of K usually depends on the size of the block cipher.

The input N is the 15-L octet size nonce. The value of the nonce is always unique within a given encryption key K range. The ability to have duplicate nonces within a given superframe destroys any security that is present.

The m variable input is the actual message that needs to be encrypted and sent. The length restriction on this field is $0 \leq l(m) < 2^{8L}$. The reason for this restriction is to make sure that the message will be encoded on an L-octet length field. Input a refers to the additional authentication data. The $l(a)$ octet field size input is not encrypted; however, it is authenticated. Usually this information is used to authenticate headers, keys, and messages. If there are no additional authentication data, then the length of this field is zero.

9.4.2.3 Data Authentication. In order to authenticate the data, CBC-MAC is used to authenticate a block sequence of B_0 to B_n. Each block is formatted with $l(m)$, a nonce, and flags. The $l(m)$ is the most significant field.

For the fields, the reserved bit is used for program implementation and so, until a future time, is set to 0. The Adata bit is equal to 0 if $l(a) = 0$. If $l(a) > 0$, then the Adata bit is set to 1. The M input field is encoded as $(m-2)/2$. The L input field takes on values of 2–8. If $l(a) > 0$, then authentication data are added. Once this happens, the authentication blocks have $l(a)$ and the field is constructed as follows: It is encoded as 2 octets if $0 < l(a) < 2^{16} - 2^8$, as 6 octects if $2^{16} - 2^8 \leq l(a) < 2^{32}$, and as 10 octets if $2^{32} \leq l(a) < 2^{64}$. The blocks that are used to encode a are performed on the $l(a)$ string. The result is 16-octet blocks. These blocks contain zeros as fillers if necessary. The authentication block rendering for additional authentication blocks can be seen in Fig. 9.8. The $l(a)$ is first, followed by the first octets of a, then the next octets of a, followed by the final octets of a, and ending with a zero.

Once the authentication blocks have been added, the message blocks must be formed. The ordering for the message blocks is a result of the CBC-MAC computed with the following equations:

$$X_1 := E(K, B_0) \tag{9.1}$$

$$X_{i+1} := E(K, X_i \oplus B_i) \quad \text{for } i = 1, \ldots, n \tag{9.2}$$

$$TT := \text{first} - M - \text{octets}(X_{n+1}) \tag{9.3}$$

where the function E is the block cipher encryption and T is the value of the integrity code obtained by XORing B_n with X_n. It is then encrypted with the block cipher.

0	Final octets of a	Next octets of a	First octets of a	$l(a)$
B_k		B_{k-1} to B_2	B_1	

FIGURE 9.8 Format for authentication block ordering for additional authentication blocks.

9.4.2.4 Encryption. Encryption on a message is done in the CTRq mode. The format for the encryption blocks A_i are counter i, nonce N, and flags. In this case, i would be the most significant octet.

The block ordering for the encryption would contain $S_1, \ldots, S_{n-k}$. It is important to note that S_0 is not used in the block ordering for the encryption. To authenticate the encryption, the value U is computed by encrypting T with block S_0. The value is then truncated whatever length that is desired. The equation looks like this:

$$U := T \oplus \text{first} - M - \text{octets}(S_0) \tag{9.4}$$

9.4.2.5 Output. The output for an encrypted message consists of the encrypted message itself as well as the authentication value U that was obtained in the encryption process by Equation (9.4).

9.4.2.6 Decryption. To decrypt an encrypted message, the K, N, a, and c variables are needed, where, again, K is the encryption key and N is the nonce value. The additional authenticated data are represented by the variable a. The variable c is the encrypted and authenticated message.

The decryption process starts by finding the integrity code value, T. Once T has been found, it is checked against the CBC-MAC value of the message and additional authentication data. If the T value is not correct, then the process stops. At this point all we know is that T is incorrect.

9.4.2.7 Restrictions. Any implementation of encryption is limited to a total number of 2^{61} block cipher encryption operations. This allows a number close

TABLE 9.2 Symbols Used

Variable Name	Description
A	Additional authenticated data
A_i	Counter block to generate key stream
B_i	Input block for CBC-MAC
c	Ciphertext
K	Block cipher key
L	Number of octets in length field
m	Message to be encrypted and sent
M	Number of octets in authentication field
N	Nonce
S_i	Block of encryption key stream
T	Unencrypted authentication tag
U	Encrypted authentication tag
X_i	Intermediate value of CBC-MAC

to 2^{64} without ever going over. If a receiver does not expect to decrypt the same message twice, this limit needs to be implemented.

Before any information is given to the other layers of the network model, the DEV verifies the CBC-MAC. If the CBC-MAC is not verified, all information that was received is destroyed.

9.4.2.8 Symbols Used. This last section contains Table 9.2, which lists the symbols used in the symmetric-key cryptography used by the IEEE 802.15.3 HR-WPAN. This was included at the end of this chapter to clarify any variables that were used that were not explicitly explained.

9.5 CONCLUSION

Many of the aspects of the three PS modes in IEEE 802.15.3 are desirable. One could see, depending upon the number of devices, the type and amount of data, and the battery life of some portable DEVs, how one could create a hybrid of these modes to successfully create a piconet that would both save energy and optimize data transfers. With the little bit of knowledge known about Bluetooth, this standard seems to be more simplified yet as dynamic as IEEE 801.15.1. The only disadvantage that can be seen is that the energy consumption, even using PS modes, would be slightly higher than Bluetooth. Just as the trade-offs described in this chapter could determine mode selections, the same could be said of trade-offs using HR-WPAN versus other WPANs.

ACKNOWLEDGMENTS

This work was partially supported by the U.S. National Science Foundation (NSF) under grants DUE-0633445, CNS-0716211, and CNS-0737325 as well as the Texas Advanced Research Program under grant 003581-0006-2006.

REFERENCES

1. LAN/MAN Standards Committee of the IEEE Computer Society. Part 15.3: "Wireless medium access control (MAC) and physical layer (PHY) specifications for high rate wireless personal area networks (WPAN)," IEEE, New York, Feb. 2003.
2. Y. Xiao, "MAC layer issues and throughput analysis for the IEEE 802.15.3a UWB," *Dynamics of continuous, discrete and impulsive systems—An International Journal for Theory and Applications (Series B). Special Issue: Ultra-Wideband (UWB) Wireless Communications* 12(3), 443–462 (2005).
3. Y. Xiao and X. Shen, "Adaptive ACK schemes of the IEEE 802.15.3 MAC for the ultra-wideband system," *Proc. of IEEE Consumer Communications and Networking Conference 2006*, Las Vegas, Nevada.

4. Y. Xiao, X. Shen, and H. Jiang, "Optimal ACK schemes of the IEEE 802.15.3 MAC for the ultra-wideband system," *IEEE J. Sel. Areas Commun.* 24(4), 836–842 (2006).
5. W. Stewart, Y. Xiao, B. Sun, H. Chen, and S. Guizani, "Security issues in the IEEE 802.15.3 WPANs," *Proc. of IEEE GLOBECOM 2006*, San Francisco, CA.
6. W. Stewart, Y. Xiao, B. Sun, and H. Chen, "Security mechanisms and vulnerabilities in the IEEE 802.15.3 wireless personal area networks," *International Journal of Wireless and Mobile Computing, Special Issue on Security of Computer Network and Mobile Systems* 2(1), 14–27 (2007).

CHAPTER 10

PERFORMANCE EVALUATION AND OPTIMIZATION OF IEEE 802.15.3 PICONETS

ZHANPING YIN and VICTOR C. M. LEUNG

10.1 INTRODUCTION

Recent advances in wireless networking technologies have brought in a new era of pervasive computing with ubiquitous network connectivity. The rapid proliferation of digital consumer electronic devices, especially those supporting high bandwidth multimedia applications, is fueling an increasing demand for wireless connectivity solutions that support very high data rates with quality-of-service (QoS) guarantees at very low costs and with very low power consumption. However, cost-effective connection of low power mobile devices to each other and to the Internet while sustaining a high level of networking performance and service quality remains a technological challenge.

In the past several years, the IEEE 802.11 family (802.11a/b/g) of wireless local area network (WLAN) standards, commercially known as WiFi, has been a great success. The baseline IEEE 802.11 medium access control (MAC) protocol, the distributed coordination function, employs carrier sensing multiple access with collision avoidance (CSMA/CA), which is designed for bursty packet data and is not suitable for delay-sensitive real-time multimedia streams. The recent extension to IEEE 802.11 MAC, i.e., 802.11e, is a step forward to support QoS by differentiating streams with different priorities. However, the generally higher power consumption of WLAN devices makes the technology less suitable for battery-powered handheld devices.

Wireless personal area networks (WPANs) offer a new frontier of wireless networking, which is made possible by the emergence of low power and low cost devices for wireless communications over short distances. A personal

Emerging Wireless LANs, Wireless PANs, and Wireless MANs. Edited by Y. Xiao and Y. Pan

operating space is a space about a person, whether stationary or in motion, that typically extends up to 10 m in all directions. Unlike most WLANs, which operate in the infrastructure mode to connect devices to access points, most WPANs connect their devices (DEVs) in an ad hoc manner.

Bluetooth is the earliest standard for WPANs, and its physical and MAC layer specifications have been incorporated in the IEEE 802.15.1 standard. Nowadays, Bluetooth technology is widely deployed to interconnect personal digital assistants, mobile phones, headsets, computer peripheries, etc. However, the data rate of Bluetooth [1 Mbps at the physical (PHY) layer] is too low to support high bandwidth multimedia applications. Within the IEEE 802.15 standard family, the 802.15.3 standard [1] that was approved in June 2003 has been developed specifically for high data rate (not less than 20 Mbps) WPANs to provide low power and low cost solutions addressing the needs of time-critical and large-file-transfer applications, such as transfer of digital images and multimedia streaming. Besides supporting a high data rate, the standard also enables QoS guarantees for real-time voice and video applications. An amended version of this standard, 802.15.3b [2], that was approved in December 2006 enhances the original standard to improve its efficiency while preserving backward compatibility. In addition, this amendment corrects several errors and clarifies some ambiguities in the 802.15.3-2003 base standard.

Since the 2002 Federal Communications Commission (FCC) report and order [3] that permits the use of unlicensed ultra-wideband (UWB) technologies in the 3.1–10.6-GHz frequency band, research on 802.15.3 WPANs has accelerated due to the promise of UWB technologies to enable a very high data rate in excess of 100 Mbps in the PHY layer. Unfortunately, the effort of the 802.15.3a task group [4] to develop an alternate UWB-based PHY layer specification for 802.15.3 by unifying the two main proposals, namely multi-band orthogonal frequency division multiplexing (MB-OFDM) and direct-sequence UWB (DS-UWB), was unsuccessful and the task group was dissolved in January 2006. Nevertheless, there is still substantial interest in 802.15.3 WPANs. In this chapter, we will focus on the performance evaluation and optimization of 802.15.3 MAC within a piconet.

The rest of this chapter is organized as follows. Section 10.2 gives a brief introduction of 802.15.3 MAC. Section 10.3 presents an overview of existing 802.15.3 performance improvement schemes. We propose two simple and effective 802.15.3 enhancement schemes for fast peer discovery and route optimization in Sections 10.4 and 10.5, respectively. Section 10.6 concludes the chapter.

10.2 IEEE 802.15.3 WPAN

An IEEE 802.15.3 WPAN is typically configured as a piconet in which communications are confined between DEVs located within the small area of a personal space. Unlike the IEEE 802.15.1 protocol for Bluetooth, which employs a master node to forward all traffic among slave nodes within a piconet, the 802.15.3 MAC supports peer-to-peer communications among DEVs within the same piconet. Timing and data transmissions in the piconet are based on the superframe, which consists of three parts: the beacon, the optional contention access period (CAP), and the channel time allocation period (CTAP). A piconet is formed when a DEV, acting as the piconet coordinator (PNC), begins transmitting beacons that define the start of the superframes. All DEVs within radio coverage of the PNC can then associate with it to form a piconet. The 802.15.3 piconet is formed without preplanning in an ad hoc manner for only as long as the piconet is needed, and its coverage area is given by the radio range of the beacons transmitted by the PNC.

The beacons transmitted by the PNC provide timing for synchronization of DEVs within the piconet. The PNC further performs admission control, allocates network resources, and manages power save requests. Real-time and large-volume data transmissions are managed in a connection-oriented manner by pre-allocating time slots in the CTAP.

The CTAP is composed of channel time allocations (CTAs), including management CTAs (MCTAs), which are used for commands, isochronous streams, and asynchronous data connections. Channel access in a CTAP is based on time division multiple access (TDMA) such that all CTAs have guaranteed start times and durations, thus enabling both power saving and QoS guarantees. The PNC controls channel access by assigning CTAs to individual DEVs or groups of DEVs. CTA assignments as well as other management information are sent by the PNC to all DEVs within the piconet using the beacons. In 802.15.3b, a method to relinquish unused time in a CTA is included to allow another DEV to transmit data. In addition, the amendment allows multicast connections by assigning device identifications (DEVIDs) to group addresses. It also defines an additional data frame, the logical link control/subnetwork access protocol (LLC/SNAP), which allows multiple protocols to share a single data connection.

The CAP is used by DEVs to communicate commands and/or asynchronous data using CSMA/CA as the MAC protocol, similar to IEEE 802.11 WLANs. The length of a CAP and the types of data and commands sent over it are dynamically determined by the PNC. A PNC may choose to use MCTAs instead of CAPs for sending command frames. MCTAs are used for communications between the DEVs and the PNC, and thus only command frames to or from the PNC can be sent in MCTAs. In the base 802.15.3 standard, slotted Aloha is employed as the MAC protocol for accessing the MCTAs. Due to the inefficiency of slotted Aloha access, 802.15.3b specifies use of CSMA/CA in all contention periods (CPs) in each superframe, which include the CAP and contention access CTAs, consisting of association MCTAs, association CTAs, open MCTAs, and open CTAs.

10.3 OVERVIEW OF EXISTING PERFORMANCE ENHANCEMENT MECHANISMS

Designed for wireless multimedia applications, the IEEE 802.15.3 standard clearly defines the MAC functionalities and frame/superframe structures. However, how to allocate CTA slots to the streams is not defined. Therefore, most recent research on 802.15.3 is concerned with scheduling to guarantee QoS, taking into account the multimedia traffic characteristics, such as burstiness and peak-to-average ratio of data rates.

In [5], it is proposed to add a byte to the MAC header to update the instantaneous queue size of each stream so that the PNC is aware of the instantaneous loads of all data streams. Thus the PNC may schedule the CTA time for each stream using an appropriate scheduling method, such as earliest deadline first (EDF) and shortest remaining processing time (SRPT) [5, 6]. Fairness is further considered in Fair-SRPT [7]. It is proposed [8] to address the disadvantages and limitations of instantaneous queue size update by allocating one MCTA to each active stream to facilitate stream traffic load feedback; furthermore, it is found that providing the feedback MCTA at the end of each superframe gives better performance than allocating the feedback MCTA at the beginning of each superframe. A feedback-assisted CTA (FACTA) scheme [9] has also been proposed, in which DEVs send their channel time request (CTRq) at the end of each superframe.

The enhanced SRPT (ESRPT) resource reservation algorithm [10] applies SRPT in two steps for the first MAC service data unit (MSDU) and remaining MSDUs, respectively. In [11], an $M/M/c$ queuing model is used which differentiates streams by assigning different priorities for scheduling. An application-aware MAC scheme is introduced in [12], in which the maximum sizes of I, P, and B frames in a group of pictures are estimated by the source DEV before sending CTA requests to the PNC. In [13], rate-adaptive acknowledgment (RA-ACK) is used to let the PNC choose the data rate for the next transmissions during CTA allocation. A dynamic MAC scheduling method for MPEG-4 (Moving Pictures Experts Group) traffic is presented in [14] to maximize the total network throughput and minimize the delay between each pair of DEVs. Several MAC scheduling schemes are compared in [15], and a scheduling scheme is proposed to determine the allocated CTA dynamically according to the variations of frame sizes.

All the above scheduling approaches require active (control frames in minislots) or passive (implicit by overhearing) feedback of traffic parameters, such as load, frame size, and type. The PNC then applies various scheduling algorithms to dynamically assign CTAs in the superframe. Thus, they all require a dynamic superframe structure which imposes significant frame exchange overhead and hence incurs a high energy cost. In [16], the advantages and disadvantages of the static and dynamic algorithms are analyzed, and a hierarchical superframe formation algorithm is proposed to combine the benefits of the static and dynamic algorithms.

All the scheduling methods discussed above are based on a single shared channel and assume that only one stream can be sent at any given time. Other research work has taken into account of spatial channel reuse and multiple channels in 802.15.3 WPANs. For example, maximum traffic (MT) scheduling [17] is proposed to allow simultaneous transmissions following a graph coloring approach, and two heuristic scheduling algorithms with polynomial time complexity are proposed in [18] to schedule concurrent transmissions in UWB-based 802.15.3 WPANs. The relative locations between communicating DEVs are considered in [19] for channel scheduling to allow parallel transmissions within a piconet. Due to the small piconet size, simultaneous transmissions require complex power control and scheduling schemes. Thus, the additional complexity is hard to be justified by the benefits realized. In fact, simultaneous transmissions are more appropriate for impulse radio-based time-hopping (TH) UWB [20] as each link uses a different TH sequence, but the low data rate of TH-UWB makes it more suitable for 802.15.4 low data rate WPANs than 802.15.3 high data rate WPANs. However, it is an interesting research area to study the performance of WPANs operating simultaneously over multiple channels [21] and methods to connect large numbers of piconets to form large-scale scatternets.

Unlike scheduling, some schemes are proposed to enable other DEVs to utilize the idle times in the allocated CTAs through active carrier sensing. Enhanced CAP [22] considers a static superframe structure and reuses the sleep CTAs as CAPs at the cost of a higher energy consumption because all active DEVs have to listen to all CTAs. Similarly, in the method proposed in [23], the PNC determines whether a CTA is idle and, if so, broadcasts a cancellation message so that other DEVs can contend for access to reuse the idle time in a CTA. A CTA sharing protocol named VBR-MCTA proposed in [24] enables the sharing of CTAs in the same group by giving unused time slots of one variable bit rate (VBR) stream to another flow that requires peak rate allocation.

Among the different acknowledgment methods specified in 802.15.3, delayed acknowledgment (Dly-ACK) is designed to be used specifically for real-time streams in CTA transmissions, and it can reduce the MAC layer overhead and improve channel utilization. Thus, the behavior and optimization of Dly-ACK have been studied extensively [25–28]. An adaptive Dly-ACK scheme is proposed in [25] for both transmission control protocol (TCP) and user datagram protocol (UDP) traffic with two enhancement mechanisms. The delay performance of the Dly-ACK scheme is analyzed in [26] with a dynamic burst size method for performance improvement. An application-aware Dly-ACK method is introduced in [27], which includes a frame-based dynamic Dly-ACK burst size adjustment and a minimum sequence sending up algorithm. Optimal No-ACK, immediate acknowledgment (Imm-ACK), and Dly-ACK mechanisms in contention-free CTA and contention-based CAP have also been studied [28].

The CAP saturation throughput is analyzed in [1, 7]. Since CAP is used mainly for command frames, the saturation assumption is far from reality.

Other studies have taken higher layer protocols, such as TCP, into account. Simulation results for TCP and real-time flows under various MAC operating parameters are given in [29]. The work in [30] aims to make TCP transmissions more energy efficient with dynamic superframe durations.

All the performance enhancement schemes discussed above are designed for WPAN operations within a piconet and assume all DEVs within the piconet can communicate with each other in a peer-to-peer manner. However, full piconet connectivity cannot be guaranteed with only direct peer-to-peer connectivity. Depending on the locations of the associated DEVs, some DEV pairs may be out of radio range of each other (referred as unreachable pairs in the sequel), and as a result, direct peer-to-peer connection is unavailable between them. How to discover peers [31] that do not have a direct connection between them is an open issue that is not addressed in the 802.15.3 standard.

Furthermore, most of the algorithms assume a fixed data rate for all connections and aim to allocate more data streams efficiently by channel time scheduling. However, most contemporary wireless systems, including 802.15.3, incorporate a PHY layer supporting multiple data rates that may be chosen from among a predefined set of modulation parameters based on the channel conditions. The scheduling method in [13] uses RA-ACK to adapt to the instantaneous channel condition. More simply, in the absence of propagation artifacts, the data rate can adapt to the transmission distance between two DEVs. In general, multiple connection paths may exist between two DEVs, including possibly the direct peer-to-peer connection and multihop connections through other DEVs within the same piconet. The direct connection, if available, may not be the best route when rate adaptation is taken into account. This presents an interesting engineering design problem for 802.15.3 WPANs on intrapiconet path optimization [32].

10.4 THIRD-PARTY HANDSHAKE PROTOCOL FOR EFFICIENT PEER DISCOVERY

In this section, we analyze the limitations of peer-to-peer intrapiconet communications as specified in the current 802.15.3 standards [1, 2], particularly the issues related to peer discovery, and present a novel peer discovery scheme called the third-party handshake protocol (3PHP) [31].

Since an 802.15.3 piconet, defined as the set of DEVs synchronized to and controlled by a common PNC, supports ad hoc communications between peer DEVs, peer discovery is crucial to its operation. The DEVs should be able to obtain information about the services and capabilities of other DEVs in the piconet at any time by exchanging information discovery commands and responses. Particularly, peer information is needed before a source DEV can send any data to a destination DEV or generate a CTRq to the PNC for CTA allocations.

All data frames in an 802.15.3 piconet are exchanged directly between DEVs in a peer-to-peer manner. Therefore, if the necessary peer information is not available, the corresponding DEVs need to execute the peer discovery procedure before actual data transmissions. The source DEV should first send a PNC information request command to the PNC to find out if the destination DEV exists in the piconet and, if so, it should then send a probe request command to the destination DEV to gather the peer communication information such as the data rate. Note that all these commands are exchanged by CSMA/CA access in the CAP. The standard peer discovery process is given in Fig. 10.1.

In a successful peer discovery frame exchange sequence following the current standard [1, 2], the minimum delay is achieved if both the PNC and the destination DEV send the corresponding responses immediately, i.e., after the Imm-ACK and a short interframe space (SIFS). However, if the destination DEV is outside the radio coverage of the source DEV, it will not receive the probe request command from the source and the source will not receive the

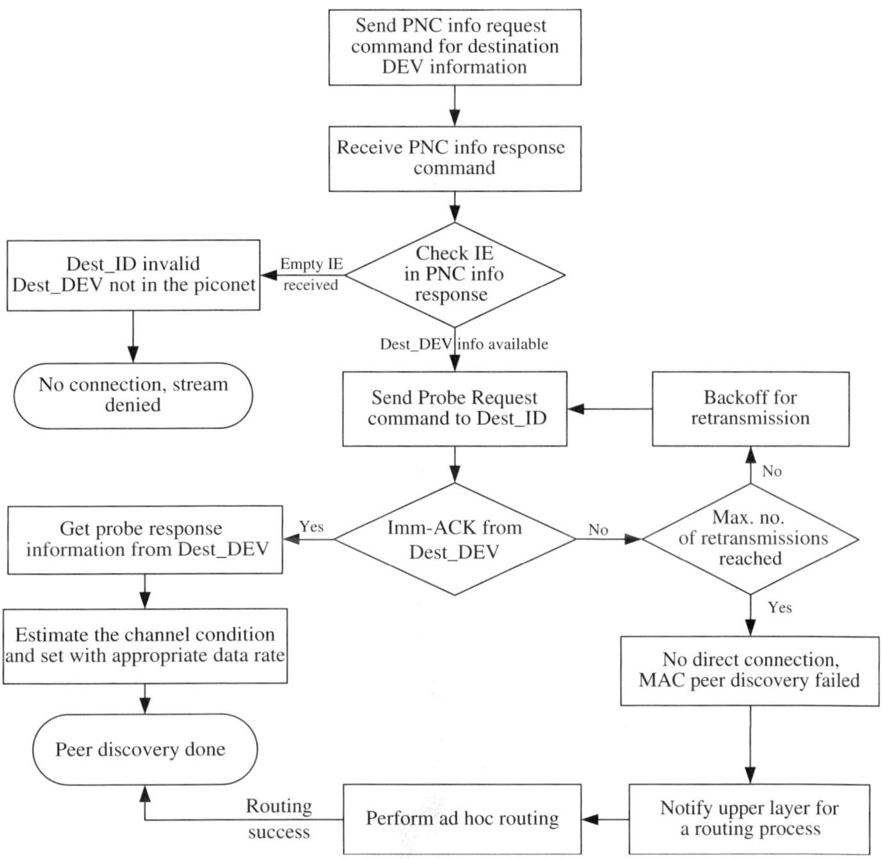

FIGURE 10.1 Standard peer discovery process.

Imm-ACK. Even worse, the sender cannot distinguish the out-of-range condition from a collision. Therefore, it will assume that a collision has occurred and begin the backoff retransmission process until the specified maximum number of retries is reached. In this case the MAC layer peer discovery ends in a failure, and the piconet fails to establish connectivity between the peer DEVs.

When peer discovery fails, the MAC layer shall notify the network layer accordingly, and the latter may initiate a route discovery process. Routing protocols, such as ad hoc on-demand vector routing and dynamic source routing (DSR), use packet flooding for route discovery in multihop ad hoc networks where no central controller, such as the PNC, is present. Route request (RREQ) packets are broadcast with no acknowledgments and no retransmissions. Because all RREQ and route response (RRES) packets are sent via contention access, collisions may occur with other commands. Failed deliveries of RREQ/RREP packets are recovered at the source by repeating the route discovery process. The transmission costs of route discovery packets are also higher than MAC command frames as the former have more overheads and may need to include the path information in the payload, as in DSR.

Any DEVs between the source and destination may become an intermediate node if it successfully delivers the RREQ to the destination. If a route is discovered by network layer routing, each hop needs to request a CTA slot separately from the PNC with a unique stream number in the MAC layer. Thus, the PNC will consider each hop as an independent traffic stream, and there is no coordination among these hops that belong to the same connection. For instance, a failure in an intermediate hop breaks the connection, but the PNC assumes that an independent stream is terminated and keeps allocating CTAs for the other hops until it is eventually notified by all participating DEVs, and another route discovery is needed to reroute the traffic. Thus the use of network layer routing between unreachable DEV pairs within a piconet is inefficient and time consuming.

On the other hand, for a successful peer discovery sequence, the exchange of PNC information request and response commands seems redundant. However, without it, the same backoff retransmission problem as discussed above happens if the destination DEV is not associated with the PNC.

Since a piconet is confined by the radio coverage of the PNC, a two-hop connection through another DEV in the piconet, such as the PNC, can always satisfy the connectivity between any pair of DEVs in the same piconet. In this case, a MAC layer forwarding method is preferable over the network layer routing procedure described above. With MAC layer forwarding, the PNC knows that these hops belong to a single connection, and thus it can better manage the corresponding CTAs by, e.g., adjusting downstream CTAs to match upstream allocations, rerouting the traffic if the intermediate node fails, and releasing all CTAs for this stream if either the source or destination terminates the session.

For fast and efficient peer discovery, we propose a 3PHP [31], which incorporates MAC layer forwarding capabilities. With the active involvement of the PNC, costly network layer intrapiconet routing is eliminated, and full piconet connectivity is guaranteed. The proposed 3PHP works as follows.

As the PNC information request and response exchange is redundant if the destination DEV is reachable, we propose that the source DEV sends only a probe request command to the destination. If the destination DEV receives the command, it returns an Imm-ACK after as SIFS and then the probe response command as in the standard protocol operations [1, 2]. At the same time, the third party, namely the PNC, actively monitors these frame exchanges and intervenes on behalf of the destination DEV if necessary. Upon receiving a probe request, the PNC checks the destination ID (Dest_ID) field in the MAC header. If the Dest_ID is not associated in the piconet, the PNC will send an Imm-ACK to the source DEV after the SIFS, followed by a PNC information command with an empty information element (IE) to notify the source that the destination does not exist in the piconet. Otherwise, instead of ignoring the probe request frame, the PNC waits for the Imm-ACK from the destination DEV. If no Imm-ACK arrives after a backoff interframe space (BIFS), which is the sum of an SIFS and a clear-channel assessment detect time (CCADetectTime), the PNC realizes that the destination DEV is not responding to the source, most likely because it is out of radio range. The PNC should then immediately send an Imm-ACK to the source, followed by a PNC information command with the route information, which is decided by a route optimization algorithm given in the next section in this chapter.

Clearly, with the proposed 3PHP, any peer discovery process requires only one round of frame exchange, which is approximately half the time of the peer discovery procedure used in the standard method [1, 2]. Thus, 3PHP fully utilizes the broadcasting nature of wireless transmissions, combining the advantages of centralized control with ad hoc communications to achieve maximum peer discovery efficiency with virtually no extra cost. It also prevents the undesirable problem of futile backoff retransmissions between unreachable DEV pairs. Figure 10.2 illustrates the frame exchange sequence in the worst case scenario with 3PHP, where the dotted lines represent commands/responses that are not completed as the source and destination DEVs are out of each other's radio range. Moreover, the MAC layer forwarding method incorporated into 3PHP guarantees full piconet connectivity and finds the best route with minimum transmission time for traffic delivery. Thus it totally eradicates the need for network layer routing within the piconet and benefits system performance greatly.

Peer discovery is done on demand when a connection is to be set up where no peer information is available or when an existing link is broken. Therefore, the reduced peer discovery time brings a fast response to connection establishment. Moreover, when link outage happens, 3PHP provides prompt route recovery and session reestablishment to the higher layer, which is essential to real-time streaming applications.

The proposed scheme is fully compatible with the standard [1, 2] and requires minimal modifications, which include the Imm-ACK exchange

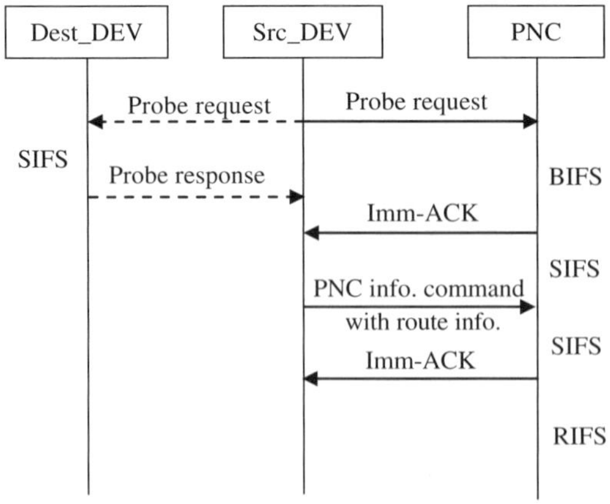

FIGURE 10.2 3PHP frame exchange sequence when destination DEV is unreachable.

discussed above. At the PNC, besides actively monitoring the commands exchanged over the piconet, the only addition is a new route IE that the PNC appends to the reply PNC information response to indicate the explicit MAC forwarding information when the source cannot directly reach the destination. Moreover, as defined in the standard [1, 2], during the CAP, DEVs resume backoff after the channel has been idle for the BIFS. Thus, with the proposed 3PHP method, when peer DEVs are unreachable, the Imm-ACK from the PNC can potentially collide with other commands. To overcome this problem, DEVs should defer for a slightly longer time, e.g., a retransmission interframe space (RIFS) before backoff resumes.

The detailed analysis of peer discovery time, including the time of routing and backoff in the case of unreachable pairs, can be found in [31]. Figure 10.3 presents the simulation results for peer discovery delays versus the contention access collision probability p under different conditions over an MB-OFDM UWB PHY [33]. Due to the extra frame exchange for the PNC information commands in the standard method, the peer discovery delays between DEVs within range are 100 μs higher than those of 3PHP over the full range of values of p. However, for unreachable DEV pairs, the standard method fails after an unproductive backoff process, delays of which are indicated by the second curve from the top in Fig. 10.3, and extra delays shown by the third curve from the top in Fig. 10.3 are incurred for network layer routing. In this case the mean successful peer discovery time, given by the top curve in Fig. 10.3, is 9–10 times that of 3PHP, given by the bottom curve in Fig. 10.3. Clearly, a successful MAC layer peer discovery is much more efficient than network layer routing. At the same time, due to the multiple hops and the RREQ broadcasting, network layer route discovery has a much higher failure probability [31] than that of the MAC layer peer discovery. Thus, not only is the 3PHP method more efficient but it is also much more reliable than the standard method, which

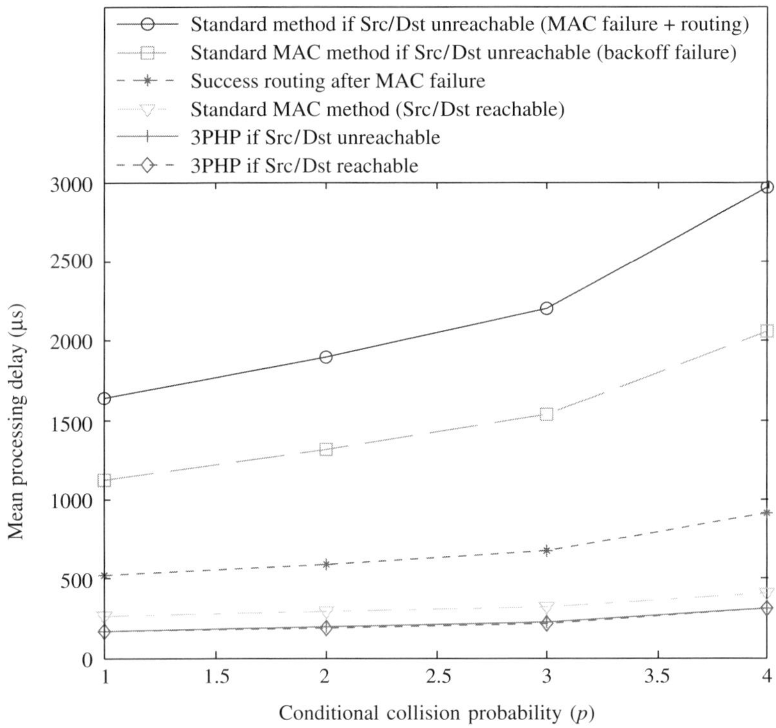

FIGURE 10.3 Peer discovery delay versus conditional collision probability.

needs to resort to network layer routing between unreachable DEVs in the same piconet.

The standard MAC layer peer discovery methods cannot guarantee full connectivity between DEVs within a piconet through direct peer-to-peer connections if the piconet operates with a radius larger than half of the maximum transmission distance. The coverage radius of a piconet can be varied by the PNC by reducing the transmit power of the beacons. Analytical and simulation results [31] show that up to 41.3% of intrapiconet DEV pairs may not be able to communicate with each other directly if the piconet radius equals the maximum transmission distance when DEVs are uniformly distributed.

Define the piconet coverage ratio r/R as the ratio between the piconet radius r and the maximum transmission distance R. The expected piconet peer discovery times for both 3PHP and the standard method (including the network layer route discovery if needed) versus the piconet coverage ratio are given in Fig. 10.4. When the piconet coverage ratio increases, the peer discovery times for 3PHP remain almost unchanged, but those of the standard method increase significantly due to an increasing probability of no direct connection between a randomly chosen pair of DEVs. For instance, at

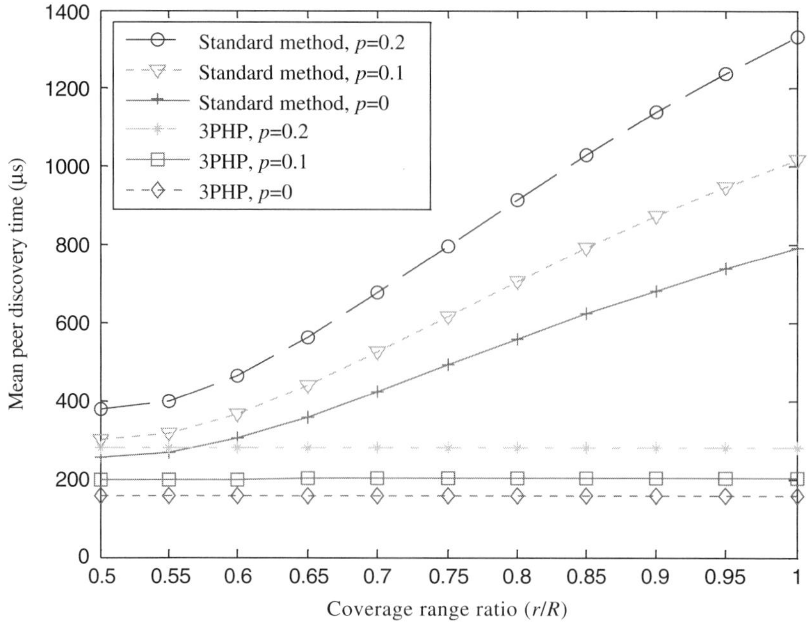

FIGURE 10.4 Piconet peer discovery time (standard vs. 3PHP).

contention access collision probability $p = 0.1$, the average peer discovery time with 3PHP is almost constant at around 217 µs, but that of the standard method increases from 316 to 1097 µs when the piconet coverage range ratio is increased from 0.5 to 1. Therefore, 3PHP is not only more efficient but also more stable and robust compared to the standard method.

10.5 INTRAPICONET ROUTE OPTIMIZATION

In this section, we propose an intrapiconet route optimization method [32] with application awareness, which takes into account multirate support in the 802.15.3 PHY layer.

Many contemporary wireless systems, including IEEE 802.15.3 WPANs, employ a multirate PHY layer that includes a set of modulation and coding parameters which may be combined to support a number of distinct data rates. A rate adaptation method or algorithm is employed to select the transmit data rate that is appropriate for the given channel condition in order to maximize the throughput of the wireless link. In 802.15.3, the beacons, command frames, and PHY and MAC headers are sent at the base rate to maximize robustness, and the frame payloads are sent with the highest data rate that achieves the minimum required link quality (called achievable data rate in what follows for simplicity). Therefore, the carrier-sensing range between DEVs is determined by the base rate, and the PNC can adjust the size of a piconet by changing the beacon power.

Depending on the locations of the associated DEVs in the piconet, some DEVs may not be able to directly communicate in a peer-to-peer manner even if they are all under the control of the same PNC. For unreachable DEVs, an intrapiconet connection can be established by PNC forwarding, e.g., using the 3PHP presented in the last section. This is good enough for exchanging commands which are sent at the base rate, and it would be the optimal choice if all wireless links between DEVs operate at the same data rate. However, data frames may be sent using any of the data rates supported by the variable-rate PHY layer. Therefore a multihop route through some intermediate DEVs with shorter links and higher achievable data rates between successive DEVs may attain a higher throughput than either a longer single-hop direct connection between reachable DEV pairs or a double-hop route via PNC forwarding between unreachable DEV pairs. Hence we can optimize frame forwarding within a piconet by considering DEVs that are closer in distance to the source and destination as potential intermediate nodes for frame forwarding.

In an 802.15.3 piconet, the CTAP in each superframe is shared using TDMA, and thus the total resource utilized by a multihop connection is simply the sum of CTAs of each individual hop. To maximize the piconet capacity, the intrapiconet route should be optimized to minimize the total transmission time, which in turn maximizes the effective data throughput and reduces the energy consumption.

In general, shortest path algorithms (SPAs) can be used for route optimization. However, as they are designed for wireline networks, SPAs such as open shortest path first (OSPF), require all routers to maintain a routing table based on the link state information of the network. This is not practicable in wireless systems where terminals employ half-duplex transmissions and link bandwidths vary with distance, mobility, and channel interference. Maintaining link state information at each node would require frequent information updates and excessive overhead. Because the PNC has up-to-date information about all DEVs in the piconet and full control of CTAs, a centralized route optimization at the PNC is more efficient and realizable.

During a frame transmission, the physical preamble and header are transmitted at the base rate, and only the payload is sent with the achievable data rate. Moreover, interframe spaces (IFSs) and acknowledgments are inserted as necessary. Thus, transmission of each data frame can incur substantial overheads. Define the effective CTA rate (S_{CTA}) as the data rate seen by the upper layer during a CTA, which is contributed by the frame payload, counted only once for the same data frame forwarded over a multihop connection. Due to IFSs and base rate transmission of physical preamble and PHY/MAC headers, S_{CTA} is much lower than the corresponding PHY rate and varies widely for different applications even if the same physical link is used. As a result, an optimal route between two DEVs for one application may not necessarily be the best for another due to different traffic parameters and protocol interactions. Therefore, we need to enhance route optimization with application awareness [32].

The frame payload transmission rate is chosen based on the receive signal strength indicator (RSSI) measured during the reception of a PHY preamble. Due to the extremely low transmit power of WPAN devices, the achievable data rate drops dramatically when the distance increases. Since only one DEV can transmit at a time during the CTAP, for simplicity the multirate PHY layer can be modeled as a function of distance d as follows if interference is neglected:

$$\text{Rate}(d) = S_i \quad \text{for } R_{i+1} < d \leq R_i \quad 1 \leq i \leq n \tag{10.1}$$

where R_i is the transmission range of data rate S_i that meets a given bit error rate objective and S_1 and S_n are the minimum and maximum data rates, respectively. Thus, $R_{n+1} = 0$ and R_1 is the maximum transmission distance.

The traffic parameters of a given application as seen by the MAC layer can be expressed as

$$\text{app} = (\text{size, ack, } k, m) \tag{10.2}$$

where size is the MAC frame payload size in bytes; ack is the ACK policy for the application stream; k is the number of blocks, each consisting of m data frames, required by the application in a CTA; and m is the number of data frames in a block. Thus, $m = 1$ for No-ACK and Imm-ACK and $m \geq 1$ when Dly-ACK is used. The PNC allocates CTA time based on the traffic parameters. Various scheduling algorithms can be used to dynamically change the CTA length in each superframe with instantaneous queue status information [5–9]. Here we consider the CTA as a constant represented by CTA(i, app), where i is the index for the frame transmission rate S_i. In practice this constant can represent the mean CTA length required to support the specific application.

Although the link rate is modeled as a function of link distance in Equation (10.1), the actual link distance is difficult to determine without costly location determination capabilities. However, we may assume that when two DEVs establish communication, they will adapt their transmissions to utilize the achievable data rate. It will then be possible for the PNC to learn the achievable data rate between reachable DEV pairs by monitoring data frames exchanged between peer DEVs. The PNC can store the achievable data rates between DEV pairs in an $N \times N$ rate matrix (RM), where N is the total number of DEVs within the piconet including the PNC, and use this information for route optimization purposes. As defined by the standard, each DEV is assigned a unique DEVID in the piconet, which is used in all frame headers instead of the real MAC address. The DEVID for the PNC is 0. Therefore, rate information can be stored in the PNC with a relatively small memory space.

For a lower rate link, a multihop connection via higher rate links may consume less transmission time. Consider a link between DEV1 and DEV2 with distance D and data rate S_l and a DEV3 that has distances D_1 and D_2 from DEV1 and DEV2 with data rates S_i and S_j, respectively, as shown in Fig. 10.5. Without loss of generality, assume $D_1 \geq D_2$ so that $S_i \leq S_j$. For a given application, the two-hop connection is more bandwidth efficient if the sum of the CTAs for the two hops with S_i and S_j is smaller than the CTA required for the direct link with S_l, i.e.,

$$\mathrm{CTA}(i, \mathrm{app}) + \mathrm{CTA}(j, \mathrm{app}) < \mathrm{CTA}(l, \mathrm{app}) \tag{10.3}$$

In other words, a link can be optimized by a multihop connection if and only if the reduction in payload transmission time is greater than the time consumed by the additional overhead. Thus, given rate index i, j, l, the frame payload size threshold (P_{th}) for a specific traffic type can be derived. No optimization is possible if the frame size is smaller than the threshold for that kind of traffic. Provided Equation (10.3) is satisfied, a link with distance D can be optimized with two links of rates S_i and S_j if there is a DEV within the optimization region shown as the shadowed area in Fig. 10.5. Obviously, the link optimization probability decreases as D increases since the optimization region decreases. The link optimization probability increases with the number of DEVs as the chance that a DEV is in the overlapped area increases. Thus, the optimization probability for links employing a specific data rate depends on the density of DEVs and the distribution of the distance D for that data rate. The detailed analysis of the link optimization probability can be found in [32].

Based on the rate information stored in the RM, the PNC can determine the optimal route for a given application stream between any DEVs using a shortest path routing algorithm, such as Dijkstra's algorithm. Because S_{CTA}

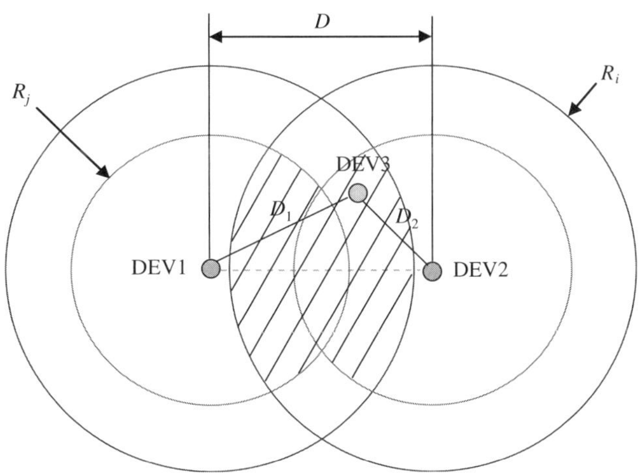

FIGURE 10.5 Link optimization with two higher rate links.

varies dramatically for different traffic parameters, the link cost has to be modified dynamically. Such dynamic adjustments would not be feasible in fully distributed systems, but they would be easy to implement with centralized control as in 802.15.3. In general, let $\psi = (i, k, l, \ldots, z, j)$ be the DEVs in a route between DEV_i and DEV_j; e.g., $\psi = (i, j)$ represents the nonoptimized direct link between DEV_i and DEV_j with rate RM_{ij} and route $\psi = (i, k, j)$ consists of an intermediate node DEV_k. Let "hop" be a link between DEV_a and DEV_b within ψ with the link rate index $\mathrm{RateID}_{\mathrm{hop}} = \mathrm{RM}_{ab}$. Then the optimal route ψ is a route that minimizes the total transmission time of all hops along the route. Thus, the AASP algorithm is proposed to find the best path ψ as follows:

$$\mathrm{CTA}_{\psi}(\mathrm{app}) = \sum_{\mathrm{hop} \in \text{all hops in } \psi} \mathrm{CTA}(\mathrm{RateID}_{\mathrm{hop}}, \mathrm{app}) \tag{10.4}$$

$$\mathrm{AASP}(i,j) = \min_{\psi} (\mathrm{CTA}_{\psi}(\mathrm{app})) \tag{10.5}$$

In Equation (10.4), the link cost (transmission time) for each hop CTA $(\mathrm{RateID}_{\mathrm{hop}}, \mathrm{app})$ depends on the application parameters and can be recomputed for each new application.

In standard 802.15.3 operations, the source DEV calculates the time unit (TU) size based on the application parameters and requests a CTA by sending a CTRq to the PNC with the number of TUs and TU size so that the PNC can efficiently allocate channel time. However, the TU size alone does not provide enough information about the traffic stream for the PNC to decide if route optimization is possible. Thus, we propose some minor modifications to the CTRq to enable integration of the AASP route optimization algorithm into the MAC protocol. Instead of sending the TU information only, the CTRq is expanded to provide the application parameters, including payload size, ACK method, block size, and number of blocks in a TU. With this information, a DEV can use the CTRq to request for a CTA to any destination DEV in the piconet even if the destination DEV is not directly reachable, and the PNC can choose the optimal route using the AASP algorithm based on the current RM. The PNC then computes the TU and CTA length accordingly, includes the explicit MAC forwarding information in the channel time response command returned to the source DEV and all DEVs involved in the optimized route, and broadcasts the CTA reservations in the following beacon frames.

Extensive simulations have been performed with the MB-OFDM PHY layer [33]. To evaluate the effectiveness of the AASP method, we define the link optimization ratio (LOR) for direct (single-hop) links with a specific data rate as the fraction of such links that can be optimized by the AASP. For unreachable DEV pairs that may be connected via PNC forwarding, the LOR is defined as the fraction of such links that can be optimized by forwarding over DEVs other than the PNC. In terms of the effective CTA rate S_{CTA}, a higher LOR leads to a higher S_{CTA}.

For unreachable DEV pairs, PNC forwarding is not always optimal. Figure 10.6 presents the LORs for unreachable pairs. It shows that with the same acknowledgment policy and payload, for any given piconet size, the LOR increases with the number of DEVs N due to the increased availability of potential multi-hop routes; e.g., with piconet radius r set to the maximum transmission range of 17 m, the LOR is 14.9% with 5 DEVs, increasing to 62.1% and 77.9% with 20 and 40 DEVs, respectively. Moreover, applications with a larger payload and employing Dly ACK achieve higher LORs due to reduced frame overhead ratios. Figure 10.6 also shows that the LOR and rate enhancements are lowest at $r \approx 12$ m because the PNC is more likely to be an optimum forwarding DEV between unreachable DEVs in a medium-sized piconet. Translated to throughput, a higher LOR results in a higher effective link throughput. For a further example of the effectiveness of the AASP, even

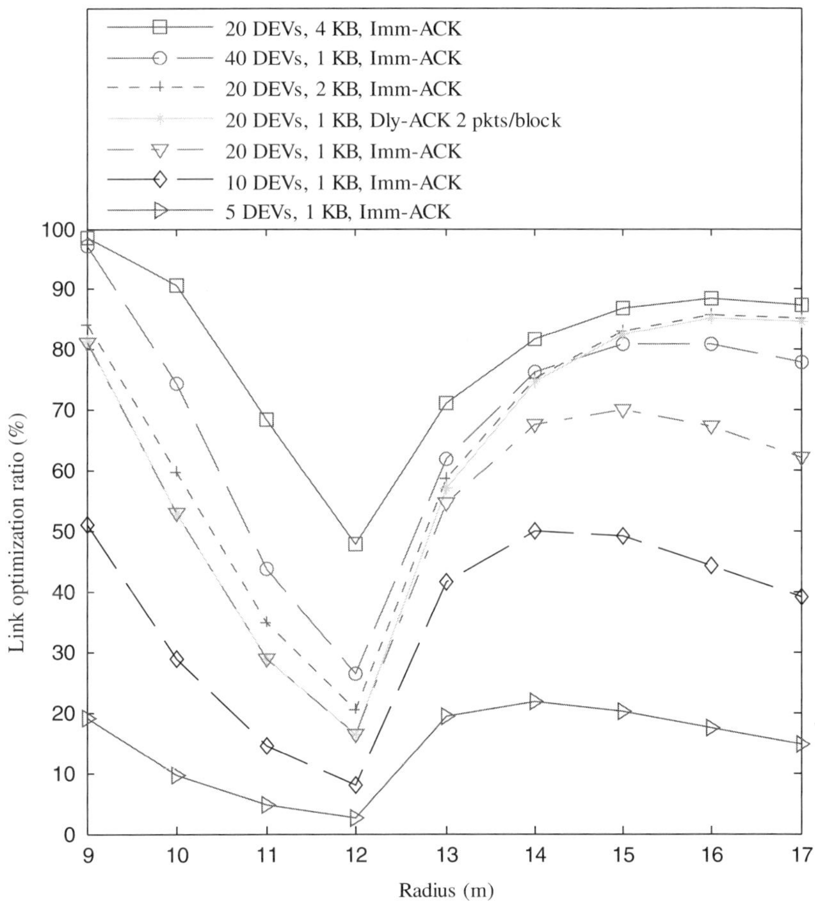

FIGURE 10.6 LOR for unreachable DEV pairs.

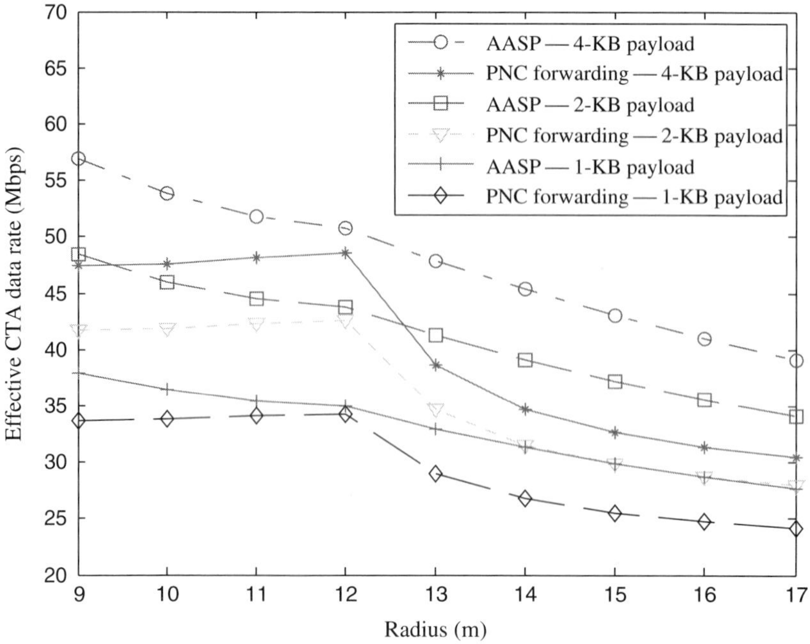

FIGURE 10.7 Effective CTA rates for unreachable DEV pairs (Imm-ACK, 20 DEVs).

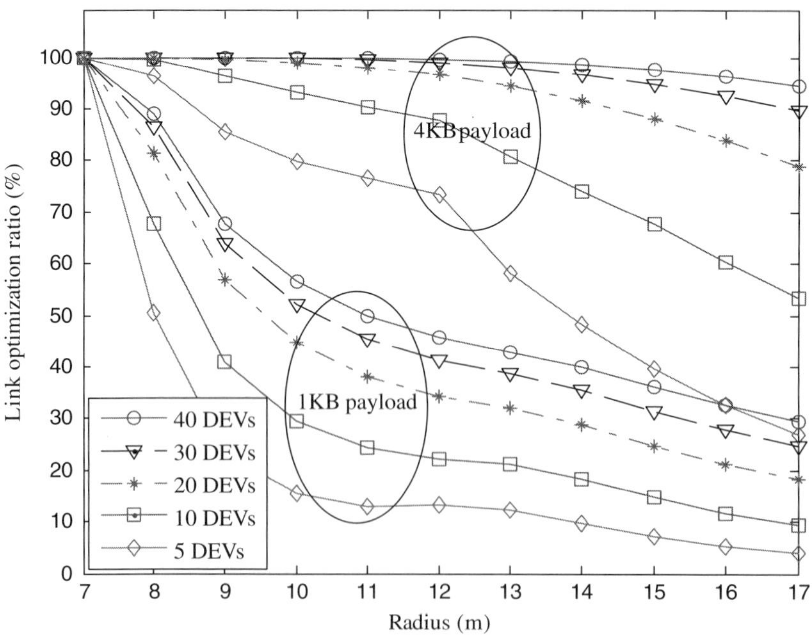

FIGURE 10.8 LOR for 53.3-Mbps links with Imm-ACK.

with Imm-ACK and 20 DEVs only, S_{CTA} values for 1- and 4-kilobyte payloads are increased by 14.4% and 28%, respectively, over PNC forwarding between unreachable DEV pairs when $r = 17\,\text{m}$ (Fig. 10.7).

Between directly reachable DEVs, a significant portion of low rate links can be optimized. As an example, Fig. 10.8 presents the LOR results for 53.3-Mbps links at different piconet size and number of DEVs in the piconet. Again, the LOR increases with the number of DEVs N and the payload size. However, the LOR decreases as the piconet size is increased if N is held constant, since a larger piconet has a lower density of DEVs, which reduces the possibility of route optimization.

The results clearly show that the proposed AASP algorithm yields very high optimization ratios, which increase with frame payload size and the density of DEVs in the piconet. This brings significant improvements in effective data rates to unreachable DEV pairs and low rate links, which in turn increases the system capacity and minimizes the energy consumption.

10.6 CONCLUSION

We have presented an extensive survey of different performance improvement methods for 802.15.3 MAC, which mainly address the scheduling of CTA assignments to provide QoS guarantees. As existing work in the literature fails to address the issues of providing full connectivity to all the DEVs within a piconet and optimization of intrapiconet connections, we have also presented our recent proposals that address these important issues.

The first issue arises due to the fact that not all DEV pairs under the coverage of the same piconet are within the radio range of each other, resulting in failed MAC layer peer discovery and excessive delays and resource usage if the network layer is called upon to complete the task. Taking advantage of the centralized control of an 802.15.3 WPAN by the PNC, we have proposed a novel 3PHP for fast and reliable peer discovery and connection reestablishment with full connectivity support. The 3PHP scheme achieves much faster peer discovery in all cases using only a single round of control frame exchange. For DEV pairs in the piconet that are not directly reachable, 3PHP provides connectivity via a simple, efficient, and reliable MAC layer forwarding method.

The second issue arises as the rate-adaptive PHY layer of 802.15.3, which enables data payloads in MAC frames to be transmitted at the highest data rate that satisfies the link quality requirement, makes it possible to optimize connections between DEVs within the piconet via multihop MAC layer forwarding. We have developed an AASP route optimization algorithm in the PNC to provide the best MAC layer forwarding routes with application awareness by self-learning the achievable data rates between DEVs. Intrapiconet route optimization using AASP significantly increases the system performance in terms of reducing transmission time and increasing effective data rate without extra cost and overhead, which can translate to power savings and increases in piconet capacity.

The various performance enhancement algorithms presented in this chapter have been proposed to address different problems with regard to the operations of IEEE 802.15.3 piconets. However, several of these algorithms can be jointly deployed, e.g., the AASP route optimization can be used simultaneously with any of the proposed scheduling algorithms. The design of an IEEE 802.15.3 WPAN should consider these different issues and incorporate the appropriate performance enhancement algorithms so that piconet performance can be optimized.

ACKNOWLEDGMENTS

This work was supported by a grant from Bell Canada under the Bell University Laboratories program and by the Canadian Natural Sciences and Engineering Research Council under grant CRDPJ 320552-04.

REFERENCES

1. IEEE 802.15.3-2003, "Wireless medium access control (MAC) and physical layer (PHY) specifications for high rate wireless personal area networks (WPANs)," IEEE, New York, Sept. 2003.
2. IEEE 802.15.3b-2005, "Wireless medium access control (MAC) and physical layer (PHY) specifications for high rate wireless personal area networks (WPANs), amendment 1: MAC sublayer," IEEE, New York, May 2006.
3. FCC's UWB first report and order, FCC02-48A1, Federal Communications Commission, Washington, DC, Feb. 14, 2002.
4. IEEE 802.15 WPAN high rate alternative PHY task group 3a (TG3a), Jan. 2006, available: http://www.ieee802.org/15/pub/TG3a.html.
5. R. Mangharam and M. Demirhan, "Performance and simulation analysis of 802.15.3 QoS," IEEE 802.15-02/297r1, IEEE, New York, July 2002.
6. A. Torok, L. Vajda, A. Vidacs, and R. Vida, "Techniques to improve scheduling performance in IEEE 802.15.3 based ad hoc networks," paper presented at the 2005 IEEE Global Telecommunications Conference (IEEE GLOBECOM'05), Vol. 6, Saint Louis, Missouri, USA, Dec. 2005, pp. 3523–3528.
7. R. Mangharam, M. Demirhan, R. Rajkumar, and D. Raychaudhuri, "Size matters: Size-based scheduling for MPEG-4 over wireless channels," paper presented at the SPIE & ACM MMCN'04, Vol. 3020, San Jose, CA, Jan. 2004, pp. 110–122.
8. X. Liu, Q. Dai, and Q. Wu, "Scheduling algorithms analysis for MPEG-4 traffic in UWB," paper presented at the IEEE VTC'04-Fall, Vol. 7, Sept. 2004, pp. 5310–5314.
9. S. M. Kim and Y.-J. Cho, "Scheduling scheme for providing QoS to real-time multimedia traffics in high-rate wireless PANs," *IEEE Trans. Consumer Electron.* 51(4), 1159–1168 (2005).

10. X. Liu, Q. Dai, and Q. Wu, "An improved resource reservation algorithm for IEEE 802.15.3," paper presented at the IEEE ICME'06, Toronto, Canada, July 2006, pp. 589–592.
11. R. Zeng and G. S. Kuo, "A novel scheduling scheme and MAC enhancements for IEEE 802.15.3 high-rate WPAN," paper presented at the IEEE WCNC'05, Vol. 4, New Orleans, LA, Mar. 2005, pp. 2478–2483.
12. S. H. Rhee, K. Chung, Y. Kim, W. Yoon, and K. S. Chang, "An application-aware MAC scheme for IEEE 802.15.3 high-rate WPAN," paper presented at the IEEE WCNC'04, Vol. 2, Atlanta, GA, Mar. 2004, pp. 1018–1023.
13. B.-S. Kim, Y. Fang, and T. F. Wong, "Rate-adaptive MAC protocol in high-rate personal area networks," paper presented at the IEEE WCNC'04, Vol. 3, Atlanta, GA, Mar. 2004, pp. 1394–1399.
14. D. Tarchi, R. Fantacci, and G. Izzo, "Multimedia traffic management in IEEE 802.15.3a wireless personal area networks," paper presented at the IEEE ICC'06, Vol. 9, Istanbul, Turkey, June 2006, pp. 3941–3946.
15. M. Wang and, G. S. Kuo, "Dynamic MAC scheduling scheme for MPEG-4 based multimedia services in 802.15.3 high-rate networks," paper presented at the IEEE VTC'05-Fall, Vol. 3, Dallas, TX, Sept. 2005 pp. 1559–1563.
16. L. Vajda, A. Torok, K.-J. Youn, and S.-D. June, "Hierarchical superframe formation in 802.15.3 networks," paper presented at the IEEE ICC'04, Vol.7, Paris, France, June 2004, pp. 4017–4022.
17. Y. H. Tseng, E. H.-K. Wu, and G.-H. Chen, "Maximum traffic scheduling and capacity analysis for IEEE 802.15.3 high data rate MAC protocol," paper presented at the IEEE VTC'03-Fall, Vol. 3, Orlando, FL, Oct. 2003, pp. 1678–1682.
18. K.-H. Liu, L. Cai, and X. Shen, "Performance enhancement of medium access control for UWB WPAN," paper presented at the IEEE Globecom'06, San Francisco, CA, Dec. 2006.
19. S. B. Kodeswaran and A. Joshi, "Using location information for scheduling in 802.15.3 MAC," paper presented at the IEEE Broadnets'05, Vol. 1, Boston, MA, Oct. 2005, pp. 668–675.
20. X. Shen, W. Zhuang, H. Jiang, and J. Cai, "Medium access control in ultra-wideband wireless networks," *IEEE Trans.Vehic. Technol.* 54, 1663–1677 (2005).
21. A. Rangnekar and K. M. Sivalingam, "Multiple channel scheduling in UWB based IEEE 802.15.3 networks," paper presented at the IEEE Broadnets'04, San Jose, CA, Oct. 2004, pp. 406–415.
22. J. E. Kim, Y.A. Jeon, S. Lee, and S. S. Choi, "ECAP: An enhancement of the IEEE 802.15.3 MAC via novel scheduling scheme," paper presented at the IEEE VTC'06-Spring, Vol. 3, Melbourne, Australia, May 2006, pp. 1313–1317.
23. E. Kwon, D. Hwang, and J. Lim, "An idle timeslot reuse scheme for IEEE 802.15.3 high-rate wireless personal area networks," paper presented at the IEEE VTC'05-Fall, Vol. 2, Dallas, TX, Sept. 2005, pp. 715–719.
24. K. W. Chin and D. Lowe, "A novel IEEE 802.15.3 CTA sharing protocol for supporting VBR streams," paper presented at the IEEE ICCCN'05, San Diego, CA, Oct. 2005, pp. 107–112.

25. H. Chen, Z. Guo, R. Yao, and Y. Li, "Improved performance with adaptive Dly-ACK for IEEE 802.15.3 WPAN over UWB PHY," *IEICE Trans. Fund. Electron.* E88-A(9), 2364–2372 (2005).

26. H. Chen, Z. Guo, R. Y. Yao, X. Shen, and Y. Li, "Performance analysis of delayed acknowledgment scheme in UWB-based high-rate WPAN," *IEEE Trans. Vehic. Technol.* 55(2), 606–621 (2006).

27. W. Yu, X. Liu, Y. Cai, and Z. Zhou, "An application-aware delayed-ACK for video streaming over IEEE 802.15.3 WPANs," paper presented at the VTC'06-Spring, Vol. 3, Melbourne, Australia, May 2006, pp. 1318–1322.

28. Y. Xiao, X. Shen, and H. Jiang, "Optimal ACK mechanisms of the IEEE 802.15.3 MAC for ultra-wideband systems," *IEEE J. Sel. Areas Commun.* 24(4), 836–842 (2006).

29. K. W. Chin and D. Lowe, "Simulation study of the IEEE 802.15.3 MAC," paper presented at the ATNAC, Sydney, Australia, Dec. 2004.

30. S.-Y. Hung, P.-Y. Chuang, Y.-H. Tseng, E.H-K. Wu, and G.-H. Chen, "Energy efficient TCP transmission for IEEE 802.15.3 WPAN," paper presented at the IEEE PIMRC'06, Helsinki, Finland, Sept. 2006, pp. p 1–6.

31. Z. Yin and V. C. M. Leung, "Third-party handshake protocol for efficient peer discovery and route optimization in IEEE 802.15.3 WPANs," ACM/Springer *J. Mobile Network Appl.* 11(5), 681–695 (2006).

32. Z. Yin and V. C. M. Leung, "IEEE 802.15.3 intra-piconet route optimization with application awareness and multi-rate carriers," paper presented at the ACM IWCMC'06, Vancouver, BC, July 2006, pp. 851–856.

33. A. Batra et al., "Multi-band OFDM physical layer proposal for IEEE 802.15 task group 3a," IEEE P802.15-04/0493r0, IEEE, New York, Sept. 2004.

CHAPTER 11

PERFORMANCE ANALYSIS OF MB-OFDM UWB SYSTEMS

CHRIS SNOW, LUTZ LAMPE, and ROBERT SCHOBERG

11.1 INTRODUCTION

Ultra-wideband (UWB) radio has recently been popularized as a technology for short-range, high data rate communication and locationing applications (cf., e.g., [1]). The IEEE 802.15 standardization group, responsible for wireless personal area networks (WPANs), organized Task Group 3a to develop an alternative physical layer based on UWB signaling. There were two main contenders for this standard: a multiband frequency hopping orthogonal frequency division multiplexing (OFDM) proposal known as *MB-OFDM* and a code division multiple access (CDMA) based technique. Unfortunately, the 3a Task Group disbanded without finalizing a standard.

This work is concerned with the MB-OFDM system, which is supported by the WiMedia Alliance,[1] and which has been standardized by the European Computer Manufacturers Association (ECMA) [2–4] and also adopted for use in Wireless USB.[2] MB-OFDM is a conventional OFDM system [5] combined with bit-interleaved coded modulation (BICM) [6] for error prevention and frequency hopping for multiple access and improved diversity. The signal bandwidth is 528 MHz, which makes it a UWB signal according to the definition of the U.S. Federal Communications Commission (FCC) [7], and hopping between three adjacent frequency bands is employed for first generation devices [2–4]. Thus, MB-OFDM is a rather pragmatic approach for UWB transmission, which builds upon the proven BICM-OFDM concept.

[1] See http://www wimedia.org/.

[2] See http://www.usb.org/developers/wusb/.

Emerging Wireless LANs, Wireless PANs, and Wireless MANs. Edited by Y. Xiao and Y. Pan

11.1.1 Objectives

Our objective here is to study the suitability and to analyze the (potential) performance of MB-OFDM for UWB transmission. We approach the performance analysis problem from several different angles:

- *Information theoretic*. We calculate the channel capacity and cutoff rate of MB-OFDM systems for UWB channels.
- *Communication theoretic*. We present two methods of analytically approximating the error rate of coded MB-OFDM systems.
- *Practical*. We present simulation results for MB-OFDM.

Our investigations rely on the quasi-static stochastic time domain UWB channel model developed under IEEE 802.15 [8]. We first analyze this channel model in the frequency domain and extract the relevant statistical parameters that affect the performance of OFDM-based transmission. In particular, the amount of diversity available in the wireless channel as a function of the signal bandwidth is examined.

As appropriate performance measures for coded communication systems, we next discuss the capacity and cutoff rate limits of MB-OFDM systems for UWB channels. In this context, since one limiting factor of performance in practical and especially in wideband MB-OFDM systems is the availability of high quality channel state estimates, the effect of imperfect channel state information (CSI) at the receiver is specifically addressed.

We then present an alternative approach to system performance analysis, namely (semi-)analytical approximation of the MB-OFDM system error rate. Classical bit error rate (BER) analysis techniques [6, 9] are not applicable to the MB-OFDM system due to the short-length channel-coded packet-based transmission and because of the quasi-static nature of the channel. Motivated by the considerations mentioned above, we develop two analytical methods to obtain BER performance over ensembles of channel realizations. The two methods are best suited to different types of analysis, as will be discussed in detail below.

After applying the methods of analysis mentioned above to the study of the MB-OFDM system performance, we end the chapter by proposing system performance enhancements through the application of capacity-approaching Turbo and repeat-accumulate (RA) codes and by using OFDM bit loading. These specific techniques were chosen because of their potential for improved system performance without requiring substantial changes to other portions of the MB-OFDM system, nor requiring major increases in complexity.

This chapter unifies in a tutorial form some of the main results of our previous studies of MB-OFDM, without elaborating on all the details and extensions. The interested reader is referred to [10, 11].

11.1.2 Notation

Table 11.1 introduces the notation used in this chapter.

TABLE 11.1 Notation

Symbol	Meaning
$\mathbf{x}$	a column vector
X	a matrix
$(\cdot)^T$	matrix transpose
$(\cdot)^H$	Matrix Hermitian transpose
$(\cdot)^*$	Complex conjugate
diag($\mathbf{x}$)	a matrix with the elements of vector $\mathbf{x}$ on the main diagonal
$\mathbb{E}(\cdot)$	the expectation of a random variable
$\Pr\{\cdot\}$	the probability of some event
$Q(\cdot)$	the Gaussian Q-function [9]
$\oplus$	the element-wise XOR operation
$\boldsymbol{I}_\rho$	the identity matrix of dimension $\rho \times \rho$
$\mathbf{0}_{\rho \times 1}$	the all-zero column vector of length ρ
det $(\cdot)$	the determinant of a matrix

11.2 MB-OFDM SYSTEM MODEL

In this section, we first introduce the transmitter model of the MB-OFDM system according to the standard [2–4]. We then describe the considered receiver structure, for which we adopt a conventional state-of-the-art architecture including channel estimation based on pilot symbols. Discussion of the UWB channel model is postponed until Section 11.3.

11.2.1 Transmitter

The block diagram of the MB-OFDM transmitter is shown in Fig. 11.1. A total of eight data rates (from 53.3 to 480 Mbps) are supported by the use of different code puncturing patterns as well as time and/or frequency repetition.

11.2.1.1 Channel Coding. Channel coding in MB-OFDM consists of classical BICM [6] with a punctured maximum free distance rate $\frac{1}{3}$ constraint length 7 convolutional encoder. Puncturing patterns defined in [2] provide rates of $\frac{1}{3}$, $\frac{1}{2}$, $\frac{5}{8}$, and $\frac{3}{4}$.

A multistage block-based channel interleaver is used (see [2–4] for details). The channel interleaver length (300, 600, or 1200 coded bits) depends on the spreading factor (see Section 11.2.1.3).

11.2.1.2 Modulation. The interleaved coded bits are mapped to quaternary phase shift keying (QPSK) symbols using Gray labeling, which carry $R_m = 2$ bits per symbol. This can be regarded as equivalent to binary modulation.

11.2.1.3 Spreading. After modulation, modulated symbols are optionally repeated in time (in two consecutive OFDM symbols) and/or frequency (two

FIGURE 11.1 MB-OFDM transmitter.

tones within the same OFDM symbol), reducing the effective code rate by a factor of 2 or 4 and providing an additional spreading gain for low data rate modes.

The spreading can equivalently be represented as a lower rate convolutional code with repeated generator polynomials. This alternative representation will be used in Section 11.5 when we consider analytical error rate approximations for coded MB-OFDM.

We denote the effective code rate (after puncturing and optional repetition) by R_c.

11.2.1.4 Framing and Transmission.

After the (optional) spreading, groups of 100 data symbols are used to form OFDM symbols with $N_t = 128$ tones. The time domain signal is generated via inverse fast Fourier transform (IFFT) and and zero padding of 37 samples is applied.

The radio frequency (RF) transmit signal hops to a different 528-MHz subband after each OFDM symbol. For first-generation devices, three bands are used, with center frequencies at 3.432, 3.960, and 4.448 GHz (see [2–4] for more details). We assume without loss of generality that hopping pattern 1 of [2–4] is used, meaning the subbands are hopped in order. As a result we can consider MB-OFDM as an equivalent $N = 384$ subcarrier OFDM system, with $N_d = 300$ data-carrying subcarriers. The remainder of the subcarriers are used for synchronization and as guard tones.

Transmission is organized in packets of varying payload lengths. Each packet header contains two pilot OFDM symbols (all tones are pilots) per frequency band, which are used at the receiver to perform channel estimation (see Section 11.2.2.1).

11.2.2 Receiver

We adopt a conventional state-of-the-art receiver structure for the MB-OFDM receiver, as depicted in Fig. 11.2. We assume perfect timing and frequency synchronization have been established. We also assume the zero padding (ZP) is longer than the delay spread of the channel impulse response and neglect noise effects in the overlap-and-add operation at the receiver [12].

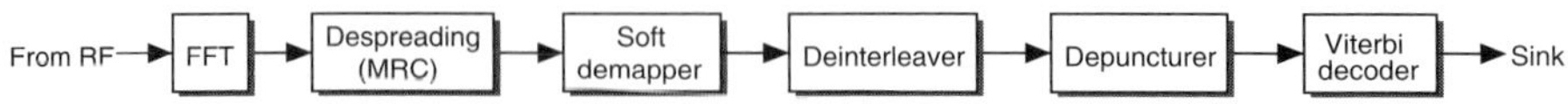

FIGURE 11.2 MB-OFDM receiver.

Due to the synchronization and sufficient ZP length, the OFDM subcarriers remain orthogonal [5]. Thus, after overlap-and-add and FFT, and considering all three bands jointly, we see an equivalent N-dimensional frequency nonselective vector channel, with subcarrier gains

$$\mathbf{h} = [\mathbf{h}_1\ \mathbf{h}_2\ \mathbf{h}_3] \triangleq [h_1\ h_2\ \ldots\ h_N]$$

where $\mathbf{h}_b (1 \leq b \leq 3)$ denotes the N_t-dimensional frequency nonselective vector channel of the bth subband.

The transmitted symbols

$$\mathbf{x}[k] \triangleq [\mathbf{x}_1[k]\ \mathbf{x}_2[k]\ \mathbf{x}_3[k]]^{\mathrm{T}}$$

of the kth composite OFDM symbol pass through the fading channel $\mathbf{H} = \mathrm{diag}(\mathbf{h})$, and the length-$N$ vector of received symbols

$$\mathbf{y}[k] \triangleq [\mathbf{y}_1[k]\ \mathbf{y}_2[k]\ \mathbf{y}_3[k]]^{\mathrm{T}}$$

(after the FFT) is given by

$$\mathbf{y}[k] = \mathbf{H}\mathbf{x}[k] + \mathbf{n}[k] \tag{11.1}$$

where

$$\mathbf{n}[k] \triangleq [\mathbf{n}_1[k]\ \mathbf{n}_2[k]\ \mathbf{n}_3[k]]^{\mathrm{T}}$$

is a vector of independent complex additive white Gaussian noise (AWGN) variables with variance $\mathcal{N}_0$. Note that $\mathbf{x}_b[k]$, $\mathbf{y}_b[k]$, and $\mathbf{n}_b[k]$ $(1 \leq b \leq 3)$ denote the transmitted symbols, received symbols, and AWGN noise variables for the bth subband.

We denote the energy per modulated symbol by

$$E_s = R_c R_m E_b$$

where E_b is the energy per information bit.

The channel estimation, diversity combining, demapping, and decoding are described in the following.

11.2.2.1 Channel Estimation. For the purposes of channel estimation, the MB-OFDM packet header structure includes $P = 2$ pilot OFDM symbols for each frequency band [2–4]. For a more general treatment, we let P be a design parameter. The responses in different frequency bands can be estimated separately.

We consider a standard channel estimation technique: least-squares error (LSE) estimation of the time domain channel impulse response (CIR), using the pilot symbols in the packet header. We apply LSE instead of minimum mean-square error (MMSE) estimation because it does not require assumptions regarding the statistical structure of the channel correlations. Furthermore, it has been shown that LSE and MMSE estimation perform almost equally well for cases of practical interest [13].

The LSE estimator exploits the fact that the CIR is shorter than the zero padding added to each OFDM symbol. We denote the assumed CIR length by L_t. Starting from Eq. (11.1), the frequency domain vector channel estimate for band b ($1 \leq b \leq 3$) can be represented as (cf., e.g., [13])

$$\hat{\mathbf{H}}_b = \mathbf{H}_b + \mathbf{E}_b \tag{11.2}$$

where $\mathbf{H}_b \triangleq \mathrm{diag}(\mathbf{h}_b)$, and the channel estimation error vector

$$\mathbf{E}_b = \mathbf{F}_{N_t \times L_t} \mathbf{F}^H_{N_t \times L_t} \cdot \frac{1}{P} \sum_{k=1}^{P} \mathbf{x}_b^H[k] \mathbf{n}_b[k] \tag{11.3}$$

is independent of $\mathbf{H}_b$ and zero-mean Gaussian distributed with correlation matrix

$$\begin{aligned} \mathbf{R}_{\mathbf{E}_b \mathbf{E}_b} &= \mathbf{F}_{N_t \times L_t} \mathbf{F}^H_{N_t \times L_t} \left(\frac{\mathcal{N}_0}{P^2} \sum_{k=1}^{P} \mathbf{x}_b^H[k] \mathbf{x}_b[k] \right) \mathbf{F}_{N_t \times L_t} \mathbf{F}^H_{N_t \times L_t} \\ &= \mathbf{F}_{N_t \times L_t} \mathbf{F}^H_{N_t \times L_t} \mathcal{N}_0 / P \end{aligned} \tag{11.4}$$

In Eqs. (11.3) and (11.4), $\mathbf{F}_{N_t \times L_t}$ denotes the normalized $N_t \times L_t$ FFT matrix with elements $\mathbf{e}^{-j\mu\nu 2\pi/N_t}/\sqrt{N_t}$ in row μ and column ν. For the last step in Eq. (11.4) we assumed the use of constant modulus pilot symbols, which holds for the MB-OFDM system. We observe from Eqs. (11.2) and (11.4) that the LSE channel estimate is disturbed by correlated Gaussian noise with variance

$$\sigma_E^2 = \frac{L_t}{N_t P} \mathcal{N}_0 = \eta \mathcal{N}_0 \tag{11.5}$$

where the channel estimation quality parameter is given by

$$\eta \triangleq \frac{L_t}{N_t P} \tag{11.6}$$

and is independent of the band number b. Because of interleaving, the effect of correlation is negligible, and it is common to ignore it for signal detection in order to keep complexity low. We will refer to parameter η in Eq. (11.5) when evaluating the performance of MB-OFDM with imperfect CSI in Sections 11.6.2 and 11.6.4. In the remainder of this chapter, we assume a maximum impulse response length of $L_t = 32$.

11.2.2.2 Diversity Combining, Demapping, and Decoding. Maximum-ratio combining (MRC) [9] in the case of time and/or frequency spreading (see Section 11.2.1.3 and [2–4]) and demapping in the standard BICM fashion [6] are performed based on the channel estimator output $\hat{\mathbf{H}}_b$. The resulting "soft" bit metrics are deinterleaved and depunctured. Standard Viterbi decoding results in an estimate of the originally transmitted information bits.

11.3 UWB CHANNEL MODEL

For a meaningful performance analysis of MB-OFDM, we consider the channel model developed under IEEE 802.15 for UWB systems [8]. In this section, we will study the channel model in detail, in order to:

1. Extract the channel parameters relevant for the performance of OFDM-based UWB systems.
2. Examine whether the design of MB-OFDM is adequate to exploit the channel characteristics.
3. Quantify the impact of the different UWB channel types on system performance.
4. Possibly enable a classification of the UWB channel model into more standard channel models used in communication theory.

11.3.1 IEEE 802.15.3a Channel Model

The IEEE 802.15.3a UWB channel model is a stochastic time domain model. The channel impulse response is a Saleh–Valenzuela model [14] modified to fit the properties of measured UWB channels. Multipath rays arrive in clusters with exponentially distributed cluster and ray interarrival times. Both clusters and rays have decay factors chosen to meet a given power decay profile. The ray amplitudes are modeled as lognormal random variables, and each cluster of rays also undergoes a lognormal fading. To provide a fair system comparison, the total multipath energy is normalized to unity. Finally, the entire impulse response undergoes an "outer" lognormal shadowing. The channel impulse response is assumed time invariant during the transmission period of several packets (see [8] for a detailed description).

Four separate channel models (CM1–CM4) are available for UWB system modeling, each with arrival rates and decay factors chosen to match a different usage scenario.[3] The four models are tuned to fit 0–4 m line-of-sight (LOS), 0–4 m non-LOS, 4–10 m non-LOS, and an "extreme non-LOS multipath channel," respectively. The means and standard deviations of the outer

[3] We note that more recently additional models have been introduced for use by IEEE 802.15 Task Group 4a [15].

lognormal shadowing are the same for all four models. The model parameters can be found in [8, Table 2].

For the remainder of this work, we consider only channels CM1–CM3, where the assumption that the zero padding is longer than the delay spread of the channel impulse response holds (some realizations of channel CM4 violate this assumption). However, we note that the CM4 performance is very similar to that of CM3.

Because of the quasi-static nature of the channel, we must consider the performance of MB-OFDM over ensembles of channel realizations. As we will discuss later, the number N_c of realizations we use must be sufficiently large to capture the true behavior of the channel model. This is a problem when performing system simulations since the simulation program must be run separately for all N_c channel realizations, which leads to high computational complexity (see also Section 11.6.3 for a discussion of the computational requirements).

11.3.2 Frequency Domain Channel Description

As mentioned in Section 11.2, we consider a stochastic frequency domain channel description, that is, we include transmitter IFFT and receiver FFT into the channel definition and consider realizations of $\mathbf{H}$ in Eq. (11.1).

From Eq. (11.1) we observe that the OFDM transmit signal experiences a frequency nonselective fading channel with fading along the frequency axis. Thus, the outer lognormal shadowing term is irrelevant for the fading characteristics as it affects all tones equally. Hence, the lognormal shadowing term is omitted in the following considerations. Denoting the lognormal term by G, we obtain the corresponding *normalized* frequency domain fading coefficients as

$$h_i^n = \frac{h_i}{G} \tag{11.7}$$

and note that $\mathbf{H}^n \triangleq \text{diag}([h_1^n \cdots h_N^n])$.

In Section 11.6, we show performance results for ensembles of channel realizations. In this case, we denote by $\bar{E}_s$ and $\bar{E}_b$ the average energy per symbol and average energy per information bit, respectively. These quantities are obtained by averaging the instantaneous energies E_s and E_b over the distribution of the lognormal shadowing G [8].

11.3.3 Marginal Distribution

The first parameter of interest is the marginal distribution of h_i^n, that is, the probability density function (pdf) $p(h_i^n)$.

First, we note that the frequency domain coefficient h_i^n is a zero mean random variable since the time domain multipath components are zero mean

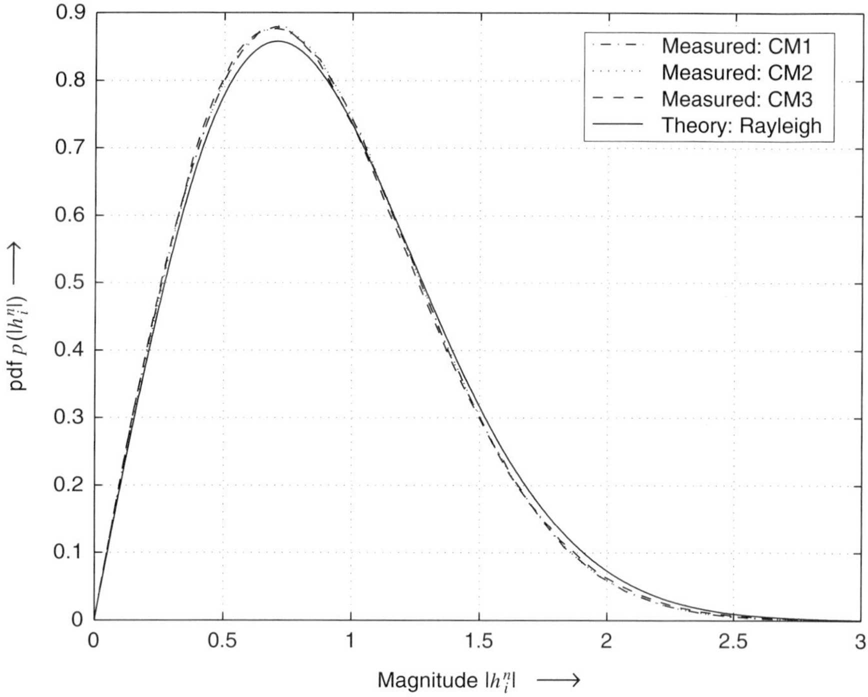

FIGURE 11.3 Distributions of normalized channel magnitude $|h_i^n|$ for channel types CM1–CM3. For comparison: Rayleigh distribution with same variance.

quantities. Furthermore, we have observed that h_i^n is, in good approximation, circularly symmetric complex Gaussian distributed, which is explained by the fact that h_i^n results from the superposition of many time domain multipath components. Since these multipath components are mutually statistically independent, the variance of h_i^n is independent of the tone index i.

Figure 11.3 shows measurements of the pdfs $p(|h_i^n|)$ of the magnitude frequency domain gain $|h_i^n|$ for the different channel models CM1–CM3, obtained from 10,000 independent realizations of channel model. As can be seen, the experimental distributions agree well with the exact Rayleigh distribution of equal variance, which is in accordance with the statements above. We note that similar conclusions regarding the frequency domain gains were also obtained in [16].

11.3.4 Correlation

The findings in the previous section indicate that the OFDM signal effectively experiences a (classical) frequency nonselective Rayleigh fading channel (along the OFDM subcarriers). Therefore, knowledge of the second-order channel

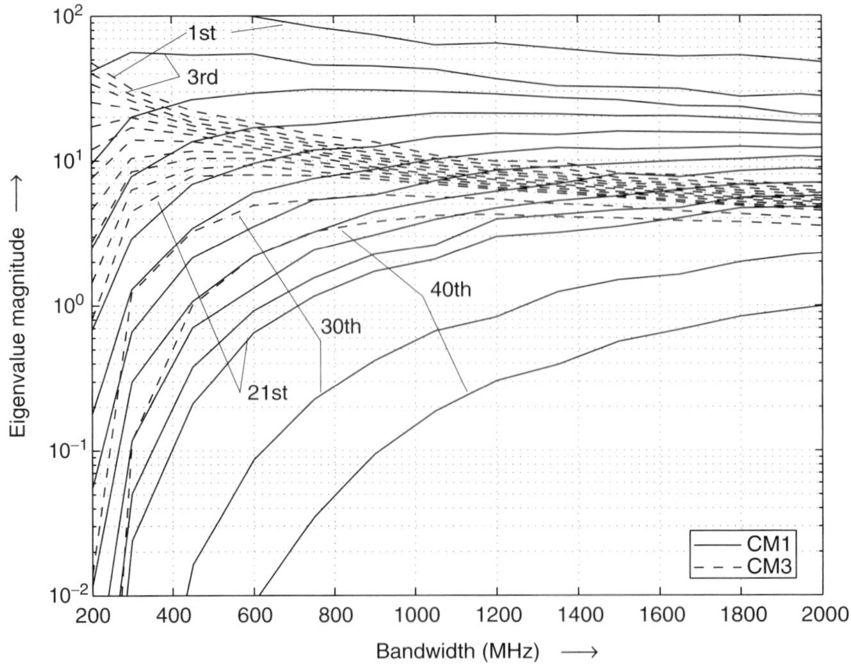

FIGURE 11.4 First 40 ordered eigenvalues of correlation matrix $\mathbf{R}_{\mathbf{H}^n\mathbf{H}^n}$ (every second from 1st to 21st, and the 30th and 40th).

statistics, that is, the correlation between different fading coefficients h_i^n and h_j^n, $i \neq j$, is important for the design and assessment of diversity techniques such as coding, interleaving, and frequency hopping, which are envisioned in the MB-OFDM system. Since coding is performed over all bands, we consider all three bands jointly.

As an appropriate figure of merit we examine the ordered eigenvalues of the $N \times N$ correlation matrix $\mathbf{R}_{\mathbf{H}^n\mathbf{H}^n}$ of $\mathbf{H}^n$. Figure 11.4 shows the first 40 ordered eigenvalues (every second from 1st to 21st, and the 30th and 40th) of the measured $\mathbf{R}_{\mathbf{H}^n\mathbf{H}^n}$, which has been obtained from averaging over 1000 channel realizations, as a function of the total employed signal bandwidth. We only show results for channel models CM1 and CM3, which constitute the two extreme cases as the corresponding impulse responses have the least (CM1) and most (CM3) independent multipath components. The respective curve for model CM2 lies in between those for CM1 and CM3.

From Fig. 11.4 we infer the following conclusions:

1. By increasing the bandwidth of the OFDM signal, the diversity order of the equivalent frequency domain channel, that is, the number of the significant nonzero eigenvalues of $\mathbf{R}_{\mathbf{H}^n\mathbf{H}^n}$, is improved since, generally,

more time domain multipath components are resolved. However, a 1500 MHz total bandwidth provides already ≥ 40 (CM3) and ≥ 30 (CM1) strong diversity branches. This indicates that the 528 MHz bandwidth and three-band frequency hopping of MB-OFDM is a favorable compromise between complexity and available diversity.

2. Since the system, comprising the convolutional code (see Section 11.2.1.1) with free distance ≤ 15 (depending on the puncturing) and spreading factor 1, 2, and 4, can at best exploit diversities of order 15, 30, and 60, respectively, bandwidths of more than 500 MHz per band would only be beneficial for the lowest data rate modes, and then only for very low error rates. Similar considerations apply to concatenated codes (e.g., Turbo and RA codes as considered in Section 11.7), which do not exceed convolutional codes with spreading in terms of free distance.
3. Though CM3 provides higher diversity order than CM1, the latter appears advantageous for high data rate modes with code puncturing due to its larger first ordered eigenvalues.

In summary, we conclude that, given a particular realization of the lognormal shadowing term G, the equivalent frequency domain channel $\mathbf{H}^n = \mathbf{H}/G$ is well approximated by a Rayleigh fading channel with relatively high "fading rate," which increases from CM1 to CM3.

11.4 CHANNEL CAPACITY AND CUTOFF RATE

The purpose of this section is to quantify potential data rates and power efficiencies of OFDM-based UWB transmission. Of particular interest here are:

1. The channel capacity and cutoff rate, which are widely accepted performance measures for coded transmission using powerful concatenated codes and convolutional codes, respectively
2. The influence of the particular channel model (CM1-CM3)
3. The effect of imperfect channel estimation on these measures

It is important to note that the capacity and cutoff rate discussed here are *constellation constrained*, that is, they are calculated assuming a given input constellation with uniform input probabilities. While the MB-OFDM standard specifies QPSK modulation, we consider arbitrary constellations for the extensions in Section 11.7, that is, general BICM-OFDM schemes.

Since coding and interleaving are limited to single realizations of lognormal shadowing, we focus on the notion of *outage probability*, that is, the probability that the instantaneous capacity and cutoff rate for a given channel realization $\mathbf{H}$ fall below a certain threshold. These theoretical performance measures will be compared with the other analysis techniques in Section 11.6.

In Section 11.4.1, we review the instantaneous capacity and cutoff rate expressions for BICM-OFDM. The required conditional pdf of the channel output is given in Section 11.4.2. We will present extensions to these expressions in Section 11.7.2.2, when we consider OFDM bit loading as a possible extension for improved MB-OFDM performance.

11.4.1 Capacity and Cutoff Rate Expressions

The instantaneous capacity in bits per complex dimension of an N_d tone BICM-OFDM system in a frequency-selective quasi-static channel is given in [17] (by extending the results of [6]) as

$$C(\mathbf{H}) = m - \frac{1}{N_d} \sum_{\ell=1}^{m} \sum_{i=1}^{N_d} \mathbb{E}_{b,y_i} \left\{ \log_2 \left(\frac{\sum_{x_i \in \mathcal{X}} p\left(y_i | \hat{h}_i, x_i\right)}{\sum_{x_i \in \mathcal{X}_b^\ell} p\left(y_i | \hat{h}_i, x_i\right)} \right) \right\} \tag{11.8}$$

where

m number of bits per symbol ($m = 2$ for QPSK)

χ signal constellation

$\mathcal{X}_b^\ell$ set of all constellation points $x \in \mathcal{X}$ whose label has value $b \in \{0,1\}$ in position ℓ

$p(y_i|\hat{h}_i, x_i)$ pdf of channel output y_i for given input x_i and channel estimate $\hat{h}_i$.

Similarly, we can express the instantaneous cutoff rate in bits per complex dimension as (cf., e.g., [6, 17])

$$R_0(\mathbf{H}) = m - m \log_2[B(\mathbf{H}) + 1] \tag{11.9}$$

with the instantaneous Bhattacharya parameter ($\bar{b}$ denotes the complement of b)

$$B(\mathbf{H}) = \frac{1}{mN_d} \sum_{\ell=1}^{m} \sum_{i=1}^{N_d} E_{b,y_i} \left\{ \sqrt{\frac{\sum_{x_i \in \mathcal{X}_{\bar{b}}^\ell} p\left(y_i | \hat{h}_i, x_i\right)}{\sum_{x_i \in \mathcal{X}_b^\ell} p\left(y_i | \hat{h}_i, x_i\right)}} \right\} \tag{11.10}$$

11.4.2 Conditional pdf

In order to calculate capacity and cutoff rate, we require the conditional pdf $p(y_i|\hat{h}_i, x_i)$. In the case of perfect CSI we have $\hat{h}_i = h_i$, and $p(y_i|\hat{h}_i, x_i)$ is a Gaussian pdf with mean $h_i x_i$ and variance $\mathcal{N}_0$.

We now obtain $p(y_i|\hat{h}_i, x_i)$ for the more realistic case of imperfect CSI assuming the application of LSE channel estimation as described in Section 11.2.2.1, According to the results of Section 11.3.3, and since channel estimation is performed for one realization G of the lognormal shadowing term, we further assume zero-mean circularly symmetric Gaussian distributed channel coefficients h_i with variance $\sigma_H^2 = G^2$ [see Eq: (11.7)]. This means that $\hat{h}_i$ is also zero-mean Gaussian distributed with variance

$$\sigma_{\hat{H}}^2 = \sigma_H^2 + \sigma_E^2 \tag{11.11}$$

[see Eqs. (11.2) and (11.5)]. Let μ be the correlation between h_i and $\hat{h}_i$,

$$\begin{aligned} \mu &= \frac{\mathbb{E}_{h_i,\hat{h}_i}\{h_i\hat{h}_i^*\}}{\sigma_H \sigma_{\hat{H}}} \\ &= \sqrt{\frac{\sigma_H^2}{\sigma_E^2 + \sigma_H^2}} \\ &= \sqrt{\frac{\gamma}{\gamma + \eta}} \end{aligned} \tag{11.12}$$

where η is defined in Eq. (11.6), and

$$\gamma = \frac{\sigma_H^2}{\mathcal{N}_0}$$

is the signal-to-noise ratio (SNR). Then, we can arrive via algebraic manipulations at (cf., e.g., [18])

$$p(y_i|\hat{h}_i, x_i) = \frac{1}{\pi[\mathcal{N}_0(\eta\mu^2 + 1)]} \exp\left(- \frac{|y_i - x_i\hat{h}_i\mu^2|^2}{\mathcal{N}_0(\eta\mu^2 + 1)} \right) \tag{11.13}$$

The Gaussian density in Eq: (11.13) implies that the system with imperfect CSI can be seen as a system with perfect CSI at an equivalent SNR of

$$\gamma_e = \frac{\mathbb{E}_{\hat{h}_i}\left\{|\hat{h}_i|^2\right\}\mu^4}{\mathcal{N}_0(\eta\mu^2 + 1)} = \frac{\gamma}{\eta(1 + 1/\gamma) + 1} \tag{11.14}$$

We note that in the high SNR regime the loss due to estimation error reaches a constant value of

$$\frac{1}{\eta + 1}$$

11.5 ERROR RATE APPROXIMATIONS

As mentioned in Section 11.3.1, simulation-based approaches to obtain MB-OFDM system performance are very time consuming as a result of the quasi-static nature of the UWB channel, which necessitates simulating the system over a large number of channel realizations. There are well-known techniques for bounding the performance of convolutionally encoded transmission over many types of fading channels, for example, [6, 9]. However, such classical BER analysis techniques are not applicable to MB-OFDM systems for several reasons. First, the short-length channel-coded packet-based transmissions are nonideally interleaved, which results in nonzero correlation between adjacent coded bits. Second, and more importantly, the quasi-static nature of the channel limits the number of distinct channel gains to the (relatively small) number of OFDM tones. This small number of distinct channel gains must not be approximated by the full fading distribution for a valid performance analysis, as would be the case in a fast-fading channel. We also note that, due to the relative complexity of this channel model, even BER analysis for uncoded UWB systems has proven challenging [19].

In the quasi-static channel setting it is also often of significant interest to obtain the outage BER performance, that is, the minimum expected BER performance after excluding some percentage of the worst-performing channel realizations [20, Section III. C-2]. Until now, one has had to resort to intensive numerical simulations in order to obtain the BER performance for each channel realization, and hence obtain the outage BER performance [4].

Motivated by the considerations mentioned above, we have developed two analytical methods to obtain outage as well as average BER performance over ensembles of channel realizations [11]. We assume perfect frequency synchronization and perfect channel state information at the receiver for both methods. The two methods are best suited to different types of analysis, as will be discussed in detail below.

The remainder of this section is organized as follows:

1. We develop a method for approximating the BER of coded multicarrier systems on a *per-realization* basis (method I). This method is most suitable for obtaining the outage BER but can also be used to obtain the average BER performance (Section 11.5.3).
2. For quasi-static channels with correlated Rayleigh-distributed subcarrier channel gains, we present an alternative method (method II) to directly and efficiently obtain the average BER performance (Section 11.5.4).

Both methods are based on considering the set of error vectors, introduced in Section 11.5.1, and the pairwise error probability (PEP) of an error vector, given in Section 11.5.2.

11.5.1 Error Vectors

Consider a convolutional encoder initialized to the all-zero state, where the reference (correct) codeword is the all-zero codeword. We construct all L input sequences, which cause an immediate deviation from the all-zero state (i.e., those whose first input bit is 1) and subsequently return the encoder to the all-zero state with an output Hamming weight of at most $w_{\max}$.

We let:

- $\mathscr{E}$ be the set of all vectors $\mathbf{e}_\ell$ $(1 \leq \ell \leq L)$ representing the output sequences (after puncturing) associated with these input sequences, that is, $\mathscr{E} = \{\mathbf{e}_1, \mathbf{e}_2, \ldots, \mathbf{e}_L\}$.
- l_ℓ be the length of $\mathbf{e}_\ell$ (the number of output bits after puncturing).
- a_ℓ be the Hamming weight of the input associated with $\mathbf{e}_\ell$.

Note that the choice of $\omega_{\max}$ governs the value of L (i.e., once the maximum allowed Hamming weight is set, the number of error events L is known).

We term $\mathbf{e}_\ell$ an "error vector" and $\mathscr{E}$ the set of error vectors. The set $\mathscr{E}$ contains all the low weight error events, which are the most likely deviations in the trellis. As with standard union-bound techniques for convolutional codes [9], the low weight terms will dominate the error probability. Hence, it is sufficient to choose a small $w_{\max}$—the MB-OFDM code of rate $R_c = \frac{1}{2}$ [2–4] has a free distance of 9 after puncturing, and choosing $w_{\max} = 14$ (resulting in a set of $L = 242$ error vectors of maximum length $l = 60$) provides results that are not appreciably different from those obtained using larger $w_{\max}$ values.

We obtained $\mathscr{E}$ by modifying an algorithm for calculating the convolutional code distance spectrum [21] in order to store the code output sequences (i.e., the error vectors $\mathbf{e}_\ell$) in addition to the distance spectrum information.

11.5.2 Pairwise Error Probability for an Error Vector

We assume that the transmitter selects $R_c R_m N_d$ random message bits for transmission, denoted by $\mathbf{b} = [b_1\, b_2 \ldots b_{R_c R_m N_d}]^{\mathrm{T}}$. The vectors $\mathbf{c}$ and $\mathbf{c}^\pi$ of length $L_c = R_m N_d$ represent the bits after encoding/puncturing and after interleaving, respectively. The bits $\mathbf{c}^\pi$ are then modulated using QPSK on each subcarrier, and the resulting N_d-modulated symbols are denoted by $\mathbf{x} = [x_1\, x_2 \ldots x_{N_d}]^{\mathrm{T}}$. At the receiver, we denote the estimate of the original bitstream, produced by the Viterbi decoder, as $\hat{\mathbf{b}} = [\hat{b}_1\, \hat{b}_2 \cdots \hat{b}_{R_c R_m N_d}]^{\mathrm{T}}$.

We now consider error events starting in a given position i of the codeword $(1 \leq i \leq L_c)$. For a specific error vector $\mathbf{e}_\ell$ $(1 \leq \ell \leq L)$, form the full error codeword

$$\mathbf{q}_{i,\ell} = [\underbrace{0\,0 \cdots 0}_{i-1}\ \underbrace{\mathbf{e}_\ell}_{l_\ell}\ \underbrace{0\,0 \cdots 0}_{L_c - l_\ell - i + 1}]^{\mathrm{T}} \tag{11.15}$$

of length L_c by padding $\mathbf{e}_\ell$ with zeros on both sides as indicated above. Given the error codeword $\mathbf{q}_{i,\ell}$ and given that codeword $\mathbf{c}$ is transmitted, the competing codeword is given by

$$\boldsymbol{\upsilon}_{i,\ell} = \mathbf{c} \oplus \mathbf{q}_{i,\ell} \tag{11.16}$$

The decoder employs a standard Euclidean distance metric. Letting $\mathbf{z}_{i,\ell}$ be the vector of QPSK symbols associated with $\boldsymbol{\upsilon}^{\pi}_{i,\ell}$ (the interleaved version of $\boldsymbol{\upsilon}_{i,\ell}$), and recalling that $\mathbf{x}$ is the modulated symbol vector corresponding to the original codeword $\mathbf{c}$, the PEP for the ℓth error vector starting in the ith position, that is, the probability that $\boldsymbol{\upsilon}_{i,\ell}$ is detected given that $\mathbf{c}$ was transmitted, is given by

$$\text{PEP}_{i,\ell}(\mathbf{H}) = \Pr\left\{ \|\mathbf{y} - \mathbf{Hx}\|^2 > \|\mathbf{y} - \mathbf{Hz}_{i,\ell}\|^2 | \mathbf{H} \right\} \tag{11.17}$$

11.5.3 Per-Realization Performance Analysis (Method I)

In this section, we obtain an approximation of the BER $P(\mathbf{H})$ for a particular channel realization $\mathbf{H}$. For simplicity, we refer to this method as Method I. As noted above and discussed in more detail in Section 11.5.3.3, the main strength of this method is the ability to obtain the outage BER.

11.5.3.1 Pairwise Error Probability. The PEP for an error vector $\mathbf{e}_\ell$ $(1 \le \ell \le L)$ with the error event starting in a position i $(1 \le i \le L_c)$ is given by Eq: (11.17). For a given $\mathbf{H}$, we obtain the expression

$$\text{PEP}_{i,\ell}(\mathbf{H}) = Q\left(\sqrt{\frac{\|\mathbf{H}(\mathbf{x} - \mathbf{z}_{i,\ell})\|^2}{\mathcal{N}_0}} \right) \tag{11.18}$$

11.5.3.2 Per-Realization BER. The corresponding BER for the ℓth error vector, starting in the ith position, is given by

$$P_{i,\ell}(\mathbf{H}) = a_\ell \, \text{PEP}_{i,\ell}(\mathbf{H}) \tag{11.19}$$

Summing over all L error vectors, we obtain an approximation of the BER for the ith starting position as

$$P_i(\mathbf{H}) = \sum_{\ell=1}^{L} P_{i,\ell}(\mathbf{H}) \tag{11.20}$$

We note that Eq. (11.20) can be seen as a standard truncated union bound for convolutional codes (i.e., it is a sum over all error events of Hamming weight less than $\omega_{\max}$). We also note that we can tighten this bound by limiting P_i to a maximum value of $\frac{1}{2}$ before averaging over starting positions [22].

TABLE 11.2 Pseudocode for Method I

Method I — Final BER is P (for given $\mathbf{H}$)

```
P :=0
for i := 1 to L_c do
  P_i := 0
  for ℓ := 1 to L do
    form q_{i,ℓ} as per Eq. (11.15)
    form υ_{i,ℓ} as per Eq. (11.16)
    form υ^π_{i,ℓ} and z_{i,ℓ} from υ_{i,ℓ}
    calculate PEP_{i,ℓ} as per Eq. (11.18)
    calculate P_{i,ℓ} as per Eq. (11.19)
    P_i :=P_i+P_{i,ℓ}
endfor
P := P+min(1/2, P_i)
endfor
P :=P/L_c
```

Finally, since all starting positions are equally likely, the BER $P(\mathbf{H})$ can be written as

$$P(\mathbf{H}) = \frac{1}{L_c} \sum_{i=1}^{L_c} \min\left[\frac{1}{2}, \sum_{\ell=1}^{L} P_{i,\ell}(\mathbf{H})\right] \tag{11.21}$$

Table 11.2 contains pseudocode to calculate $P(\mathbf{H})$ according to Eq. (11.21).

11.5.3.3 Average and Outage BER. The average BER can be obtained by averaging Eq. (11.21) over a (large) number N_c of channel realizations, where the ith channel realization is denoted by $\mathbf{H}_i$ ($1 \leq i \leq N_c$), as

$$P = \frac{1}{N_c} \sum_{i=1}^{N_c} P(\mathbf{H}_i) \tag{11.22}$$

As mentioned previously, method I also readily lends itself to the consideration of the outage BER, a common measure of performance for packet-based systems operating in quasi-static channels [20]. The outage BER[4] provides a measure of the minimum performance that can be expected of the system given a specified X% outage rate and is often employed in UWB system performance studies [4].

[4] An alternative measure of outage is the *outage probability*, that is, the probability that the BER exceeds some nominal value BER_0 in an OFDM block. The outage probability can also obtained given the per-realization BER in Eq. (11.21).

We evaluate Eq. (11.21) for a set of N_c channel realizations denoted by

$$\mathscr{H} = \{\mathbf{H}_i, 1 \leq i \leq N_c\}$$

The worst-performing $X\%$ of realizations are considered in outage, and those channel realizations are denoted by $\mathscr{H}_{\text{out}}$. Denoting the remaining $(100-X)\%$ of channel realizations by $\mathscr{H}_{\text{in}}$, the outage BER is given by

$$P_{\text{out}} = \max_{\mathbf{H}_i \in \mathscr{H}_{\text{in}}} P(\mathbf{H}_i) \tag{11.23}$$

11.5.4 Average Performance Analysis (Method II)

In this section, we propose a method based on knowledge of the frequency domain channel correlation matrix, which can be used directly in order to obtain the average BER performance of coded multicarrier systems. The advantage of this method is that it allows for simple and direct evaluation of the average BER, without the need to evaluate the BER of many different channel realizations as in method I; cf., Eq. (11.22). For simplicity, we refer to this method as method II.

For this method we will explicitly assume that the elements of $\mathbf{h}$ are Rayleigh distributed and have known correlation matrix Σ_{hh} (in practice, Σ_{hh} can be obtained from actual channel measurements or can be numerically estimated by measuring the correlation over many realizations of a given channel model; cf. Section 11.3.4). As noted in Section 11.3, the UWB channel models for MB-OFDM communication satisfy this assumption.

11.5.4.1 Average PEP. Noting that only the $\rho_{i,\ell}$ nonzero terms of $(\mathbf{x}-\mathbf{z}_{i,\ell})$ in Eq. (11.17) contribute to the PEP (and suppressing the dependence of ρ on i and ℓ for notational clarity), we let $\mathbf{x}'$, $\mathbf{z}'_{i,\ell}$, $\mathbf{H}' = \text{diag}(\mathbf{h}')$, and $\mathbf{n}'$ represent the transmitted symbols, received symbols, channel gains, and AWGN noises corresponding to the ρ nonzero entries of $(\mathbf{x}-\mathbf{z}_{i,\ell})$, respectively, and form $\Sigma_{h'h'}$ by extracting the elements from Σ_{hh}, which correspond to $\mathbf{h}'$.

Letting

$$\mathbf{D} = \text{diag}(\mathbf{x}' - \mathbf{z}'_{i,\ell})$$

be the diagonal matrix of nonzero entries and

$$\mathbf{g} = \mathbf{H}'(\mathbf{x}' - \mathbf{z}'_{i,\ell}) = \mathbf{D}\mathbf{h}'$$

we have

$$\mathbb{E}(\mathbf{g}) = 0_{\rho \times 1} \tag{11.24}$$

$$\mathbb{E}\left(\mathbf{g}\mathbf{g}^{H}\right) = \mathbf{R}_{gg} = \mathbf{D}\Sigma_{\mathrm{h'h'}}\mathbf{D}^{H} \tag{11.25}$$

that is, the distribution of $\mathbf{g}$ is zero-mean complex Gaussian with covariance matrix $\mathbf{R}_{gg}$.

We would like to obtain the average $\overline{\mathrm{PEP}}_{i,\ell}$ for the ℓth error vector, starting in the ith position. Rewriting Eq: (11.7) including only the contributing terms, we obtain

$$\overline{\mathrm{PEP}}_{i,\ell} = \Pr\left\{ \left\| \mathbf{y}' - \mathbf{H}'\mathbf{z}'_{i,\ell} \right\|^2 - \|\mathbf{y}' - \mathbf{H}'\mathbf{x}'\|^2 < 0 \right\} \tag{11.26}$$

From Eq. (11.26) and following [23, Eq. (7)], we can write the average PEP for the ℓth error vector starting in the ith position as

$$\overline{\mathrm{PEP}}_{i,\ell} = \frac{1}{\pi} \int_0^{\pi/2} \left[\det\left(\frac{2E_s \mathbf{R}_{gg}}{\mathcal{N}_0 \sin^2\theta} + \mathbf{I}_\rho \right) \right]^{-1} d\theta \tag{11.27}$$

11.5.4.2 Average BER. Given the average PEP according to Eq. (11.27), the corresponding BER for the ℓth error vector, starting in the ith position, is given by

$$\bar{P}_{i,\ell} = a_\ell \, \overline{\mathrm{PEP}}_{i,\ell} \tag{11.28}$$

Summing over all L error vectors, the BER for the ith starting position can be written as

$$\bar{P}_i = \sum_{\ell=1}^{L} \bar{P}_{i,\ell} \tag{11.29}$$

Finally, since all starting positions are equally likely to be used, the average BER $\bar{P}$ can be written as

$$\bar{P} = \frac{1}{L_c} \sum_{i=1}^{L_c} \bar{P}_i = \frac{1}{L_c} \sum_{i=1}^{L_c} \sum_{\ell=1}^{L} \bar{P}_{i,\ell} \tag{11.30}$$

Table 11.3 contains pseudocode to calculate $\bar{P}$ according to Eq. (11.30). Note that, since $\bar{P}_{i,\ell}$ in Eq. (11.30) is already averaged over $\mathbf{H}$, we cannot upper bound it by $\frac{1}{2}$ as we did in Eq. (11.21) for method I. This implies that the result for method II may be somewhat looser than that for method I (see also Section 11.6.3).

11.6 RESULTS AND DISCUSSION

In this section, we present numerical results from the capacity and cutoff rate analysis (Section 11.4) and the error rate approximations (Section 11.5), as well as from system simulations. We discuss the suitability of the MB-OFDM system for UWB communication based on comparison with theoretic measures of performance.

TABLE 11.3 Pseudocode for Method II

Method II	Final BER is $\bar{P}$
	$\bar{P}$:=0
	for i :=1 to L_c do
	for ℓ :=1 to L do
	form $q_{i,\ell}$ as per Eq. (11.15)
	form $\upsilon_{i,\ell}$ as per Eq. (11.16)
	form $\upsilon^{\pi}_{i,\ell}$ and $z_{i,\ell}$ from $\upsilon_{i,\ell}$
	form $x'_{i,\ell}$, $z'_{i,\ell}$, h', $\mathcal{J}'$ and
	compute $D := \text{diag}(x' - z'_{i,\ell})$
	compute $g = Dh'$ and $R_{gg} := D\Sigma_{h'h'}D^H$
	calculate $\overline{\text{PEP}}_{i,\ell}$ as per Eq. (11.27)
	calculate $\bar{P}_{i,\ell}$ as per Eq. (11.28)
	$\bar{P} := \bar{P} + \bar{P}_{i,\ell}$
	endfor
	endfor
	$\bar{P} := \bar{P}/L_c$

11.6.1 Capacity and Cutoff Rate

We evaluated Eqs. (11.8) and (11.9) via Monte Carlo simulation using 1000 realizations of each UWB channel model CM1–CM3. To keep the figures legible, we present representative results for CM1 and CM3 only. The performance of CM2 (not shown) is between that of CM1 and CM3 (cf. also Section 11.3.4). For comparison we also include results for independent and identically distributed (i.i.d.) Rayleigh fading on each tone and an outer lognormal shadowing term identical to that of the UWB models (labeled as "Rayleigh + LN").

First, we consider the case of perfect CSI. Figure 11.5 shows the outage capacity $\Pr\{C(\mathbf{H}) < R\}$ (left) and cutoff rate $\Pr\{R_0(\mathbf{H}) < R\}$ (right) as a function of the threshold rate R for $10\log_{10}(\bar{E}_s/\mathcal{N}_0) = 5$ and 10 dB, respectively.

It can be seen that both capacity and cutoff rate for the UWB channel models are similar to the respective parameters of an i.i.d. Rayleigh fading channel with additional lognormal shadowing. In fact, the curves for CM3, which provides the highest diversity (see Section 11.3.4), are essentially identical to those for the idealized i.i.d. model. The high diversity provided by the UWB channel also results in relatively steep outage curves, which means that transmission reliability can be considerably improved by deliberately introducing coding redundancy. This effect is slightly more pronounced for the capacity measure relevant for more powerful coding. On the other hand, the effect of shadowing, which cannot be averaged out by coding, causes a flattening toward low outage probabilities ≤ 0.1. In the high outage probability range we note that CM1 is slightly superior to CM3, which is due to the large dominant eigenvalues of CM1 identified in Section 11.3.4.

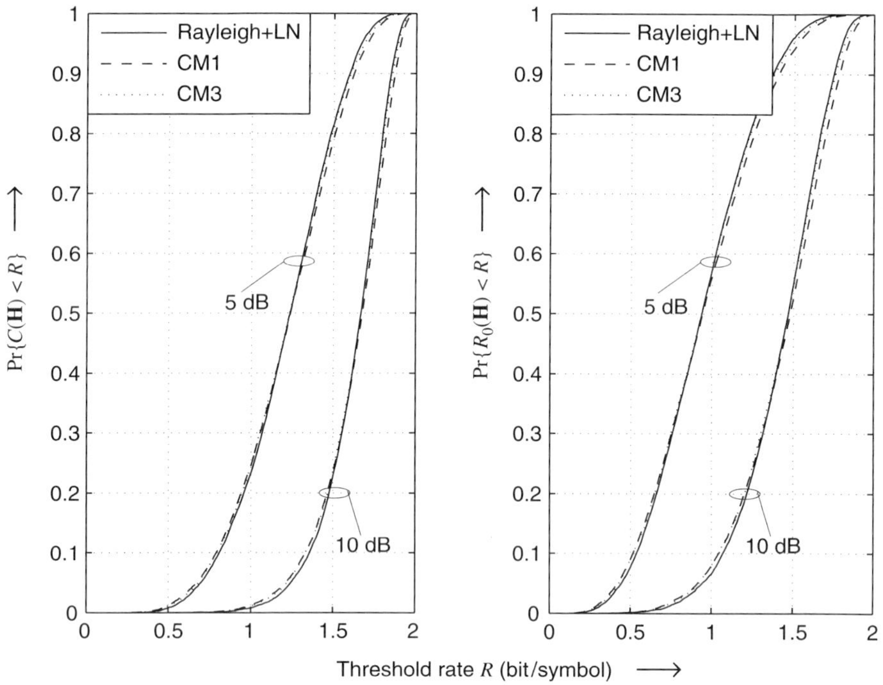

FIGURE 11.5 Outage probability for $10\log_{10}(\bar{E}_s/\mathcal{N}_0) = 5$ and 10 dB and perfect CSI. (*Left*) Outage capacity. (*Right*) Outage cutoff rate.

In Fig. 11.6 we consider the 10% outage[5] capacity and cutoff rate as a function of the SNR $10\log_{10}(\bar{E}_s/\mathcal{N}_0)$. Again we note the close similarity between the UWB channel models and the i.i.d. Rayleigh fading channel with lognormal shadowing. A comparison of the capacity with the corresponding cutoff rate curves indicates that decent gains of 2.5–3 dB in power efficiency can be anticipated by the application of more powerful capacity approaching codes such as Turbo or RA codes instead of the convolutional codes proposed in [2–4], which usually perform in the vicinity of the cutoff rate. We will return to this issue in Section 11.7.

11.6.2 Effects of Imperfect CSI

Figure 11.7 shows the SNR loss due to LSE channel estimation according to Eq. (11.14) with various values of η. For reference, the MB-OFDM system uses $P = 2$, $N_t = 128$, and so choosing a value of $L_t = 32$ (less than the length of the zero padding) results in a value of $\eta = 0.125$ from Eq. (11.6).

[5] We note that 10% outage is a typically chosen value for UWB systems and the considered channel model [2–4].

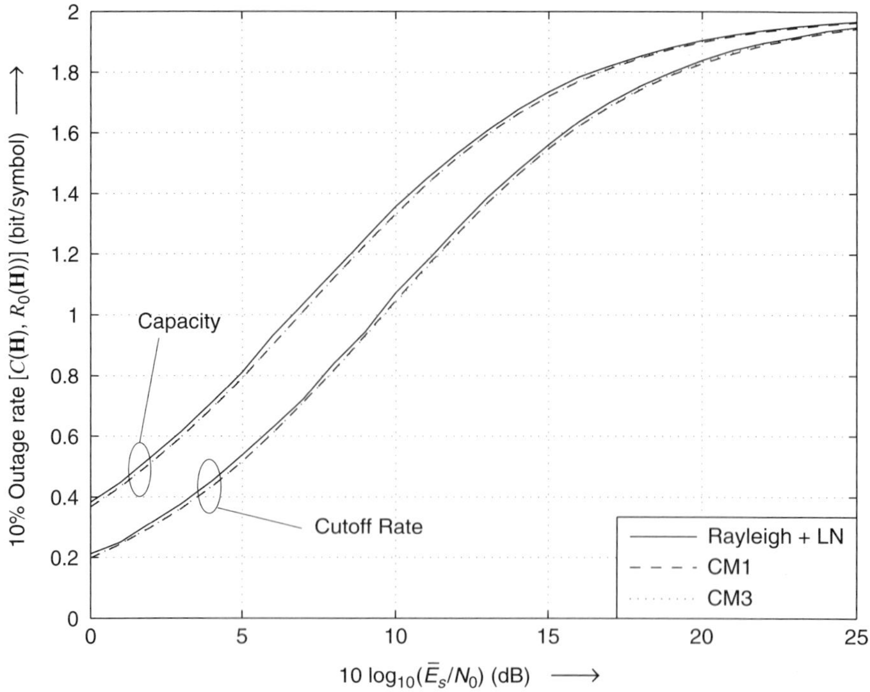

FIGURE 11.6 A 10% outage capacity and cutoff rate for perfect CSI.

We can see from Fig. 11.7 that the performance penalty 10 $\log_{10}(\gamma/\gamma_e)$ due to imperfect CSI is about 0.5 dB in the range of interest for the MB-OFDM system. The actual loss in $\bar{E}_s/\mathcal{N}_0$ is slightly different since γ in Fig. 11.7 is for a fixed lognormal shadowing, and the actual $\bar{E}_s/\mathcal{N}_0$ loss must be obtained by averaging over the lognormal pdf. However, we can see from Fig. 11.7 that the SNR loss is relatively constant for relevant values of γ, which (since the lognormal shadowing has a 0-dB mean), results in $E_s/\mathcal{N}_0$ loss of approximately 10 $\log_{10}(\gamma/\gamma_e)$. Reducing the channel estimation overhead to $P = 1$ ($\eta = 0.25$) could be an interesting alternative for short packets, as the additional loss is only about 0.5 dB (in terms of required energy per information bit $\bar{E}_b$ the loss is even smaller). Further reduction of pilot tones is not advisable as the gains in throughput are outweighed by the losses in power efficiency.

11.6.3 Error Rate Approximations

In this section, we present numerical results for methods I and II introduced above. We focus on the particular case of MB-OFDM operating in the CM1 UWB channel. For method II we include the effect of "outer" lognormal

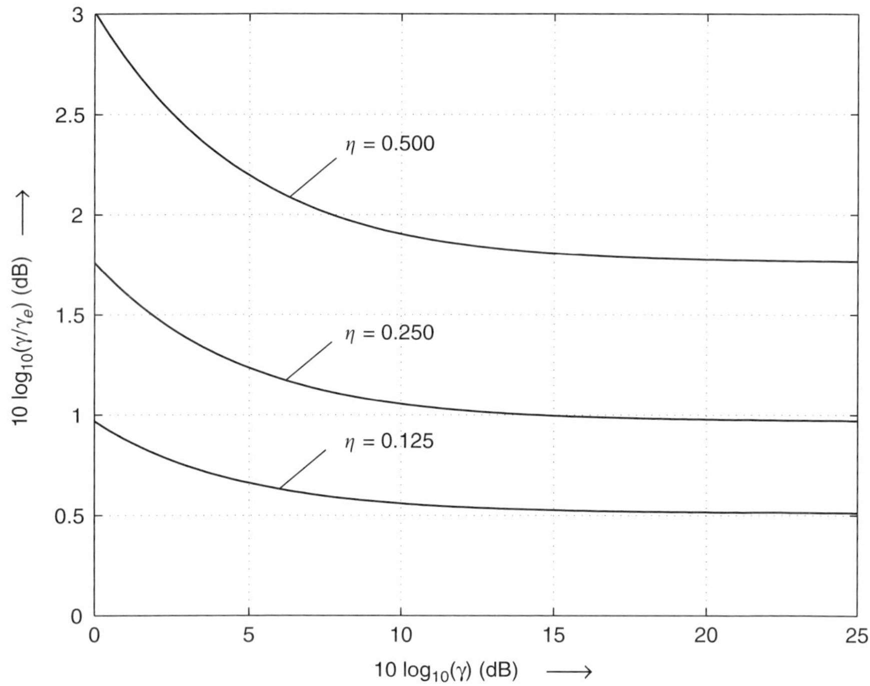

FIGURE 11.7 Loss in SNR due to LSE channel estimation with different η according to Eq. (11.14).

shadowing by numerically integrating the results of Eq. (11.30) over the appropriate lognormal distribution.

In Fig. 11.8, we present the 10% outage BER as a function of $\bar{E}_b/\mathcal{N}_0$ obtained using method I (lines), as well as simulation results (markers) for different code rates and modulation schemes using a set of 100 UWB CM1 channel realizations with lognormal shadowing. We can see that method I is able to accurately predict the outage BER for a variety of different code rates, with a maximum error of less than 0.5 dB. It is also important to note that obtaining the method I result requires significantly less computation than is required to obtain the simulation results for all 100 UWB channel realizations. For example, it took about 15 min to obtain one of the analytical curves of Fig. 11.8 (using a short MATLAB program), while it took approximately 48 h to obtain the corresponding simulation results using a hand-optimized C++ MB-OFDM simulator on the same computer (with two Intel Xeon 3-GHz processors).

Figure 11.9 illustrates the average BER as a function of $\bar{E}_b/\mathcal{N}_0$ for code rates $R_c = \frac{1}{2}$ and $\frac{3}{4}$ using two approaches: method I with an average over 10,000 channel realizations (dashed lines) and the direct average from method II (solid lines). As expected, the two methods are in close agreement at low BER.

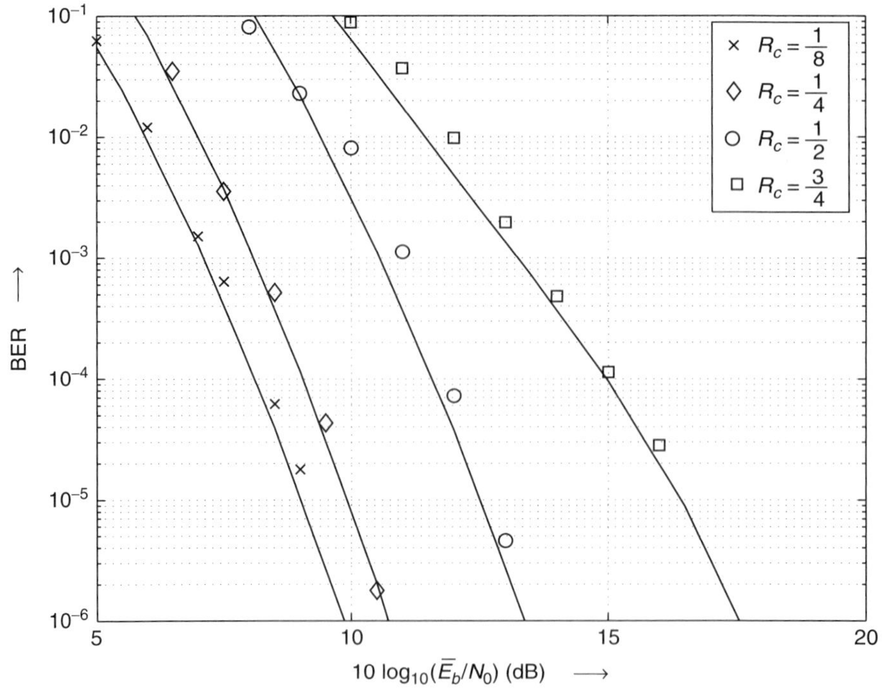

FIGURE 11.8 A 10% outage BER vs. $10\log_{10}(\bar{E}_b/\mathcal{N}_0)$ (dB) from method I (lines) and simulation results (markers) for different code rates. UWB CM1 channel. Code rates $\frac{1}{4}$ and $\frac{1}{8}$ include spreading (see Section 11.2.1.3).

The deviation between the two results at higher BER is due to (a) the loosening effect of the averaging of method II over the lognormal distribution, and (b) the fact that method I is somewhat tighter due to the upper bounding by $\frac{1}{2}$ in Eq. (11.21).

11.6.3.1 A Caution to System Designers. We should note that 100 channel realizations (standard for MB-OFDM performance analysis [4]) may not be sufficient to accurately capture the true system performance. Figure 11.10 (solid lines) shows the average BER with respect to $\bar{E}_b/\mathcal{N}_0$ for four different sets of 100 UWB CM1 channel realizations, obtained via method I. For comparison, the average performance obtained via method II is also shown (bold solid line). We can see that the average system performance obtained using sets of only 100 channel realizations depends greatly on the specific realizations chosen. Similarly, Fig. 11.10 illustrates the 10% outage BER with respect to $\bar{E}_b/\mathcal{N}_0$ for four different sets of 100 UWB CM1 channel realizations, obtained via method I (dashed lines). For comparison the 10% outage BER

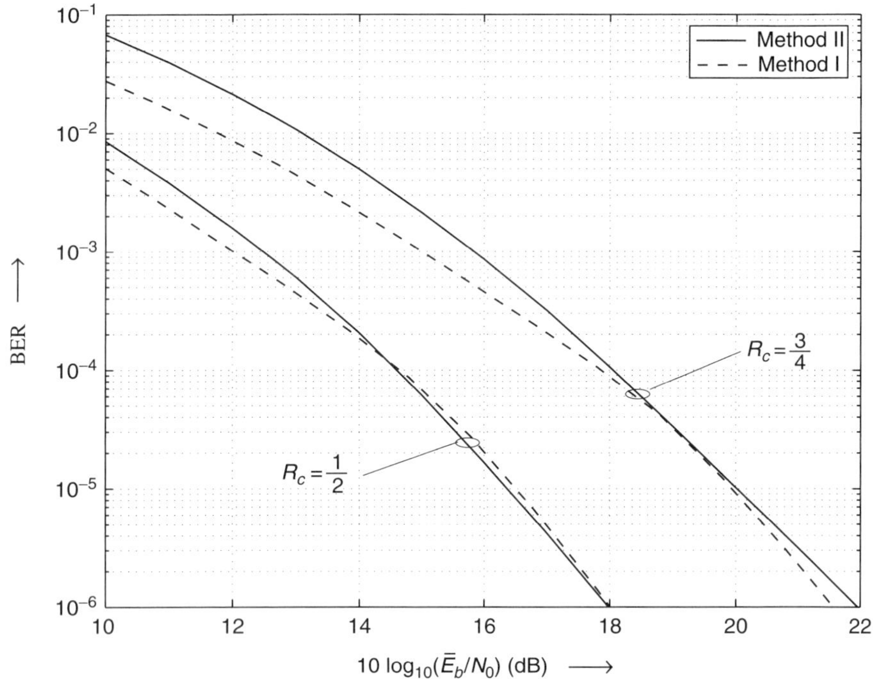

FIGURE 11.9 Average BER vs. $10\log_{10}(\bar{E}_b/\mathcal{N}_0)$ (dB) for code rates $R_c = \frac{1}{2}$ and $\frac{3}{4}$. Solid lines: Direct average from method II. Dashed lines: method I with an average over 10,000 channel realizations. UWB CM1 channel.

obtained using a set of 1000 realizations is also shown (bold dashed line). We see that the outage BER curves, while less variable than the average BER curves, are still quite dependent on the selected channel realization set. (See also the Λ marker and error bar of Fig 11.11.)

Based on the results above, it seems that performance evaluation for systems operating in quasi-static channels using only small numbers of channel realizations may be prone to inaccurate results. This is one of the main strengths of the two methods presented in Section 11.5: The performance can easily be evaluated over any number of channel realizations (method I), or the average performance can be directly obtained (method II), without resorting to lengthy simulations.

11.6.4 Simulation Results

In this section, we examine channel CM1 with four different transmission modes with data rates of 80, 160, 320, and 480 Mbps corresponding to 0.25, 0.50, 1.00, and 1.50 bit/symbol, respectively, and to code rates (R_c) of $\frac{1}{8}$, $\frac{1}{4}$, $\frac{1}{2}$, and

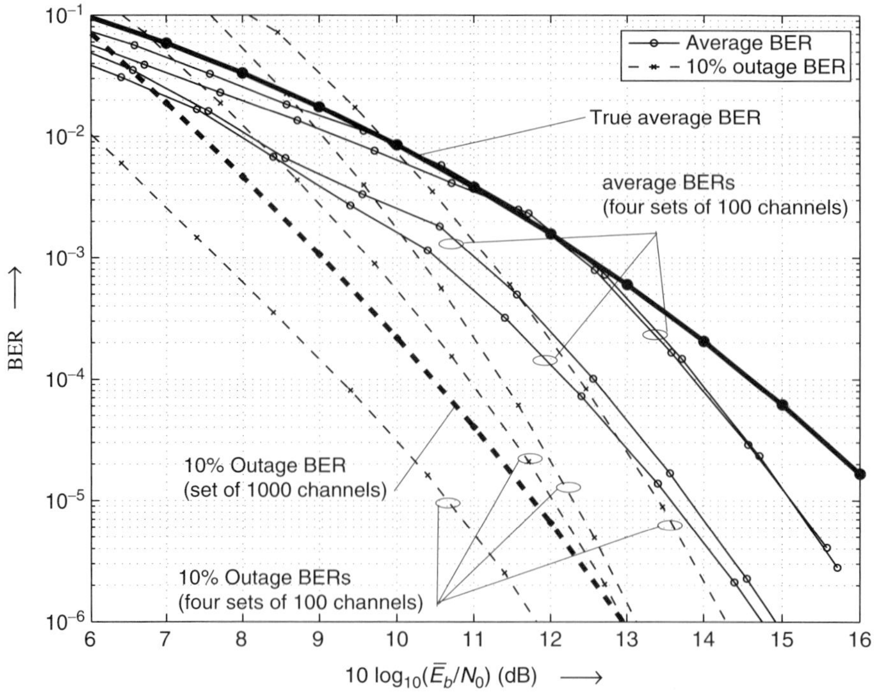

FIGURE 11.10 Average BER (solid lines) and 10% outage BER (dashed lines) vs. $10\log_{10}(\bar{E}_b/\mathcal{N}_0)$ (dB) for four different sets of 100 channels using method I. For comparison: average BER from method II (bold solid line), and 10% outage BER for a set of 1000 channels (bold dashed line). UWB CM1 channel, $R_c = \frac{1}{2}$.

$\frac{3}{4}$, respectively. In the simulations, detection is performed with perfect CSI as well as with LSE channel estimation using $\eta = 0.125$. The simulation results presented in these two sections are the worst-case $10\log_{10}(\bar{E}_s/\mathcal{N}_0)$ values required to achieve BER $\leq 10^{-5}$ for the best 90% of channel realizations over a set of 100 channels (i.e., they are simulation results corresponding to 10% outage).

Figure 11.11 (markers) shows SNR points when using convolutional codes (as in the MB-OFDM system), together with the corresponding 10% outage cutoff rate curves. We observe that the simulated SNR points are approximately 3–4 dB from the cutoff rate curves, which is reasonable for the channel model and coding schemes under consideration. These results (a) justify the relevance of the information-theoretic measure and (b) confirm the coding approach used in MB-OFDM. More specifically, the diversity provided by the UWB channel is effectively exploited by the chosen convolutional coding and interleaving scheme.

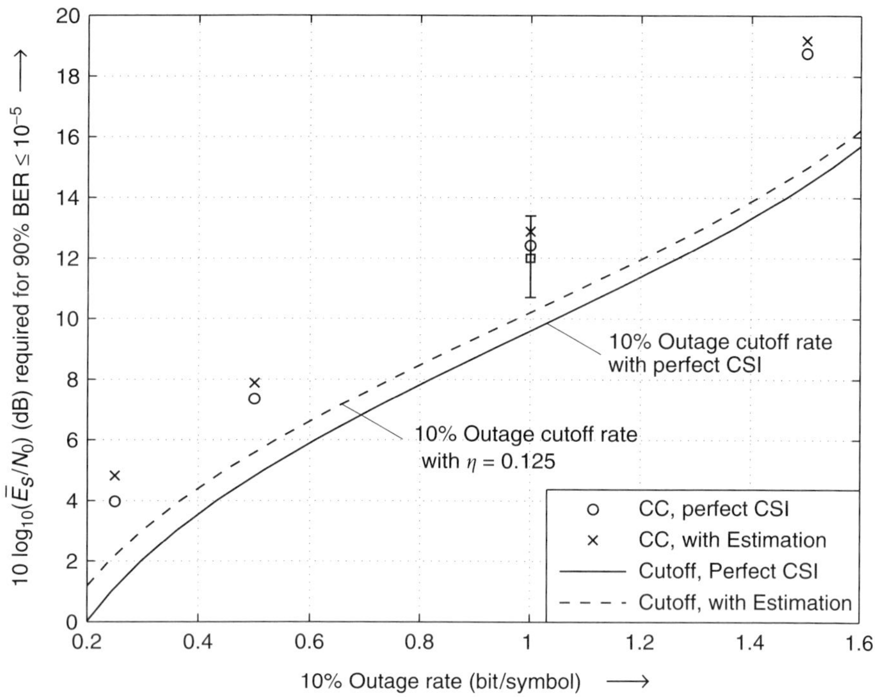

FIGURE 11.11 $10\log_{10}(\bar{E}_s/\mathcal{N}_0)$ required to achieve BER$\leq 10^{-5}$ for the 90% best channel realizations using convolutional codes (CC) (markers). For comparison: 10% outage cutoff rate (lines). Channel model CM1 and LSE channel estimation.

Furthermore, the system with LSE channel estimation performs within 0.5–0.7 dB of the perfect CSI case as was expected from the cutoff rate analysis [see also the discussion in Section 11.6.2 on the relationship between the loss $10\ \log_{10}(\gamma/\gamma_e)$ and the $10\log_{10}(\bar{E}_s/\mathcal{N}_0)$ loss].

The Λ marker and error bar for 1.00 bit/symbol in Fig 11.11 indicate the range of possible 10% outage values from different sets of channel realizations (cf. Fig. 11.10) for perfect CSI. We can see again that, depending on the particular set of channel realizations used, the results obtained can be somewhat variable.

11.7 EXTENSIONS FOR PERFORMANCE ENHANCEMENT

In this section we consider several techniques to improve the performance of MB-OFDM systems. Our objective herein is not to provide an exhaustive list of potential performance enhancements, but rather to illustrate some examples, and, in addition, to apply some of the tools of analysis developed in the previous section to these examples. In particular, we consider powerful coding schemes and OFDM bit loading.

11.7.1 High Performance Turbo and RA Codes

We propose the use of Turbo codes [24] in order to improve the system power efficiency and more closely approach the channel capacity. We have adopted the generator polynomials and interleaver design of the 3rd Generation Partnership Project (3GPP) [25], due to their excellent performance for the code lengths considered as well as reasonable interleaver memory storage requirements. For low data rates, the time/frequency spreading technique of MB-OFDM is retained.

We would like to maintain compatibility with the MB-OFDM channel interleaver by having each coded block fit into one channel interleaver frame, as is done with the convolutional codes used in the standard. Keeping the block lengths short also reduces the memory requirements and decoding delay at the receiver. However, to maintain compatibility at the lowest data rates would require a Turbo code interleaver length of only 150 or 300 bits. Due to the poor distance properties and resultant performance degradation associated with short-length Turbo codes, at low data rates we consider both MB-OFDM-compliant block lengths and longer blocks of 600 input bits (the same length as used without spreading).

The limited length of the MB-OFDM channel interleaver motivates the consideration of serially concatenated codes, where the interleaver is positioned between the constituent encoders and thus has a longer length. We consider nonsystematic regular RA codes [26] due to their simplicity and good performance for the required code lengths. The time/frequency spreading mechanism described above is discarded, and low rate RA codes ($R_c = \frac{1}{4}$ or $\frac{1}{8}$) are used. The interleaver between the repeater and accumulator is randomly generated (no attempt is made to optimize its performance).

Convolutionally coded schemes use a soft-input Viterbi decoder to restore the original bit stream, requiring a decoding complexity of 64 trellis states searched per information bit. Turbo-coded schemes are decoded with 10 iterations of a conventional Turbo decoder using the log-domain BCJR algorithm [27], with a complexity of roughly $10 \times 2 \times 2 \times 8 = 320$ trellis states searched per information bit (i.e., 10 iterations of two 8state component codes, and assuming that the BCJR algorithm is roughly twice as complex as the Viterbi algorithm due to the forward-backward recursion). RA decoding is performed by a Turbo-like iterative decoder, using a maximum of 60 iterations and an early exit criterion, which, at relevant values of SNR, reduces the average number of decoder iterations to less than 10 [28]. We note that the per-iteration decoding complexity of the RA code is less than that of the Turbo code (since only a 2-state accumulator and a repetition code are used), making the total RA decoder complexity slightly more than the convolutional code but less than the Turbo code. The increased decoder complexities of the Turbo and RA codes, compared to the convolutional code, are reasonable considering the performance gains they provide (see Section 11.7.3.2).

11.7.2 Bit Loading

In this section we introduce the use of bit-loading schemes for MB-OFDM and extend the information-theoretic results of Section 11.4 to encompass them.

11.7.2.1 Bit-Loading Schemes. The UWB channel (see Section 11.3) is considered time invariant for the duration of many packet transmissions. For that reason, it is feasible to consider bit-loading algorithms to assign unequal numbers of bits to each OFDM subcarrier [5]. Channel state information is obtained at the transmitter, either by:

1. Exploiting channel reciprocity (if the same frequency band is used in the uplink and downlink, as in the standard), or
2. Some form of feedback (which may be required even if the same frequency band is used since reciprocity may not apply due to different interference scenarios for transmitter and receiver).

We consider loading for higher data rates, without time or frequency spreading. To illustrate the potential performance gains of bit loading, we have applied the algorithm of Chow, Cioffi, and Bingham (CCB) [29] because it loads according to the information-theoretic capacity criterion, as well as for its moderate computational complexity.

The data rates and OFDM symbol structure of MB-OFDM are maintained by loading each OFDM symbol with 200 bits. Each tone carries from 0 to 6 bits using quadrature amplitude modulation (QAM) signal constellations with gray or quasi-gray labeling (note that 6 bit/symbol corresponds to 64 QAM, which is a reasonable upper limit for modulation on a wireless channel). Due to FCC restrictions on the transmitted power spectral density [7], power loading is not used (all tones carry the same power). For the CCB algorithm, the SNR gap parameter Γ is either 6 dB (when convolutional codes are used) or 3 dB (for Turbo codes). When the algorithm is unable to determine a suitable loading, an all-QPSK loading is used; cf. [29] for details.

11.7.2.2 Clustered Bit Loading. One potential feedback-based method of bit loading is for the receiver to determine the appropriate modulation for each tone and feed the loading information back to the transmitter. To lower the feedback transmission requirements and significantly reduce the loading algorithm's computational complexity, we also propose a clustered loading scheme where clusters are formed by considering groups of D adjacent tones. We make the following modification to the CCB algorithm. We substitute Eq. (1) of [29] with:

$$b(i) = \frac{1}{D}\sum_{k=1}^{D}\log_2[1 + \mathrm{SNR}(i,k)10^{-((\Gamma+\gamma_{\mathrm{margin}})/10)}] \tag{11.31}$$

where $\mathrm{SNR}(i, k)$ = SNR of the kth tone in the ith cluster

γ_{margin} = system performance margin in dB (iteratively calculated by the CCB algorithm),

$b(i)$ = (possibly noninteger) number of bits allocated for each tone in cluster i

Using the modified algorithm to load $200/D$ bits on $100/D$ clusters provides the final integer-valued loadings $\hat{b}(i)$ for each cluster. Finally, all tones in cluster i are assigned $\hat{b}(i)$ bits (i.e., the loading inside each cluster is constant). This modification causes the CCB algorithm to load according to the mean capacity of all tones in a cluster.

11.7.2.3 Capacity and Cutoff Rate with Loading. The instantaneous capacity in bits per complex dimension of an N_d tone BICM-OFDM system using loading can be found by extending Eqs. (11.8) and (11.9) (following the methodology of [6, 17]) as

$$C(\mathbf{H}) = \bar{m} - \frac{1}{N_d}\sum_{i=1}^{N_d}\sum_{\ell=1}^{m_i} E_{b,y_i}\left\{\log_2\left(\frac{\sum_{x_i\in\mathcal{X}_i} p\left(y_i|\hat{h}_i, x_i\right)}{\sum_{x_i\in\mathcal{X}_{i,b}^{\ell}} p\left(y_i|\hat{h}_i, x_i\right)}\right)\right\} \tag{11.32}$$

where $\bar{m}$ = average number of bits/symbol ($\bar{m} = 2$ for MB-OFDM)

m_i = number of bits per symbol for the ith tone

χ_i = signal constellation for the ith tone

$\mathcal{X}_{i,b}^{\ell}$ = set of all constellation points $x \in \mathcal{X}_i$ whose label has the value $b \in \{0,1\}$ in position ℓ.

Similarly, we can express the instantaneous cutoff rate for bit-loading systems in bits per complex dimension as

$$R_0(\mathbf{H}) = \bar{m} - \bar{m}\log_2[B(\mathbf{H}) + 1] \tag{11.33}$$

with the instantaneous Bhattacharya parameter

$$B(\mathbf{H}) = \frac{1}{N_d}\sum_{i=1}^{N_d}\sum_{\ell=1}^{m_i}\frac{1}{m_i}\mathbb{E}_{b,y_i}\left\{\sqrt{\frac{\sum_{x_i\in\mathcal{X}_{i,\bar{b}}^{\ell}} p(y_i|\hat{h}_i, x_i)}{\sum_{x_i\in\mathcal{X}_{i,b}^{\ell}} p(y_i|\hat{h}_i, x_i)}}\right\} \tag{11.34}$$

11.7.3 Performance Results

We illustrate the potential performance improvements through the use of the methods proposed above by considering the capacity and cutoff rate in Section

11.7.3.1. In Section 11.7.3.2, we then compare those results with system simulations. Finally, in Section 11.7.3.3, we illustrate the improvements in range that are possible with these extensions.

11.7.3.1 Capacity and Cutoff Rate. We examine the capacity and cutoff rate of systems employing the CCB loading algorithms. We evaluated Eqs. (11.32) and (11.33) via Monte Carlo simulation as discussed in Section 11.6.1. Figure 11.12 (lines) shows the 10% outage capacity and cutoff rates for the CM1 channel using the CCB loading algorithms. (The markers in this figure will be discussed in Section 11.7.3.2). It should be noted that $\bar{E}_s$ is not adjusted to account for tones carrying 0 bits because we assume operation at FCC transmit power limits [7], precluding the reallocation of power from unused tones to other subcarriers (which would put the transmit power spectral density beyond the allowed limits). We also do not adjust for the overhead associated with the feedback of loading information from the receiver to the transmitter. For high rates, the CCB loading algorithm provides a gain of several decibels in capacity and in cutoff rate compared to the unloaded case; and this gain grows with increasing rate and $\bar{E}_s/\mathcal{N}_0$.

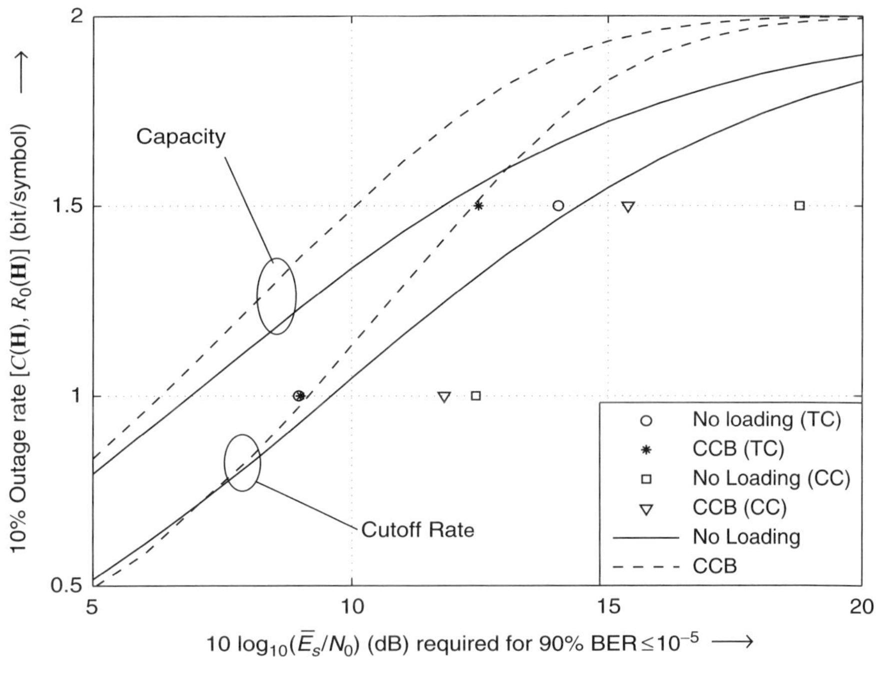

FIGURE 11.12 10% outage capacity and cutoff rate with and without loading for CM1 (lines). $10\log_{10}(\bar{E}_s/\mathcal{N}_0)$ required to achieve BER $\leq 10^{-5}$ for the 90% best channel realizations using convolutional and Turbo codes, with and without loading (markers).

We next consider the application of clustered loading using the modified CCB algorithm as described in Section 11.7.2.2. Figure 11.13 shows the 10% outage capacity (solid lines) and cutoff rate (dashed lines) for various values of cluster size D, for channels CM1 and CM3. Also included for comparison are the nonclustered loading ($D = 1$) and unloaded (all-QPSK) curves. As the cluster size D increases, the attainable rates decrease because the modulation scheme chosen for each cluster is not optimal for all tones in the cluster. This loss is slightly more pronounced for the cutoff rate than for the capacity, which indicates that when using clustered loading we should expect more performance degradation with convolutional codes than with Turbo codes (see also Section 11.7.3.2). The performance degradation with increasing cluster size is higher for CM3 than for CM1, which can be predicted from the correlation matrix results of Section 11.3.4. Specifically, we note from Fig. 11.4 that the frequency responses of adjacent subcarriers are more correlated (fewer significant eigenvalues) in CM1 and less correlated (more significant eigenvalues) in CM3. The less correlated the tones of a cluster are, the higher the average

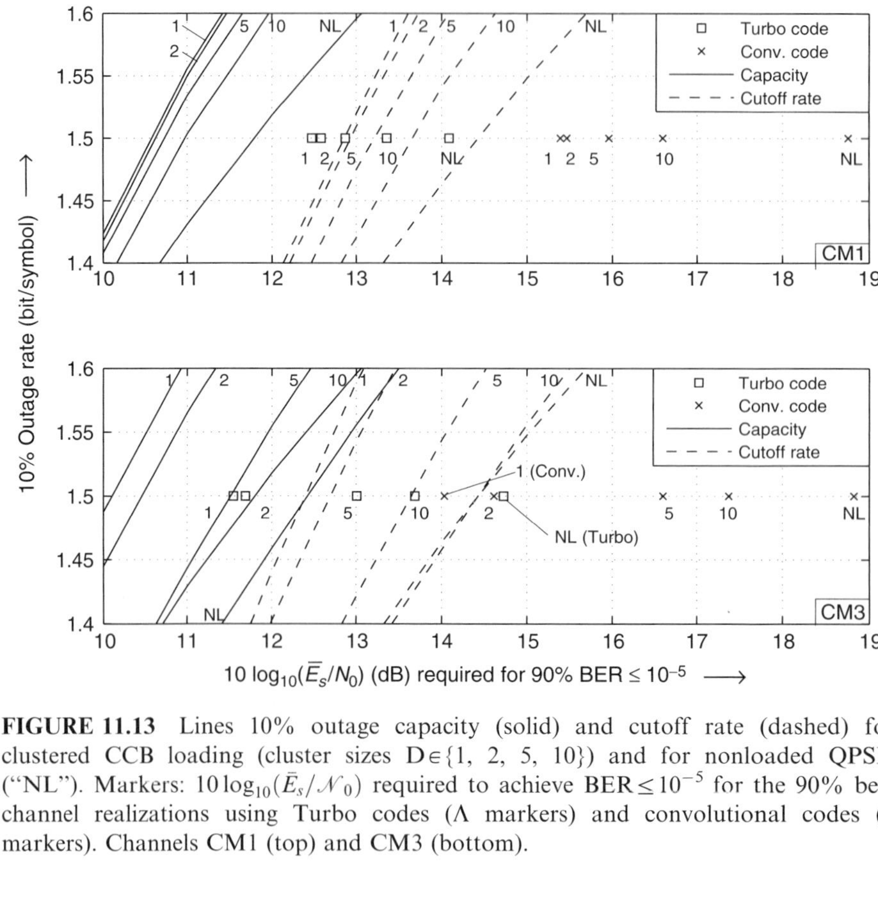

FIGURE 11.13 Lines 10% outage capacity (solid) and cutoff rate (dashed) for clustered CCB loading (cluster sizes D∈{1, 2, 5, 10}) and for nonloaded QPSK ("NL"). Markers: $10\log_{10}(\bar{E}_s/\mathcal{N}_0)$ required to achieve BER$\leq 10^{-5}$ for the 90% best channel realizations using Turbo codes (Λ markers) and convolutional codes (x markers). Channels CM1 (top) and CM3 (bottom).

mismatch between the optimal modulation for each tone (i.e., that chosen by the nonclustered loading algorithm) and the fixed modulation chosen for the cluster. The higher average mismatch on CM3 results in lower performance when clustered loading is applied.

11.7.3.2 Simulation Results. We next consider the Turbo and RA coding schemes. Figure 11.14 (markers) shows the simulation results for Turbo and RA codes on channel CM1 with perfect CSI, as well as the convolutional code results for comparison. We also show the corresponding 10% outage capacity and cutoff rate curves. Turbo codes give a performance gain of up to 5 dB over convolutional codes and perform within 2.5 dB of the channel capacity, depending on the rate. At rates of 0.25 and 0.50 bit/symbol, Turbo code interleaver sizes compatible with the channel interleaver design of MB-OFDM (the "std" points) incur a performance penalty of 1–2 dB compared with the longer block length ($K = 600$) points. Repeat-accumulate codes have a performance roughly 1 dB worse than the long-block-length Turbo codes, but the RA codes are both (a) compatible with the MB-OFDM channel interleaver and

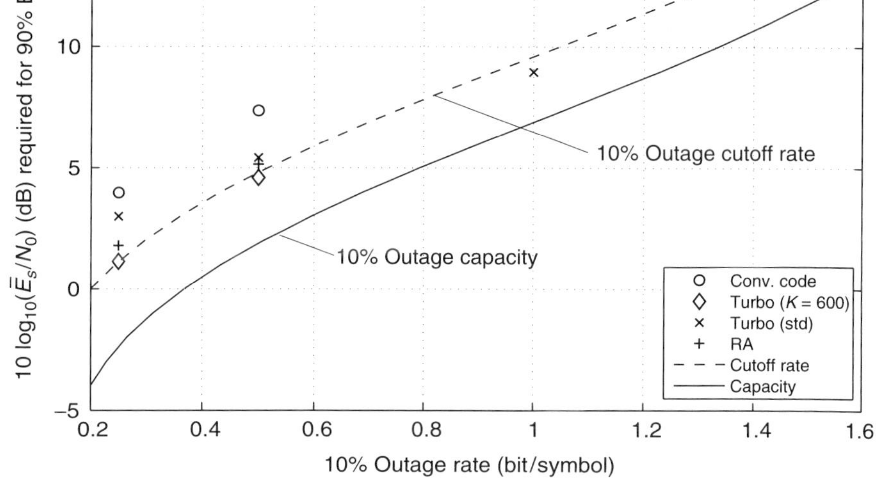

FIGURE 11.14 $10\log_{10}(\bar{E}_s/\mathcal{N}_0)$ required to achieve BER $\leq 10^{-5}$ for the 90% best channel realizations using Turbo codes, RA codes, and convolutional codes (markers). For comparison: 10% outage capacity and cutoff rate (lines). Channel model CM1 and perfect CSI.

(b) less complex to decode. They are thus a good candidate for low rate MB-OFDM transmission.

Figure 11.12 (markers) shows the simulation results for Turbo codes and for convolutional codes, using the CCB loading algorithm on channel CM1 with perfect CSI. At 1.00 bit/symbol and using convolutional codes, we see a performance gain of less than 1 dB using CCB loading. Performance using Turbo codes at 1.00 bit/symbol is relatively constant regardless of loading. However, at 1.50 bits/symbol we see gains of approximately 1.5 dB for Turbo codes and almost 4 dB for convolutional codes when CCB loading is used. Finally, we note that at 1.50 bits/symbol the system employing CCB loading and Turbo codes is approximately 6 dB better than the unloaded convolutionally coded system and performs within approximately 2.5 dB of the channel capacity.

In Fig. 11.13 (markers) we consider the performance of clustered loading with Turbo codes and with convolutional codes for 1.50 bits/symbol on the CM1 and CM3 channels with perfect CSI. As predicted by information-theoretic analysis, clustered loading incurs a performance penalty with increasing cluster size D. We note that Turbo codes suffer a smaller performance degradation (relative to $D = 1$) than convolutional codes because the more powerful Turbo code is better suited to handle the mismatched modulation (as discussed in Section 11.7.3). The performance degradation is larger for CM3 due to that channel model's lower correlation between adjacent subcarrier frequency responses and resultant larger loading mismatch. However, even $D = 10$ loading provides performance gains for both channels and code types. Cluster size $D = 2$ is a good trade-off point for both Turbo and convolutional codes, allowing for feedback reduction by a factor of 2 with losses of approximately 0.1 dB for CM1 and 0.4 dB for CM3. Cluster sizes as large as $D = 5$ could be used with Turbo codes, depending on the required power efficiency and expected channel conditions.

TABLE 11.4 Power Efficiency Gains and Range Increases Available Using Some of the Extensions Considered, Compared to the MB-OFDM Standard[a]

System	$10\log_{10}(\bar{E}_s/N_0)$	Gain (dB)	% range increase
CC, no loading (standard)	18.76	—	—
CC, CCB loading	15.38	3.38	47
CC, $D=2$ clustered loading	15.47	3.29	6
TC, no loading	14.09	4.67	71
TC, CCB loading	12.48	6.28	106
TC, $D=2$ clustered loading	12.58	6.18	103

[a]Channel CM1, rate 1.50 bits/symbol (480 Mbps), path loss exponent $d=2$. $10\log_{10}(\bar{E}_s/\mathcal{N}_0)$ values are those required to achieve BER $\leq 10^{-5}$ for the 90% best channel realizations. (CC: convolutional code, TC: Turbo code).

11.7.3.3 Range Improvements from Turbo Codes and Loading. Table 11.4 lists the gains in required $10\log_{10}(\bar{E}_s/\mathcal{N}_0)$ and percentage range increases on channel CM1 for various combinations of the extensions we have proposed. We assume a path loss exponent of $d = 2$, as in [4]. We can see that bit loading alone provides up to 47% increase in range, Turbo codes without loading provide a 71% increase, and the combination of Turbo codes and loading allows for a 106% increase in range. Furthermore, the use of clustered loading with $D = 2$ only reduces these range improvements by 1–3% over the nonclustered case, while providing reduced-rate feedback and lower computational complexity.

11.8 CONCLUSIONS

In this chapter, we have analyzed the MB-OFDM system for UWB communication from the information-theoretic, communication-theoretic, and practical points of view. We have shown that the UWB channel model developed under IEEE 802.15 is seen by OFDM systems in the frequency domain as Rayleigh fading with additional shadowing. The 528 MHz signal bandwidth chosen for MB-OFDM essentially captures the diversity provided by the UWB channel. As a result, we have found that the information-theoretic limits of the UWB channel are similar to those of a perfectly interleaved Rayleigh fading channel with shadowing.

We have presented two methods for evaluating the performance of coded MB-OFDM operating over frequency-selective, quasi-static, nonideally interleaved fading channels. The realization-based method (method I) estimates the system performance for each realization of the channel and is suitable for evaluating the outage performance. Method II, based on knowledge of the correlation matrix of the Rayleigh-distributed frequency domain channel gains, allows for direct calculation of the average system performance over the ensemble of quasi-static fading channel realizations. The results in Section 11.6 demonstrate that the proposed methods of analysis provide an accurate measure of the system performance and allow for much greater flexibility than simulation-based approaches.

The BICM-OFDM scheme used in MB-OFDM performs close to the outage cutoff rate measure and is thus well suited to exploit the available diversity. The application of stronger coding, such as Turbo codes or repeat-accumulate codes, improves power efficiency by up to 5 dB, depending on the data rate. Bit-loading algorithms provide additional performance gains for high data rates, and a simple clustering scheme allows for reduced-rate feedback of loading information depending on the channel conditions and required power efficiency. Finally, a simple LSE channel estimator has been shown to enable performance within 0.5–0.7 dB of the perfect CSI case for the MB-OFDM system.

ACKNOWLEDGMENTS

The completion of this research was made possible thanks to Bell Canada's support through its Bell University Laboratories R&D program and the Natural Sciences and Engineering Research Council of Canada (Grant CRDPJ 320 552), and with the support of a Canada Graduate Scholarship.

REFERENCES

1. S. Roy, J. Foerster, V. Somayazulu, and D. Leeper, "Ultrawideband radio design: The promise of high-speed, short-range wireless connectivity," *Proc. IEEE* 92(2), 295–311 (2004).
2. ECMA, "Standard ECMA-368: High rate ultra wideband PHY and MAC standard, available: http://www.ecma-international.org/publications/standards/Ecma-368.htm, Dec. 2005.
3. IEEE P802.15, "Multiband OFDM physical layer proposal for IEEE 802.15 Task Group 3a, document No. P802.15-03/268r3, IEEE, New York, Mar. 2004.
4. A. Batra, J. Balakrishnan, G. Aiello, J. Foerster, and A. Dabak, "Design of a multiband OFDM system for realistic UWB channel environments," *IEEE Trans. Microwave Theory Technol.* 52(9), 2123–2138 (2004).
5. J. Bingham, *ADSL, VDSL, and Multicarrier Modulation*, Wiley, New York, 2000.
6. G. Caire, G. Taricco, and E. Biglieri, "Bit-interleaved coded modulation," *IEEE Trans. Inform. Theory* 44(3), 927–946 (1998).
7. Federal Communications Commission (FCC), Revision of Part 15 of the Commissions rules regarding ultra-wideband transmission systems, First Report and Order, ET Docket 98-153, FCC 02-48, adopted February 14, 2002, released Apr. 22, 2002.
8. A. F. Molisch, J. R. Foerster, and M. Pendergrass, "Channel models for ultrawideband personal area networks," *IEEE Wireless Commun. Mag.* 10(6), 14–21 (2003).
9. J. G. Proakis, *Digital Communications*, 4th ed., McGraw-Hill, New York, 2001.
10. C. Snow, L. Lampe, and R. Schober, "Performance analysis and enhancement of multiband OFDM for UWB communications," *IEEE Trans. Wireless Commun.* 6(6), 2182–2192 (2007).
11. C. Snow, L. Lampe, and R. Schober, "Error rate analysis for coded multicarrier systems over quasi-static fading channels," *IEEE Trans. Commun.* 55(9), 1736–1746 (2007).
12. B. Muquet, Z. Wang, G. Giannakis, M. de Courville, and P. Duhamel, "Cyclic prefixing or zero padding for wireless multicarrier transmissions?" *IEEE Trans. Commun.* 50(12), 2136–2148 (2002).
13. X. Cai and G. B. Giannakis, "Error probability minimizing pilots for OFDM with M-PSK modulation over Rayleigh-fading channels," *IEEE Trans. Vehic. Technol.* 53(1), 146–155 (2004).
14. A. Saleh and R. Valenzuela, "A statistical model for indoor multipath propagation," *IEEE J. Sel. Areas Commun.* SAC-5(2), 128–137 (1987).
15. A. Molisch, D. Cassioli, C.-C. Chong, S. Emami, A. Fort, B. Kannan, J. Karedal, J. Kunisch, H. Schantz, K. Siwiak, and M. Win, "A comprehensive standardized model for ultrawideband propagation channels," *IEEE Trans. Antennas Propagat.* 54(11), 3151–3166 (2006).

16. M.-O. Wessman, A. Svensson, and E. Agrell, "Frequency diversity performance of coded multiband-OFDM systems on IEEE UWB channels," In *Proc. IEEE Vehic. Tech. Conf., Fall* (*VTC*), Vol. 2, Sept. 2004, pp. 1197–1201.
17. A. Ekbal, K.-B. Song, and J. M. Cioffi, "Outage capacity and cutoff rate of bit-interleaved coded OFDM under quasi-state frequency selective fading," in *Proc. IEEE Global Telecomm. Conf.* (*GLOBECOM*), Vol. 2, Dec. 2003, pp. 1054–1058.
18. J. Wu, C. Xiao, and N. C. Beaulieu, "Optimal diversity combining based on noisy channel estimation," in *Proc. IEEE Intl. Conf. Comm.* (*ICC*), Vol. 1, Paris, June 2004, pp. 214–218.
19. J. A. Gubner and K. Hao, "An exact computable formula for the average bit-error probability of the IEEE 802.15.3a UWB channel model," in *Proc. IEEE Intl. Conf. on Ultra-Wideband*, Zurich, Sept. 2005, pp. 142–146.
20. E. Biglieri, J. Proakis, and S. Shamai, "Fading channels: information-theoretic and communications aspects," *IEEE Trans. Inform. Theory*. 44(6), 2619–2692 (1998).
21. M. Cedervall and R. Johannesson, "A fast algorithm for computing distance spectrum of convolutional codes," *IEEE Trans. Inform. Theory* 35(6), 1146–1159 (1989).
22. E. Malkamäki and H. Leib, "Evaluating the performance of convolutional codes over block fading channels," *IEEE Trans. Inform. Theory* 45(5), 1643–1646 (1999).
23. V. V. Veeravalli, "On performance analysis for signaling on correlated fading channels," *IEEE Trans. Commun.* 49(11), 1879–1883 (2001).
24. C. Berrou and A. Glavieux, "Near optimum error correcting coding and decoding: Turbo-codes," *IEEE Trans. Commun.* 44(10), 1261–1271 (1996).
25. "3G technical specification: Multiplexing and channel coding," technical report 25.212 v6.2.0, 3GPP, 2004.
26. D. Divsalar, H. Jin, and R. J. McEliece, "Coding theorems for 'Turbo-Like' codes," In *Proc. 36th Allerton Conf. on Communications, Control, and Computing*, 1998, pp. 201–210.
27. J. Hagenauer, E. Offer, and L. Papke, "Iterative decoding of binary block and convolutional codes," *IEEE Trans. Inform. Theory* 42(2), 429–445 (1996).
28. D. J. MacKay, "Gallager codes—Recent results, technical report, University of Cambridge, available: http://www.inference.phy.cam.ac.uk/mackay/.
29. P. S. Chow, J. M. Cioffi, and J. A. Bingham, "A practical discrete multitone transceiver loading algorithm for data transmission over spectrally shaped channels," *IEEE Trans. Commun.* 43(2/3/4), 773–775 (1995).

CHAPTER 12

DISTRIBUTED SOLUTION FOR RESOURCE ALLOCATION IN ULTRA-WIDEBAND WIRELESS PANs

HAI JIANG, KUANG-HAO LIU, WEIHUA ZHUANG, and XUEMIN (SHERMAN) SHEN

12.1 INTRODUCTION

The future wireless personal area networks (PANs) are expected to support high data rate multimedia applications, such as home networking, gaming, imaging, and the like. Ultra-wideband (UWB) has emerged as a promising technology for the high speed services in a short range, especially after the U.S. Federal Communications Commission's (FCC) approval of the frequency band from 3.1 to 10.6 GHz to be used by UWB indoor applications. A UWB system has a 10-dB bandwidth larger than 500 MHz, or has a 10-dB fractional bandwidth larger than 20%. UWB signals can be obtained in two ways: pulse based [1, 2] and multiband orthogonal frequency division multiplexing (MB-OFDM) based [3]. In pulse-based UWB, information bits are sent via pulses with very short duration (say a nanosecond). Time hopping (TH) or direct sequence (DS) are two typical modulation methods. On the other hand, an MB-OFDM system applies OFDM and frequency hopping. UWB technology has many promising merits such as high transmission rate, low power/interference, and localization capability [4].

In UWB wireless PANs, to achieve desired quality of service (QoS) such as reception quality, delay/jitter bound, and throughput, the radio resources (e.g., transmission power, time, and rate) should be managed effectively to achieve spectrum efficiency [5]. The resource allocation for wireless local area networks (LANs) or ad hoc networks has been extensively investigated in the past decade. However, the unique characteristics of UWB pose new challenges and

Emerging Wireless LANs, Wireless PANs, and Wireless MANs. Edited by Y. Xiao and Y. Pan

provide new opportunities as well in resource allocation design, which should be tackled in UWB system development.

- The most popular resource allocation protocols for wireless LANs or ad hoc networks are random access based, such as carrier sense multiple access (CSMA) and its variants. In a neighborhood, only one transmission is allowed; otherwise, collisions will happen. Such *single-channel* protocols are not suitable for UWB wireless PANs because the inherent spread spectrum in UWB allows two or more simultaneous transmissions in a neighborhood as long as different pseudorandom sequences are applied. In such a *multichannel* case in UWB wireless PANs, the transmissions of two nodes only generate interference to each other. In the literature, the IEEE 802.15.3 protocol is designed for wireless PANs. However, it is still for a single-channel case.
- The precise synchronization in UWB reception requires a long acquisition time [6–8]. The acquisition can be obtained by the preamble that is prior to the data payload. The preamble duration in a UWB system is usually in the range from tens of microseconds to tens of milliseconds [9]. As a comparison, a narrowband system usually has an acquisition time on the order of microseconds. Therefore, for very high speed UWB wireless PANs, the preamble will occupy a large portion of the transmission time, generate significant overhead, and thus severely degrade the spectrum efficiency [10].
- UWB devices usually have very limited power supply. In a UWB wireless PAN, it may not be appropriate to have a fixed central controller, which is usually power demanding. Thus, resource allocation should be performed in a distributed manner. At each node, only local information and limited information exchanged from other nodes are available, thus making resource allocation more complicated.
- Pulse-based UWB is capable of positioning/ranging because of the small time domain resolution. Resource allocation can benefit from such location information.

In the following, effective distributed resource allocation for UWB wireless PANs is presented to address the preceding challenges and take advantage of the preceding capability. Since traditional OFDM resource allocation approaches can be applied in MB-OFDM–based wireless PANs, pulse-based UWB wireless PANs are the focus of this chapter. Although TH-UWB is considered in the following, the approach can be extended to DS-UWB as well.

12.2 SYSTEM MODEL

Consider a pulse-based TH-UWB wireless PAN that supports peer-to-peer single-hop communications with low mobility. Binary pulse position modulation

(PPM) is applied. The maximum transmission power of each node is P^{max}. Pulse-based UWB reception can be immune to multipath fading because of the very short pulse transmissions [11–13]. Therefore, it is assumed that there is no multipath fading [14], and the power at the receiver is attenuated due to path loss, that is, the channel gain from link i's transmitter to link j's receiver can be represented as $h_{ij} = K \cdot d_{ij}^{-\theta}$, where K and θ are constants, and d_{ij} is the distance from link i's transmitter to link j's receiver. As UWB technology is capable of positioning/ranging, it is assumed that each node has the location information of any other node in the wireless PAN.

Multiple access in TH-UWB can be achieved by assigning unique time-hopping codes to different links. Each node is assigned a unique receiving code. The receiving code of the destination is used for any peer-to-peer transmission. Hence, each node only needs to monitor its own receiving code for the desired coming traffic [15].

By proper code assignment, collisions among simultaneous transmissions can be avoided. However, the interference among them should also be managed. It has been shown that the total interference from a large number of links can be approximated as Gaussian noise [2]. Therefore, for a TH-UWB wireless PAN with N active links, the received signal-to-interference-plus-noise ratio (SINR) of link i can be represented as

$$\text{SINR}_i = \frac{P_i h_{ii}}{R_i(\eta_i + T_f \sigma^2 \sum_{j=1, j\neq i}^{N} P_j h_{ji})} \qquad i = 1, \ldots, N \tag{12.1}$$

where P_i denotes the average transmission power of link i's transmitter, R_i link i's bit rate, η_i the background noise energy plus non-UWB interference energy, T_f the pulse repetition time, and σ^2 a parameter depending on the shape of the pulse [16].

In the wireless PAN, each link is provided with a certain level of reception quality. In other words, for link i, the following inequality should hold:

$$\frac{P_i h_{ii}}{R_i(\eta_i + T_f \sigma^2 \sum_{j=1, j\neq i}^{N} P_j h_{ji})} \geq \gamma_i \tag{12.2}$$

where γ_i is the required SINR value for link i.

An equivalent form of the inequality is

$$R_i \leq \frac{P_i h_{ii}}{\gamma_i(\eta_i + T_f \sigma^2 \sum_{j=1, j\neq i}^{N} P_j h_{ji})} \tag{12.3}$$

The inequality gives the maximum achievable bit rate of link i with the constraint of SINR value γ_i. It is assumed that an adaptive rate can be achieved by changing the processing gain, for example, via adapting the number of pulses for each symbol and/or the maximum time-hopping shift, or via adaptive channel coding. On the other hand, for each link, the maximum achieved bit rate should not exceed $R^{\max} = 1/T_f$ as the processing gain should be at least 1. Therefore, for link i the achievable bit rate is

$$\min\left\{\frac{1}{T_f}, \frac{P_i h_{ii}}{\gamma_i(\eta_i + T_f \sigma^2 \sum_{j=1, j\neq i}^{N} P_j h_{ji})}\right\}$$

12.3 POWER/INTERFERENCE CONTROL

In a UWB wireless PAN, the interference among the links should be managed so that each link can be guaranteed the minimum SINR value. Generally, power control can be applied to manage the interference levels in the network, targeting at the guaranteed reception quality of each link. Power control can be executed in two manners: global or incremental [17]. A global power control approach reassigns the power levels of all the links in the network at any time when there is a change in link activities, for example, when a new link is admitted or an existing link is completed. This may lead to very large computation burden and overhead. Therefore, global power control is suitable only for a centralized network with powerful central controllers that have global information of all the links. On the other hand, upon a new link arrival, the incremental power control approach assigns the link a power level and keeps the power levels of all existing active links unchanged. With a much lower control overhead, this power control strategy is more suitable for UWB wireless PANs under consideration.

An incremental power control approach is based on a *maximum sustainable interference (MSI)* concept, which is also referred to as the *interference margin* [16]. The MSI of a link denotes the additional tolerable interference at the link's receiver while not violating the SINR requirement, that is, for link i,

$$\frac{P_i h_{ii}}{R_i(\eta_i + T_f \sigma^2 \sum_{j=1, j\neq i}^{N} P_j h_{ji} + \mathrm{MSI}_i)} = \gamma_i \tag{12.4}$$

which leads to

$$\mathrm{MSI}_i = \frac{P_i h_{ii}}{\gamma_i R_i} - \eta_i - T_f \sigma^2 \sum_{j=1, j\neq i}^{N} P_j h_{ji} \tag{12.5}$$

The MSI values of all the links are updated upon each new link admission and should be nonnegative in order to keep the reception quality of all the links. For multiple access in a multichannel environment, each active link periodically

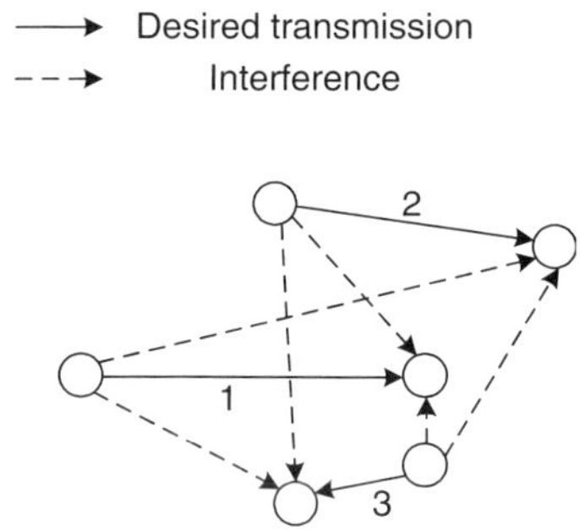

FIGURE 12.1 Near-sender-blocking problem in MSI-based scheme.

announces its MSI value over a control channel. If a link's MSI is honored by all the neighboring links, its transmission rate and reception quality can be guaranteed. When a call request for a new link arrives at a node, according to its local measurement of interference and noise levels and MSI information of other links, the node determines whether or not it is feasible to join the network while keeping reception quality of other links. If so, the new link is admitted [16].

It can be seen that the preceding MSI-based scheme uses a kind of circuit-switching channel reservation. Each link reserves a code channel, and a new link is required not to violate the QoS of existing links. However, it may suffer a severe *near-sender-blocking problem*, as demonstrated in Fig. 12.1. At the beginning, there are two links, links 1 and 2, each with transmission power $P^{\max}$ at the sender. Subsequently link 3 becomes active. As the sender of link 3 is close to the receiver of link 1, it may generate significant interference to link 1's receiver. Thus the MSI of link 1 may be largely reduced (even to zero). Then it is very difficult for the network to accommodate another link because of no sufficient MSI at link 1, even though links 2 and 3 have large MSIs.

Further, in the MSI-based scheme, some information exchanges are needed such as the MSI values, location, and transmission power of the active links. However, it is difficult for a link to obtain complete and updated information due to possible collisions of the exchange messages.

An effective solution to address the near-sender-blocking problem is to use a temporally exclusive mechanism. As shown in Fig. 12.1, if links 1 and 3 transmit in different periods, the significant reduction of link 1's MSI can be avoided. Accordingly, we propose a distributed scheme for the UWB wireless PAN with multichannel transmissions. The time frame structure and resource allocation algorithm of the proposed scheme are detailed in the following, respectively.

12.4 TIME FRAME STRUCTURE AND MESSAGE EXCHANGES

The time frame structure is shown in Fig. 12.2. Each frame starts with a beacon, followed by M time slots. The beacon is to indicate the beginning of a frame.

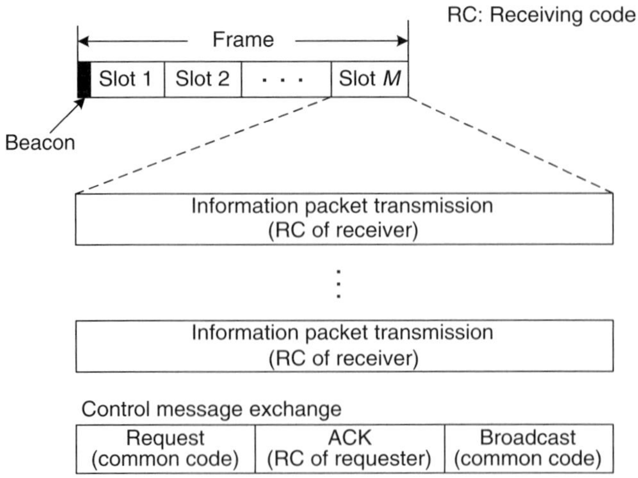

FIGURE 12.2 Time frame structure.

In each slot, multiple simultaneous transmissions are supported for data packets and control message exchanges. For effective control message exchanges, among all the active senders in a slot, one is selected to act as *slot head* (the selection procedure will be discussed later). The head is responsible to collect information at this slot and broadcast to potential new senders during its period of duty. As shown in Fig. 12.2, the control message exchanges are performed in parallel with other information packet transmissions. The exchange procedure includes three phases: request phase for a new caller to send a request by a prespecified request common code; ACK phase for the slot head to acknowledge the successful reception of a new call request; and broadcast phase for the slot head to broadcast (by a prespecified broadcast common code) the MSI, location, and transmission power information of the existing links in the slot. It can be seen that because of UWB's capability in supporting multiple simultaneous transmissions, the time frame structure does not need an extra control slot for channel request as those in traditional single-channel protocols [18].

Consider a UWB wireless PAN with no active nodes at the beginning. When a node has traffic to send, it first detects the beacon. If no beacon is detected, the node assumes that all the channels are idle, transmits a beacon, picks up the first one or more slots (depending on its rate request), and acts as the head for the slots. In each of the slots, the node not only transmits to its receiver (using the receiver's receiving code) but also transmits the location, MSI values, and transmission power information of all the active links in the slot via the broadcast common code. The beacon is always transmitted by the head of the first slot.

When another node (a potential sender) has a call request for a new link, it first listens to the broadcast channel in all the slots and collects the MSI values, location, and transmission power information of each active link in each slot. It then selects (based on criterion discussed in Section 12.5) a slot as its target slot and sends a request at the request phase of the slot. There are two possible outcomes of the request: (1) If the request is received successfully by the slot head, the slot head responds with an ACK via the requester's receiving code. Upon the reception of the ACK, the requested link will be accommodated in the slot, and its sender can transmit at the same slot of subsequent frames until the call is completed. The new sender also becomes the new slot head in order not to pose all computation complexity on a single head. As the broadcast message from the old head contains all required information for a head, there is no extra overhead for information transfer between the old and new heads. The new head updates the MSI values of all the existing links and broadcasts the MSI values, location, and transmission power information of all the links (including itself). (2) If no ACK is received (e.g., when at least one other node also sends a request simultaneously) for a new link request, the call sender resends the request at the slot of the following frames with a probability p until an ACK is received.

12.5 RESOURCE ALLOCATION

12.5.1 Resource Definition

Defining the "resource" is a challenging task for UWB wireless PANs with peer-to-peer connections. Because the interference level at each receiver can be quite different, it may not be feasible to find a global resource definition for the whole UWB wireless PAN. In the following, the resource is defined from the viewpoint of each receiver based on the concept of MSI.

From the MSI_i given in (12.5), it can be seen that when R_i takes the value of the maximum achievable rate given by (12.3) (if feasible), the MSI_i value is equal to 0. Link i achieves its maximum MSI value when there is no other active link, that is,

$$\text{MSI}_i^{\max} = \frac{P_i h_{ii}}{\gamma_i R_i} - \eta_i \tag{12.6}$$

For active link i, denote ΔMSI_{ji} as its MSI reduction due to the activity of link j, which is given by

$$\Delta\text{MSI}_{ji} = T_f \sigma^2 P_j h_{ji} \tag{12.7}$$

Hence, the MSI of link i can be rewritten as

$$\text{MSI}_i = \text{MSI}_i^{\max} - \sum_{j=1, j\neq i}^{N} \Delta\text{MSI}_{ji} \tag{12.8}$$

From (12.8), to guarantee the QoS of link i (i.e., a nonnegative MSI_i), the summation of all the $\Delta\mathrm{MSI}_{ji}$ over j should be bounded. From the linear constraint, we define the amount of resources for link i as the $\mathrm{MSI}_i^{\max}$. The $\Delta\mathrm{MSI}_{ji}$ in Eq. (12.8) represents the amount of resources consumed by link j from the perspective of link i, and the MSI_i is the amount of available resources at link i.

12.5.2 Slot Selection and Power/Rate Allocation

In the following, the criterion is discussed regarding how a link determines from which slots it requests service and with what power and rate levels it transmits in order to meet its rate requirement and not to degrade the QoS of existing links.

Consider link i with target rate R_i^t. Let $\Omega_i \subseteq \{1, 2, \ldots, M\}$ denote the set of slots (in each frame) from which link i has gained services, and R_i^k denote the transmission rate of link i at slot k when $k \in \Omega_i$. Then the effective rate of link i, that is, the total achieved rate of link i in the system, is given by

$$R_i^e = \sum_{k \in \Omega_i} R_i^k (1 - \mathbf{1}_{\bar{\Omega}_i}\{k-1\}\xi) \frac{T_{\text{slot}}}{T_{\text{frame}}} \tag{12.9}$$

where T_{frame} and T_{slot} are the length of a frame and a time slot as shown in Fig. 12.2, respectively, ξ is the acquisition overhead ratio (i.e., the fraction of time for acquisition in a slot if acquisition is needed), and the indication function

$$\mathbf{1}_{\bar{\Omega}_i}\{k-1\} = \begin{cases} 1 & k-1 \in \bar{\Omega}_i \quad \text{or} \quad k-1 = 0 \\ 0 & \text{otherwise} \end{cases} \tag{12.10}$$

When slot $k-1$ is serving link i, the reacquisition overhead in slot k is not necessary.

When a new call request for link i arrives, by checking whether there exists at least one idle slot, the sender node selects the first idle slot k and transmits with maximum power $P_i^k = P^{\max}$ and rate R_i^k so as to achieve the target R_i^t:

$$R_i^k : R_i^t = R_i^k (1 - \xi) \frac{T_{\text{slot}}}{T_{\text{frame}}} \tag{12.11}$$

that is, $R_i^k = R_i^t / [(1-\xi)(T_{\text{slot}}/T_{\text{frame}})]$. If $R_i^t / [(1-\xi)(T_{\text{slot}}/T_{\text{frame}})] > R^{\max}$, we set $R_i^k = R^{\max}$. Thus, when the target rate R_i^t cannot be achieved through slot k, we set R_i^k to be the maximum achievable rate $R^{\max}$ at a slot. If no idle slot exists, the call sender of link i selects one among all the M slots. For the slot selection, let Q_i^k denote a penalty function of the resource consumption by link i at slot k,

U_i^k (a utility function) the gain of link i in its effective rate, and C_i^k (a cost function) the overall penalty versus utility for link i at slot k. To achieve efficient resource utilization, the slot with the minimal cost function is chosen, as discussed in the following.

Let $N(k)$ denote the number of existing links within slot $k \in \{1, \ldots, M\}$, and P_j^k and R_j^k denote the transmission power and rate, respectively, for an active link j at slot k. The MSI value (i.e., the amount of available resources) of active link j at slot k is given by

$$\mathrm{MSI}_j^k = \frac{P_j^k h_{jj}}{\gamma_j R_j^k} - \eta_j - T_f \sigma^2 \sum_{l=1, l \neq j}^{N(k)} P_l^k h_{lj} \tag{12.12}$$

Under the hypothesis that the new call request (for link i) selects slot k for service, let p_i^k and r_i^k denote the transmission power and rate of link i at slot k, respectively.[1] The power p_i^k should be constrained by the MSIs of existing active links at slot k [16], that is,

$$p_i^k = \min\left\{ P^{\max}, \min_{1 \leq j \leq N(k)} \left(\frac{\mathrm{MSI}_j^k}{T_f \sigma^2 h_{ij}} \right) \right\} \tag{12.13}$$

If $p_i^k = 0$, link i cannot be accommodated at slot k as it will violate other links' MSIs. Thus, we set the cost function of link i at slot k as $C_i^k = \infty$. If $p_i^k > 0$, we set r_i^k so as to achieve the target rate R_i^t, that is,

$$r_i^k : R_i^t = r_i^k (1 - \xi) \frac{T_{\mathrm{slot}}}{T_{\mathrm{frame}}} \tag{12.14}$$

where r_i^k is bounded by $R^{\max}$. This setting is feasible only if the resulting MSI of link i at slot k is nonnegative, that is,

$$\frac{p_i^k h_{ii}}{\gamma_i r_i^k} - \eta_i - T_f \sigma^2 \sum_{j=1}^{N(k)} P_j^k h_{ji} \geq 0 \tag{12.15}$$

If (12.15) cannot be satisfied (i.e., the SINR of link i at slot k is not high enough to support r_i^k), we set $C_i^k = \infty$.

Under the hypothesis that link i is admitted in slot k, the MSI reduction of an existing active link j $[1 \leq j \leq N(k)]$ at slot k due to the admission of link i (or equivalently, the amount of link j's available resources that will be consumed by

[1] We use capital letters P and R to denote the actual transmission power and rate values already allocated at a slot and use the lowercase p and r to denote those in hypothesis.

link i) is given by

$$\Delta\text{MSI}_{ij}^{k} = T_f \sigma^2 p_i^k h_{ij} \tag{12.16}$$

The penalty function Q_i^k should reflect the resource consumption by link i if it is accommodated at slot k. One intuitive way to define Q_i^k is to aggregate all the ΔMSI_{ij}^k, $1 \leq j \leq N(k)$. However, it is also desired to differentiate existing links at slot k with different MSI_j^k values. In general, with the same value of ΔMSI_{ij}^k, the MSI reduction of an existing link j with small MSI_j^k should cause a large penalty as the link j may become the bottleneck for subsequent new link accommodation. Therefore, the penalty induced by admitting link i at slot k is defined by

$$Q_i^k = \sum_{j=1}^{N(k)} \frac{\Delta\text{MSI}_{ij}^k}{\text{MSI}_j^k} \tag{12.17}$$

To reflect the increase of effective rate achieved from admitting link i in slot k, the utility function is defined by

$$U_i^k = r_i^k (1 - \xi) \frac{T_{\text{slot}}}{T_{\text{frame}}} \tag{12.18}$$

We use a heuristic cost function C_i^k to deal with the overall penalty versus gained effective rate, that is, $C_i^k = Q_i^k / U_i^k$.

For the call request of new link i, if all the cost functions at the M slots are infinity, the call will be dropped; otherwise, the target slot k^* is chosen by

$$k^* = \arg \min_{1 \leq k \leq M} C_i^k \tag{12.19}$$

with the power level $P_i^{k^*} = p_i^{k^*}$ and rate level $R_i^{k^*} = r_i^{k^*}$. Once the target slot k^* is chosen, the source node sends a request at the request phase of slot k^* and expects to receive an ACK. After that, $\Omega_i = \{k^*\}$.

It is possible that the effective rate of link i in Ω_i calculated by (12.9) is less than its rate requirement R_i^t. In such circumstances, link i needs to request services at more slots (not in Ω_i) until its rate requirement is satisfied or it is degraded.[2] The sender first tries idle slots. If there is no idle slot, a procedure similar to that described above is executed, except for the determination of r_i^k and the utility calculation. Combined with previous allocated rates in other

[2] We term a call degraded if it is admitted into one or more slots but its rate requirement is not satisfied (as it cannot gain more service from other slots).

slots (in Ω_i), the r_i^k is the value that satisfies the target rate requirement, given by

$$r_i^k : \sum_{l \in \Omega_i} R_i^l (1 - \mathbf{1}_{\overline{\Omega}_i}\{l-1\}\xi) \frac{T_{\text{slot}}}{T_{\text{frame}}} + r_i^k (1 - \mathbf{1}_{\overline{\Omega}_i}\{k-1\}\xi) \frac{T_{\text{slot}}}{T_{\text{frame}}} + R_i^{k+1} \mathbf{1}_{\Omega_i}\{k+1\}\xi \frac{T_{\text{slot}}}{T_{\text{frame}}} = R_i^t \tag{12.20}$$

Similarly, if $r_i^k > R^{\max}$, we set $r_i^k = R^{\max}$. The second term on the left side of (12.20) means the effective rate at slot k, and the third term represents the increase of effective rate at slot $k+1$ if it already provides service to link i (because of the reduction of the acquisition overhead at slot $k+1$). Accordingly, the utility is given by

$$U_i^k = r_i^k (1 - \mathbf{1}_{\overline{\Omega}_i}\{k-1\}\xi) \frac{T_{\text{slot}}}{T_{\text{frame}}} + R_i^{k+1} \mathbf{1}_{\Omega_i}\{k+1\}\xi \frac{T_{\text{slot}}}{T_{\text{frame}}} \tag{12.21}$$

From the utility function, it can be seen that consecutive slot allocation is favored, and thus the negative effect of long acquisition time can be mitigated.

In (12.20), it is possible that $r_i^k < 0$. This happens when the target rate can be achieved by the increase of effective rate at slot $k+1$. In this case, a low r_i^k is assigned to maintain the link activity at slot k.

The above procedure continues until the rate requirement of link i is met or link i is degraded.

12.6 PERFORMANCE EVALUATION

Simulation is carried out to evaluate the performance of the proposed scheme and to compare it with the MSI-based scheme [16]. The experimental wireless PAN is set up as follows. A number of stationary nodes are uniformly distributed in a two-dimensional 100-m × 100-m square. The call arrival to the whole network is a Poisson process with rate λ, and each call duration is exponentially distributed with mean value equal to μ. For the simplicity of presentation, each call arrival is assigned a sender and a receiver, both independently and uniformly located in the square (thus referred to as *uniform topology*). The time frame structure, as illustrated in Fig. 12.2, has a fixed period of 30 ms, which is further divided into a beacon (with duration of 5 ms) and five time slots (each with duration of 5 ms). For comparison, a similar time frame structure is implemented in the MSI-based scheme except that there is only one time slot (with duration of 25 ms) in a frame. Our proposed control message exchange procedure is also used in the MSI-based scheme. In addition, the acquisition time in both schemes is the same, which leads to the acquisition overhead ratio in our scheme five times that in the MSI-based

TABLE 12.1 Parameters Used in Performance Evaluation

Parameters	Values
θ	2.4
K	1
η	2.568×10^{-21} W/Hz
γ	7 dB
T_f	100 ns
σ^2	1.9966×10^{-3}
$P^{\max}$	0.5 mW
R_i^t	2 Mbps

scheme. For presentation simplicity, the acquisition time in both schemes is measured by the acquisition overhead ratio ξ in our scheme. Other relevant parameters used in the simulation are given in Table 12.1. All simulation results are obtained by averaging over 2000 calls.

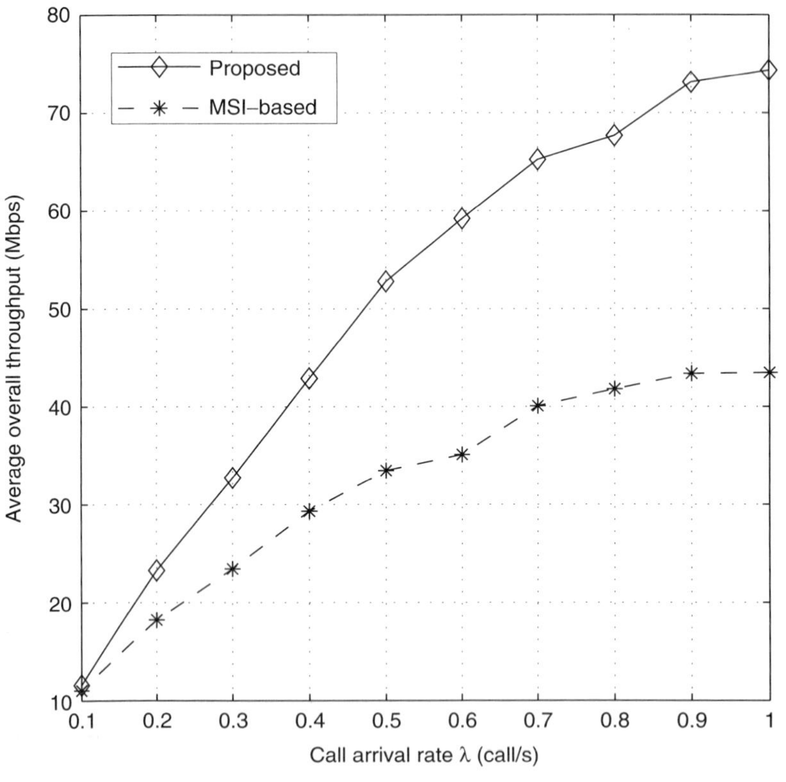

FIGURE 12.3 Throughput over uniform topology.

12.6.1 Throughput

The achieved average overall throughput versus the call arrival rate λ is shown in Fig. 12.3 with $\xi = 0.5$ and $\mu = 60$ s. It is observed that our scheme achieves up to 70% increase in the system throughput over that of the MSI-based scheme. The gain becomes more significant as the call arrival rate λ increases. This is because our scheme can solve the near-sender-blocking problem by temporal exclusion. Therefore, the system employing our scheme can admit more calls. Although the use of multiple slots in our scheme may have the risk of aggravating the overhead due to acquisition, the operation explained in Section 12.5 tends to regulate each active node acquiring consecutive slots and thus reducing the impact of long acquisition time.

12.6.2 Call Dropping Probability

A call is dropped if its admission may damage the reception of existing calls, or equivalently there are no sufficient resources available at the instant that the sender requests an admission. It is assumed that there is no retry, which may be added to the scheme for implementation purposes. Figure 12.4 shows that our scheme provides a relatively small likelihood of call dropping than the

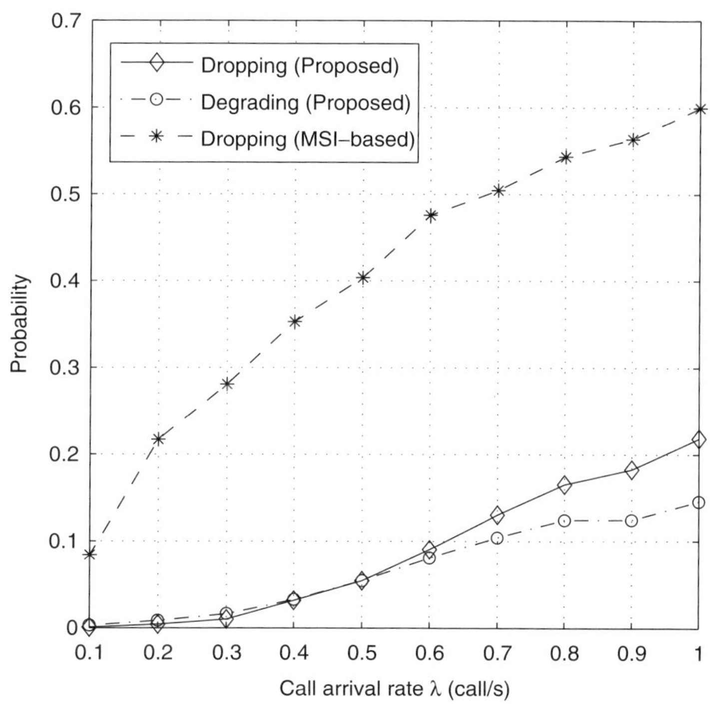

FIGURE 12.4 Probability of call being dropped or degraded over uniform topology.

MSI-based scheme. It is further observed that some calls may be admitted but unable to acquire sufficient resources to satisfy their QoS requirements (i.e., degraded). As the offered load increases, more calls are likely to be degraded. Note that there is no degraded call in the MSI-based scheme (with only one slot in each frame) to avoid the bottleneck effect.

12.6.3 Power Consumption

Besides the throughput and call dropping probability, another concern of the resource allocation in the UWB wireless PAN is the power consumption. Figure 12.5 shows the average transmission power consumption of each admitted call. Here the energy consumed in sending the control messages is ignored. It can be seen that our scheme can reach a higher throughput with much lower average power consumption. Furthermore, when the traffic load increases, the average power consumption slightly decreases. This can be explained as follows. When the call arrival rate is small, almost all the existing active links at each slot have sufficient MSI values. Thus, a new link is very likely to transmit with maximum power $P^{\max}$ at its serving slots. When the call arrival rate increases (after the threshold $\lambda = 0.5$ call/s in the example), there are

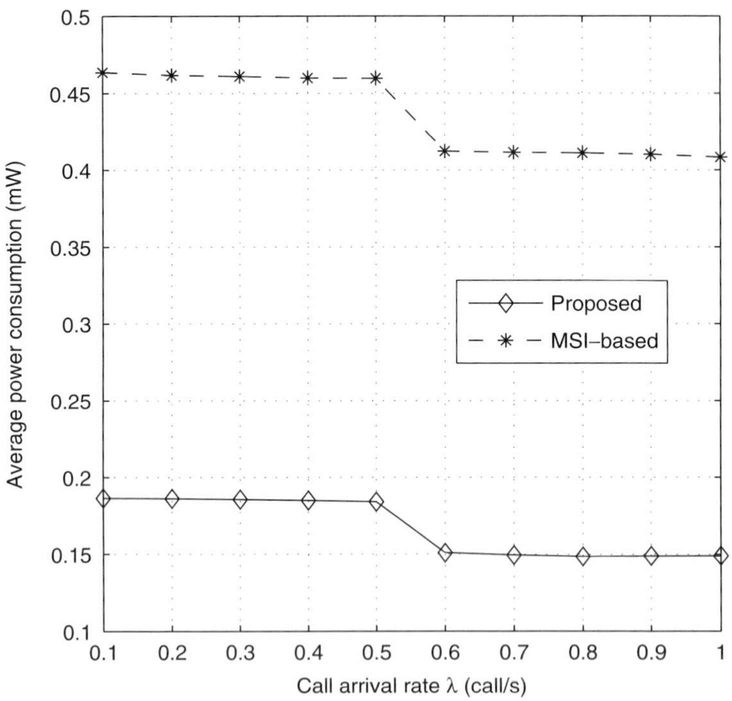

FIGURE 12.5 Average power consumption over uniform topology.

more active links at a slot, resulting in insufficient MSI values of some links. It is likely that the senders of new calls are not allowed to transmit with $P^{\max}$ due to the constraints of MSIs of existing active links. Thus, the average power consumption slightly decreases.

12.6.4 Performance in a Clustered Topology

Consider a clustered UWB wireless PAN where the 100-m $\times$ 100-m square is equally partitioned into four regions, and each node only communicates with nodes in the same region. A node can hear any other node's transmission as long as it tunes to the code used and the received SINR exceeds a threshold. The beacon and broadcast messages can be heard by all nodes, the request can be heard by the slot head, and the acknowledgment for the request can be heard by the requester. Call arrivals are distributed in different regions so that calls are equally located in the experimental area. Figures 12.6–12.8 illustrate the behavior of the proposed scheme compared to the MSI-based scheme over the clustered topology. Similar to the result of the uniform topology, Fig. 12.6 shows that both schemes achieve nearly the same throughput when the traffic density is low. As the call arrival rate increases, the throughput improvement

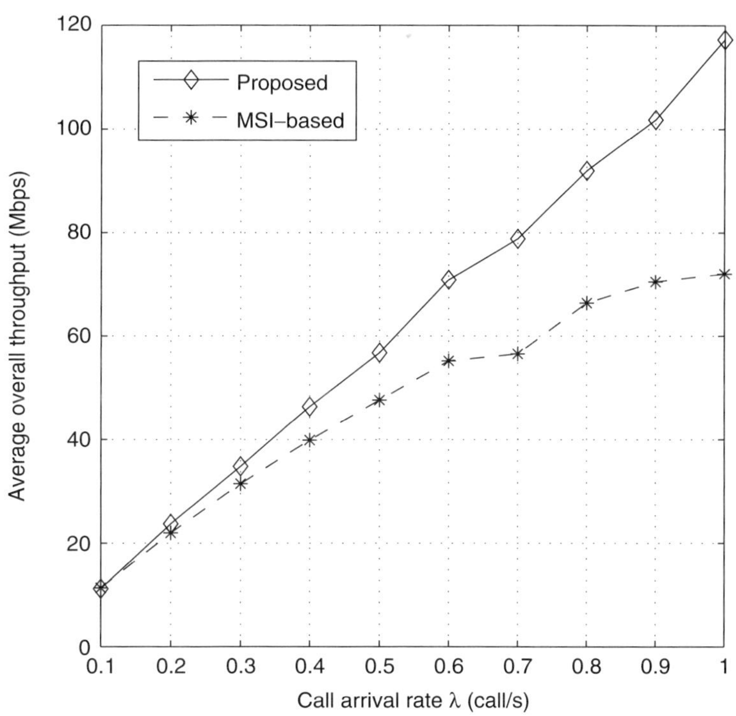

FIGURE 12.6 Throughput over clustered topology.

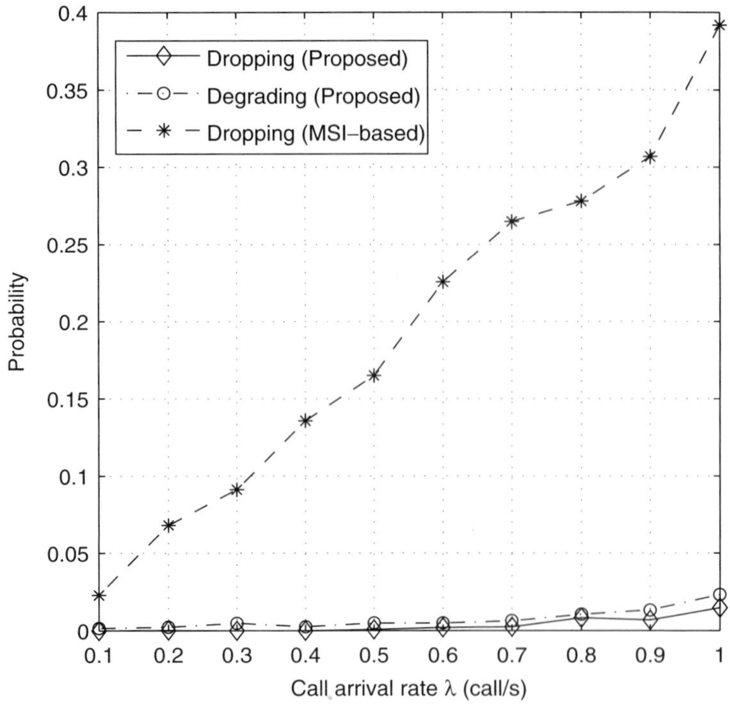

FIGURE 12.7 Probability of call being dropped or degraded over clustered topology.

achieved by our scheme is more significant. In terms of the throughput and call dropping/degrading probability, it can be seen from Figs. 12.3, 12.4, 12.6, and 12.7 that both schemes perform better in the clustered topology than in the uniform topology. The power consumption shown in Fig. 12.8 reveals the similar tendency to that of the uniform topology.

To further compare the network capacity with the uniform and clustered topologies, Table 12.2 gives the system throughput, dropping and degrading probabilities, and average number of active links per slot in our scheme with different topologies. It can be seen that, with the uniform topology, $\lambda = 1$ call/s makes the network saturated with dropping probability of 22%. When λ increases to 2 calls/s, the throughput increase is relatively small. However, with the clustered topology, $\lambda = 1.5$ calls/s makes the network close to saturation, with 6% dropping probability. This is because the path gain of each link in the clustered topology is likely larger than that in the uniform topology, which in turn increases the MSI of each link. Thus, more links can be admitted in a slot with the clustered topology than with the uniform topology, as seen in Table 12.2. This implies that routing at the network layer should be jointly designed with the radio resource allocation at the link layer. For a link with the sender and receiver separated far away, a single hop transmission may lead to a

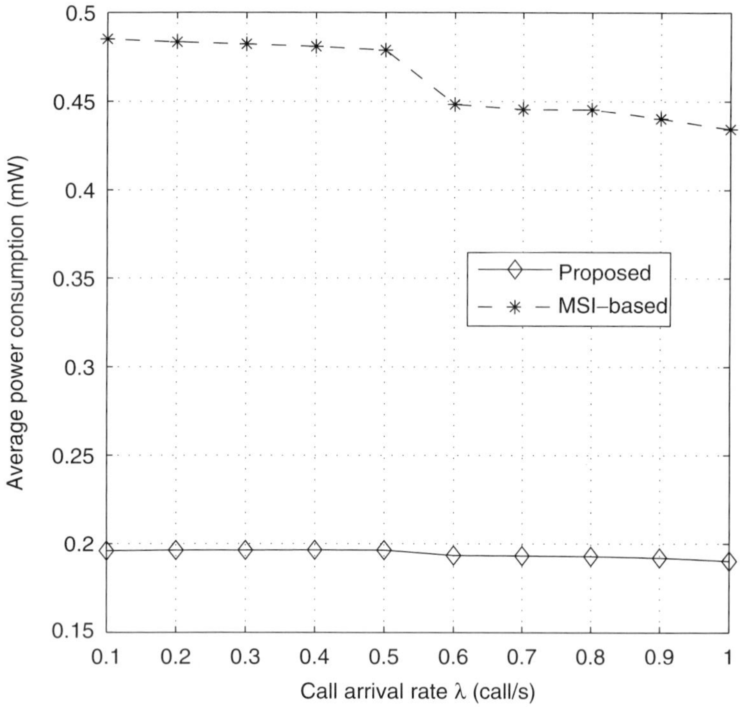

FIGURE 12.8 Average power consumption over clustered topology.

small delay, at the cost of the possibility of its becoming a bottleneck due to the small MSI, while multihop transmissions can keep a large MSI for each hop, at the cost of a large delay due to the multihop link. How to achieve an appropriate trade-off is an interesting issue.

TABLE 12.2 Capacity Comparison of Uniform and Clustered Topologies

Topology	λ (call/s)	Throughput (Mbps)	Dropping Prob. (%)	Degrading Prob. (%)	Average Link Number per Slot
Uniform	2.0	92	43	16	18
	1.5	84	37	14	17
	1.0	75	22	14	17
Clustered	2.0	170	16	11	32
	1.5	156	6	7	31
	1.0	118	1	2	25

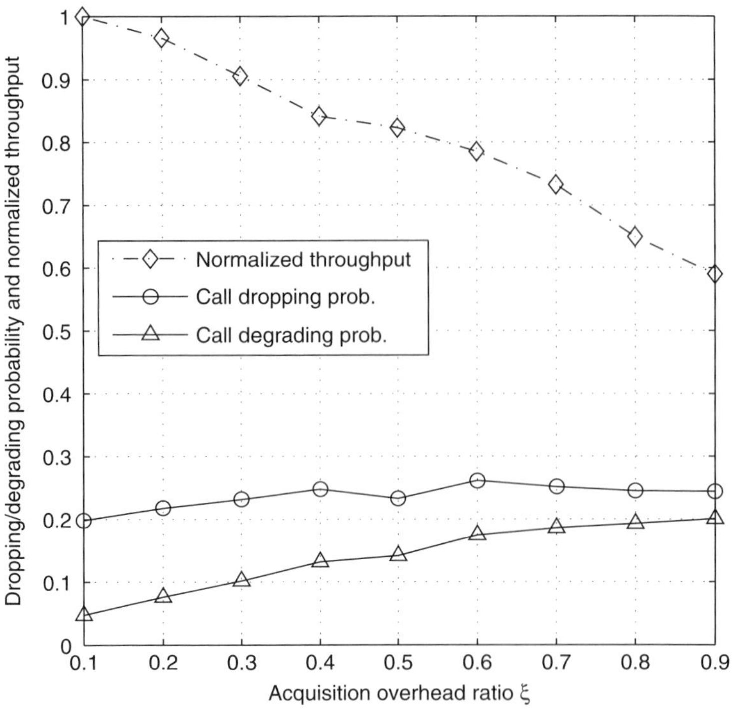

FIGURE 12.9 Call dropping/degrading probability and normalized throughput (with respect to $\xi = 0.1$ case) versus acquisition overhead ratio ξ in our proposed scheme.

12.6.5 Effects of Acquisition Overhead

The impacts of acquisition overhead on the overall network performance, such as call dropping and degrading probabilities, and normalized throughput (with respect to $\xi = 0.1$) are shown in Fig. 12.9 with $\lambda = 1$ call/s. As expected, a higher acquisition overhead ratio tends to reduce the achievable throughput. On the other hand, if a transmission encounters a higher acquisition overhead ratio, the effective transmission time is reduced, which leads to an increasing number of calls without transmission rate requirement satisfaction. This illustrates the increase of call degrading probability as the acquisition overhead ratio increases. From Fig. 12.9, when the acquisition overhead ratio increases from 0.1 to 0.9 (i.e., the transmission efficiency significantly decreases from 90% to 10% when acquisition is needed), the overall throughput only decreases by approximately 40%. This is because our proposed scheme favors transmissions of a link at consecutive slots.

12.7 CONCLUSION

Peer-to-peer communications in UWB wireless PANs can be affected by the near-sender-blocking problem. An effective solution to address the issue is to use a temporally exclusive mechanism. Our proposed solution can achieve temporal exclusion and distribute computation burden of admission and resource allocation to the admitted nodes. The effect of long acquisition time in UWB transmissions is mitigated as well.

REFERENCES

1. W. Zhuang, X. Shen, and Q. Bi, "Ultra-wideband wireless communications," *Wireless Commun. Mobile Comput.* 3(6), 663–685 (2003).
2. M. Z. Win and R. A. Scholtz, "Ultra-wide bandwidth time-hopping spread-spectrum impulse radio for wireless multiple-access communications," *IEEE Trans. Commun.* 48(4), 679–691 (2000).
3. A. Batra, J. Balakrishnan, G. R. Aiello, J. R. Foerster, and A. Dabak, "Design of a multiband OFDM system for realistic UWB channel environment," *IEEE Trans. Microwave Theory Techniques* 52(9), 2123–2138 (2004).
4. R. C. Qiu, H. Liu, and X. Shen, "Ultra-wideband for multiple-access communications," *IEEE Commun. Mag.* 43(2), 80–87 (2005).
5. H. Jiang, W. Zhuang, X. Shen, and Q. Bi, "Quality-of-service provisioning and efficient resource utilization in CDMA cellular communications," *IEEE J. Sel. Areas Commun.* 24(1), 4–15 (2006).
6. S. Roy, J. R. Foerster, V. S. Somayazulu, and D. G. Leeper, "Ultra-wideband radio design: The promise of high-speed, short range wireless connectivity," *Proc. IEEE* 92(2), 295–311 (2004).
7. R. L. Peterson, R. E. Ziemer, and D. E. Borth, *Introduction to Spread Spectrum Communications*, Prentice Hall, Englewood Cliffs, NJ, 1995.
8. Y. Ma, F. Chin, B. Kannan, and S. Pasupathy, "Acquisition performance of an ultra wide-band communications system over a multiple-access fading channel," in *Proc. IEEE Conference on Ultra Wideband Systems and Technologies* (*UWBST2002*), Baltimore, MD, May 2002, pp. 99–103.
9. S. Aedudodla, S. Vijayakumaran, and T. F. Wong, "Rapid ultra-wideband signal acquisition," in *Proc. IEEE Wireless Communications and Networking Conference* (*WCNC'04*), Atlanta, GA, Mar. 2004, pp. 1148–1153.
10. J. Ding, L. Zhao, S. R. Medidi, and K. M. Sivalingam, "MAC protocols for ultra-wide-band (UWB) wireless networks: Impact of channel acquisition time," in *Proc. SPIE*—Vol. 4869: *Emerging Technologies for Future Generation Wireless Communications*, Boston, MA, Nov. 2002, pp. 97–106.
11. F. Ramirez-Mireles, "On the performance of ultra-wide-band signals in Gaussian noise and dense multipath," *IEEE Trans. Vehic. Technol.* 50(1), 244–249 (2001).
12. K. W. Lam, Q. Li, L. Tsang, K. L. Lai, and C. H. Chan, "On the analysis of statistical distributions of UWB signal scattering by random rough surfaces based on Monte Carlo simulations of Maxwell equations," *IEEE Trans. Antennas Propagat.* 52(12), 3200–3206 (2004).
13. M. Z. Win and R. A. Scholtz, "On the robustness of ultra-wide bandwidth signals in dense multipath environments," *IEEE Commun. Lett.* 2(2), 51–53 (1998).
14. B. Radunovic and J.-Y. Le Boudec, "Optimal power control, scheduling, and routing in UWB networks," *IEEE J. Sel. Areas Commun.* 22(7), 1252–1270 (2004).

15. E. S. Sousa and J. A. Silvester, "Spreading code protocols for distributed spread-spectrum packet radio networks," *IEEE Trans. Commun.* 36(3), 272–281 (1988).
16. F. Cuomo, C. Martello, A. Baiocchi, and F. Capriotti, "Radio resource sharing for ad hoc networking with UWB," *IEEE J. Sel. Areas Commun.* 20(9), 1722–1732 (2002).
17. S. Lal and E. S. Sousa, "Distributed resource allocation for DS-CDMA-based multimedia ad hoc wireless LAN's," *IEEE J. Sel. Areas Commun.* 17(5), 947–967 (1999).
18. C. Zhu and M. S. Corson, "A five-phase reservation protocol (FPRP) for mobile ad hoc networks," In *Proc. IEEE INFOCOM'98*, San Francisco, CA, Mar.–Apr. 1998, pp. 322–331.

PART IV

IEEE 802.15.4 AND 802.15.5 WIRELESS PANs

CHAPTER 13

IEEE 802.15.4 MEDIUM ACCESS CONTROL AND PHYSICAL LAYERS

YANG XIAO, MICHAEL J. PLYLER, MING LI, and FEI HU

13.1 INTRODUCTION

The IEEE 802.15.4 specification [1] defines the medium access control (MAC) and physical (PHY) layers targeted for low rate wireless personal area networks (LR-WPANs) using short-distance applications with communication networks of low power consumption and low cost, particularly short-range applications such as wireless sensor networks and residential/industrial setting networks. Applications of IEEE 802.15.4 include light control systems, environmental and agricultural monitoring, consumer electronics, energy management and comfort functions, automatic meter reading systems, industrial applications, and alarm and security systems [2].

This chapter surveys the IEEE 802.15.4 MAC and PHY layers [1] and is organized as follows. Section 13.2 gives a short overview. Section 13.3 introduces personal area network (PAN) functionality. Section 13.4 introduces the 802.15.4 frame formats. Section 13.5 introduces the 802.15.4 MAC command frame formats. The importance of Sections 13.3–13.5 will be seen in its different methods, managements, security, and hardware devices in the LR-WPAN.

13.2 A SHORT OVERVIEW

The IEEE 802.15.4 specification favors low cost and low power LR-WPANs for a wide variety of applications requiring short-range distances. Low power consumption is one of the major design requirements in the IEEE 802.15.4 specification to maximize battery life, assuming that the amount of data transmitted is small and transmissions are infrequent [2–4]. The frame structure is designed with minimal overhead.

Emerging Wireless LANs, Wireless PANs, and Wireless MANs. Edited by Y. Xiao and Y. Pan

This section gives a short overview of IEEE 802.15.4, including its basic component devices, network topology, physical layer, and MAC layer.

13.2.1 Devices

The PAN coordinator (PANC) is the principal controller of a PAN. It controls the network and defines the parameters of the network. An IEEE 802.15.4 network has exactly one PANC. There are two types of devices present in the specification that communicate together to form different network topologies: full-function device (FFD) and reduced-function device (RFD). An FFD is capable of operating as a coordinator or device and implementing the complete protocol set. An RFD operates with a minimal implementation of the IEEE 802.15.4 protocol. An RFD can only connect to an FFD, whereas an FFD can connect to both FFDs and RFDs. The FFD that acts as a PANC is the main controller of the network and can initiate a communication, terminate it, and route it around the network. At the physical level, an FFD and an RFD distinguish themselves based on their capabilities of hardware platforms. An RFD can perform a logical role of end device with extremely simple applications such as a light sensor and a lighting controller, whereas an FFD can take up the roles of coordinator and a router.

13.2.2 Network Topology

The RFDs and FFDs combine together to form two types of network topologies: star topology and peer-to-peer topology, shown in Fig. 13.1. In

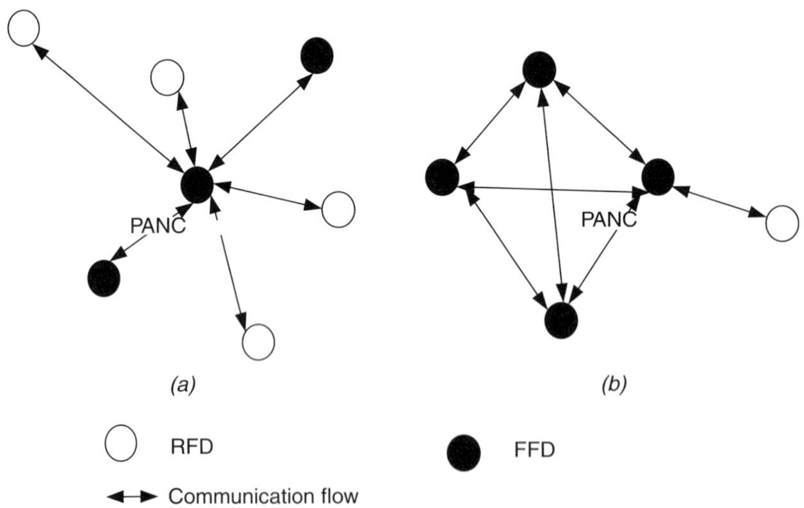

FIGURE 13.1 Network topologies: (*a*) star; (*b*) peer to peer.

the star topology, the PANC acts as the initiation point for the network and other FFDs and RFDs connect to it. Communications are performed between RFDs/FFDs and the PANC, which is in charge of managing all the star functionality. In the peer-to-peer topology, every FFD can communicate with other FFDs, including a PANC. Peer-to-peer topology allows more complex network formations to be implemented, e.g., ad hoc and self-configuring networks. Each PANC has a unique identifier or the link key through which other devices can communicate with each other.

13.2.3 Physical Layer

The IEEE 802.15.4 specification supports two PHY options based on direct-sequence spread spectrum (DSSS), which allows the use of low cost digital integrated circuit (IC) realizations [2]. The PHY layer adopts the same basic frame structure for low duty cycle, low power operations, except that the two PHY layers adopt different frequency bands: low band (868–915 MHz) and high band (2.4 GHz). The low band adopts binary phase shift key (BPSK) modulation and operates in the 868-MHz band in Europe. It offers 1 channel with a raw data rate of 20 kbps.I. In the 915-MHz industrial, scientific, and medical (ISM) band in North America it offers 10 channels with a raw data rate of 40 kbps [1, 2]. The low band adopts offset quadrature phase shift key (O-QPSK) modulation, operates in the 2.4–2.483-GHz band, and is a part of the ISM band, which is available almost worldwide, and has 16 channels with channel spacing of 5 MHz and a raw data rate of 250 kbps. The PHY layer uses a common frame structure, containing a 32-bit preamble, a frame length, and 2–127 bytes of payload field.

13.2.4 Medium Access Control

The IEEE 802.15.4 MAC layer is used for reliable and single-hop communication among the devices, providing access to the physical channel for all types of transmissions and appropriate security mechanisms. MAC uses acknowledged frame delivery, performs frame validation, maintains network synchronization, controls the association/disassociation, administers device security, and schedules the time slots.

The specification allows the optional use of a superframe structure for applications requiring dedicated bandwidth with guaranteed delay. The PANC defines the format of the superframe, which includes a beacon frame, the contention access period (CAP), and the contention-free period (CFP). The total length of the CAP and the CFP is 16 equally sized time slots. The time slots for the CFP are called guaranteed time slots (GTSs) and are administered by the PANC. The CAP adopts the carrier sense multiple-access/collision avoidance (CSMA/CA) mechanism.

13.3 PAN FUNCTIONALITY

The functionality of IEEE 802.15.4 encompasses the following topics: superframe structure, data transfer, frame structure, power consumption, robustness, and security.

It is apparent that there are many functional protocols for the IEEE 802.15.4 standard to follow. However, there has been enough slack given to enable the application programmer to make application-specific adjustments for the functionality and management of the network.

13.3.1 Superframe Structure

In IEEE 802.15.4, it is optional for the network to use the superframe structure. For networks that use the superframe structure, the format of the superframe is given by the coordinator. As you can see in Fig. 13.2, the superframe is bounded by two network beacons. These beacons are sent by the coordinator. The frame is divided into 16 equally sized slots. The beacon is transmitted in the first of these slots for each frame.

If the coordinator chooses not to use the superframe, the beacon transmission can be turned off and is not used for the PAN. This is usually only helpful for PANs that are operating on peer-to-peer topologies. In the other case, star topologies, the beacons are used to synchronize the devices (DEVs), identify the PAN, and describe the structure of the superframe.

Devices that would like channel time during the CAP compete with each other using a slotted CSMA/CA channel access algorithm. Any communication that takes place must be finished before the next beacon.

An 802.15.4 superframe can have an active and an inactive portion in the structure, shown in Fig. 13.3. The active portion is comprised of the CAP and CFP. During the inactive portion, the coordinator does not talk with the PAN. This allows the DEVs to enter a low power mode. Since most of the DEVs are battery powered, this is very important to the life of the network.

If there are certain transmissions that require a given amount of network bandwidth, the PANC can assign GTSs. These GTSs will make up the CFP portion of the active section of the superframe. This comes after the CAP but

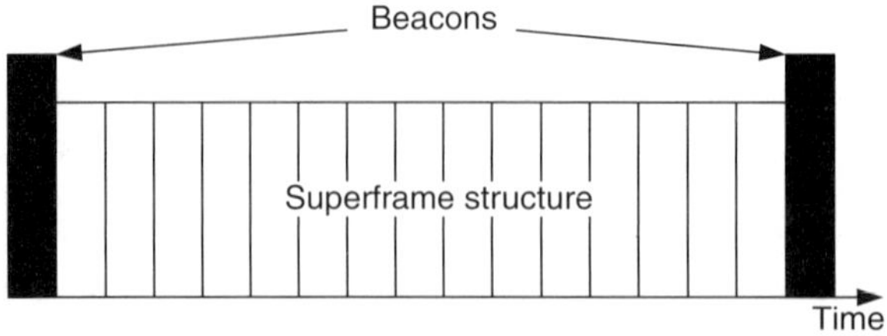

FIGURE 13.2 Superframe structure between two network beacons.

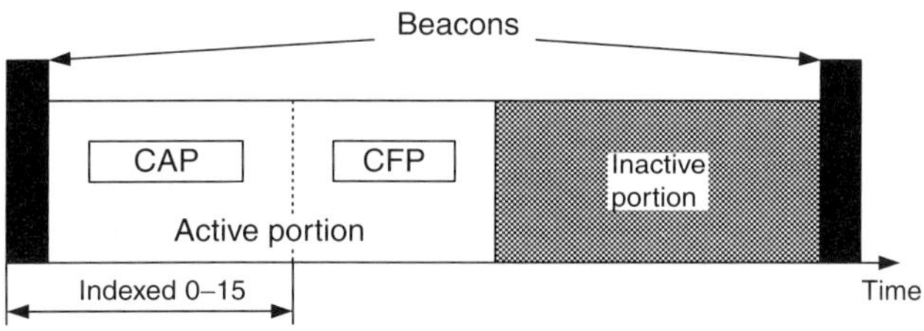

FIGURE 13.3 Superframe structure showing active and inactive portions of frame.

ends before the next beacon. The PANC can allocate up to seven GTSs. It is possible for a single GTS to occupy more than one of these time slots. Essentially two GTSs can be given to one network DEV and function as one GTS, or one duration of time. Although GTSs can be assigned, a good portion of the network is given to the CAP. Both the CAP and CFP sizes are dynamic and can be altered by the PANC based upon what it needed by the DEVs.

The important thing to remember about these periods is that contention-based transmission and the GTSs must complete their transmissions before the beginning of the next period. If the DEV is not able to complete the transmission before this time, the transmission will not commence.

13.3.2 Data Transfer

There are three different types of data transfers: transfer to a coordinator, from a coordinator, and between peer-to-peer DEVs. In the first type, the DEV transmits the data. In the second type, the DEV receives the data. The last type of data transfer involves at least two DEVs communicating with each other. With star topologies only the first two types are used. In a peer-to-peer network topology all three types can be used, but it is more common for the peer-to-peer communication to take place. The mechanisms for each of these types of data transfer depend upon whether or not the network supports a beacon.

Beacon-enabled networks have the ability to use personal computer (PC) peripheral DEVs. If these types of DEVs are not needed, then the beacon can be turned off for normal operations. Operations such as network association require beacons.

13.3.2.1 DEV to Coordinator. When a DEV wants to send data to a coordinator in a beacon-enabled network, it first waits for a network beacon. Once a beacon is found, the DEV synchronizes to the superframe. It sends data to the coordinator using the slotted CSMA/CA method. Once the coordinator acknowledges the reception with an acknowledgment (ACK) frame, the transmission is complete. Fig. 13.4*a* shows how this process takes place in steps.

For non-beacon-enabled networks, the DEV will transmit the data to the coordinator using unslotted CSMA/CA. If the coordinator receives the

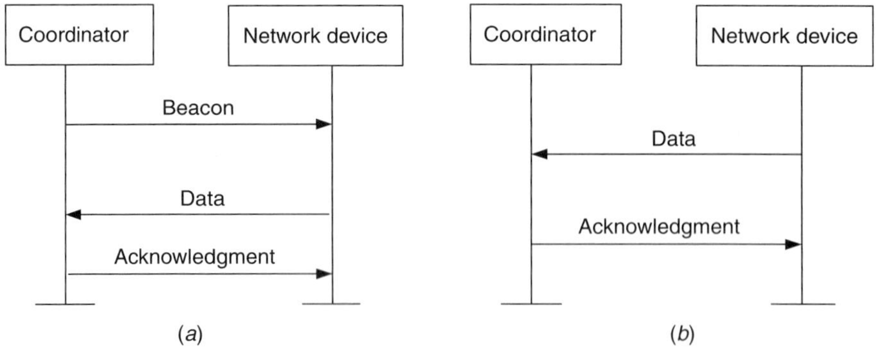

FIGURE 13.4 Relationship between DEV and coordinator (*a*) with beacons and (*b*) without use of beacons.

transmission, it will send an ACK frame. Figure 13.4*b* represents this transaction between the DEV and the coordinator.

13.3.2.2 Coordinator to DEV. When a coordinator wants to send data to a DEV in a beacon-enabled network, it indicates in the beacon that the data are pending for the DEV. The DEV listens for the beacon. Once it receives it and sees that data are pending, the DEV sends a MAC command, using slotted CSMA/CA, requesting the data. The coordinator acknowledges the request with an ACK frame. Then the coordinator sends the data using slotted CSMA/CA. Once the transmission has been successfully received by the DEV, it sends an ACK frame. After the coordinator receives the ACK, it will remove the data from the list of pending transmissions. This can be seen in Fig. 13.5*a*.

In a non-beacon-enabled network, the coordinator stores data until the appropriate DEV requests the data. A DEV can contact the coordinator by

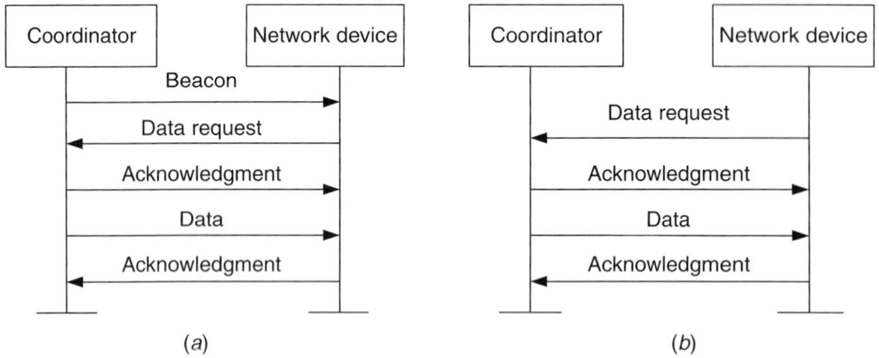

FIGURE 13.5 Relationship between coordinator and DEV: (*a*) with beacons and (*b*) without use of beacons.

sending a MAC request using unslotted CSMA/CA. The coordinator acknowledges the request with an ACK frame. If data are pending for the given DEV, the coordinator sends the data to the DEV using unslotted CSMA/CA. If there are no data for the DEV, then the coordinator transmits a zero-length frame. The DEV interprets this as no data. In Fig. 13.5*b*, it is shown that the DEV acknowledges this transmission with an ACK frame.

13.3.2.3 Peer to Peer. Within a DEV's given radio sphere, every DEV can communicate with any other DEV. To make this take place effectively, DEVs need to either be receiving always or be synchronized to each other. If the DEVs were receiving constantly, they would transmit using CSMA/CA. The DEVs need to be synchronized. This is an important approach to the functionality of this topology and communication, but it is beyond the scope of the 802.15.4 standard. Most of these methods are left to the application programmer.

13.3.3 Frame Structure

The frame structures of the 802.15.4 PAN have minimal complexity, yet they are robust enough to send data on a noisy channel. There are four basic frame types for this network: the beacon sent by the coordinator, the data frame for all data transfers, an ACK frame for confirming receipts of data, and the MAC command frame that is used for MAC layer management entity (MLME). Each protocol layer adds to the structure of the frame by placing specific headers and footers.

13.3.3.1 Beacon Frame. The beacon frame originates in the MAC layer. The rest of this section will refer to Fig. 13.6. The MAC service data unit (MSDU) contains the frame controls, beacon sequence number (BSN), source

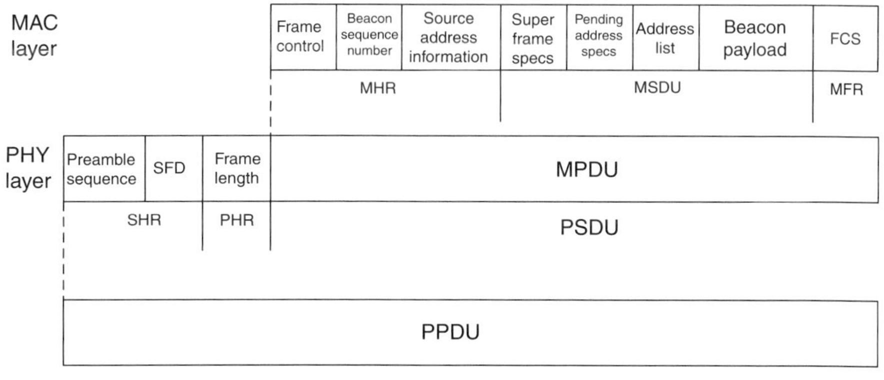

FIGURE 13.6 Beacon structure.

address information, superframe specification, pending address information, address list, beacon payload, and frame check sequence (FCS).

The MSDU has a MAC header (MHR) and a MAC footer (MFR). The MHR, MSDU, and MFR together make up the MAC protocol data unit (MPDU).

Once the header and footer are appended to the MSDU, the MPDU is passed from the MAC layer to the PHY layer. In the PHY layer, the MPDU is called the PHY beacon packet or the PHY service data unit (PSDU). A PHY header (PHR) is added to the PSDU. The PHR contains the length of the PSDU in octet measurement. This is prefixed with a synchronization header (SHR) which contains the start of the frame delimiter (SFD) fields and a preamble sequence. The PSDU, PHR, and SHR form the PHY protocol data unit (PPDU).

13.3.3.2 Data Frame. The data frame, shown in Fig. 13.7, is made up of the data payload. This is passed to the MAC layer. It is called the MSDU. The MHR and the MFR are added to the MSDU, which is passed to the PHY layer. In the PHY layer, the PHR and SHR are added to the PSDU. Together these form the PHY data packet. This is also called the PPDU.

13.3.3.3 ACK Frame. This frame structure, shown in Fig. 13.8, works similarly to the previous two, except that the MPDU is comprised of only the MHR and MFR. The MPDU is passed to the PHY layer where the SHR and PHR are added to the PSDU, forming the PPDU.

13.3.3.4 MAC Command Frame. The structure of the MAC command frame, shown in Fig. 13.9, originates in the MAC layer. The MSDU for the MAC command frame contains the command type and the command payload. The MHR contains the frame control, data sequence number (DSN), and

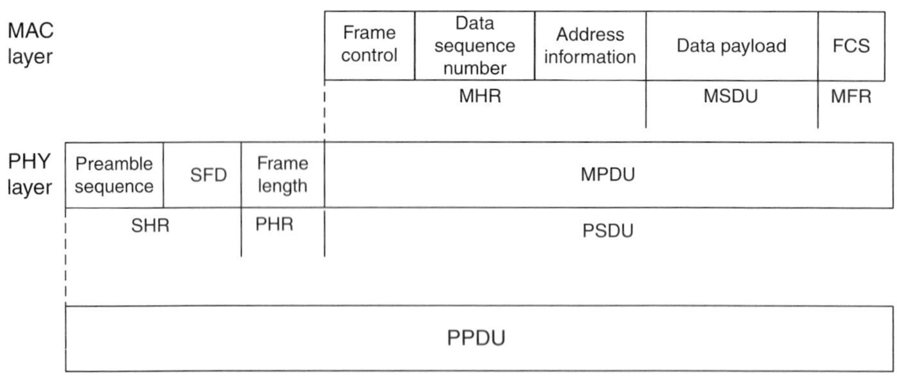

FIGURE 13.7 Data frame.

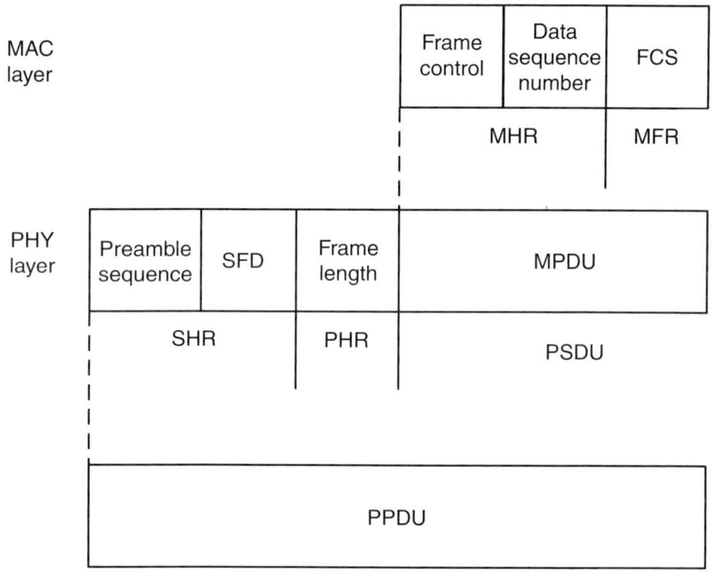

FIGURE 13.8 ACK frame.

address information. The MPDU is passed to the PHY layer. Here the SHR and PHR are added to the PSDU. This makes up the PPDU.

13.3.4 Power Consumption

In most applications that use this standard, the majority of the DEVs will be battery powered. It is highly impractical to replace or try to charge these batteries over a short period of time. Because of this, power consumption is an

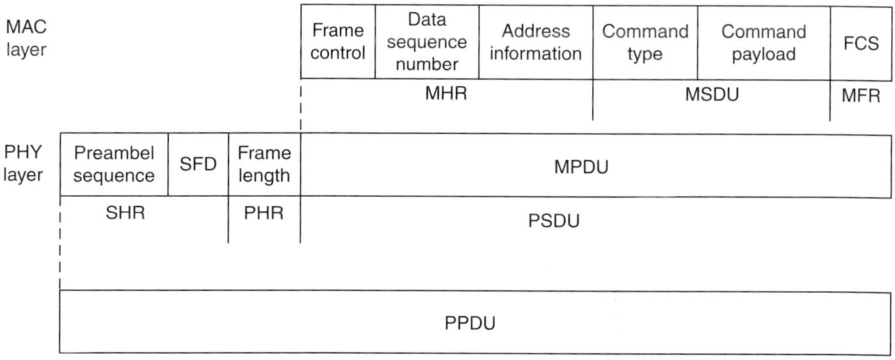

FIGURE 13.9 MAC command.

important aspect of this standard. Limited power supply availability was the driving force behind this standard.

PHY layer implementations that require other application-dependent considerations are outside the scope of the 802.15.4 standard. However, the standard was set up in favor of battery-operated DEVs. The standard has set up the transmission frames in such a way that the DEV will spend most of its life in a sleep state.

Since the majority of the implementations and mechanisms are not specified, it allows the application designer to implement a program-specific utility to handle PHY implementation.

13.3.5 Robustness

There are several different tools that the 802.15.4 standard uses to ensure the robustness of its data transfers: CSMA/CA channel access mechanism, frame ACK, and data verification. Others can be implemented. However, these are the ones specified by the standard.

13.3.5.1 CSMA/CA Algorithm. This algorithm is used before any data and MAC command frames. It is used for basically everything except beacon frames or frames sent in the CFP part of the active frames. ACK frames do not use CSMA/CA. It is also possible for data to be quickly sent right after an ACK. In this case, the DEVs do not have to use CSMA/CA to access the channel.

If the network is using beacons, the MAC layer uses the slotted CSMA/CA for transmissions. If beacons are not used, the unslotted algorithm is used. This is also true if a beacon is not able to be located for a period of time by the DEVs in the PAN.

Both types of CSMA/CA use backoff periods. One of these periods is equal to a variable called aUnitBackoffPeriod. The remainder of this section refers to Fig. 13.10.

13.3.5.1.1 Slotted CSMA/CA. The boundaries of all the DEVs in the network are aligned with boundaries given by the PANC. These are aligned with the beacon transmission. Synchronization of all of the DEVs takes place here. System clocks, timers, and counters are also synchronized.

The transmission from a slotted CSMA/CA access can only start at the end of a backoff period. In a slotted CSMA/CA mechanism, each DEV maintains up to three variables: NB, CW, and BE.

NB is the number of times the CSMA/CA backs off while trying to access the channel. This value is reinitialized to zero at the start of every transmission attempt.

CW is the contention window length. This is the number of backoff periods that have to be cleared in the channel before a transmission can start. This helps

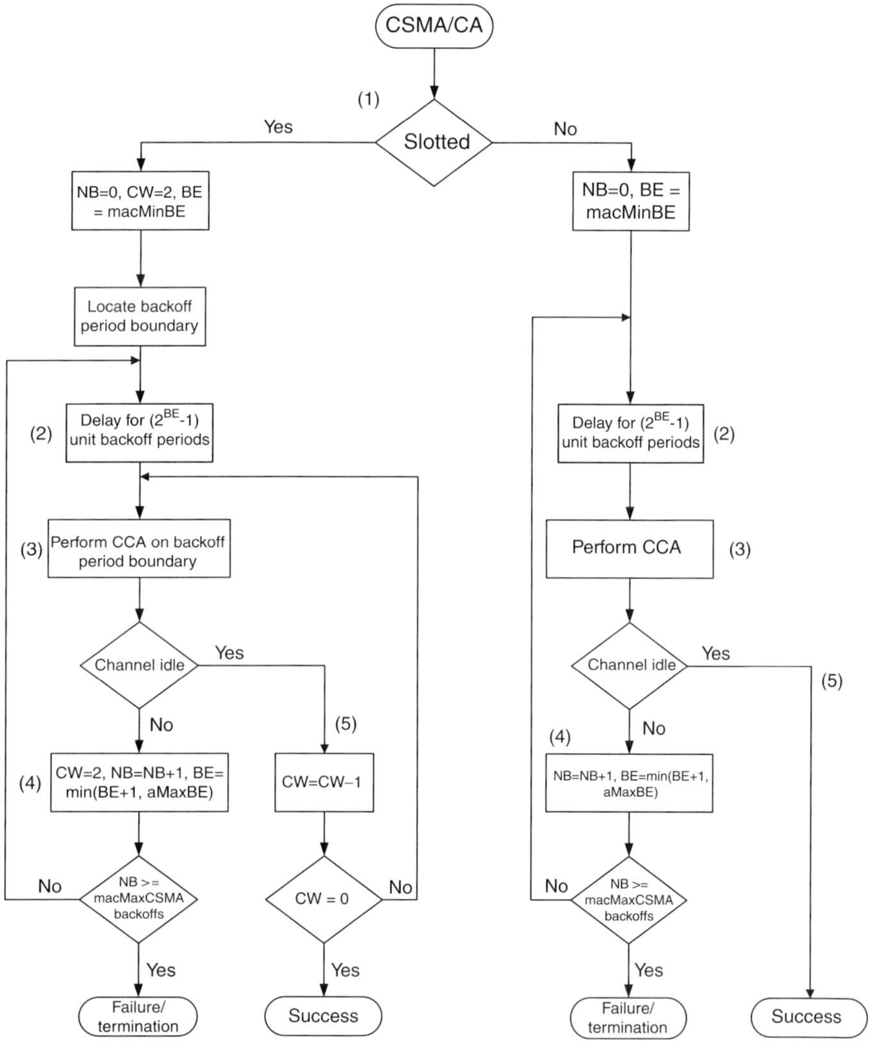

FIGURE 13.10 CSMA/CA algorithm for channel access used by IEEE 802.15.4.

avoid collisions. The CW value is set to 2. It is reset to this number every time the channel is busy. Only slotted CSMA/CA uses the CW variable.

BE is the backoff component. This is linked to how many backoff periods the DEV waits before trying to access a channel. If this value is ever zero, the CA is disabled during the first iteration of the algorithm. During this backoff period, even though a DEV is enabled, it discards any frames that it might have received.

In Fig. 13.10, the five steps for the slotted algorithm use these three variables. In step 1, the NB, CW, and BE variables are initialized. In this step, the boundary of the next period is also found. In step 2, the DEV delays for a number of random backoff periods. The range of the random numbers is somewhere between zero and $2^{BE}-1$. Step 3 starts from the time the DEV waits until the end of the backoff period in order to request a clear-channel assessment (CCA). The MAC layer makes sure that the transmission can be completed in the time that will be given. If the DEV can proceed, it will go on to the fifth step. Otherwise, the DEV must wait until the next CAP in the next superframe and perform the evaluation process over again. If the channel was not idle, step 4 is implemented. In this step the backoff process is used. If NB is less than macMaxCSMABackoffs, then the process goes back to step 2. Otherwise, the algorithm terminates due to not being able to find a clear channel. If the channel at step 3 was found idle, then the DEV makes sure that the contention window has expired and then commences its transmission. To make sure that the contention window has expired, the MAC layer decrements CW by 1 and then checks to see if it is equal to zero. If it is equal to zero, then after a backoff period the DEV transmits its data. If it is not equal to zero, the DEV goes back to step 3 of the algorithm.

13.3.5.1.2 Unslotted CSMA/CA. This portion of the CSMA/CA algorithm follows the same basic process as the slotted algorithm. However, in the unslotted CSMA/CA, the backoff periods of a DEV are not related to any other DEV backoff periods. In step 1 of the unslotted algorithm, NB and BE are initialized. From this point forward, the algorithm works the same as the slotted CSMA/CA.

13.3.5.2 Frame Acknowledgment. There are two ways in which all data transmissions are acknowledged. A successful data transfer and validation of data or command frames can be confirmed with an ACK frame. If the receiving DEV cannot or does not receive the frames, there is no ACK.

If the sending DEV does not receive an ACK after a certain period of time, it is assumed that it was not received and is sent again. After several tries, the sender can stop the transmission. If the ACK bit is not set, this means that an ACK will not be sent for a successful transmission, and it is assumed by the DEVs that all transmissions are successful.

13.3.5.3 Data Verification. To help in detecting bit errors in data transmissions, 802.15.4 uses FCSs, which employ a 16-bit International Telecommunications Union, Telecommunication Standardization Sector (ITU-T), cyclic redundancy check (CRC) in order to protect every data frame. This is inserted in the final field of the transmission block.

13.3.6 Security

Although it is the intent of this standard to implement a number of applications that provide interoperability among DEVs using it, there is a baseline of security services. Some of these services include the ability to have an access control list and the ability to use cryptography to protect the data frames [4, 5].

The security features that this standard provides are determined by the higher layers which manage when security is to be used and to provide the key materials that are necessary. Most of these higher level functions are beyond the bounds of the 802.15.4 standard.

13.3.6.1 Security Services. Some of the security services that are provided are access control, data encryption, frame integrity, and sequential freshness. It is assumed that the data that most of these services use are obtained, stored, and used in a secure manner. This standard recommends that established security practices be used.

Access control allows a DEV to select the DEVs with which it wishes to communicate. The DEV should keep a list of the DEVs with which it will communicate.

Data encryption can be done on beacon payloads, command payloads, and data payloads. This is typically done by using a key shared by a certain group of DEVs.

Frame integrity works with the data encryption to ensure that frames being received are actually coming from the group of DEVs with the cryptographic key.

Sequential freshness is a service that will reject frames that have been replayed. When frames are read, a freshness value is given to the frame. This value is compared to identical frames in order to tell which one is the most up-to-date frame.

13.3.6.2 Security Modes. Based upon the security mode, the MAC layer provides security services. There are three modes: unsecured, asynchronous connectionless (ACL), and secured.

In the unsecured mode, there are no security services.

In the ACL mode the upper layer provides the services to the DEV. This mode essentially makes sure that the sending and receiving DEVs belong to the same cryptographic key group.

If a DEV is in the secure mode, the MAC layer can provide any of the services from the previous section. These services are access control, data encryption, frame integrity, and sequential freshness.

13.4 FRAME FORMATS

This section contains information concerning the format of the MAC frame. Each frame contains a MHR, which contains the frame control, sequence number, and address information; a MAC payload (MSDU), which is variable length and contains information about the specific frame type; and the MFR, which contains the FCS.

Although this chapter takes the stance that the IEEE 802.15.4 superframe structure is less complex and more simplified than other wireless network protocols, it is evident that it is only "less complex" in relative terms. Once the general format is understood, it does seem less complex; however, for small, often temporary DEVs to have so much programming going into them, it would be expensive. These DEVs will provide a great way to harvest information.

13.4.1 MAC Frame Format

The MAC frame format is made up of a number of fields in a specific order. All of the frames are transmitted by the PHY layer, left to right, as most of the diagrams in this chapter will show. For these cases, the leftmost bit is transmitted first. All of the fields within each frame are numbered in a typical computer science fashion. If you have k bits for a field, they are indexed 0, ..., $k-1$. If any of these fields are longer than an octet, then the fields are broken down into a lower order and a higher order bit sequence. This means that the fields cover more than one octet and are broken down this way to keep them in sequence.

13.4.1.1 General MAC Frame. The general MAC frame contains a MHR, MAC payload, and MFR. Fig. 13.11 shows the frame field order and how many octets are reserved for each of these fields. In the figure, we observe that the MFR is composed of 2 bits for the FCS; the payload length varies; and the MHR contains the source address, source PAN ID, destination address, destination PAN ID, sequence number, and two octets for the frame control number.

13.4.1.2 Frame Control Field. The frame control field, 16 bits in length, is the first part of a MAC frame that is transmitted. This field contains information about the type of frame, the addressing fields that specify other DEVs, and other control flags. The frame control field, shown in Fig. 13.12,

Octets: 2	1	0/2	0/1/2/8	0/2	0/1/2/8	x	2
Frame control	Sequence number	Destination PAN identifier	Destination address	Source PAN identifier	Source address	Frame payload	Frame check sequence
		Addressing fields					
MHR						MAC payload	MFR

FIGURE 13.11 General MAC frame format.

Bits

0–2	3	4	5	6–9	10–11	12–13	14–15
Frame type	Security enabled	Frame pending	ACK required	Reserved	Destination addressing mode	Reserved	Source addressing mode

FIGURE 13.12 Frame control field subfields.

contains the frame type, whether or not security is enabled, a frame pending bit, ACK request bit, reserved bits, destination addressing mode bits, two more reserved bits, and the source addressing information.

The frame type subfield is 3 bits in length and is set to one of four values: 000 if it is a beacon frame, 001 if it is a data frame, 010 if it is an ACK frame, and 011 if it is a MAC command frame. The bit sequences 100–111 are reserved for application implementers.

The security subfield is only 1 bit in length. If the frame is using cryptography, the bit is set to 1. If the frame is not protected by the MAC layer, it is set to 0. The cryptographic keys that are used are stored in the MAC personal area network information base (PIB). These are for the security settings for the current frame. Any of the security services that are used in the frame are defined by the security suite that is used.

The third subfield of Fig. 13.12 is the frame pending subfield. This field is used, and the bit is set to 1, if there are any other data frames that the sender has to give to the receiver. If this bit is set, the receiver will request the sender to transmit the data. The sender will then send any pending data. If there are no pending data, the bit is set to 0. This subfield is only used for beacon-enabled networks during the CAP. It can also be used by DEVs in any non-beacon-enabled network.

The ACK request subfield is also 1 bit in length. The job for this part of the MAC frame is to specify whether or not an ACK is required from the receiver. If this bit is set to 1, an ACK frame will be sent once the transmission has been deemed valid. If the bit is set to 0, an ACK will not be sent no matter whether the data have been correctly received or not.

The next subfield, the destination addressing mode subfield, is 2 bits in length. The bits will be set as follows: 00 if the PAN ID and address field do not exist, 01 if the address field is a short address, 10 if the address field is a 16-bit short address, and 11 is the mode value for the 64-bit address field for the extended addresses. By chance, if the subfield is equal to zero, not specifying whether if it is a beacon or ACK frame, then the source addressing mode field is nonzero. This means that the frame will be directed to the coordinator.

The source addressing mode is the same as the destination addressing mode subfield except that it deals with the source or sender of the transmissions. It uses the same mode values and descriptions as the destination addressing mode subfield: 00 if the PAN ID and address field do not exist, 01 if the address field is a short address, 10 if the address field is a 16-bit short address, and 11 is the mode value for the 64-bit address field for the extended addresses.

13.4.1.3 Sequence Number Field. This field specifies a unique sequence identifier for the frame. This identifier is 8 bits in length. For the beacon frames, this specifies a BSN to keep up with the contiguous beacons that could be sent. For the beacon frames, every coordinator stores the BSN in a variable called macBSN. This value is initialized to a randomly generated number. It is the job of the coordinator to take macBSN, copy it to the sequence number field, and then increment it by one number after each beacon frame is transmitted.

For MAC command frames or ACK frames, this number is a DSN. This is stored in a variable called macDSN. This is done in order not to increment but to match this value to prove the validity of the frames for data transfer. The algorithms used to produce these random numbers are not part of this standard.

For a DEV that does not request an ACK, the sending DEV should increment the macDSN variable by 1 after the transmission is sent. If the DEV does expect an ACK, then the DEV increments macDSN by 1, waits aAckWaitDuration amount of time, retransmits the data, and then increment macDSN by 1 again.

13.4.1.4 Destination PAN ID Field. This field is 16 bits in length. It specifies the PAN identifier of the receiving DEV. This field uses the value of 0xffff to represent a broadcast ID that will be valid for all DEVs within the PAN. This, of course, only happens if the DEVs are on the same channel. If the destination addressing mode subfield is set to 0, then this subfield is not used.

13.4.1.5 Destination Address Field. The size of this field is dynamic based on the address value of the destination address. This specifies the address of the receiving DEV. The 8-bit value 0xff signifies the broadcast address for the DEVs on the same channel. If the destination addressing mode subfield is 0, then this subfield is not used.

13.4.1.6 Source PAN ID Field. This field is used to represent the PAN ID of the sender of the frames. The PAN ID number is something that it determined during the PAN association. It is possible for this number to change to resolve DEV conflicts. However, in most cases, it remains unchanged.

13.4.1.7 Source Address Field. Depending upon the destination address size, this field size can be 8, 16, or 64 bits in length. This specifies the address of the DEV that is sending the frames. This field is not used if the previous fields are not used or if the source addressing mode subfield is set to 0.

13.4.1.8 Frame Payload Field. Depending upon the destination address size, this field size can be 8, 16, or 64 bits in length. This specifies the address of

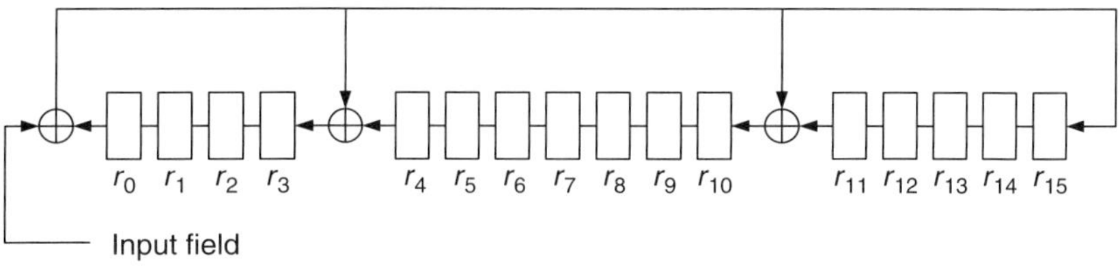

FIGURE 13.13 Frame check sequence implementation.

the DEV that is sending the frames. This field is not used if the previous fields are not used or if the source addressing mode subfield is set to 0.

13.4.1.9 Frame Check Sequence. This field is 16 bits in length. This contains the CRC code. The FCS is calculated as

$$G_{16}(x) = x^{16} + x^{12} + x^5 + 1 \tag{13.1}$$

This is a standard generator polynomial of sixteenth degree. Figure 13.13 shows how the FCS is calculated for the transmission with the following algorithm: First we set $M(x) = b_0x^{k-1} + b_1x^{k-2} + \cdots + b_{k-2} + b_{k-1}$; then we multiply $M(x)$ by x^{16}. This gives us $x^{16}M(x)$. If we divide $x^{16}M(x)$ modulo 2 by the generator polynomial of $G_{16}(x)$, then we get the remainder $R(x) = r_0x^{15} + r_1x^{14} + \cdots + r_{14}x + r_{15}$, which is the remainder coefficients used for the FCS field.

Figure 13.13 shows the general implementation of the CRC-16 generator polynomial. In the figure, the remainder register is first initialized to 0. Then the MHR and payload are shifted into the divider by the order of transmission. Once all of the data fields have been shifted into the divider, the remainder register will contain the FCS. The FCS is appended so that the r_0 bit is transmitted first.

13.4.2 Individual Frame Format

The IEEE 802.15.4 standard defines four types of frames: beacon, data, acknowledgment, and MAC command frames.

13.4.2.1 Beacon. Figure 13.14*a* shows the format of the beacon frame. It contains an FCS in the MFR; a beacon payload, address list, pending address specification, and superframe specification subfield in the MAC payload; and the addressing fields, sequence number, and frame control in the MHR.

13.4.2.1.1 MAC Header. The MHR field contains the addressing fields, sequence number, and frame control. The addressing fields contain the PAN

Frame control	Sequence number	Addressing fields	Superframe spec.	Pending address spec.	Address list	Beacon payload	FCS
MHR			MAC payload				MFR

(*a*)

Frame control	Sequence number	Addressing fields	Data payload	Frame check sequence
MHR			MAC payload	MFR

(*b*)

FIGURE 13.14 (*a*) Beacon frame format. (*b*) Data frame format.

ID field and the source address field. The addressing field information is typically for the information of the DEV that is sending the data transmission.

13.4.2.1.2 Superframe Specification Field. The superframe specification field is 16 bits in length and is divided into six subfields: beacon order, superframe order, final CAP slot, a reserved subfield, PANC subfield, and an association permit subfield.

The beacon order is the first of the 4 bits of the frame. It specifies the interval at which the beacon will be transmitted. The variable BO keeps up with the beacon order. The beacon interval is calculated as aBaseSuperframeDuration $\times 2^{BO}$ symbols, where $0 \leq BO \leq 14$.

The next subfield is the superframe order. It is 4 bits in length as well. This subfield specifies the time for which the superframe will be active. Therefore, the DEV will know how long it will be enabled to receive data. In any PAN, the coordinator only transmits to the DEVs during an active superframe. Thus this subfield is important. This subfield uses a variable called SO for the superframe order. Since we already know that we operate on 15 time slots, once this variable reaches 15 or grows larger than the BO, the active portion of the superframe has ceased.

The final CAP slot subfield is the next 4-bit field. This lets the DEV know which slot is the last slot in the CAP transmission period.

The PANC subfield indicates to the PAN whether or not a beacon will be transmitted by the coordinator.

The last subfield of the superframe specification field is the association permit. This bit coincides with a variable called macAssociationPermit. If this Boolean variable is set to true, indicating that the coordinator is accepting associations from DEVs to the PAN, then this association bit is set to 1. If the PANC is not accepting associations to the PAN, then macAssociationPermit is set to false and the association permit subfield is set to 0.

13.4.2.1.3 Pending Address Specification. The pending address specification field is formatted with a number of short addresses pending subfield, reserved subfield, and number of extended addresses pending subfield, ending with another 1-bit reserved subfield. The number of short addresses pending subfield is used to indicate to the PAN the number of 8–16-bit addresses in the address list of the beacon. The number of extended addresses pending subfield works the same way for 64-bit extended addresses.

13.4.2.1.4 Address List Field. The actual number of addresses for devices in the address list field is dependent upon the number given by the pending address specification field of the beacon frame. The address list contains a list of the addresses of DEVs that have pending transmissions.

There is only enough space to have a total of seven pending addresses at any one time. The addresses can be made up of short or extended DEV addresses. However, all of the short addresses will be first in the address list.

13.4.2.1.5 Beacon Frame Payload Field. The payload for the beacon is optional. If it is used, it can have up to MaxBeaconMSDULength number of octets. If this variable is zero, then the field is not used. If there is a number greater than zero, then the number of octets will be put in this field. The number is kept in a variable called macBeaconMSDU.

Since there can be security imposed on these frames, if the security bit is set, then the payload field is processed with the security suite that is indicated. When this happens, the DEV processes the payload corresponding to macCoordExtendedAddress. If there is a payload present in this frame, the DME must first process the data contained in the superframe specification and address list fields. If there is no payload, the MAC layer begins processing the information in the superframe specification and address list fields.

13.4.2.2 Data Frame Format. The data frame, Fig. 13.14*b*, conforms to the general MAC frame by having frame control, sequence number, and addressing field subfields within the MHR; a data payload subfield as part of the MAC payload; and an FCS for the MFR.

13.4.2.2.1 Data Frame MHR. The MHR contains the frame control field, sequence number field, and addressing fields. The addressing fields are made up of the destination PAN ID and address field and the source PAN ID address fields.

The frame control field contains the data that indicate this frame to be a data frame.

The sequence number field keeps up with the frame numbers in a variable called macDSN. Once data transmission and acknowledgment have been made, the macDSN value is incremented by 1 by the DEV.

13.4.2.2.2 Data Payload Field. The data payload field is dynamic in size dependent upon how much information has been passed to the MAC layer by the higher layers of the network model.

If there is security indicated for these frames, then the data are processed with the current security suite for macCoordExtendedAddress. This happens if the destination address field is not filled in.

For incoming data frames, if the security bit is 0, the data payload subfield will pass the data directly to the higher layers. If the security bit is set to 1, then the data payload is decrypted according to the security suite specified, and the information is passed on to the higher layers of the network model.

13.4.2.3 ACK Frame. The ACK frame, Fig. 13.15*a*, has three basic parts: the frame control subfield and the sequence number subfield as part of the MHR and the FCS, which is part of the MFR.

The frame control field specifies the type of frame. The sequence number field indicates the sequence number that was received by the DEV. This can be used to retransmit only the frame number that was lost if there was a transmission data reception error.

13.4.2.4 MAC Command Frame Format. The MAC command frame, Fig. 13.15*b*, is made up of the MHR, MAC payload, and MFR. The MHR is made up of the frame control field, the sequence number field, and the addressing fields. The MAC payload is comprised of the command frame identifier and the command payload.

The frame control field contains the value of the MAC command frame. This is the type of frame. The sequence number field has a variable called macDSN. Once data transmission, reception, and acknowledgment have taken place, the macDSN is incremented by 1. The addressing fields contain the source and destination addresses of the DEVs.

Frame control	Sequence number	Frame check sequence
MHR		MFR

(*a*)

Frame control	Sequence number	Addressing fields	Command frame ID	Command payload	Frame check sequence
MHR			MAC payload		MFR

(*b*)

FIGURE 13.15 (*a*) ACK frame format. (*b*) MAC command frame format.

The command frame identifier field indicates which MAC command is used. The different types of commands are beyond the scope of this chapter.

The command payload field contains the actual MAC command. If the command frame is security enabled, then the payload is processed with the current security suite provided by the destination address field. If this field is not present, then macCoordExtendedAddress indicates the type of security to use. If security is set on an incoming frame, then it is processed by the DEV before it is passed along to the higher layers.

13.5 MAC COMMAND FORMATS

The command frames that the IEEE 802.15.4 standard deals with are association request, association response, disassociation notification, data request, PAN ID conflict notification, orphan notification, beacon request, coordinator realignment, GTS request, and GTS allocation. There are also command frame identifiers that are reserved for later commands to be added by the program implementer.

An FFD is capable of transmitting and receiving all types of commands. The RFD can only transmit association requests, disassociation notifications, data requests, and orphan notifications. RFDs can only receive association responses, disassociation notifications, and coordinator realignment commands. For DEVs operating in a PAN without a beacon, commands can be sent at any time. If beacons are used, commands can only be sent during the CAP.

13.5.1 Association and Disassociation

These commands let DEVs, FFDs, and RFDs associate or disassociate from a PAN.

13.5.1.1 Association Request. With the association request command, a DEV has the ability to associate itself with a PANC. It is not possible for an associated DEV to use this command. It should only be used by DEVs that want to join the PAN. It is required that both types of network DEVs are able to send this command.

The association request command format is broken up into three main parts: the MHR fields, command frame identifier, and a capability information field.

13.5.1.1.1 MAC Header. The MHR fields contain a source addressing mode subfield, destination addressing mode subfield, security-enabled subfield, frame pending subfield, and destination PAN ID field. The source addressing mode subfield for an association request is set to 3. This indicates that it will be using the 64-bit extended addressing.

If security is used for the frame, then the security-enabled field is set to 1. The security that is used corresponds to the destination address. If the security enabled field is set to 0, then no security is used.

For the association request command, the frame pending subfield is set to 0 and is basically ignored. Once this happens, the acknowledgment frame subfield will be set to 1.

The destination PAN ID field is used for PAN information when the DEV is trying to associate with the network. It contains the address in the beacon that was transmitted by the PANC.

13.5.1.1.2 Capability Information Field. The capability information field contains an alternate PANC, DEV type, power source, receiver on when idle, reserved, security capabilities, and allocate address subfield.

The alternate PANC subfield is the field that indicates whether or not a DEV is capable of being a PANC. If the DEV is capable, the bit is set to 1. If not, it is set to 0.

If a DEV is an FFD, the DEV type subfield is set to 1. If the DEV is an RFD, then the bit is set to 0. The next subfield is the power source subfield. This is 1 bit in length. If the DEV is receiving mains power, then the bit is set to 1. The power source subfield will be set to 0 if a DEV is running on any other kind of power.

The next 3 bits are the receiver on when idle subfield. This field is set to 1 when the DEV will still be able to receive frames during idle time periods.

Following the receive frame during idle subfield is a 2-bit reserved subfield. This will be used by the manufacturer or program implementer. After this subfield is the security capability subfield. This bit is set to 1 if the DEV can send and receive MAC commands using a security suite. If a security suite is not used, then the bit is set to 0.

The last subfield of the capability information field is the allocate address subfield. This bit is set if the DEV would like the PANC to assign it a short address. If this subfield contains a 0, then the PANC will continue to use the DEV's 64-bit extended address.

13.5.1.2 Association Response Command. This command allows the PANC to respond back to an association request. This command is only sent to the DEV if the DEV is trying to associate with the PAN; the association response command format can be seen in Fig. 13.16. All DEVs can receive these types of commands. RFDs are not capable of transmitting it. The association response command format contains an MHR field, a command frame ID, a short address, a coordinator short address, and an association status field.

MHR. The MHR contains a destination addressing mode subfield, source addressing mode subfield, security-enabled subfield, frame pending

Octets: 23	1	2	2	1
MAC header fields	Command frame ID	Short address	Coordinator short address	Association status

FIGURE 13.16 Association response command format.

subfield, acknowledgment request subfield, destination and source PAN ID fields, and destination and source address fields. If security is set for this frame, it uses the proper security suite that is indicated by the destination address.

Short Address Field. This field is 16 bits in length and is used to assign a DEV a short address. This assignment is given by the coordinator. If the DEV was not associated with the PAN, the short address field is set to $0 \times 00\text{ff}$. The associate status field has the reason why the DEV was not associated. If an association was made, the short address is used by the PANC to communicate to the newly associated DEV. If a DEV is not given a short address, the 64-bit address is used.

Coordinator Short Address Field. This field is 16 bits in length. It contains the PANC's short address. If an 8-bit short address is used, then the DEV will set the most significant octets to 0. If a 16-bit address is used, then the entire 16-bit field is used by the DEVs.

Association Status Field. The last field, the association status field, is 8 bits and contains values indicating the type of association made or the reason an association was not made. If the association status is set to 0×00, this means that it is reserved for the MAC primitive enumeration value. If 0×01 is inputted, then the PAN was at capacity and could not process and accept the association. If the association status is 0×02, then the PAN access was denied. These are just a few examples of association request failures.

13.5.1.3 Disassociation Notification Command. A disassociation notification command can be sent by the PANC or a member DEV. All DEVs must be able to send and receive this command. The disassociation notification command format contains an MHR field, a command frame identifiers field, and a disassociation reason field.

MHR. In the MHR for the disassociation notification command, the destination addressing mode and source addressing mode subfields are set to 3. If the disassociation notification command uses security, the security-enabled subfield will be set to 1, and the security suite will correspond to the destination address field. The variable macPAN ID is contained in the destination and source PAN ID. If a PANC wants a DEV to disassociate from the PAN, it will set the destination address field to the 64-bit address of the DEV to be disassociated. When a DEV wants to disassociate from the network, it will set the value of the destination address field to macCoordExtendedAddress.

Disassociation Reason Field. This 8-bit field contains the reason that a disassociation took place. These reasons can vary from the coordinator wishing the DEV to leave the PAN to the DEV wishing to leave the PAN by itself. Most of the time, these reasons are associated with channel clarity and resource availability.

13.5.2 Data Request Command

This command is sent when a DEV wants information from a PANC. The format of this frame consists of an MHR field and a command frame ID field. All DEVs must be capable of transmitting this command. An FFD must be able to receive it as well.

In a beacon-enabled network, this command is sent once the macAutoRequest value is set to true. A beacon must have also been sent to the DEV, indicating that it had pending data for it.

A DEV also has the capability of sending the command after an acknowledgment to a request command. This could happen after events such as an association or after a DEV's request for GTS channel time allocation.

13.5.2.1 MAC Header. In the MHR field, there are many subfields. The destination addressing mode subfield is for the data request command. If the source addressing mode subfield is set to 3, it indicates that there is no short address being used for the DEV. Any other setting for this subfield suggests that the coordinator is accepting short addressing.

If security is enabled, then the security suite indicated by macCoordExtendedAddress is used. If security is not enabled, then the security-enabled subfield is set to 0.

13.5.3 Orphan Notification Command

This command, transmitted by all DEVs, is used by a DEV that has lost its beacon. This also is sent when a DEV is out of synchronization from the PANC. The orphan notification command format contains an MHR field and a command frame identifier field.

The source addressing mode subfield is set to 3 for the orphan notification command subfield. If security is enabled, the security is based on the value of the variable macCoordExtendedAddress. If security is not used, then the security-enabled field is set to 0.

For these command frames, no pending frame or acknowledgment requests are sent. The last field of the orphan notification command format is the command frame identifier. This tells the DEV what type of frame is being sent.

13.5.4 Beacon Request Command

The beacon request command is used by a DEV to locate beacons. Specifically, these DEVs are trying to locate a coordinator within its area of transmission. An RFD does not have to be able to send a beacon request command. The beacon request command has an MHR field and a command frame identifier.

The MHR has a destination addressing mode subfield, source addressing mode subfield, frame pending subfield, acknowledgment request subfield, destination PAN ID subfield, and destination address field.

13.5.5 Coordinator Realignment Command

This command is only sent by a coordinator. It is not a requirement for an RFD to be able to transmit this frame. This type of command is sent after the PANC receives an orphan notification command. The coordinator realignment command is sent straight to the orphaned DEV. The command frame contains the MHR field, command frame identifier, PAN identifier, logical channel field, and short address field. Figure 13.17 is the representation for the coordinator realignment command format.

MHR. The MHR for the coordinator realignment command consists of the destination addressing mode subfield, source addressing mode subfield, security-enabled subfield, frame pending subfield, acknowledgment request subfield, destination PAN ID subfield, and destination address field. The MHR for the coordinator realignment command is the same as the other commands. The settings and descriptions are similar as well.

PAN ID Field. This field is 16 bits in length. The purpose of this field is to keep track of the ID that the coordinator will use in its transmissions.

Logical Channel Field. This field is 8 bits in length. The purpose of this field is to keep track of the channel that the coordinator will use for its transmissions.

Short Address Field. The short address field is the last field in the coordinator realignment command frame. It is 16 bits in length. If this field is sent to an orphan DEV, it will contain the short address that the DEV will use to operate within the PAN.

Octets: 16 or 23	1	2	1	2
MAC header fields	Command frame ID	PAN ID	Logical channel	Short address

FIGURE 13.17 Coordinator realignment command.

13.5.6 GTS Request Command

The GTS request command is one of the commands used to manage GTSs. It is used when a DEV requests GTSs to allocate to it. This command is also used by a DEV to request the deallocation of GTSs or allocation acknowledgments. This command is used by DEVs that have a valid address. The GTS request command frame contains the MHR field, command frame ID field, GTS control field, and GTS characteristics field.

MHR. The MHR contains the destination addressing mode subfield, source addressing mode subfield, security subfield, frame pending subfield, acknowledgment request subfield, source PAN ID field, and source address field. The fields and subfields all work similarly to the other command frame types.

GTS Control Field. The GTS control field is the third field of the GTS request command frame. This field is 8 bits in length and contains the request type subfield, reserved subfield, and GTS ID subfield. The request type subfield contains the specific type of GTS request that is being transmitted. This 2-bit subfield uses the values 00 to represent a GTS allocation, 01 to represent a GTS deallocation, 10 to represent a GTS confirmation, and 11 for the application programmer. The last 3 bits of the GTS control field is the GTS ID subfield. This specifies the ID of the GTS to which the command is referring. Remember, there can only be to seven GTSs per frame CFP.

GTS Characteristics Field. The GTS characteristics field is composed of the GTS length field, reserved field, and GTS direction field. The information contained within the frame is the number of superframe slots being used for GTSs. The GTS characteristic format also gives the direction of the data frame transmissions by DEVs.

13.5.7 GTS Allocation Command

The last command type used by the IEEE 802.15.4 PAN is the GTS allocation command. This command is used in one of two ways: to respond to a GTS request or to allocate or deallocate a GTS to a DEV. The GTS allocation command frame includes ID fields, GTS specification fields, and GTS status field. The MHR and command frame ID fields work the same as the previous commands.

The GTS specification field is 16 bits in length and contains the GTS ID, GTS direction, GTS start slot, GTS length, and reserved subfield.

The GTS ID subfield refers to a specific GTS that is being sent to the DEVs by the PANC. For transmissions where a GTS is allocated, this subfield contains the ID number for future transmission about a certain GTS. If a GTS is not allocated, the bits will be set to 0.

The GTS direction bit is set to 1 for receiving and 0 for transmitting.

The next 4 bits, the GTS start slot, identify the starting position of the GTS.

The next bits of the GTS specification field contain the length of the GTS that has been allocated. If a GTS has not been allocated, this field will contain the number of slots that remain in the CAP that can be used for GTSs.

13.6 CONCLUSIONS

This chapter has introduced the MAC/PHY layer specifications and characteristics in IEEE 802.15.4 (Zigbee standard lower layers). We point out that there are still lots of research issues to be addressed in those two layers. For instance, the current MAC layer does not include an accurate time synchronization function for sensor network cases, although higher layers could achieve certain synchronization accuracy. It also does not consider asymmetric wireless links [i.e., the forward link and backward link have different radio ranges and bit error rates (BERs)]. A multi-interface transceiver (for multi-radio switch) can further make the design challenging. Although it supports time division multiple access (TDMA) very well in the MAC layer, code division multiple access (CDMA) technology could be used to avoid communication interferences among different neighborhoods. However, the CDMA specifications are missing in the current standard.

ACKNOWLEDGMENT

This work is partially supported by the U.S. National Science Foundation (NSF) under grants CNS-0716211 and CNS-0716455.

REFERENCES

1. LAN/MAN Standards Committee of the IEEE Computer Society, Part 15.4, "Wireless medium access control (MAC) and physical layer (PHY) specifications for low rate wireless personal area networks (LR-WPAN)," IEEE Computer Society, Oct. 2002.
2. Zigbee Alliance, www.zigbee.org.
3. I. Howitt and J. A. Gutierrez, "IEEE 802.15.4 low rate-wireless personal area network coexistence issues," *Proc. IEEE Wireless Commun. Network. Conf.* 4(1), 1481–1486 (2003).
4. Y. Xiao, S. Sethi, H.-H. Chen, and B. Sun, "Security services and enhancements in the IEEE 802.15.4 wireless sensor networks," *Proc. IEEE GLOBECOM* 24(1), 411–415 (2005).
5. Y. Xiao, H. Chen, B. Sun, R. Wang, and S. Sethi, "MAC security and security overhead analysis in the IEEE 802.15.4 wireless sensor networks," *EURASIP J. Wireless Commun. Network.* 2006 (2006), Article ID 93830, 12 pages, doi:10.1155/WCN/2006/93830.

CHAPTER 14

PERFORMANCE ANALYSIS FOR IEEE 802.15.4 WIRELESS PERSONAL AREA NETWORKS

HSUEH-WEN TSENG, YU-KAI HUANG, and AI-CHUN PANG

14.1 INTRODUCTION

Recently, wireless sensor networks have received tremendous attention from both academia and industry. With the advance of technologies for micro-sensors, wireless networking and embedded processing, wireless sensor networks are now widely tested and deployed for different application domains [1, 2]. The existing applications include environmental monitoring, industrial sensing and diagnostics, health care, and data collecting for battlefield awareness. Most of the applications are developed by using low rate, short-distance, and low cost wireless technologies. Among the well-known wireless personal area network (WPAN) specifications, ultra wideband (i.e., IEEE 802.15.3) is designed for high rate WPANs [3]. Bluetooth (i.e., IEEE 802.15.1) supports various applications, such as wireless headsets of home appliances and computer peripherals, and provides quality of service (QoS) transmissions, especially for audio traffics [4]. As low cost and low power consumption are considered, IEEE 802.15.4 emerges as a good alternative for WPANs [5].

IEEE 802.15.4 targets ultra-low complexity, cost, and power for low rate wireless connectivity among inexpensive, portable, and moving devices [6]. An IEEE 802.15.4 system can operate in one of the three frequencies. In 2.4-GHz ISM (industrial, scientific, and medical) band, 16 channels are supported and the transmission rate is 250 kbps. In 915-MHz ISM band, 10 channels are supported with a 40-kbps transmission rate. The European 868-MHz band provides 1 channel with a 20-kbps transmission rate. The specifications of the physical (PHY) layer and medium access control (MAC) layer for IEEE 802.15.4 are defined in [5]. Specifically, the IEEE 802.15.4 physical technology

Emerging Wireless LANs, Wireless PANs, and Wireless MANs. Edited by Y. Xiao and Y. Pan

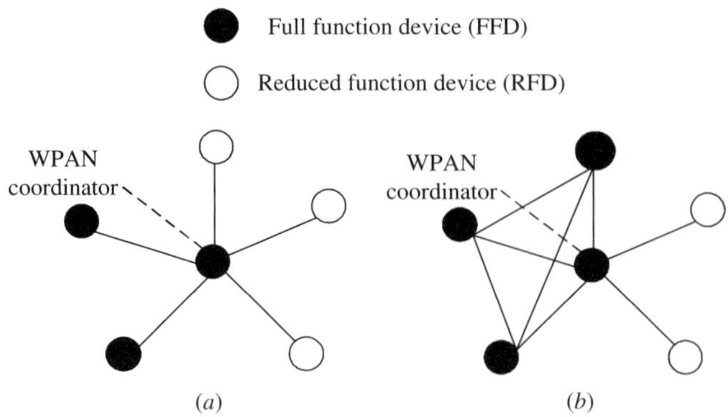

FIGURE 14.1 IEEE 802.15.4 network topologies: (*a*) star topology; (*b*) peer-to-peer topology.

adopts binary phase shift keying (BPSK) and O-QPSK, and its MAC design follows the CSMA/CA (carrier sense multiple-access/collision avoidance) contention-based mechanism.

Based on the data-processing capabilities, two types of devices are provided in IEEE 802.15.4: (1) reduced-function device and (2) full-function device (FFD). These devices constitute a network, and a coordinator equipped with the FFD capability is responsible for organizing and managing the network. In IEEE 802.15.4, both star and peer-to-peer topologies are supported. In the star topology shown in Fig. 14.1*a*, the communication is established between end devices and a single central controller (i.e., coordinator). In the peer-to-peer topology shown in Fig. 14.1*b*, a device could communicate with any other devices within its transmission range. Multihop routing is allowed in the peer-to-peer topology, and routing paths could be dynamically updated. This topology provides more complex network formations such as mesh networking.

14.1.1 Related Work and Motivation

Previous work for IEEE 802.15.4 focused on analytical and simulation modeling for the existing MAC specifications. Gang et al. [7] conducted simulation-based performance evaluation for IEEE 802.15.4. Jelena et al. [8] derived the probability distribution of access delay and calculated the throughput of a beacon-enabled IEEE 802.15.4 network. Zheng and Lee [6] investigated whether IEEE 802.15.4 is fit for ubiquitous networking. Golmie et al. [9] evaluated the performance of IEEE 802.15.4 for medical applications in terms of goodput, delay, and packet loss. Jelena et al. [8] pointed out the bottlenecks in the MAC design of IEEE 802.15.4 systems, and several solutions were proposed.

A considerable amount of contention overhead results from iterative backoff operations of a standard IEEE 802.15.4 CSMA/CA mechanism, especially

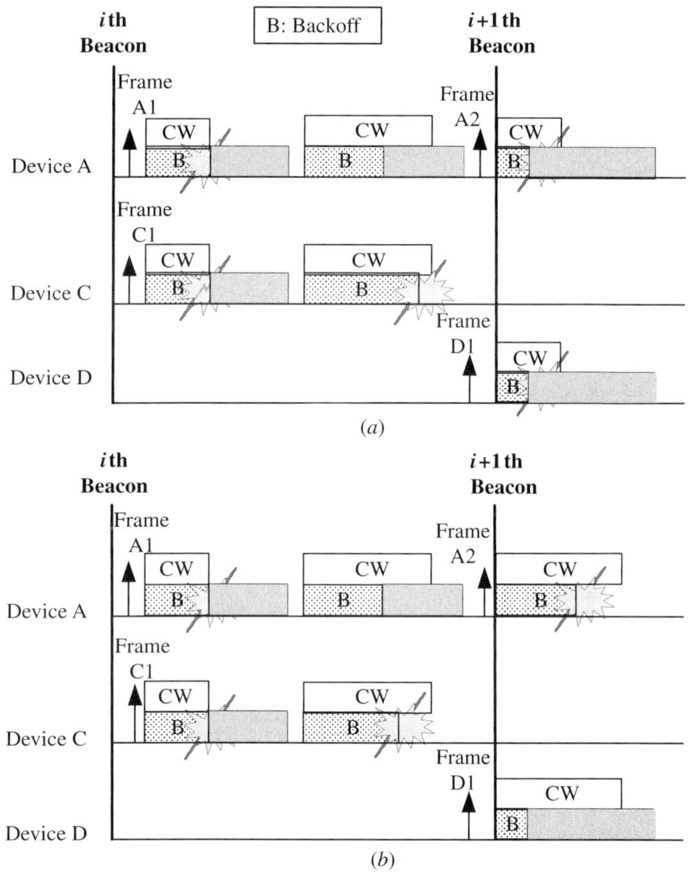

FIGURE 14.2 Backoff flows for (*a*) IEEE 802.15.4 and (*b*) our MBS.

when the system traffic load is heavy. Figure 14.2*a* illustrates the backoff flow for a standard IEEE 802.15.4 CSMA/CA mechanism. When the data frames A1 and C1, respectively, arrive at Devices A and C in the *i*th superframe, the two end devices randomly select a backoff time based on CW_{min}.[1] If the collision occurs (e.g., due to the same backoff time selected by devices A and C), the size of the contention window (CW) is doubled, and the contention process is repeated until one of the devices successfully occupies the channel. As shown in Fig. 14.2*a*, device A obtains the channel access in the *i*th superframe and successfully tranits its data frame. Then when the next superframe [i.e., the $(i+1)$th superframe] starts, the window size will be reset to CW_{min}. Suppose that devices A and D intend to send the frames in the $(i+1)$th superframe. The two devices probably select the same backoff time due to the

[1] In IEEE 802.15.4, $CW_{min} = 2^3$.

small CW, and the collision may occur again. When the network load is heavy, the serious contention could not be resolved within a narrow backoff window, which leads to the increase of the number of collisions, and hence the performance degradation. If a broad contention window is initially used in the $(i+1)$th superframe, the collisions could be reduced, and the devices have higher opportunities to successfully transmit their data frames. However, when the network load is light, a large contention window causes the reduction of the network utilization. Also, the frame transmission delay may be raised because of a relatively large backoff period to determine a specific device that could access the channel.

On the other hand, some data transfers might be time critical (such as sensor/meter applications that issue requests at a constant interval) in WPAN. In order to support time-critical data transfers generated by repetitive low latency applications, IEEE 802.15.4 provides a guaranteed time slots (GTS) mechanism to allocate a specific duration within a superframe for frame transmissions. Although the dedicated bandwidth could guarantee the reliability and performance of data deliveries, the abusing of dedicated resources might also result in the exclusion of other transmissions. The data transmission problem is further complicated by the first-come-first-serve (FCFS) GTS allocation policy [5] because of the lack of scheduling flexibility to respond to the network workload and application needs in low latency data delivery. Starvation is even possible for devices with low data transmission frequencies due to a fixed timer maintained in IEEE 802.15.4 for GTS deallocation.

Thus, how to adequately and efficiently provide a GTS allocation scheme with low latency and fairness is a very challenging problem. Among the work related to this problem, Zheng and Lee [6] did a feasibility study for IEEE 802.15.4 standard over ubiquitous networks. Lu et al. [7] worked on the energy-cost analysis of IEEE 802.15.4 beacon-enabled and non-beacon-enabled transmission modes. A performance analysis of IEEE 802.15.4–based body area networks (BANs) for medical sensors was presented by Timmons and Scalon [10], and the system throughput and the probability distribution of access delay are derived for a beacon-enabled WPAN [11]. An adaptive algorithm [12] for beacon interval adjustment in IEEE 802.15.4 star topology networks was proposed. Kim et al. [13] developed an off line real-time message-scheduling algorithm based on the GTS parameters, such as the length of a beacon interval. Although the performance analysis for IEEE 802.15.4 was investigated extensively, little work has been done on the problems of IEEE 802.15.4 GTS allocation. We must point out that many existing polling algorithms, for example, those for IEEE 802.11 contention-free period (CFP) [14–16], cannot be applied to IEEE 802.15.4 GTS allocation due to the extremely low power consumption of IEEE 802.15.4–based wireless devices and the scarce bandwidth of IEEE 802.15.4 networks (compared to that of IEEE 802.11).

In this chapter, we present two efficient transmission schemes to improve IEEE 802.15.4 transmission performance. A memorized backoff scheme (MBS) dynamically adjusts the CW based on the traffic load in contention access period (CAP). The adaptive GTS allocation (AGA) scheme considers the low latency and fairness issue for data transmitted in CFP. The rest of the chapter is organized as follows. Section 14.2 presents the transmission procedures of CAP and CFP of IEEE 802.15.4 standard. Then we describe the MBS and AGA schemes and the performance evaluation of these schemes in Section 14.3. Section 14.4 is the conclusion.

14.2 IEEE 802.15.4 TRANSMISSION PROCEDURE

An IEEE 802.15.4 network can operate in either a beacon-enabled mode or a non-beacon-enabled mode. In the non-beacon-enabled mode, a device can send data at any time based on CSMA/CA. On the contrary, in the beacon-enabled mode, a coordinator broadcasts a beacon frame periodically to end devices for network synchronization and association. Following the beacon, devices could transmit their data based on the superframe structure specified in the received beacon frame. To ease system management, most sensor networks adopt the beacon-enabled mode for sensor interconnections [17].

Figure 14.3 shows a superframe structure for the beacon-enabled network. A superframe consists of an active period and an inactive period. All devices including a coordinator and several end devices operate in the active period and enter a sleeping phase in the inactive period. Two parameters, beacon order (BO) and superframe order (SO) set by the coordinator, determine the lengths of an active and an inactive periods. They are, respectively equal to 48×2^{SO} UnitBackoffPeriods (UBPs) and $48 \times (2^{BO} - 2^{SO})$ UBPs, where UBP is a basic time unit (i.e., 20 symbol periods[2]) used for backoff. In the beginning of an

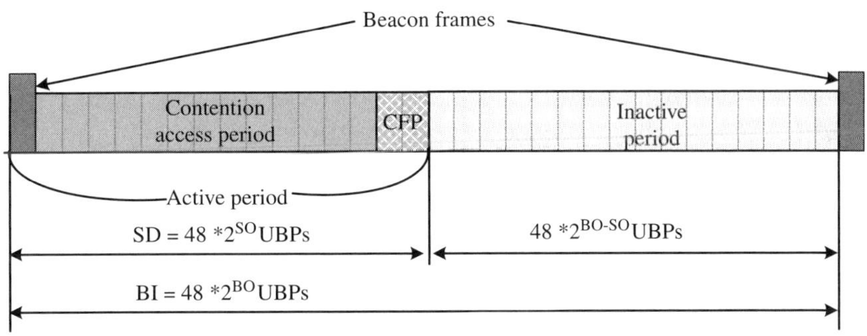

FIGURE 14.3 Superframe structure.

[2] A symbol period is defined as the time that four data bits are transmitted. In IEEE 802.15.4, its length approximates to 16 ns.

active period, a beacon frame is sent from the coordinator to the end devices. The beacon frame includes the information for timing synchronization, system configuration, a list of the end devices that have to receive data frames from the coordinator, and so on. The remaining active periods are divided into two parts. The first part is CAP, and the second part is CFP. In CAP, the end devices equally access the medium by using CSMA/CA. On the other hand, the slots in CFP are reserved for some specific end devices assigned by the coordinator. The coordinator is also in charge of the adjustment for the lengths of CAP and CFP based on traffic loads and request types.

14.2.1 IEEE 802 15.4 CAP Transmission Procedure

With the superframe structure described above, Fig. 14.4 illustrates an example of the CAP transmission procedure for IEEE 802.15.4–based networks. Assume that Device B would like to transmit a data frame to Device A, and the following steps are executed:

Step 1 (*S1*). First, device B randomly selects a backoff time according to the predefined CW (i.e., the default minimum contention window—CW_{min}) and counts it down to zero.

Step 2 (*S2*). After the countdown process, device B checks the channel condition by executing two clear channel assessments (CCAs). If the channel is determined to be free, a data frame will be transmitted to the coordinator. Upon receipt of the data frame, the coordinator delays for a predefined T_{ack} duration and then issues an ACK frame to respond to device B. On the contrary, if the channel has been occupied by other devices, the size of the CW for device B will be doubled, and steps 1 and 2 will be executed again. The contention process will be repeatedly executed until the data frame is successfully transmitted or the maximal retry count is reached (in this case, the frame will be dropped). Note that once the maximum CW value (CW_{max}[3]) defined in IEEE 802.15.4 is reached, the value is retained for the following contentions of this data frame.

Step 3 (*S3*). In the beginning of the next superframe (i.e., the *i*th superframe), the coordinator broadcasts a beacon frame to inform device A that a data frame at the coordinator is destined for device A. After receiving the beacon frame, device A repeats steps 1 and 2 to send data request to the coordinator. When device A receives the ACK frame (corresponding to a data request) from the coordinator (step 3 in Fig. 14.4), it will be awake for the duration of maximum frame response time[4] to wait for the receipt of the data frame from the coordinator.

[3] In IEEE 802.15.4, $CW_{max} = 2^5$.

[4] In IEEE 802.15.4, aMaxFrameResponseTime = 1220 symbol periods.

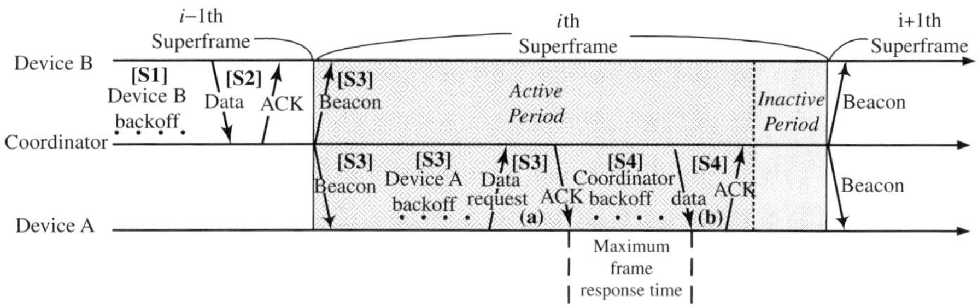

FIGURE 14.4 Example of IEEE 802.15.4 data transmission procedure.

Step 4 (*S4*). Assume that in step 3, the coordinator successfully receives the data request frame. Then following CSMA/CA, the coordinator transmits the data frame back to device A. Upon receipt of the data frame, device A responds with an ACK frame.

14.2.2 IEEE 802 15.4 CFP Transmission Procedure

The IEEE 802.15.4 standard defines the use of CFP for devices requiring dedicated bandwidth. The PAN coordinator is responsible for the GTS allocation and determines the length of the CFP in a superframe. Basically, the CFP length depends on the GTS requests and the current available capacity in the superframe. Provided that there is sufficient capacity in a superframe, the maximum number of GTSs that the PAN coordinator can allocate in the superframe is seven. The GTS direction relative to the data flow from the device that owns the GTS is specified as either *transmit* or *receive*. The transmit GTSs are used for transmitting data from devices to the PAN coordinator, and the downlink frames from the PAN coordinator to devices are delivered over the receive GTSs.

The device that requests new GTS allocation sends a GTS request command to the PAN coordinator during the CAP. Upon receipt of the GTS request command, the PAN coordinator first checks if there is available capacity in the current superframe. Provided that there is sufficient bandwidth in the current superframe, the coordinator determines a device list for GTS allocation in the next superframe based on an FCFS fashion. Then the PAN coordinator includes the GTS descriptor (i.e., the device list that obtains GTSs) in the following beacon to announce the allocation information. For GTS deallocation, devices can return the GTS resources by explicitly sending a GTS deallocation request command to the PAN coordinator. However, in most cases, the PAN coordinator has to detect the activities of the devices occupying GTSs and determine when the devices stop using their GTSs. In IEEE 802.15.4, a fixed expiration timer is used to manage the GTS usage. Once the allocated GTSs would not be utilized for 2^n superframes, the PAN coordinator shall reclaim the previously allocated GTS bandwidth for those devices, where n is defined as

$$\begin{cases} n = 2^{8-\mathrm{BO}} & 0 \le \mathrm{BO} \le 8 \\ n = 1 & 9 \le \mathrm{BO} \le 14 \end{cases}$$

14.3 OUR MBS AND AGA SCHEMES FOR IEEE 802.15.4 MAC

To solve the problems of transmission efficiency of IEEE 802.15.4, we present the MBS scheme for CAP and the AGA scheme for CFP. The MBS detects the traffic load of IEEE 802.15.4 networks and to dynamically adjust the size of the CW based on the network load. In order to accurately estimate the initial value of the CW for each superframe, the exponential weight moving average (EWMA) is incorporated into our proposed MBS. Furthermore, the AGA scheme is a two-phase approach. In the classification phase, devices are assigned priorities in a dynamic fashion based on recent GTS usage feedbacks. Devices that need more attention from the coordinator are given higher priorities. In the GTS scheduling phase, GTSs are given to devices in a non decreasing order of their priorities. A starvation avoidance mechanism is presented to regain service attention for lower priority devices.

14.3.1 Memorized Backoff Scheme

Figure 14.2*b* illustrates the backoff flow for MBS. In this scheme, the CW value for the successful data delivery in the previous superframe is recorded to predict the initial value of the CW for the current superframe. The coordinator announces the initial CW value for the current superframe to end devices via the beacon frame. In Fig. 14.2*b*, device A obtains the channel access in the *i*th superframe by using a suitable CW and successfully transmits its data frame. Then as the next superframe [i.e., the $(i+1)$th superframe] starts, the size of CW will not be reset to $\mathrm{CW}_{\min}$. The coordinator informs end devices the CW value with that device A successfully transmits its data frame in the *i*th superframe. Suppose that devices A and D intend to send the frames in the $(i+1)$th superframe. The two devices probably select different backoff times due to a relatively large CW, and the probability of the data collision may decrease. To avoid the backoff window expanding too quickly in MSB, a window-shrinking operation is designed. That is, if three consecutive successful transmissions occur with the 2^k-slot CW (which implies that the CW may be too large), the initial window value for the next superframe is decreased to 2^{k-1}.

The MBS is fully compatible with the existing IEEE 802.15.4 implementations. In MBS, bits 7–9 for the frame control field in beacon and ACK frames (shown in Fig. 14.5) are used to carry the exponent of the CW value. Once the coordinator receives an ACK frame, the CW value in the ACK frame for this successful frame transmission would be maintained in the coordinator. As the next superframe starts, the coordinator announces the initial CW value to all end devices based on all CW values collected in the current superframe.

Bits: 0–2	3	4	5	6	7–9	10–11	12–13	14–15
Frame type	Security enabled	Frame pending	Ack. req.	Intra PAN	Backoff window	Dest. addressing mode	Reserved	Source addressing mode

FIGURE 14.5 Frame control field in IEEE 802.15.4.

In order to accurately estimate the initial value of the CW for each superframe, a "weighted-average" concept similar to the EWMA [18] approach is incorporated into our proposed MBS. The equation for MBS window size estimation is shown in Eq. (14.1) where E_i denotes the exponent of the CW size for the successful transmission in the ith superframe, and E_A represents the average value of E_{i-1}, E_{i-2}, and E_{i-3}. Then the predicted initial value EP_{i+1} of the CW for the $(i+1)$th superframe is a weighted combination of E_A and E_i. As shown in Eq. (14.2), the weight X depends on the difference of E_i and E_A, and the C value is set to the difference between the exponents of $\mathrm{CW}_{\max}$ and $\mathrm{CW}_{\min}$ (i.e., $C=2$ in this chapter).

$$EP_{i+1} = \lceil E_A + (1 - X)E_i \rceil \tag{14.1}$$

where

$$X = 1 - \left\lfloor \frac{\lceil |E_i - E_A| \rceil}{C} \right\rfloor \tag{14.2}$$

The simulation model for our MBS follows the IEEE 802.15.4 at the MAC layer. The input parameters are referred to the standard and listed in Table 14.1. Without loss of generality, several assumptions are made to reduce the complexity of the simulation model and described as follows:

- All end devices support 250-Kbps transmission rate.
- The coordinator is static and located at the center of simulated area.

TABLE 14.1 Input Parameters and Their Values

Parameters	Values
SO	2
BO	2
Transmission rate of data frames	250 kbps
Beacon frame	0.96 ms (240 bits)
UnitBackkoffPeriod (slot)	0.32 ms (20 symbols)
$\mathrm{T_{ack}}$ period	0.512 ms (128 bits)
Data request (length)	1.792 ms (144 bits)
ACK frame period (length)	0.16 ms (40 bits)
$\mathrm{CW_{min}}$	8 UnitBackoffPeriods (UBPs; slots)
$\mathrm{CW_{max}}$	32 UBPs (slots)
Average data frame size	9 UBPs (slots)

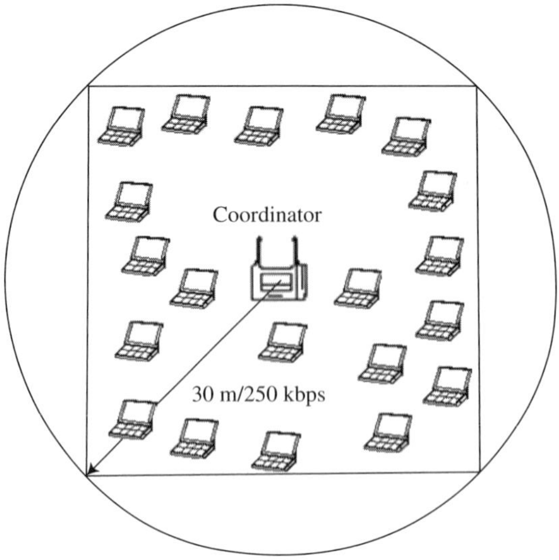

FIGURE 14.6 Scenario for simulation model.

In our simulation model, the transmission range of a coordinator is assumed to be 30 m for the transmission rate of 250 kbps. In the simulation experiments, we simulate a scenario of 20 end devices, and their initial locations are randomly assigned within the area as shown in Figure 14.6. Each simulation run lasts 320 s, and each simulation result is obtained from averaging the results of 100 independent simulations. Each end device maintains a first-in, first-out (FIFO) waiting buffer of 16 frames, and the mean frame length (i.e., $1/\mu$) is assumed to be 90 bytes (e.g., 9 UBPs at the 250-kbps transmission rate, excluding PHY and MAC headers). The network load consists of uplink transmission (from end devices to the coordinator) and downlink transmission (from the coordinator to end devices) traffic. When the network load (i.e., $M\lambda/\mu$) is less than 0.3, we define the situation is the "light" traffic load. On the other hand, when the network load is greater than 0.7, the situation is defined as the "heavy" traffic load.

Figure 14.7 shows the effects of the traffic load on goodput (GP) for MBS, MBS+EWMA, and IEEE 802.15.4 schemes. We find that MBS and MBS+EWMA schemes achieve higher GP than IEEE 802.15.4, and the peak values of the GP for MBS and MBS+EWMA schemes are much larger than that for IEEE 802.15.4. The high GP for MBS and MBS+EWMA mainly results from the decrease of the number of contentions/collisions. In other words, by using MBS and MBS+EWMA, the backoff overhead is significantly reduced and therefore the GP improves. We also observe that the performance of MBS+EWMA is better than that of MSB, which implies that the proposed EWMA approach effectively predicts the network condition and further reduces the occurrence of collisions.

Figure 14.8 shows the effects of the traffic load on the completion rate R_c for MBS, MBS+EWMA, and IEEE 802.15.4 schemes, where R_c is defined as the

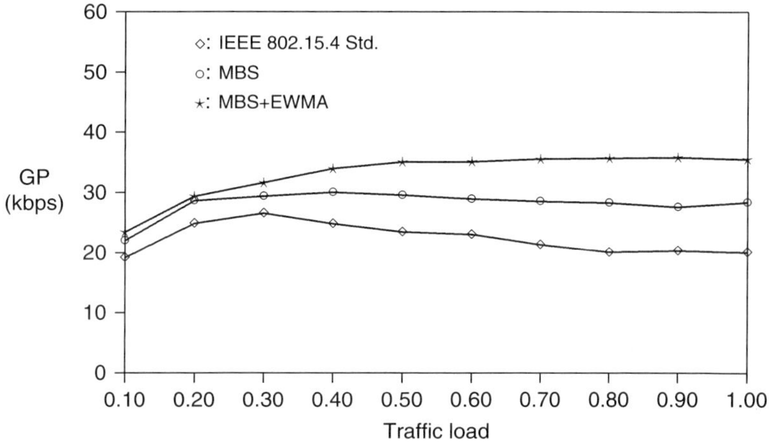

FIGURE 14.7 Effects of traffic load on goodput.

number of successful transmitted frames over the total number of transmitted frames. From this figure, we observe that for all schemes, R_c decreases as the traffic load increases. The decreasing rate is sharper for IEEE 802.15.4 than for MBS and MBS+EWMA, especially when the traffic load is heavy. The reason is that for IEEE 802.15.4, the collisions for medium contention under the heavy traffic load become severe, which significantly results in the degradation of the completion rate. From this figure, we find that MBS and MBS+EWMA have higher completion rates than IEEE 802.15.4.

Figures 14.9 and 14.10 show the effects of the traffic load on the average queueing delay D_q and average MAC delay D, respectively. These figures show

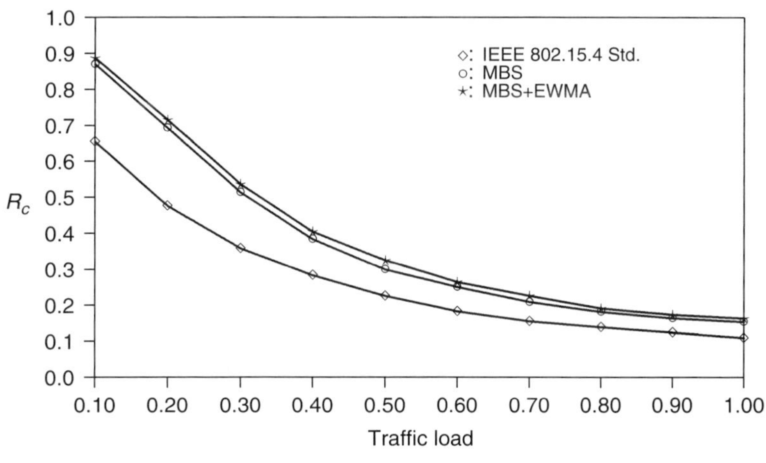

FIGURE 14.8 Effects of traffic load on completion rate.

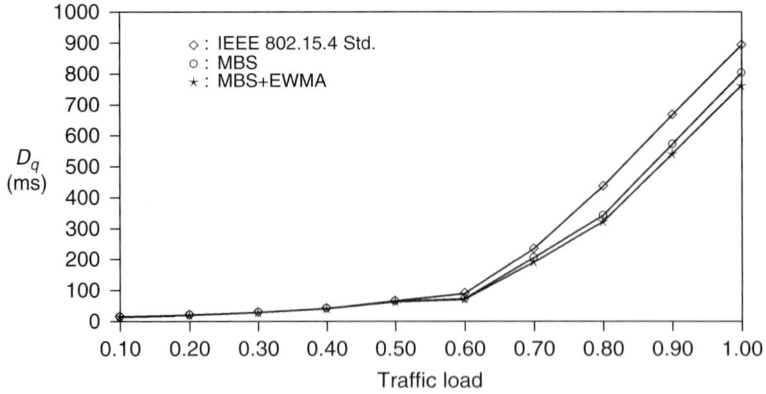

FIGURE 14.9 Effects of traffic load on average queueing delay.

the intuitive results that for MBS, MBS+EWMA, and IEEE 802.15.4 the D_q and D increase as the traffic load increases. These figures also indicate that when the traffic load is light, the curves for all schemes are insensitive to the traffic load. On the other hand, in the heavy traffic load, the average delays significantly increase as the traffic load increases, especially for IEEE 802.15.4. Since serious collision under the heavy traffic load results in longer delays in IEEE 802.15.4, the MBS and MBS+EWMA schemes are proposed. Furthermore, The D and D_q of MBS+EWMA are less than those of MBS because the contention window in the MBS+EWMA scheme can be adapted more appropriately by using the proposed EWMA approach.

Figure 14.11 shows the average number N_c of collisions occurred for each data frame prior to being successfully transmitted. A trivial result is observed

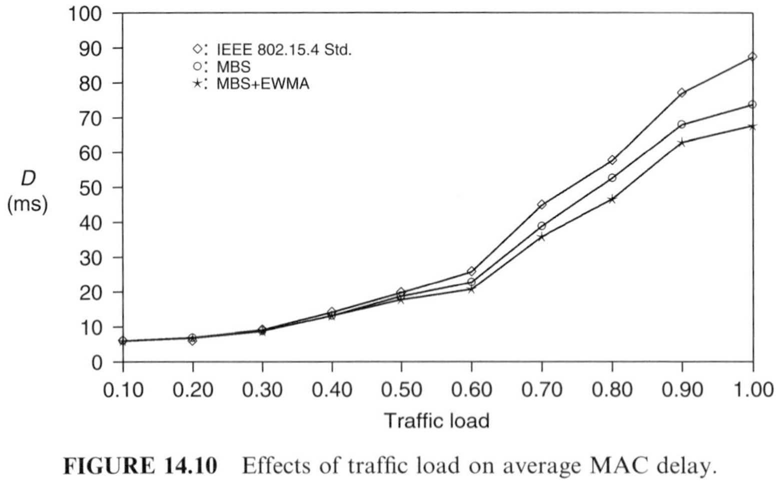

FIGURE 14.10 Effects of traffic load on average MAC delay.

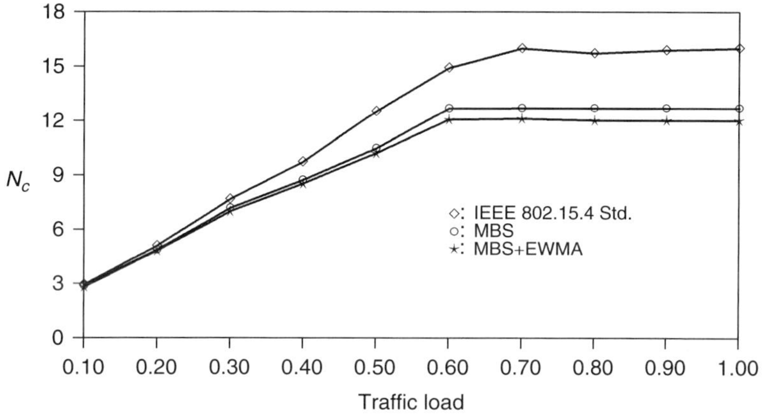

FIGURE 14.11 Effects of traffic load on the number of collisions for each data frame.

that for all schemes under investigation, N_c increases as the traffic load increases. N_c increases more significantly for the heavy traffic load than for the light traffic load, especially in IEEE 802.15.4. Furthermore, MBS and MBS+EWMA have smaller N_c values than IEEE 802.15.4. The large N_c value indicates that with a data frame arrival, the end device should try to send the data frame many times before the frame is successfully transmitted, which results in more power consumption of the end device.

14.3.2 Adaptive GTS Allocation Scheme

The objective of this section is to propose an AGA scheme for IEEE 802.15.4–based WPANs with the considerations of low latency and fairness. In IEEE 802.15.4, GTS is provided by the coordinator in a star network topology (see Fig. 14.1*a*). The communication is established between a PAN coordinator and up to 255 devices. By periodically broadcasting a beacon frame, a PAN coordinator updates its GTS descriptor to surrounding devices.

An ideal GTS allocation scheme should have a good guess for the future GTS-transmitting behaviors of devices. By using the prediction, the PAN coordinator allocates GTS resources to devices in need and reclaims the previously allocated GTSs that will not be used. Our AGA scheme is a two-phase approach. In the classification phase, devices are assigned priorities in a dynamic fashion based on recent GTS usage feedbacks. Devices that need more attention from the coordinator are given higher priorities. In the GTS scheduling phase, GTSs are given to devices in a nondecreasing order of their priorities. A starvation avoidance mechanism is presented to regain service attention for lower priority devices. Before presenting the details for the device classification and GTS scheduling phases, we define two terms, GTS hit and GTS miss, as follows.

Definition 14.1. If one device has issued a GTS request in the CAP or transmitted data within its allocated GTS to the PAN coordinator during the period of the current superframe, the device is defined to have a GTS hit. Otherwise, the device is considered to have a GTS miss.

14.3.2.1 Device Classification Phase. In this phase, each device is adaptively classified into one state maintained by the coordinator, and dynamically assigned a priority number by the coordinator based on past GTS usage feedbacks. Assume that there are N devices in an IEEE 802.15.4–based WPAN, and there are $M+1$ (0,1, ..., M) priority numbers dynamically assigned to the N devices. A large priority number represents a low priority for GTS allocation. The priority number assigned to the device n is defined as Pri_n, and then we have $0 \leq \text{Pri}_n \leq M$. In our AGA scheme, the devices with higher priorities are expected to have more recent traffic, and thus have higher probabilities to transmit their data in the coming superframe. The state and priority number of a device are internally maintained by the PAN coordinator. The maintenance of the state and priority number of each device is based on the concept for dynamic branch prediction for computer architecture design [19] and the additive increase/multiplicative decrease (AIMD) algorithm for network congestion control [18] with some improvement, and will be described as follows.

State Transition. All devices in our AGA scheme are classified into four traffic levels according to the state diagram shown in Fig. 14.12. In this figure, the four traffic levels of devices are accordingly mapped to the four states, that is, HH (high heavy), LH (low heavy), HL (high light), and LL (low light). The order of traffic levels for these states are HH > LH > HL > LL. Initially, all devices are placed in the LL state. At the end of each superframe, the PAN coordinator examines the GTS usage of all devices and then decides to which states to transmit the devices. The transition follows the solid and dashed lines in Fig. 14.12. The solid and dashed lines, respectively, represent the occurrence of a GTS hit and a GTS miss. With the state diagram, the devices with more

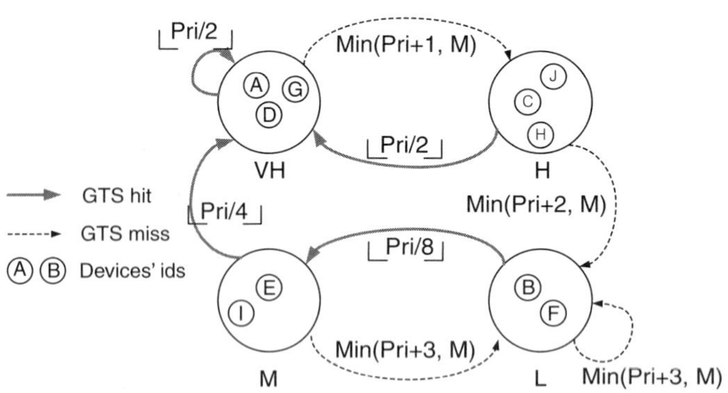

FIGURE 14.12 State diagram for our proposed AGA scheme.

frequently GTS usage will have larger probabilities to stay in heavy-traffic states (e.g., HH and LH). Also, temporarily unstable transmission behaviors of devices could be more tolerated so that the devices residing in the heaviest traffic state (i.e., HH) with an occasional transmission interruption have a second chance before being degraded to the light-traffic states. On the other hand, the devices in the LL state can be promoted to heavy-traffic states by having consecutive GTS hits.

In original IEEE 802.15.4 specification [5], the devices intending to utilize GTSs for data transmission may wait for the expiration of GTSs (i.e., the allocated GTSs that have not been used for a specific period) of high priority transmissions. This passively deal location scheme for GTS resources would result in starvation of light-traffic devices. Conversely, by using our AGA scheme, starvation of light-traffic devices can be avoided since these devices can be gradually promoted to the heavy-traffic state with the existing GTS request facility to notify the PAN coordinator for traffic-level promotion.

Priority Assignment. By using the above state diagram, the PAN coordinator can monitor the recent transmission behaviors of devices and classify the devices into proper traffic types. However, with scarce GTS resources (i.e., seven time slots) of IEEE 802.15.4–based networks, the four-state classification for devices is somewhat rough and cannot be sufficient for precisely classifying transmission behaviors of devices. Thus, the state diagram in Fig. 14.12 is further revised so that each device is dynamically assigned a priority number for GTS allocation. Upon the occurrence of GTS hit of a device, the priority number of the device is decreased by the PAN coordinator, and the priority of GTS allocation for the device upgrades. On the other hand, when GTS miss occurs to a device, the PAN coordinator increases the priority number of the device, and hence the opportunity for obtaining a GTS for the device reduces. Maintenance of priority numbers of devices depends on the transmission feedback as well as the traffic-level states of devices, and the details for maintaining priority numbers of devices are presented below.

Compared to the priority assignment by purely using AIMD [14], our scheme provides a multilevel AIMD algorithm for the updating of priority numbers. In our multilevel priority updating, the decrease/increase of a priority number of a device depends on the traffic-level state the device resides. The high priority devices with temporary interruption of GTS usage will be slightly demoted to lower priorities. On the other hand, if a low priority device starts to request the GTS service to transmit data, its priority will be greatly promoted to have GTS service as soon as possible, and starvation of such a low priority device can be avoided. What our priority assignment focus on is whether devices have continuous data to be transmitted over the GTSs. The devices with consecutive transmissions are favored by our scheme. However, a device that has idled for a period of time would be considered not to need the GTS service, and it is reasonable to greatly degrade the device's priority.

From Fig. 14.12, we can see that if the device n in state HH uses the GTS service all the time and occasionally has a GTS miss, its priority Pri_n will be increased by 1. Once device n resumes to request for the GTS service in the following superframe and then has continuous data to be transmitted, the increased priority number for device n will be exponentially halved so that the priority of device n can be "recovered" rapidly. For device k in the LL state, a similar and even more great priority promotion occurs if device k has consecutive GTS hits. On the other hand, if device k in state LL just has one GTS hit and ceases transmitting data, the degradation of the priority for device k would be more serious than that for the high traffic-level device n. Our design for device classification can prevent low priority devices from starvation, and simultaneously maintain the GTS service for heavy-traffic devices with occasional transmission interruption.

14.3.3 GTS Scheduling Phase

With the device classification phase, priorities for GTS allocation for all devices under the supervision of the PAN coordinator are determined. Next, in the GTS scheduling phase, the GTS resources are adequately scheduled and allocated to the devices. The scheduling criteria are based on the priority numbers, the superframe length depending on the BO value, and the GTS capacity in the superframe. The GTS scheduling algorithm is shown in Procedure 14.1. Assume that there are N devices in the WPAN, and P is a set including the priority numbers for the N devices. In Procedure 14.1, the PAN coordinator first checks if the GTS capacity is overloaded. In the IEEE 802.15.4 specification [5], the GTS capacity in a superframe shall meet the following two requirements:

1. The maximum number of GTS slots to be allocated to devices is seven.
2. The minimum length of a CAP shall be aMinCAPLength. The increase of the total GTS period shall not result in the reduction of the CAP length to be less than aMinCAPLength.

PROCEDURE 14.1 Device Scheduling

Assme that there are N devices in the WPAN.
$P = \{\mathrm{Pri}_1, \mathrm{Pri}_2, \ldots, \mathrm{Pri}_N\}$
$T_h = MR^{\mathrm{BO}}$ where M and R are constants.
while The GTS capacity is not overloaded **do**
    Find a device k such that $\mathrm{Pri}_k \in P$ is the minimum number of P.
If $\mathrm{Pri}_k \leq T_h$ **then**
    Device k will be scheduled in the GTS of the current superframe.
    Remove Pri_k from P.
else
    break;
end if
end while

If the requirements are met, the GTS capacity is considered not to be overloaded. Provided that there is sufficient GTS resources to accommodate more devices, lines 5–11 of the WHILE loop are executed. At each iteration of the WHILE loop, a minimum Pri_k is selected among P, and its value is compared with a threshold value T_h and T_h is defined as

$$T_h = MR^{\mathrm{BO}} \tag{14.3}$$

where R is a constant, and $0 < R < 1$.

The threshold T_h is presented here due to the consideration of the GTS traffic load. When the traffic load is light (i.e., most of the devices have high priority numbers), there is no need to allocate too many GTS resources for the devices. Too much dedicated bandwidth for GTS usage in this case leads to the resource wastage and even the degradation of overall system performance. Instead, the GTS bandwidth shall be transferred for contention-based accesses in CAP. To achieve such a goal, the PAN coordinator has to detect the workload for GTS traffic and filter unnecessary GTS allocation by using the threshold T_h. From Eq. (14.3), the value of T_h is dynamically adjusted and depends on the maximum priority number M, a constant R, and the beacon interval determined from BO. As the beacon interval increases, there are higher probabilities that many devices have requested the GTS service in the superframe. Based on our priority assignment, the devices having requested GTS are assigned small priority numbers even though they only have one request in the whole superframe. To prevent the scarce GTS resources from distributing to those devices with extremely low frequency GTS requests in such a long superframe, a more strict threshold is needed. In this case, the T_h value is set to be much smaller than M. On the other hand, in a short beacon interval, the value of T_h can be increased, and the limitation for the device selection can be more relaxed.

Based on the above discussions, the priority number of the selected device k is compared with the dynamic threshold T_h. If $\mathrm{Pri}_k \leq T_h$ (line 6), then the device k is scheduled in the GTS of the current superframe.

Then we develop a simulation model to investigate the performance of our AGA scheme. Our developed simulation follows the specification of IEEE 802.15.4 MAC protocol. Without loss of generality, several assumptions are made to reduce the complexity of the simulation model and are described as follows:

1. Only the GTS traffic is considered.
2. All GTS transmissions are successful. That is, we do not consider GTS retransmissions.
3. Only the transmit GTSs for the uplink traffic are adopted.

In the simulation model, a star topology with one PAN coordinator and N devices ($N = 5$ and 10) is adopted. Each simulation run lasts 100,000 beacon

TABLE 14.2 Input Parameters and Their Values

Parameters	Value
Frame size	128 bytes
Transmission rate	250 kbps
Network topology	Star topology
Number of devices	5 and 10
BO = SO	5
Buffer size of each devices	100
λ_h	0.3/s
λ_l	0.1/s

intervals (i.e., 49,152 seconds). The packet arrivals for each device form a Poisson stream with the interarrival rate λ. Two traffic types generated by devices are considered, heavy traffic and light traffic; and λ_h and λ_l represent, respectively, the interarrival rates for the heavy-traffic and light-traffic devices. In the simulations, we have $\lambda_h = 0.3/\text{s}$ and $\lambda_l = 0.1/\text{s}$. Such rate setting is reasonable in IEEE 802.15.4–based WPAN since IEEE 802.15.4 targets low rate wireless communications. Also, the ratio of the number of heavy-traffic devices to that of all devices is defined as v. Table 14.2 lists the input parameters for our simulation model.

As to the output measures, average packet waiting time is an important metric for our proposed AGA scheme. Furthermore, a fairness index F for packet waiting times is utilized to measure the fairness among different traffic-type devices for scheme. From [20] F is defined as

$$F = \frac{\left(\sum_{i=1}^{N} W_i\right)^2}{N \sum_{i=1}^{N} W_i^2} \tag{14.4}$$

where N is the total number of devices in the network, and W_i is the average waiting time of packets generated by the device i. In Eq. (14.4), it is clear that $0 \leq F \leq 1$. When the average waiting times for all devices are close, the F value approaches 1. On the other hand, if the variation of the W_i values becomes large, F approaches 0. Therefore, a large F implies that each device obtains the GTS bandwidth more fairly, and, probably, much starvation will not occur.

Figures 14.13 and 14.14 show the effect of v (the percentage of heavy-traffic devices) on the average packet waiting time and the fairness index F for our AGA scheme and the original scheme proposed by IEEE 802.15.4 standard. When $N = 5$, Fig. 14.13*a* indicates that as v increases (i.e., the number of heavy-traffic devices increases), both the curves for average packet waiting times of our AGA and the original schemes decrease. The reason is that for a small v, the dedicated GTS bandwidth is almost occupied by the light-traffic devices.

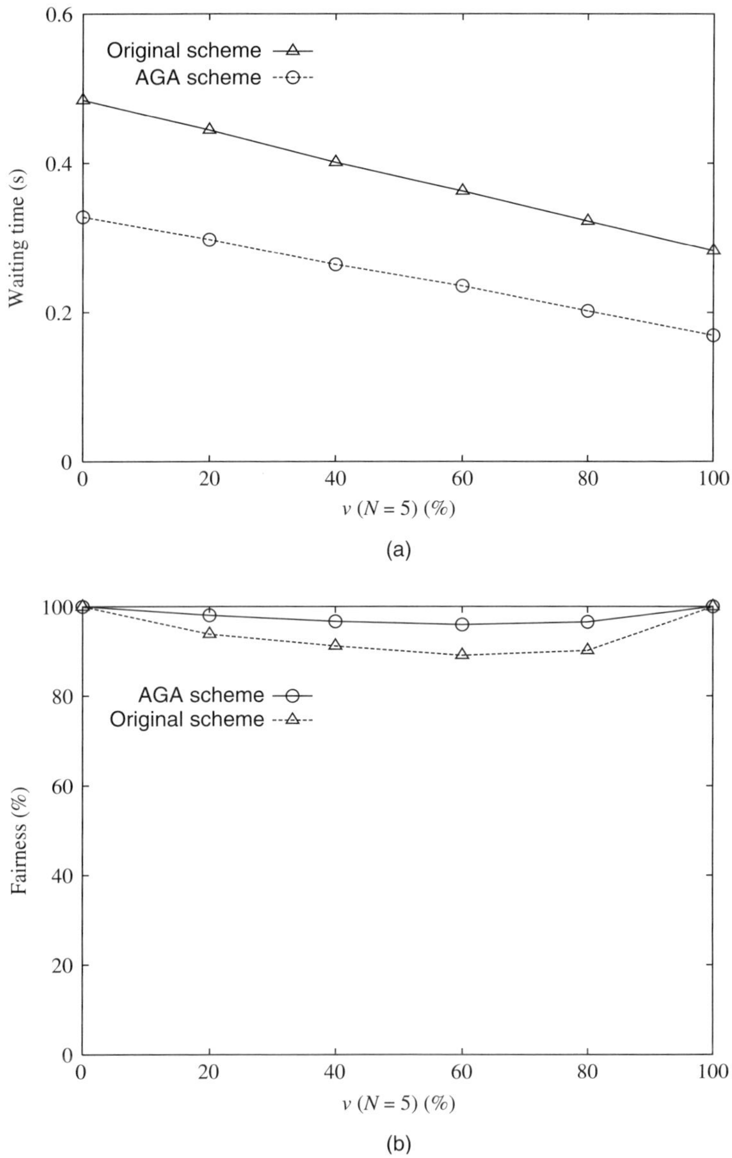

FIGURE 14.13 Effect of v on (a) average packet waiting time and (b) fairness index F when $N = 5$.

The resources allocated to the light-traffic devices would be released soon due to the low frequency packet arrivals. In this case, any packet arrival at the light-traffic devices cannot have the immediate GTS usage, which incurs the longer average waiting time. Conversely, when the number of heavy-traffic devices is

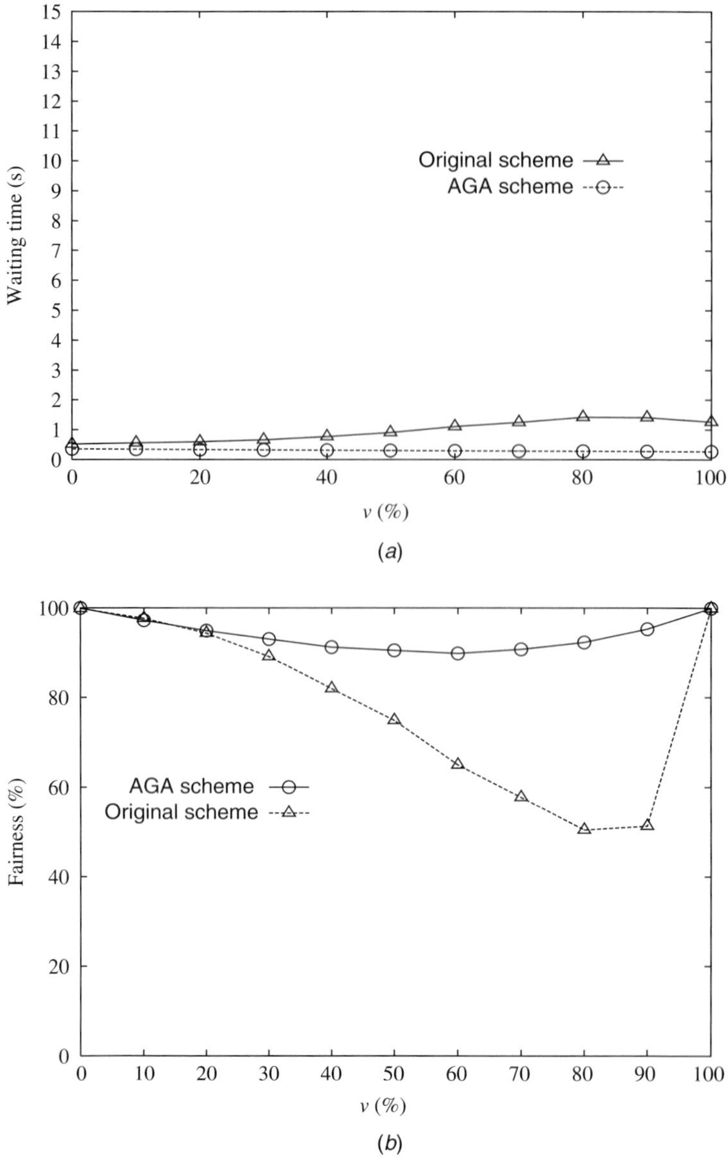

FIGURE 14.14 Effect of v on (a) average packet waiting time and (b) fairness index F when $N = 10$.

large, most packet arrivals are placed in some GTSs preallocated for the heavy-traffic devices. Specifically, with more GTS usage, a more precise prediction for our AGA scheme leads to a smaller average packet waiting time. From Fig. 14.13a, for all v under investigation, our proposed AGA scheme achieves a smaller average waiting time that the original scheme specified in IEEE 802.15.4 standard.

Based on the index *F*, Fig. 14.13*b* shows the comparison of fairness provided by our AGA scheme and the IEEE 802.15.4 original one. In Fig. 14.13*b*, we observe that *F* decreases and then increases as *v* increases, which implies that the unfairness problem comes from heterogeneity of devices. Also, the decreasing/increasing rate of *F* for AGA is smaller than that for the original scheme. In other words, our proposed scheme is equipped with the capability to provide more fair transmissions among different kinds of devices than the original IEEE 802.15.4 scheme. However, we can observe that both the scheme will not have serious unfairness when the network is sparse (i.e., when *N* is small).

A similar phenomenon for the average waiting time of our AGA scheme is observed in Fig. 14.14*a* compared with that in Fig. 14.13*a*. However, from Fig. 14.14*a*, we find out that for the original scheme, the average waiting time significantly increases and then slightly decreases when *v* increases. This is because the original scheme cannot resist the rapid workload increase with its inflexible GTS allocation presented by IEEE 802.15.4 specification. The raising of average waiting time in the original scheme results from the long-term GTS occupancy of heavy-traffic devices. On the other hand, for a dense network, our adaptive scheme provides much lower waiting times than that for a sparse network shown in Fig. 14.13*a*. The curves for the fairness index in Fig. 14.14*b* can further explain why the rapid increase of the average waiting time occurs at the original scheme. Figure 14.14*b* indicates that when $80\% \leq v \leq 90\%$, a serious unfair situation is observed in the original IEEE 802.15.4 scheme, which implies that most GTS resources are distributed to the heavy-traffic devices and starvation of the light-traffic devices may occur. However, our proposed approach can retain a small waiting time and provide more fair GTS transmissions for all devices.

14.4 CONCLUSION

IEEE 802.15.4 defines the low rate wireless transmission for personal area networks. In the IEEE 802.15.4 standard, the MAC layers are defined as CAP and CFP transmission procedures. In this chapter, we presented a memorized backoff scheme (MBS) for CAP and adaptive GTS allocation (AGA) scheme for CFP. The MBS dynamically adjusts the size of the CW based on the network load. With the EWMA approach, the CW value can be estimated more accurately. The AGA scheme considering low latency and fairness consists of two phases: device classification phase and GTS-scheduling phase. In the device classification phase, the priority for each device intending to transmit data is determined, and the GTS slots are adequately scheduled and allocated according to the priorities in the GTS-scheduling phase. Our proposed schemes can be implemented in the standard IEEE 802.15.4 MAC protocol without adding any new message type and without modifying the communicating procedure. The simulation models for these schemes were developed, and the experimental results indicated that our proposed schemes greatly outperforms the standard IEEE 802.15.4 implementations.

REFERENCES

1. B. Haowei, M. Atiquzzaman, and D. Lilja, "Wireless sensor network for aircraft health monitoring," paper presented at IEEE BroadNets 2004, 2004, pp. 748–750.
2. V. Rajaravivarma, Y. Yang, and T. Yang, "An overview of wireless sensor network and applications," *IEEE Syst. Theory* 2003, pp. 432–436.
3. "Standard for Information Technology—Part 15.3: Wireless medium access control (MAC) and physical layer (PHY) specifications for wireless personal area networks (WPANs)," IEEE, New York, 2003.
4. "Bluetooth Baseband Specification Version 1.0 B," http://www.bluetooth.com/
5. IEEE 802.15.4, "Standard for Part 15.4: wireless medium access control layer (MAC) and physical layer (PHY) specifications for low rate wireless personal area networks (LR-WPANs)," IEEE, New York, Oct. 2003.
6. J. Zheng and M. J. Lee, "Will IEEE 802.15.4 make ubiquitous networking a reality? A discussion on a potential low power, low standard," *IEEE Commun. Mag.* 42, pp. 140–146 (2004).
7. L. Gang, B. Krishnamachari, and C. Raghavendra, "Performance evaluation of the IEEE 802.15.4 MAC for low-rate low-power wireless network," paper presented at the IEEE EWCN 2004, Apr. 2004, pp. 701–706, Phoenix, Arizona.
8. M. Jelena, B. Vojislavand, and S. Shairmina, "Performance of IEEE 802.15.4 beacon enabled PAN with uplink transmissions in non-saturation mode-access delay for finite buffers," paper presented at IEEE BROADNETS 2004, 2004, pp. 416–425.
9. N. Golmie, D. Cypher, and O. Rebala, "Performance analysis of low rate wireless technologies for medical applications," *Computer Commun.* 28, 1255–1275 (2005).
10. N. F. Timmons and W. G. Scanlon, "Analysis of the performance of IEEE 802.15.4 for medical sensor body area networking," paper presented at IEEE SECON 2004, Oct. 2004.
11. J. Misic, S. Shaf, and V. B. Misic, "Performance of a beacon enabled IEEE 802.15.4 cluster with downlink and uplink traffic," *IEEE Trans. Parallel Distributed Syst.* 17(4), 361–376 (2006).
12. M. Neugebauer, J. Plonnigs, and K. Kabitzsch, "A new beacon order adaptation algorithm for IEEE 802.15.4 networks, paper presented at EWSN 2005, pp. 303–311.
13. D. Kim, M. Pham, Y. Doh, and E. Choi, "Scheduling support for guaranteed time services in IEEE 802.15.4 low rate WPAN," paper presented at IEEE RTCSA 2005, Aug. 2005.
14. X. Dong, P. Varaiya, and A. Puri, "Adaptive polling algorithm for PCF mode of IEEE 802.11 wireless LANs," *Electron. Lett.* 40, pp. 482–483 (2004).

15. Y. Kim and Y. Sun, "Adaptive polling MAC schemes for IEEE 802.11 wireless LANs," paper presented at IEEE 57th VTC, Apr. 2003.
16. B. Kim, S. W. Kim, Y. Fang, and T. F. Wong, "Linkadaptable polling-based MAC protocol for wireless LANs," paper presented at IEEE GLOBECOM 2004, Nov. 2004.
17. L. Lazos and R. Poovendran, "HiRLoc: High-resolution robust localization for wireless sensor networks," *IEEE J. Sel. Areas Commun.* 24, 233–246 (2006).
18. J. F. Kurose and K. W. Ross, *Computer Networking*, 3rd ed., Addison-Wesley, Reading, MA, 2001.
19. D. A. Patterson and J. L. Hennessy, *Computer Organization and Design*, Morgan Kaufmann, CA, 1997.
20. R. K. Jain, D. W. Chiu, and W. R. Hawe, "A quantitative measure of fairness and discrimination for resource allocation in shared computer systems," technical report TR-301, DEC, Sept. 1984.

CHAPTER 15

DATA TRANSMISSION AND BEACON SCHEDULING IN LOW RATE WIRELESS MESH PERSONAL AREA NETWORKS

JIANLIANG ZHENG

15.1 INTRODUCTION

The IEEE standard, 802.15.4 [1], defines the physical (PHY) layer and medium access control (MAC) sublayer specifications for low rate wireless mesh personal area networks (LR-WMPANs). ZigBee, an industrial alliance, has been working on the network and upper layers of LR-WMPANs [2]. And the IEEE 802.15.5 task group is currently working to provide an architectural framework for interoperable, stable, and scalable wireless mesh topologies for both low rate and high rate wireless personal area network devices. LR-WMPANs show promise to bring ubiquitous networking into our lives, at least technically [3]. They will bring many simple, originally standalone devices into networks, and thus not only open the door to an enormous number of new applications but also add value to many other existing applications.

Two major problems faced by wireless medium access are the hidden terminal (HT) and exposed terminal (ET) problems [4]. An HT problem results in collisions and an ET problem causes unnecessary delay. An HT problem is more serious to most applications, as packets may be dropped due to collisions, which reduces the network throughput. Collisions caused by an HT problem can be classified into two types. One is that a collision happens at the common destination of two or more packets. The other is that a collision happens at a node that is the destination of one of the packets involved in the collision. Here we call the first type of collision *primary collision* and the second type of collision *secondary collision*. All packets are destroyed in a primary collision,

Emerging Wireless LANs, Wireless PANs, and Wireless MANs. Edited by Y. Xiao and Y. Pan

while only packet(s) destined for the node where collision happens is destroyed in a secondary collision. Since most collisions happen between two packets, a secondary collision has a good chance to be resolved by one retransmission. In general, a primary collision needs more retransmissions to resolve. In LR-WMPANs, repeated primary collisions are likely to happen and packets have a high probability to be dropped due to the relatively short backoff period used by MAC [5].

WiFi uses request-to-send (RTS) and clear-to-send (CTS) schemes to cope with HT/ET problems [6]. Since RTS/CTS control messages themselves still suffer from ET/HT problems, the possible failure of RTS/CTS transmissions means ET/HT problems cannot be completely eliminated by the RTS/CTS scheme. And mobility can substantially reduce the reliability of the RTS/CTS scheme. Another shortcoming of the RTS/CTS scheme is that it can only handle unicast communications. The primary design goal of LR-WMPANs is low cost, low power consumption, and support of simple devices such as sensors and actuators. As a result, IEEE 802.15.4 has some distinct design features, for instance, relatively short backoff period, no RTS/CTS control frames, beacon mode and superframe structure, and orphaning and coordinator relocation. While in general IEEE 802.15.4 is a well-defined standard, an LR-WMPAN based on IEEE 802.15.4 faces some special problems, including repeated collisions, high collision probability at the beginnings of superframes, nonatomic transactions, and insufficient support of multihop beacon-enabled mesh networks [5]. To address those problems, two medium access control scheduling schemes are presented in this chapter, that is, receiver-oriented time division multiple access (ROT) and medium access scheduling midware (MASM). ROT can be used to remove all primary collisions and MASM, as a contention-free time division multiple access scheme, can be employed to eliminate both primary and secondary collisions.

15.2 RECEIVER-ORIENTED TDMA

The simple ROT scheme presented here can be used to eliminate all primary collisions. While it is possible to eliminate both primary and secondary collisions using more sophisticated time division multiple access (TDMA) schemes, nodes will suffer more serious ET problems as more time slots are generally needed. As noted above, secondary collisions can be resolved using one retransmission in most cases; the network throughput by using the simple ROT scheme is expected to be close to that by using sophisticated TDMA schemes. Transmission latency caused by retransmissions in ROT is (at least partially) compensated by the fact that less time slots are needed compared with other sophisticated TDMA schemes. Compared with a collision-free TDMA scheme, retransmissions in ROT will consume more network resource such as frequency bandwidth and energy. But fortunately, a node does not need to retransmit twice for most collisions, which may prove to be a reasonable cost paid for the simplicity of ROT. Our simulation results show that this simple scheme is very efficient for LR-WMPANs where the primary collision problem is serious. The combination of ROT and the topology-guided distributed link state (TDLS) wireless mesh routing [7] greatly improves the performance of LR-WMPANs.

In ROT, two-hop neighbor information is exchanged among nodes. To do this, each node transmits one-hop Hello messages in which information of all one-hop neighbors is included. When a node receives an Hello message from a one-hop neighbor, it adds the neighbor into its neighbor list and also calculates the time slot and slot cycle it should use to transmit packets to this neighbor. A simple way to calculate the time slot is to sort identifiers (IDs) (e.g., addresses) of all the neighbors of the neighbor from whom the Hello message comes. Based on the order its ID appears in the sorted ID list, the node knows its time slot. For example in Fig. 15.1, node 5 has four one-hop neighbors (1, 4, 6, 9). If it includes each neighbor's neighbors, the node has a view of two-hop neighbors (1 [0, 2, **5**], 4 [0, **5**, 8], 6 [2, **5**, 7, 10], 9 [**5**, 8, 10, 13]). Thus, node 5 has a time slot table (TST), (1:2:3/4:1:3/6:1:4/9:0:4), each entry of which is in the format of neighbor_id:slot_number:slot_cycle. The TST of node 5 tells that node 5 should use slot 2 (the third slot—slots are numbered from 0) (modulo slot cycle 3), 1 (modulo slot cycle 3), 1 (modulo slot cycle 4), and 0 (modulo slot cycle 4) to transmit packets to neighbors 1, 4, 6, and 9, respectively. Each time a node receives an Hello message, it will check if the TST needs to be updated. Synchronization is needed in ROT, as in any other TDMA scheme. But as we have noticed, ROT is a fully distributed receiver-oriented scheme, which means no network-wide synchronization is needed. So the synchronization problem is simply reduced to that a node should know the clock of each its

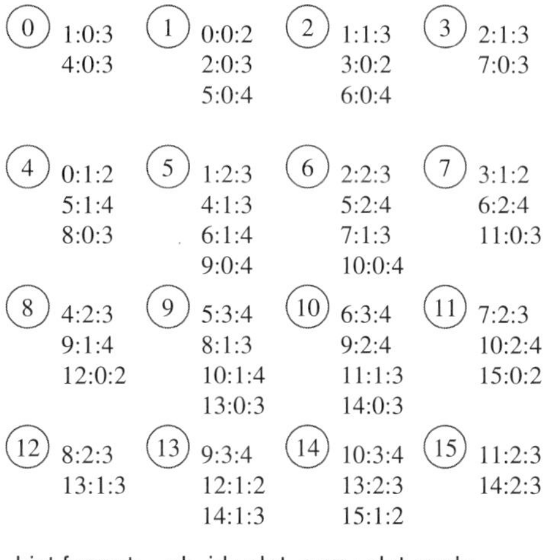

FIGURE 15.1 Example of receiver-oriented TDMA.

neighbor. Therefore, the synchronization problem can be easily solved by including the clock information in the Hello message. To compensate for clock inaccuracy and clock drift, some small guard time duration (GTD) can be added into each time slot and resynchronization should be performed before the clock drift exceeds the GTD. Note that the synchronization clock, slot number, and slot cycle used for one neighbor is independent from those for other neighbors.

While ROT eliminates all primary collisions, it suffers from secondary collisions. This is partially compensated for by the fact that ROT has a smaller time slot cycle (TSC) (equivalently, less serious ET problems and less delay) compared with a more sophisticated TDMA scheme. To see this, let us compare ROT with any TDMA scheme that is able to eliminate both primary and secondary collisions. To eliminate both primary and secondary collisions, a node cannot share a time slot with any neighbor that is within two hops away. Based on this, we can calculate the time slot reuse distance (RD) (in terms of number of nodes, or equivalently in terms of area) for the network topology given in Fig. 15.1. To reuse a time slot, two nodes must be three hops away. There are two types of neighbors that are three hops away. One is that the two nodes locate three hops away horizontally or vertically; the other is that the two nodes locate at the two opposite corners of a 1×2 rectangular area. The first case has a time slot RD 9 and the second case has a time slot RD $9/2 = 4.5$. Since each node can have the same number of three-hop neighbors for each case, the average RD is $(9 + 4.5)/2 = 6.75$. This means that at least 7 time slots are needed. This value is larger than the total number of time slots needed in ROT, which is 4 (see Fig. 15.1).

When traffic is not heavy, ROT can be combined with some existing medium access schemes such as carrier sense multiple-access with collision avoidance (CSMA-CA). For example, CSMA-CA is used for the first transmission of a packet and ROT is used for retransmissions only. In this way, CSMA-CA helps to reduce the first transmission delay that is otherwise incurred by ROT. And ROT helps to efficiently resolve collisions.

15.3 MEDIUM ACCESS SCHEDULING MIDWARE

15.3.1 Overview

TDMA schemes are in general free of HT problems. Nevertheless, when traffic load is light, TDMA results in unnecessary delay, which can be viewed as a special ET problem. The medium access scheduling midware (MASM) approach proposed here tries to eliminate HT problems and, at the same time, minimize the effect of ET problems. MASM sits on top of the MAC sublayer and works as a plug-and-play (PnP) midware (i.e., a shim sublayer). As a midware, it goes with different wireless networks and requires no or minimal modifications to existing network protocols such as MAC and routing protocols. It can be used for various purposes (e.g., data transmission, beacon scheduling, and sleeping scheduling). The following summarizes the main features of MASM:

1. Each node is assigned one or more time slots that are distinct from those of its neighbors within two hops, thus eliminating HT problems for both unicast and broadcast communications.
2. Time slot assignment is done in distributed fashion, and priority can be applied during this procedure (a node with a high priority can select slot(s) first and can require more time slots than other nodes).
3. Use multilevel scheduling (with or without using multiple channels) to minimize the effect of ET problems. Different levels can use different time slot durations. For example, multiple mini slots can be scheduled within one reserved common time slot of another level. Also different levels can use different frequency channels, if available. Multilevel approach also favors applications that need to support sleeping mode.
4. Without incurring additional delay, large time slot cycle (TSC) is used to
   a. Simplify time slot assignment.
   b. Handle unevenly distributed network topologies.
   c. Maintain fairness among nodes with the same priority.
   d. Facilitate the optimization of time slot assignment.
   e. Cope with dynamic activities (node joining, node leaving/failure, mobility, and sleeping mode).
5. The self-correcting ability enables a node to recover from the loss of synchronization due to clock drift or other problems. This feature also expedites link/node failure detection and time slot recycling as well as the handling of various dynamic activities mentioned above.

15.3.2 Basic Scheme

In the basic scheme, two-hop neighbor information, which includes time slot assignment data, is first exchanged among nodes. Then each node determines if it has the highest priority among all its neighbors that are within two hops (referred to as two-hop neighbors here) and that have not chosen a time slot. To do this, a node starts a timer if it finds it has the highest priority. If the node receives a message from any two-hop neighbor with a higher priority than itself before the timer expires, it stops the timer. When the timer expires, it selects the smallest time slot that has not been used by any its two-hop neighbor.[1]

Figure 15.2 shows two examples of time slot assignments using the basic scheme. In both examples, priority is simply determined according to node ID,

[1] An alternative way is that a node never stops a running timer; instead it checks if it still has the highest priority when the timer expires.

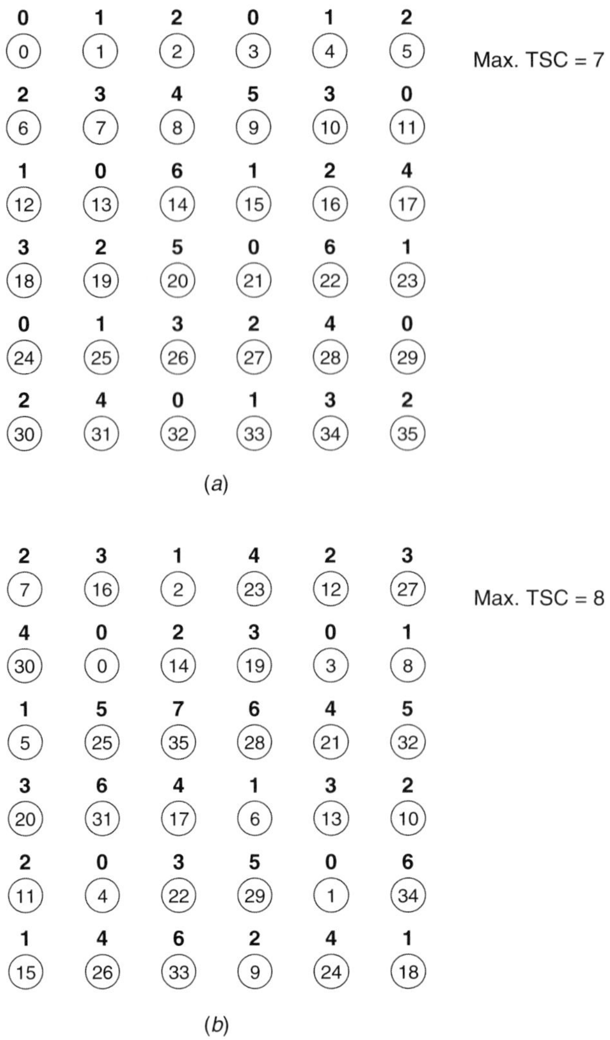

FIGURE 15.2 Examples of time slot assignment using the basic scheme: (*a*) node IDs distributed orderly; (*b*) node IDs distributed randomly.

and a smaller ID enjoys a higher priority. In Fig. 15.2*a*, node IDs are distributed orderly; and in Fig. 15.2*b*, node IDs are randomly distributed. The time slot assignment results, in terms of total number of time slots used (i.e., time slot cycle), are roughly the same for the two examples. Except node 35 in Fig. 15.2*b*, which has a time slot 7, all other nodes in Fig. 15.2*a* and 15.2*b* have a time slot between 0 and 6.

We know, from Section 15.2, the time slot reuse distance (RD) for the network topology given in Fig. 15.2 is 6.75. This means that at least 7 time slots are needed. In general, RD can be calculated as

$$\text{RD} = \sum_{i}^{k} p_i d_i \tag{15.1}$$

where k is the total number of types of three-hop neighbors a node can have, d_i is the time slot RD of type i, and p_i is the probability that a three-hop neighbor belongs to type i and it can be calculated using degree of connectivity information as follows:

$$p_i = \frac{c_i}{\sum_j^k c_j} \tag{15.2}$$

where c_i is the degree of connectivity of type i three-hop neighbors and c_j is the total degree of connectivity of all three-hop neighbors. Note that when nodes are unevenly distributed, RD varies with location. Here "unevenly distributed" means the degree of connectivity is not a constant through the whole network. A physically evenly distributed network is not a strict evenly distributed network regarding RD because the degree of connectivity around the network boundary is smaller than that in the middle of the network.

The TSC in Fig. 15.2*a* is 7, which is the number given by Eq. (15.1). However, 8 time slots are needed in Fig. 15.2*b*. We will show in Section 15.3.3 that the difference in TSC does not affect the performance of the network. Actually, for various reasons, we will intentionally use a TSC that is much larger than the value given by Eq. (15.1).

15.3.3 Advanced Scheme

There are some practical problems that are not addressed by the basic scheme presented above. In this section we study those problems and their solutions in detail. As a result, a more advanced scheme is proposed.

15.3.3.1 Time Slot Assignment and Time Slot Cycle. To determine when to transmit a packet, a node needs to know its time slot as well as the TSC. Obviously, all nodes should use the same TSC, even if they have different TSCs within the scope of two-hop neighbors. This is simply because a node normally is the two-hop neighbor of multiple nodes, and therefore a time slot (e.g., slot 3) will be calculated differently by different nodes if they use different TSCs (e.g., $n \times \text{TSC} + 3$ gives different values if different TSCs are used). From the above discussion, it is clear that each node needs to know the maximum TSC (maxTSC) that has been used in the network. However, the maxTSC depends on the node distribution, and normally it is difficult to know before the time slot assignment is done. For example, although both have the same physical topology, the maxTSC is different in Fig. 15.2*a* and 15.2*b*. One intuitive solution to let each node know the maxTSC is to exchange local TSC information after the time slot assignment is done. But this incurs additional overhead and latency. We will discuss this later and show that a better solution is available.

Another problem revealed by the fact that the maxTSC is different in Fig. 15.2*a* and 15.2*b* is that the maxTSC is not uniquely determined by the physical topology. The order by which nodes choose time slots makes a difference to the time slot assignment, including the maxTSC. From Eq. (15.1), we know the minimum slots needed for a grid topology is 6.75. Since a node has 12 two-hop neighbors for a grid topology, theoretically it may have a local TSC as large as 13 in the worst case. This means all nodes may need to use a TSC of 13 even if most of them have a local TSC of 7. One may argue that the worst case is unlikely to happen. Even so, the problem persists as it is likely that node density of the network is not uniform. If a small part of the network has a high node density, then the whole network has to use a large TSC. If the maxTSC used in the network is higher than a node's local TSC, then some slots are wasted in the basic scheme. For example, the maxTSC in Fig. 15.2*a* is 7, but node 0 has a local TSC of 4. So if only one slot is assigned to each node, some slots are wasted.

To fully utilize all available time slots, some nodes will be assigned multiple time slots. Figure 15.3 shows the final time slot assignment result for the topology given in Fig. 15.2*b*. After the first round of time slot assignment, the maxTSC is determined and announced to all nodes. Then the second round of time slot assignment begins. Different from the first round of time slot assignment in which each node needs to select the smallest possible slot, from second round on, a node may face three situations: It cannot find any unused slot; it finds some unused slot(s), but decides not to select any slot; it finds some unused slot(s) and selects one slot. In any case, the node needs to notify other nodes so that those nodes with lower priority can continue their time slot selection. Note that a node does not need to select the smallest available slot from the second round on. And, for fairness reason, a node should not try to occupy all the available slots (in above example, a node is not allowed to select more than one slot during each round). Multiple rounds may be needed before all slots are assigned.

In general, a node in a sparse area will be assigned more time slots than a node in a dense area. This will result in a smaller effective TSC for those nodes located in a sparse area. Apparently, a node's effective TSC closely reflects its local TSC. This is natural, as a node located in a sparse area should not be affected too much by another node located in a remote dense area. This result also means that which maxTSC is used is not critical in terms of time slot utilization. For example, if we triple the maxTSC from 3 to 9, then the original time slot assignment, for example, (node 0:slot 0/node 1:slot 1/node 2:slot 2) or in short (0:0/1:1/2:2), will become (0:0/1:1/2:2/0:3/1:4/2:5/0:6/1:7/2:8), which is

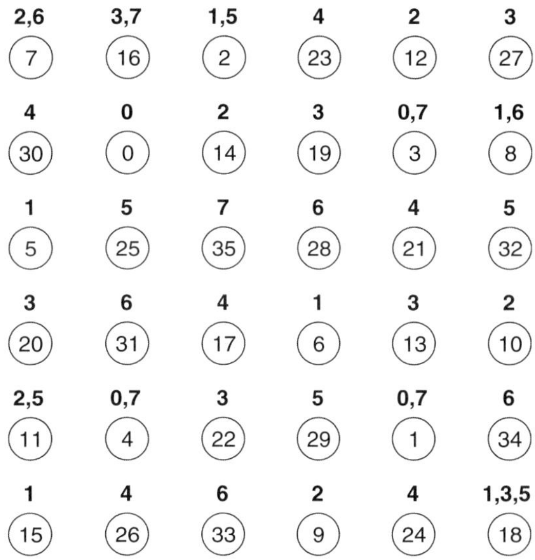

FIGURE 15.3 Full utilization of time slots.

virtually the same as before. Based on this fact, we propose to use a maxTSC that is much larger than otherwise needed. The motivation is that:

1. A predetermined maxTSC precludes the need for exchanging local TSC information to get the maxTSC after the first round of time slot assignments. As such, both overhead and delay can be reduced. When maxTSC is predetermined, a node does not need to select the smallest possible time slot in the first round of time slot assignment. It can select any available slot. This facilitates the optimization of time slot assignment (discussed later).
2. A large enough maxTSC guarantees there are enough time slots to be assigned even for a very irregular network topology (e.g., the node density is very high in some area). Yet this large maxTSC will not introduce additional delay as can be seen from the example of tripling the maxTSC.
3. A small maxTSC leads to unfairness in time slot assignment. This can be seen from Fig. 15.3. To understand this, let us compare the different results of 7 slots allocated to 6 nodes and 70 slots allocated to 6 nodes. Apparently, we can do much better in the second case in terms of fairness.
4. With a large maxTSC, dynamic activities such as node joining, node leaving/failure, mobility, and sleeping mode can be better handled. All those activities require the adjustment of time slot assignment. A large maxTSC not only makes such adjustment possible but also easier and smoother.

15.3.3.2 Acknowledgment Issue. When an immediate acknowledgment (ACK) is required for a successful data transmission, the scheme presented in Section 15.3.3.1 may fail. For example, in Fig. 15.4 nodes 17 and 21 both own time slot 4, and they are allowed to transmit data simultaneously. Since they are three hops away, data packets from these two nodes will not collide with each other. The corresponding ACK packets, although transmitted from two neighboring nodes, will not collide with each other either, as an ACK packet will not be able to reach the destination of another ACK packet. The only problem is the collision between a data packet and an ACK packet. In principle, the problem can be solved by making data transmission and ACK transmission nonoverlapping (not necessary in time). This is not a problem for a new MAC protocol. However, for an existing MAC protocol, it could be difficult to find a satisfactory solution for the problem without modifying the MAC protocol.

Most MAC schemes do not allow the ACK to be delayed, that is, the ACK should be sent within a certain short period after a data packet is received. This makes it impossible to transmit the data and its corresponding ACK in separate time slots. If not using different frequency channels or different spreading codes (as in code division multiple access) for the data transmission and the corresponding ACK transmission (Fig 15.5), then we may have to choose one of the following two solutions. One is to assign distinct time slots within three-hop neighbors instead of two-hop neighbors. The other is to separate a

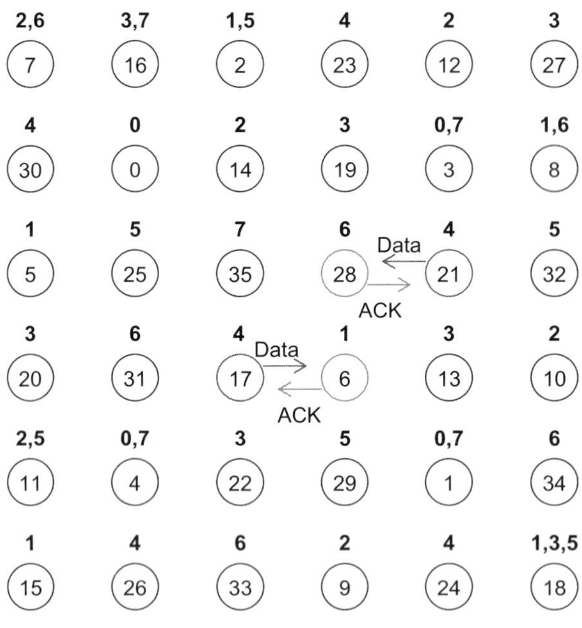

FIGURE 15.4 Acknowledgment issue.

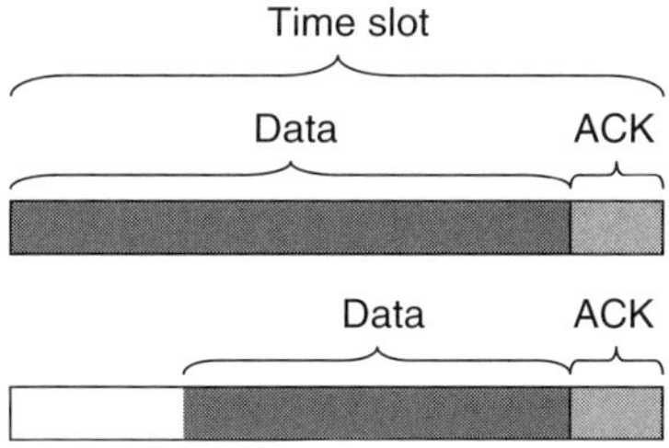

FIGURE 15.5 Nonoverlapping data transmission and ACK transmission within time slot.

time slot into two nonoverlapping parts, one for data transmission and the other for ACK transmission (Fig. 15.5). The first solution always works but may significantly reduce the bandwidth efficiency. The second solution is much better but may not work with all existing MAC protocols.

As shown in Fig. 15.5, the whole time slot is divided into two parts: one for the transmission of data, the other for the transmission of ACK. Since an ACK needs to be transmitted within a short period after a data packet is received, the transmission of the data should be scheduled in such a way that it ends at the boundary of the two parts used for transmitting data and ACK. One difficulty is that most MAC protocols employ some random backoff scheme for the purpose of medium access, which means an upper layer will not be able to accurately schedule the data transmission. Fortunately, some MAC protocols allow configuring the backoff parameters. For example, in IEEE 802.15.4 [1], an upper layer can set the minimum backoff exponent, macMinBE, in the MAC PAN information base (MPIB) to 0, thus leading to the first backoff equal to $(2^{\mathrm{macMinBE}}-1) \times \mathrm{aUnitBackoffPeriod} = 0$.

15.3.3.3 Multilevel Time Slot Assignment. One drawback of any TDMA-based scheme is that time slots are assigned regardless of the actual traffic need. As a consequence, some nodes may have more time slots than they need, while others are running out of time slots. One way to mitigate this is to use multilevel time slot assignment. Figure 15.6 illustrates a two-level time slot assignment example. Slot 0 of level 1 is reserved for all nodes. This reserved slot is further divided into several level 2 mini slots, which are assigned as usual. These mini slots can be used by nodes to borrow slots from one another. For example, a node can return its slot(s) to a slot pool (SP) if it does not have data to transmit during a certain period. The SP lists all the available time slots as well as their original owners and cycles. A node can borrow slots from the SP. The rule of borrowing slots is like that of slot assignment. That is, a node can borrow a slot listed in the SP if no two-hop neighbor is using a slot with the same number. How many level 1 slots should be reserved during each TSC depends on the specific application. This approach also favors applications that need to support sleeping mode. If a node finds out (e.g., using level 2 communications)

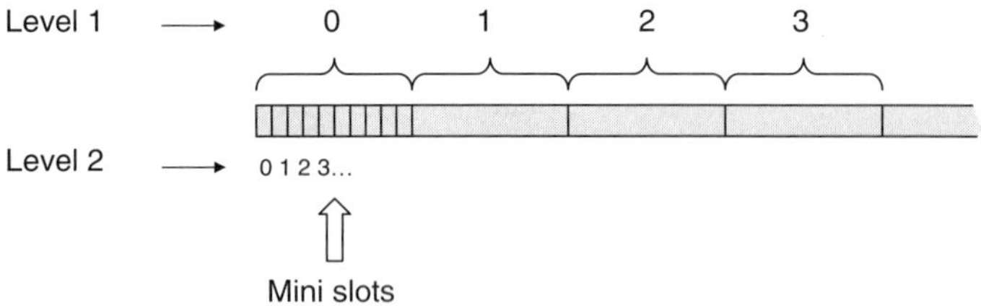

FIGURE 15.6 Multilevel time slot assignment.

that it has no data to transmit and there is no data to be transmitted from any neighbor to it for some duration in the future, it can go to sleep.

Instead of reserving slot(s) of one level for the use of another level, another approach is to use multiple channels, one for each level. In the above example, level 2 can use a channel that is different from that of level 1; and the bandwidth of the channel used by level 2 can be much smaller than that used by level 1.

15.3.3.4 Optimization of Time Slot Assignment. If time slots are allocated in such a way that each node along a traffic flow from the source to the destination owns a slot that is slightly larger than that owned by the previous hop (an example is shown in Fig. 15.7*a*), then the end-to-end packet delivery latency can be minimized. For bidirectional traffic, a node can choose some optimal slots for one direction and some others for the other direction if multiple slots can be chosen (an example is shown in Fig. 15.7*b*). If only one slot can be selected and the traffic rates of two directions are similar, then a distance of half TSC between a node's slot and the slot of its neighbor is reasonable. For instance, if the TSC is 8, then the slots of two neighboring nodes are better to have a distance of 4 slots.

In a mesh network where bidirectional traffic may flow between one node and any of its neighbors, it is difficult to find a slot that is optimal for all traffic flows. In this case, an average result should be used and a trade-off is often

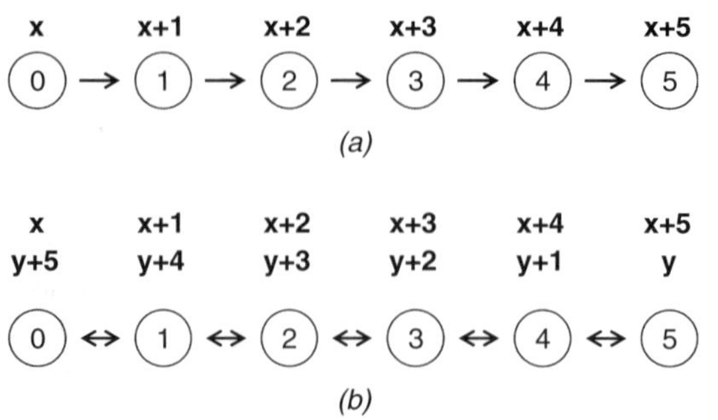

FIGURE 15.7 Optimization of time slot assignment.

needed. The situation of pure tree routing (e.g., cluster tree routing [8]) is a little better, as traffic can only go up or down along the tree. In this case, each node can select two time slots, one optimal for upward traffic and the other optimal for downward traffic. If each node can only select one slot, then the above half TSC distance rule can be applied. An alternative way is to use two channels, one for upward traffic and the other for downward traffic.

15.3.3.5 Synchronization and Self-Correcting/Adaptation. For synchronization purpose, one node's clock should be used as the "standard" clock. Practically, a powerful node such as the base station in a WiFi or the PAN coordinator in an LR-WMPAN is chosen for this purpose, but it is free to select any other node. The node with the "standard" clock first broadcasts a synchronization message in which its clock information is included. All neighbors are then synchronized with this node. Next, each neighbor tries to synchronize its neighbors by broadcasting another synchronization message, in which its synchronized clock information is included. This procedure continues until all nodes are synchronized. Due to the propagation delay and clock drift, some small GTD may be added to each time slot.

To handle clock drift problem, a resynchronization message is periodically broadcast from the node with the standard clock. This message spreads through the network and is rebroadcast during each relaying node's time slot. A node may not be able to receive and relay this broadcast message due to the loss of synchronization or other problems. To recover from loss of synchronization, each node keeps a timer, which expires after a certain number of resynchronization broadcast periods. Each time a resynchronization message is received, the timer is reset. When the timer expires, the node with a problem should stop any transmission until it receives a resynchronization message. To recover from loss of synchronization quickly, a node can also try to overhear transmissions of its neighbors and then synchronize with them. Normally, it takes much less time to overhear transmissions than wait for a resynchronization message.

The self-correcting scheme used for synchronization recovery can also be used to handle link/node failures. If a node finds that one of its neighbors has not been relaying resynchronization messages for certain duration, it concludes the link to the neighbor or the neighbor itself has failed, and it should update all its neighbors of the failure. If a failure is reported by more than one neighbor, the failure is likely to be node failure. Any node that detects a failure and/or receives a failure report will update its two-hop view and check if any time slot can be recycled.

The proposed scheme can also adapt to other dynamic activities such as node leaving, node joining, mobility, and sleeping mode. Node leaving can be handled similarly as node failure. For node joining, a joining node first asks its two-hop neighbors to send their slot information to it. Based on this, it can find out if any unused time slot is available (often true if the joining node is at the boundary of the network). If there is no unused time slot or the number of unused time slots is not enough, the joining node may request an adjustment or reassignment of time slots among its two-hop neighbors. If any node's slot assignment is changed, it should update all its two-hop neighbors so that any reclaimed time slot by the change can be reused. Note that each node has a different two-hop view. So a reclaimed time slot may be reassigned to one neighbor, but not to another. If two-hop neighbor information is updated promptly, then time slot assignment can be dynamically adjusted to reflect the effect of mobility. Another related application scenario that requires dynamic adjustment of time slots is the dense wireless sensor network where sensors need to go to sleep or may die due to energy depletion. For example, five times the number of sensors than needed at any time may be deployed in an area. Then only one fifth of the sensors are turned into normal state and all other sensors go to sleep. At the time the energy of the sensors in normal state is about to be depleted, another one fifth of the sensors are turned into normal state. This is a way to prolong the life of wireless sensor networks.

Note that multilevel time slot assignment (with or without using multiple channels) discussed above can be used to further facilitate the handling of all the dynamic activities addressed in this subsection.

15.4 CONCLUSIONS

Two medium access control scheduling schemes are presented in this chapter. The simple receiver-oriented TDMA (ROT) can be used to remove all primary collisions at a minimal addition of code size and control overhead. The medium access scheduling midware (MASM) is a contention-free TDMA scheme and is able to eliminate all collisions, that is, both primary and secondary collisions. Some special design features such as using large time slot cycle (TSC), multilevel slot assignment, optimal time slot assignment, and self-correcting ability have made MASM more robust and efficient than other TDMA schemes. As a midware, it also goes with different MAC and routing protocols.

REFERENCES

1. IEEE P802.15.4/D18, "Draft standard: Low rate wireless personal area networks," IEEE, New York, Feb. 2003.
2. ZigBee Specification, version 1.0, ZigBee Alliance, San Ramon, CA, June 2005.
3. J. Zheng and M. J. Lee, "Will IEEE 802.15.4 make ubiquitous networking a reality?: A discussion on a potential low power, low bit rate standard," *IEEE Commun. Mag.* 42(6), 140–146 (2004).
4. J. H. Schiller, *Mobile Communications*, Addison-Wesley, Reading, MA, 2000.
5. J. Zheng and M. J. Lee, "A comprehensive performance study of IEEE 802.15.4," *Sensor Network Operations*, Wiley Interscience, Hoboken, NJ, 2006, Chapter 4, pp. 218–237.

6. IEEE 802.11, "Part 11: Wireless LAN medium access control (MAC) and physical layer (PHY) specifications," IEEE, New York, Aug. 1999.
7. J. Zheng and M. J. Lee, "A resource-efficient and scalable wireless mesh routing protocol," *Special Issue of Elsevier Ad Hoc Networks Journal on Wireless Mesh Networks*, in press.
8. L. Hester, Y. Huang, A. Allen, O. Andric, and P. Chen, "neuRFon Netform: A self-organizing wireless sensor network," Paper presented at the 11th IEEE ICCCN Conference, Miami, FL, Oct. 2002.

CHAPTER 16

IMPACT OF RELIABLE AND SECURE SENSING ON CLUSTER LIFETIME IN IEEE 802.15.4 NETWORKS

JELENA MIŠIĆ

16.1 INTRODUCTION

In order to penetrate the market with cost-effective solutions for wireless sensor networks (WSNs) we need standardized low cost, low power, and short-range communication [low rate wireless personal area network (LR-WPAN)] technology. An important candidate for the application in this area is the IEEE 802.15.4 standard [1]. The 802.15.4 specification outlines some basic security services at the data link layer that can be combined with advanced techniques at the upper layers to implement a comprehensive security solution. For example, the recent ZigBee specification [2] implements a number of protocols—including security-related ones—that can be deployed in an 802.15.4 network. Given that the 802.15.4 devices are typically severely constrained in terms of their communication and computational resources, the implementation of such solutions is likely to impose a significant performance overhead. For cost effectiveness we assume that symmetric-key key establishment (SKKE) [2] is implemented over the IEEE 802.15.4 sensor cluster operating in beacon-enabled, slotted carrier sense multiple-access/collision avoidance (CSMA/CA) mode.

In this chapter we address the problem of multicluster sensor network as shown in Fig. 16.1 with integrated node sleep control and key exchange mechanism. The network is formed by three clusters interconnected in a master–slave regime wherein the coordinator of a lower cluster acts as the bridge to the upper one, and the coordinator of the topmost cluster acts as the network sink. In our previous work [3] we analyzed the impact of contention caused by the bridges and ordinary nodes on the cluster lifetimes. In this chapter we include the model of additional traffic caused by key exchanges in

Emerging Wireless LANs, Wireless PANs, and Wireless MANs. Edited by Y. Xiao and Y. Pan

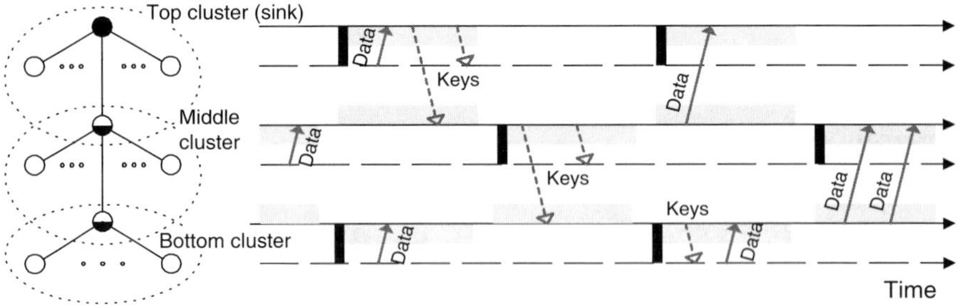

FIGURE 16.1 Network topology.

the network where each cluster has to deliver R packets per second toward the sink. All clusters are equipped with redundant sensors, which enables reduction of individual sensor duty cycle through activity management [4]. In other words, each node spends most of its time in sleep mode and wakes up only to transmit its packets. Since contention between the bridge and ordinary nodes cause nonuniform lifetimes among the clusters, we attempt to find cluster populations that will compensate bridges' activitites due to data delivery and excessive key exchanges.

Individual sensor nodes are battery operated, and their power consumption is modeled according to tmote_sky ultra-low power IEEE 802.15.4–compliant wireless sensor module [5] powered with two AA batteries. Since the coordinators/bridges have to work without ever going to sleep, their power budget is assumed to be infinite; the use of relaying nodes with larger power resources than ordinary sensing nodes has been shown to increase the useful network lifetime [6].

The chapter is organized as follows: Section 16.2 gives a brief overview of the operation of 802.15.4-compliant networks with star topology in the beacon-enabled, slotted CSMA/CA mode, followed by a review of power management techniques for 802.15.4 and basic security mechanisms provided for by the standard. As the 802.15.4 specification does not prescribe any particular key management approach, we will make use of the SKKE mechanism presented in Section 16.3. In Section 16.4 we briefly discuss bridge operation. Energy consumption for tmote_sky ultra-low power IEEE 802.15.4–compliant wireless sensor module [5], which we will consider in our modeling, is considered in Section 16.6. Section 16.5 presents derivation of the analytical model of the cluster, while Section 16.7 presents numerical results obtained from the analysis. Finally, Section 16.8 concludes the chapter.

16.2 802.15.4 BEACON-ENABLED MEDIUM ACCESS CONTROL (MAC) LAYER

The 802.15.4 networks with star topology are operated in a beacon-enabled mode where channel time is divided into superframes bounded by beacon transmissions from the personal area network (PAN) coordinator [1]. All communications in the cluster take place during the active portion of the superframe; the (optional) inactive portion may be used to switch to conserve power by switching devices to a low power mode. The standard supports 16 different frequency channels in which clusters can operate within the industrial/scientific/medical (ISM) band. Uplink channel access is regulated through the slotted CSMA/CA mechanism [1].

Data transfers in the downlink direction, from the coordinator to a node, must first be announced by the coordinator. In this case, the beacon frame will contain the list of nodes that have pending downlink packets, as shown in Fig. 16.2*b*. When the node learns there is a data packet to be received, it transmits a request. The coordinator acknowledges the successful reception of the request by transmitting an acknowledgment. After receiving the acknowledgment, the node listens for the actual data packet for the period of *aMaxFrameResponseTime*, during which the coordinator must send the data frame.

Power management consists of adjusting the frequency and ratio of active and inactive periods of sensor nodes [7, 8]. For 802.15.4 nodes it can be implemented in two ways. In the first one, supported by the standard [1], the interval between the two beacons is divided into active and inactive parts, and the sensors can switch to low power mode during the inactive period. Activity management for individual nodes can be accomplished through scheduling of their active and inactive periods. In order to avoid simultaneous activity and collisions by awakened nodes, sleep periods have to be randomized. In order to ensure fairness among the nodes, the coordinator has to periodically broadcast required event-sensing reliability (number of packets per second needed for reliable event detection) and the number of nodes that are alive.

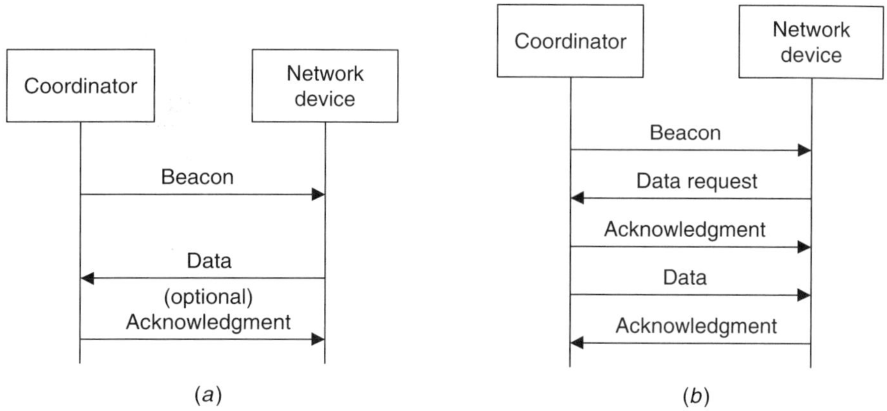

FIGURE 16.2 Data transfers in 802.15.4 PAN in beacon-enabled mode: (*a*) uplink transmission; (*b*) downlink transmission.

TABLE 16.1 Energy Consumption per Backoff Period for Various Modes of Transmitter/receiver Operation

Mode	Label	Energy Consumption per Backoff Period
Radio transmitting at 0 dBm	ω_t	15.8 μJ
Radio transmitting at −1 dBm	ω_t	15.0 μJ
Radio transmitting at −3 dBm	ω_t	13.8 μJ
Radio receiving	ω_r	17.9 μJ
Idle mode, oscillator off	ω_s	18.2 nJ

Based on that information, the node can calculate average period of sleep between transmissions. When the average sleep period is known, then some discrete random probability distribution can be used to generate individual sleep durations [9]. When the node wakes up and has a packet to transmit, it turns its receiver on in order to synchronize with the beacon. If the node's buffer is empty, it will start the new sleep. After receiving the information from the beacon, the node turns the transmitter on and starts backoff count in order to transmit the packet. After packet transmission, the node turns the receiver on in order to receive the acknowledgment. After the positive acknowledgment, the node starts the new sleep period. If the packet was not received correctly, the node has to repeat the transmission. Since the minimal beacon size is two backoff periods, we assume that an additional backoff period (10 bytes) is sufficient for transmitting information about the number of live nodes and requested event-sensing reliability. Let us denote power consumptions as ω_s, ω_r, and ω_t joules per one backoff period during sleep, receiving, and transmitting, respectively. They can be derived from typical operating conditions reported in documentation for ultra-low power IEEE 802.15.4 sensor module operating in ISM band between 2400 and 2483.5 MHz [5] and shown in Table 16.1. According to the specification of the tmote_sky module, two AA batteries are needed in order to supply voltage between 2.1 and 3.6 V.

The 802.15.4 standard specifies several security suites that consist of a "set of operations to perform on MAC frames that provide security services" [1]. Specified security services include access control lists, data encryption using prestored key, message integrity code generated using the prestored key, and message freshness protection. While these services are useful, they are by no means sufficient. In particular, procedures for key management, device authentication, and freshness protection are not specified by the 802.15.4 standard. Hence, they must be implemented on top of the 802.15.4 MAC layer.

16.3 SKKE PROTOCOL

A low cost alternative for this task with the possibility to change the symmetric keys between the nodes and the coordinator is the ZigBee protocol suite [2] developed by the ZigBee Alliance, an industry consortium working on developing a network and application programming interfaces (API) for wireless ad hoc and sensor networks. The ZigBee APIs include security extensions at different networking layers, using both symmetric and asymmetric key exchange protocols. Asymmetric key exchange protocols, which mainly rely on public key cryptography, are computationally intensive, and their application in wireless sensor networks is only possible with devices that are resource rich in computation and power and connected through high bandwidth links.

The application support sublayer of the ZigBee specification defines the mechanism by which a ZigBee device may derive a shared secret key (link key) with another ZigBee device; this mechanism is known as the symmetric-key key establishment (SKKE) protocol. Key establishment involves the coordinator and the node and should be prefaced by a trust-provisioning step in which trust information (a master key) provides a starting point for establishing a link key. The master key may be preinstalled during manufacturing, may be installed by a trust center, or may be based on user-entered data (PIN, password).

This protocol relies on keyed-hash message authentication code, or HMAC, which is a message authentication code (MAC) calculated using a cryptographic hash function in conjunction with a secret key. For the cryptographic hash function the 802.15.4 specification supports the advanced encryption standard (AES) block cipher in its basic form, while the ZigBee specification suggests the use of a modified AES algorithm with a block size of 128 bits [10]. The hash function of a data block d will be denoted as $H(d)$. The ZigBee specification suggests the use of the keyed HMAC:

$$\begin{aligned} MacTag &= HMAC(MacData) \\ &= H((MacKey \oplus opad)||H(MacKey \oplus ipad)||MacData) \end{aligned}$$

where *ipad* and *opad* are hexadecimal constants. In this chapter, we will follow the notation introduced in [2] and present the last equation in the equivalent form $MacTag = MAC_{MacKey}MacData$.

The SKKE protocol is initiated by the PAN coordinator (denoted as initiator device U) by exchanging ephemeral data (Fig. 16.3). The PAN coordinator U will generate the challenge QEU. Upon receiving the challenge QEU, the node (denoted as V) will validate it and also generate its own, different challenge QEV and send it to the PAN coordinator U.

Upon successful validation of challenges, both devices generate a shared secret based on the following steps:

1. Each device generates a $MACData$ value by concatenating their respective identifiers and validated challenges together: $MACData = U||V||QEU||QEV$.
2. Each device calculates the $MACTag$ (i.e., the keyed hash) for $MACData$ using the master key $Mkey$ as $MACTag = MAC_{Mkey}MACData$. Note

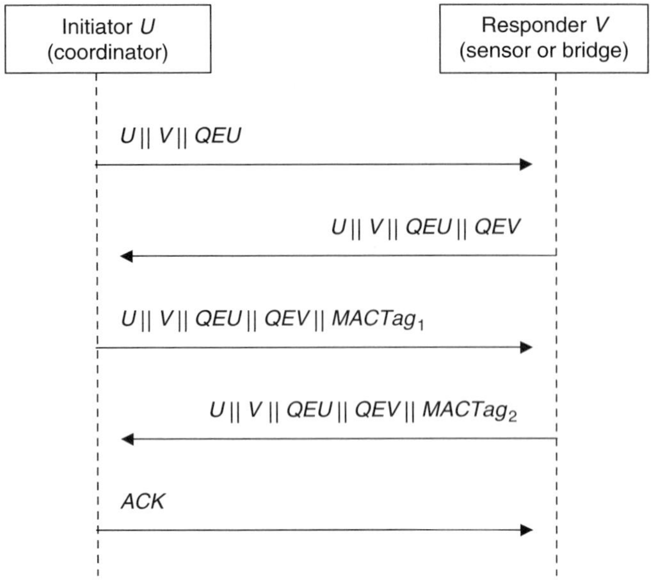

FIGURE 16.3 SKKE protocol between PAN coordinator and node.

that both devices should obtain the same shared secret $Z = MACTag$ at this time.

3. In order to derive the link key, each device generates two cryptographic hashes of the shared secret and hexadecimal numbers, that is, $Hash_1 = H(Z||01_{16}) Hash_2 = H(Z||02_{16})$. The $Hash_2$ will be the link key among two devices, while $Hash_1$ will be used to confirm that both parties have reached the same link key.

16.4 BRIDGING THE CLUSTERS

Consider the network shown in Fig. 16.1, operating in the ISM band at 2.4 GHz (other bands can be used but we don't consider them here). We assume that all clusters operate in beacon-enabled, slotted CSMA/CA mode under the control of their respective cluster (PAN) coordinators. In each cluster, the channel time is divided into superframes bounded by beacon transmissions from the coordinator [1]. All communications in the cluster take place during the active portion of the superframe, the duration of which is referred to as the superframe duration SD, as shown in Fig. 16.4.

The basic time unit of the MAC protocol is the duration of the so-called backoff period. Access to the channel can occur only at the boundary of the backoff period. The actual duration of the backoff period depends on the

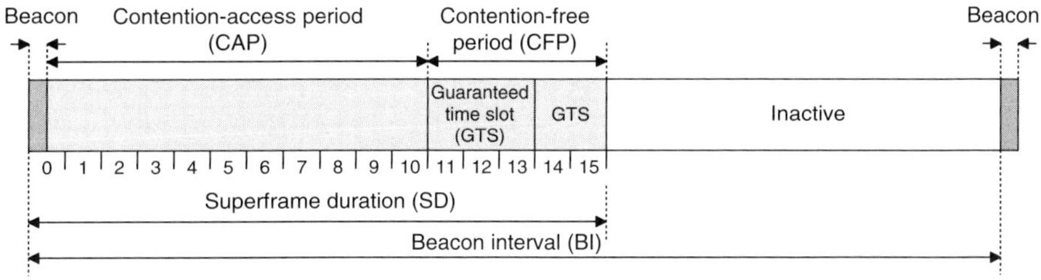

FIGURE 16.4 Composition of superframe under IEEE Std 802.15.4. (Adapted from IEEE 802.15.4-2006, "Wireless MAC and PHY specifications for low rate WPAN," IEEE, New York, 2006.)

frequency band in which the 802.15.4 wireless PAN (WPAN) is operating. Namely, the standard allows the PAN to use one of three frequency bands: 868–868.6, 902–928, or 2400–2483.5 MHz. In the two lower frequency bands, binary phase shift keying (BPSK) modulation is used, giving the data rate of 20 and 40 kbps, respectively. Each data bit represents one modulation symbol, which is further spread with the chipping sequence. In the third band, the offset quadrature phase shift keying (O-QPSK) modulation is used before spreading; in this case, four data bits comprise one modulation symbol, which is further spread with the 32-bit speading sequence. Table 16.2 summarizes the basic timing relationships in the MAC sublayer. Note that the constants and attributes of the MAC sublayer, as defined by the standard, are written in italics. Constants have a general prefix of *a*, for example, *aUnitBackoffPeriod*, while attributes have a general prefix of *mac*, for example, *macMinBE*.

TABLE 16.2 Basic Timing Relationships in MAC Sublayer

Type of Time Period	Duration	MAC Constant
Modulation symbol	1 data bit in 860- and 915-MHz bands, 4 data bits in 2.4-GHz band	N/A
Unit backoff period	20 symbols	*aUnitBackoffPeriod*
Basic superframe slot (SO = 0)	Three unit backoff periods (60 symbols)	*aBaseSlotDuration*
Basic superframe length (SO = 0)	16 basic superframe slots (960 symbols)	*aBaseSuperframeDuration* = *NumSuperframeSlots* × *aBaseSlotDuration*
(Extended) superframe duration, SD	*aBaseSuperframeDuration* $\times 2^{SO}$	*macSuperframeOrder*, *SO*
Beacon interval, BI	*aBaseSuperframeDuration* $\times 2^{BO}$	*macBeaconOrder*, BO

As shown in Table 16.2 the superframe is divided into 16 slots of equal size, each of which consists of 3×2^{SO} backoff periods. The variable SO, also known as *macSuperframeOrder*, determines the duration of the superframe; its default value of $SO = 0$ corresponds to the shortest active superframe duration of 48 backoff periods. In the ISM band, the duration of the backoff period is 0.32 ms for a payload of 10 bytes, which results in the maximum data rate of 250 kbps. The time interval between successive beacons is $BI = aBaseSuperframeDuration * 2^{BO}$, where $aBaseSuperframeDuration = 48$ backoff periods ($SO = 0$) and BO denotes the so-called *macBeaconOrder*, which can take values between 1 and 14. The duration of the inactive period of the superframe can easily be determined as $I = aBaseSuperframeDuration * (2^{BO} - 2^{SO})$. The default access mode in beacon-enabled operation is slotted CSMA/CA, with some slots optionally reserved for certain nodes.

During the inactive portion of the superframe, any device may enter a low power mode; the cluster coordinator can switch to the upper cluster in order to perform the bridging function—that is, deliver the data to the coordinator of the upper cluster. All *lower* cluster coordinators use this facility to perform the bridging function. As soon as the active part of the superframe is completed in the lower cluster, the coordinator/bridge switches to the *upper* cluster and waits for the beacon so that it can deliver the data from the lower cluster to the upper cluster coordinator/network sink.

All clusters use the CSMA/CA access, which means that the bridge has to compete for medium access with ordinary nodes in the upper cluster. As soon as the data is delivered, the bridge can return to its own cluster. This also means that, should the bridge be unable to transmit its data when the (active portion

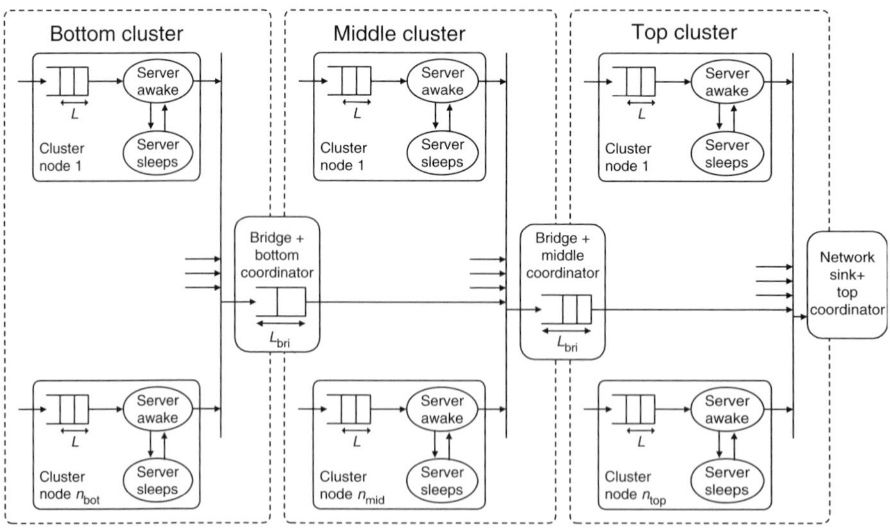

FIGURE 16.5 Queueing model of bridging process among three clusters.

of the) superframe in the upper cluster ends, it will freeze its backoff counter and leave the upper cluster. The backoff countdown will resume when the bridge returns to the upper cluster for the next superframe. Upon returning to the lower cluster, the bridge transmits the beacon, denoting the beginning of the next superframe, and the lower cluster continues to operate.

Bridge switching is schematically presented in Fig. 16.5. As can be seen, the three clusters have to operate with the same beacon interval, and the time between successive bridge visits to the "upper" cluster is therefore the same as the period between two beacons in its own, "lower" cluster. If the top and bottom clusters are far enough, that is, beyond the transmission range of each other, all three clusters may use the same radio frequency (RF) channel (the 802.15.4 standard uses 16 channels in the ISM band). Note that obtaining an increased area of coverage is the main reason for using a multicluster configuration. If the clusters are closer to each other, the top and bottom cluster may use different channels, and the middle cluster can use either of these.

16.5 ANALYTICAL MODEL FOR ORDINARY NODE IN CLUSTER WITH SKKE

In this section we will develop a Markov chain model for node behavior that includes all phases of the SKKE protocol and subsequent sleep and transmission phases. We assume that the PAN coordinator maintains a separate counter for the number of transmissions by each node. When counter value reaches threshold n_k, key update protocol is triggered. Updated keys are used to generate message authentication code. The high level Markov chain, which includes key update sleep periods followed by the transmissions, is presented in Fig. 16.6.

Furthermore, each of the steps that involves downlink transmission requires synchronization with the beacon, transmission of the uplink request packet, and transmission of the downlink packet as shown in Fig. 16.2*b*. Every transmission is implemented using slotted CSMA/CA specified by the standard [1]. Markov subchain for single CSMA/CA transmission (as the component of the Fig. 16.6) is shown in Fig. 16.7. The delay line from Fig. 16.7 models the requirement from the standard that every transmission that cannot be fully completed within the current superframe has to be delayed to the beginning of the next superframe and is shown in Fig. 16.8*a*. The probability that a packet will be delayed is denoted as $P_d = \overline{D_d}/SD$ where SD denotes duration of active superframe part (in backoff periods) and $\overline{D_d} = 2 + \overline{G_p} + 1 + \overline{G_a}$ denotes total packet transmission time including two clear channel assessments, transmission time $\overline{G_p}$, waiting time for the acknowledgment, and acknowledgment transmission time $\overline{G_a}$. The block labeled T_r denotes $\overline{D_d}$ linearly connected backoff periods needed for actual transmission.

Within the transmission subchain, the process $\{i,c,k,d\}$ defines the state of the device at backoff unit boundaries where $i \in (0..m)$ is the index of current

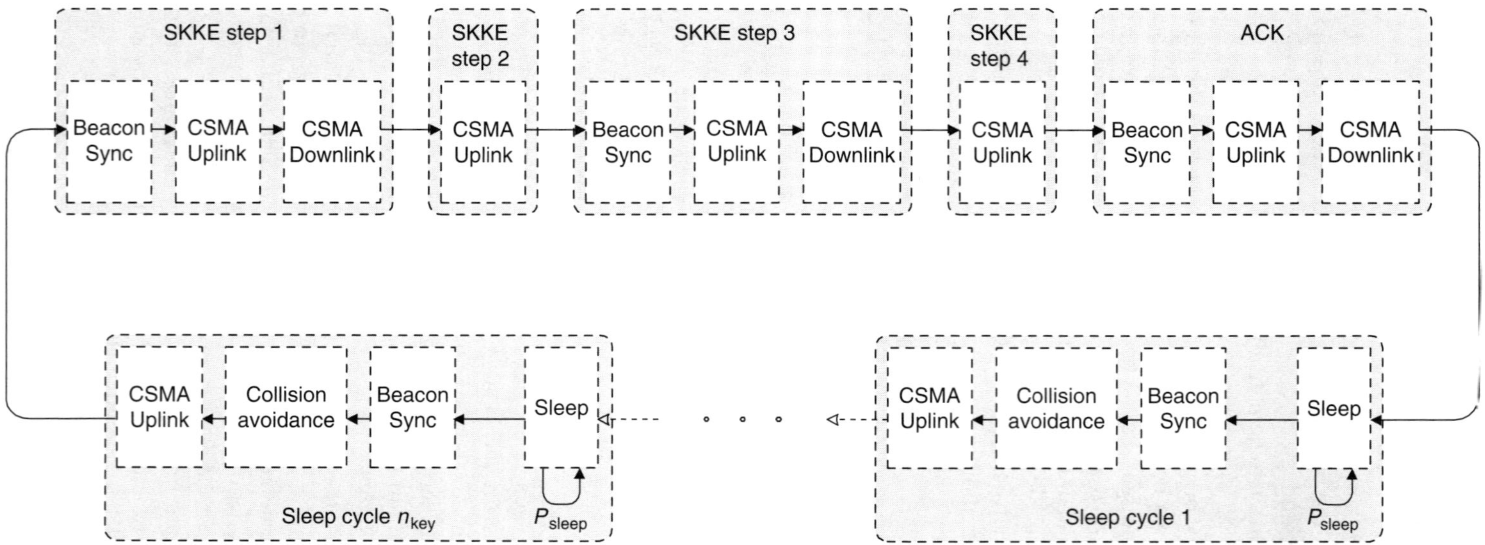

FIGURE 16.6 Markov chain for node behavior under threshold triggered key exchange.

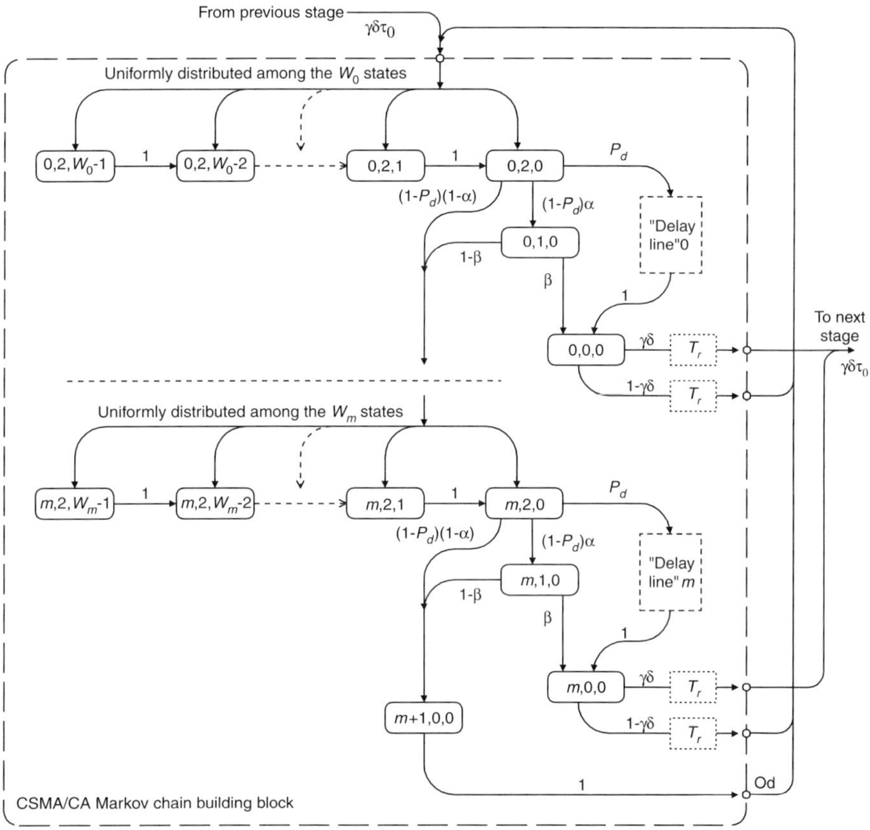

FIGURE 16.7 Markov subchain for single CSMA/CA transmission.

backoff attempt, where m is a constant defined by MAC with default value 4; and $c \in (0, 1, 2)$ is the index of the current clear channel assessment (CCA) phase. The standard prescribes two CCAs after the backoff countdown and if both are successful, trasnmission can start; and $k \in (0..W_i - 1)$ is the value of backoff counter, with W_i being the size of backoff window in ith backoff attempt. The minimum window size is $W_0 = 2^{macMinBE}$, while other values are equal to $W_i = W_0 2^{min(i,5-macMinBE)}$ (by default, $macMinBE = 3$); and $d \in (0..\overline{D}_d - 1)$ denotes the index of the state within the delay line mentioned above; in order to reduce notational complexity, it will be shown only within the delay line and omitted in other cases.

We need also to include synchronization time from the moment when the node wakes up till the next beacon, shown in Fig. 16.8*b*, as well as the uniformly distributed time needed to separate potential collisions among the nodes that wake up in the same superframe. One may argue that this separation time is not needed since CSMA/CA random backoff times will do the separation, but both request packets and data packets by awakened nodes will start backoff count

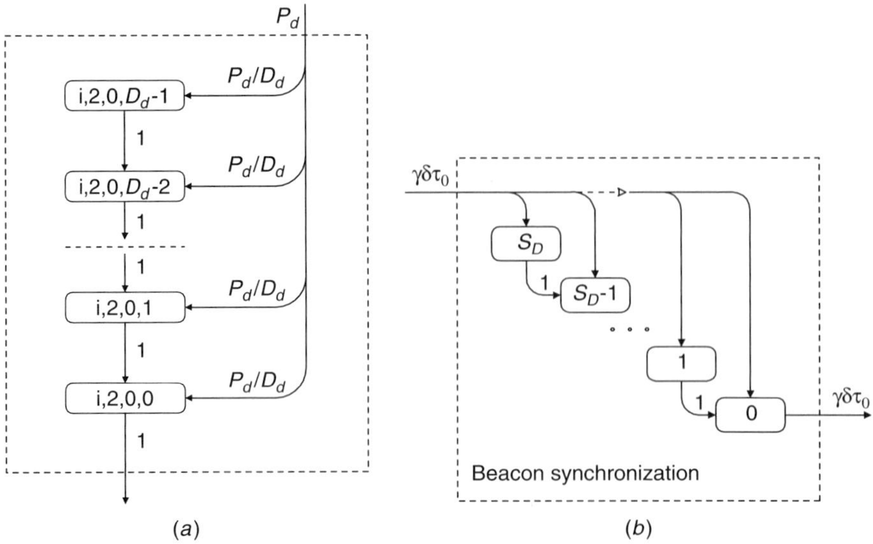

FIGURE 16.8 Delay and synchronization lines: (*a*) Markov subchain for delayed transmissions; (*b*) synchronization with beacon.

immediately after the beacon with backoff window, which has a range from 0 to 7 backoff periods. Due to the small backoff window, collisions will be likely, and we think that additional separation is needed. The collision separation line is similar to the beacon synchronization line except that the delay range is from 0 to $SD/2$ backoff periods. Synchronization with the beacon is also needed to receive the acknowledgment from the coordinator that whole SKKE transaction is completed. We assume that this acknowledgment is sent in downlink packet.

Data and key information packet sizes are assumed to be 12 backoff periods long, and therefore we assume that the probability to access the medium τ_0 as well as probabilities of transmission without the collision γ, and that the packet will not be corrupted δ have the stationary value after every transmission attempt. The same assumption holds for probabilities that the medium is idle on first and second CCA denoted with α and β, respectively.

Let us assume that the input probability to arbitrary transmission block is $\tau_0\gamma\delta$ where $\tau_0 = \sum_{i=0}^{m} x_{0,0,0}$ is the medium access probability after each packet transmission. We also assume that the medium access control layer together with the application layer will repeat transmission until thepacket is acknowledged. Therefore, the probability of finishing the first backoff phase in transmission block is equal to $x_{0,2,0} = \tau_0\gamma\delta + \tau_0(1 - \gamma\delta) = \tau_0$.

Using the transition probabilities indicated in Figs. 16.7 and 16.8*a*, we can derive the relationships between the state probabilities and solve the Markov chain. For brevity, we will omit *l* whenever it is zero and introduce the auxiliary variables C_1, C_2, C_3, and C_4:

$$\begin{aligned} x_{0,1,0} &= \tau_0(1 - P_d)\alpha = \tau_0 C_1 \\ x_{1,2,0} &= \tau_0(1 - P_d)(1 - \alpha\beta) = \tau_0 C_2 \\ x_{0,0,0} &= \tau_0[(1 - P_d)\alpha\beta + P_d] = \tau_0 C_3 \\ C_4 &= \frac{1 - C_2^{m+1}}{1 - C_2} \end{aligned} \tag{16.1}$$

Using values C_i we obtain the sum of probabilities for one transmission subchain as:

$$\begin{aligned} s_t &= \tau_0 C_4 \left[C_3(\overline{D_d} - 2) + C_1 + \frac{P_d(\overline{D_d} - 1)}{2} \right] \\ &+ \tau_0 \left[\sum_{i=0}^{m} \frac{C_2^i(W_i + 1)}{2} + C_2^{m+1} \right] \end{aligned} \tag{16.2}$$

The sum of probabilities within the beacon synchronization line is equal to $s_b = \tau_0\gamma\delta\sum_{i=0}^{SD}(i/SD) = \tau_0\gamma\delta(SD+1)/2$, and the sum of probabilities for the collision avoidance line is equal to $s_c = \tau_0\gamma\delta SD/4 + \frac{1}{2}$.

In order to model the node's sleep time, we will assume that sleep time is geometrically distributed with parameter P_{sleep}. Then the sum of probabilities of being in single sleep is equal to $s_{s1} = \tau_0\gamma\delta/((1 - P_{\text{sleep}})$. However, if the node wakes up and finds its buffer empty, it will start the new sleep. We will denote the probability of finding an empty buffer after sleep as Q_c and derive it later. The sum of probabilities of being in consecutive sleep then becomes $s_s = \tau_0\gamma\delta/[(1 - P_{\text{sleep}})(1 - Q_c)]$. if we denote the threshold value of the number of packets sent using the same key as n_k, then the normalization condition for the whole Markov chain becomes

$$3(s_b + 2s_t) + 2s_t + n_k(s_s + s_t + s_b + s_c) = 1 \tag{16.3}$$

However, the total access probability by the node is equal to the sum of access probabilities in each transaction, that is,

$$\tau = (8 + n_k)\tau_0 \tag{16.4}$$

16.5.1 Analysis of Node's Packet Queue

In order to find probability Q_c we need to consider the node's MAC layer as the $M/G/1/K$ queuing model with vacations and setup time. We assume that when the node wakes up it will transmit only one packet and go to sleep again, which is known as 1-limited scheduling [11]. A detailed model for the more general sleep policy with Bernoulli scheduling of activity period is derived in [9]. In Bernoulli scheduling, after one packet transmission, the node decides to transmit another packet with probability P_{ber} and goes to sleep with probability $1-P_{\text{ber}}$. We can apply this approach to our model using the restriction that $P_{\text{ber}}=0$. In the discussion that follows, packets are arriving to each node following the Poisson process with the rate λ. All nodes have buffers of finite capacity, L packets for an ordinary sensor node and L_{bri} packets for the two bridge/coordinators.

Consider the probability generating function (PGF) for one geometrically distributed sleep period (with parameter P_{sleep}) as

$$V(z) = \sum_{k=1}^{\infty} (1 - P_{\text{sleep}}) P_{\text{sleep}}^{k-1} z^k = \frac{(1 - P_{\text{sleep}})z}{1 - zP_{\text{sleep}}} \tag{16.5}$$

and the mean duration of the vacation is $\overline{V} = V'(1) = 1/(1 - P_{\text{sleep}})$. We also note [11] that the PGF for the number of packet arrivals to the sensor buffer during the sleep time is equal to

$$F(z) = V^*(\lambda - z\lambda) \tag{16.6}$$

where $V^*(\ )$ denotes the Laplace–Stieltjes transform (LST) of the sleep time, which (since sleep time is a discrete random variable) can be obtained by substituting the variable z with e^{-s} in the expression for $V(z)$.

A node returning from sleep (i.e., with nonempty buffer) has to synchronize with the next beacon; the synchronization time is uniformly distributed between 0 and BI$-$1 backoff periods (where BI is beacon interval), and its PGF is

$$S_1(z) = \frac{1 - z^{\text{BI}}}{\text{BI}(1 - z)} \tag{16.7}$$

When the awakened node finds the next beacon, then it has to wait for collision separation time before it starts its backoff procedure. We adopt that this time is uniformly distributed between 0 and 7 backoff periods and its PGF has the value

$$S_2(z) = \frac{1 - z^{\text{BI}/2}}{8(1 - z)} \tag{16.8}$$

The total idle time when the node is awakened then has the PGF

$$S_t(z) = S_1(z)S_2(z) \tag{16.9}$$

Its LST will be denoted as $D^*(s)$, the corresponding probability distribution function $D(x)$, and the probability density function as $d(x)$. The PGF for packet service time will be denoted as $T_t(z)$ and its probability density function will be denoted as $dt_t(x)$.

Let us now analyze the operation of the system, starting from Markov points, which include moments of packet departure and moments when the server wakes up (i.e., ends its vacation). Let $V^*(s)$ denote the LST of the vacation time, with the corresponding probability distribution function $V(x)$ and the probability density function $v(x)$.

The PGFs for the number of packet arrivals to the node's buffer during the total idle time, and packet service time, respectively, are

$$\begin{aligned} D(z) &= \sum_{k=0}^{\infty} s_k z^k = \int_0^{\infty} e^{-x\lambda(1-z)} d(x) = S_t^*(\lambda - z\lambda) \\ A(z) &= \sum_{k=0}^{\infty} a_k z^k = \int_0^{\infty} e^{-x\lambda(1-z)}\, dt_t(x) = T_t^*(\lambda - z\lambda) \end{aligned} \tag{16.10}$$

Then, the probabilities of k packet arrivals to the node's buffer during the synchronization time, packet service time, and sleep time, denoted with d_k, a_k, and f_k, respectively, can be obtained as

$$s_k = \frac{1}{k!} \frac{d^k S(z)}{dz^k}\bigg|_{z=0} \qquad a_k = \frac{1}{k!} \frac{d^k A(z)}{dz^k}\bigg|_{z=0} \qquad f_k = \frac{1}{k!} \frac{d^k F(z)}{dz^k}\bigg|_{z=0}$$

Let π_k and q_k denote the steady-state probabilities that there are k packets in the device buffer immediately upon a packet departure and after returning from vacation, respectively. Then, the steady-state equations for state transitions are

$$\begin{aligned} q_0 &= (q_0 + \pi_0) f_0 \\ q_k &= (q_0 + \pi_0) f_k + \sum_{j=1}^{k} \pi_j f_{k-j} \quad \text{for } 1 \le k \le L-1 \\ q_L &= (q_0 + \pi_0) \sum_{k=L}^{\infty} f_k + \sum_{j=1}^{L-1} \pi_j \sum_{k=L-j}^{\infty} f_k \\ \pi_k &= \sum_{j=1}^{k+1} q_j \sum_{l=0}^{k-j+1} (s_l + a_{k-j+1-l}) \quad \text{for } 0 \le k \le L-2 \\ \pi_{L-1} &= \sum_{j=1}^{L} q_j \sum_{k=L-j}^{\infty} \sum_{l=0}^{k} (s_l + a_{k-l}) \\ 1 &= \sum_{k=0}^{L} q_k + \sum_{k=0}^{L-1} \pi_k \end{aligned} \tag{16.11}$$

The probability distribution of the device queue length at the time of packet departure $\pi_i, i = 0 \cdots L-1$ and return from the sleep $q_i, i = 0 \cdots L$ can be found by solving the system of linear equations (16.11). In this manner, we obtain the probability that the Markov point corresponds to a return from the vacation and the queue is empty at that moment:

$$Q_c = \frac{q_0}{\sum_{i=0}^{L} q_i} \tag{16.12}$$

The probability distribution for the total inactive time of the node has a geometric distribution with the parameter Q_c, applied at the moments when the node returns fom sleep. The corresponding moment-generating function is

$$I^*(s) = \sum_{k=1}^{\infty} (1 - Q_c) Q_c^{k-1} V^*(s)^k = \frac{(1 - Q_c) V^*(s)}{1 - V^*(s) Q_c} \tag{16.13}$$

and the mean value is $\overline{I} = 1/[(1 - Q_c)(1 - P_{\text{sleep}})]$.

Given that there are n nodes in the cluster, the total event sensing reliability is equal to

$$R = \frac{n_k \gamma \delta \tau_0}{t_{\text{boff}}} \tag{16.14}$$

where $t_{\text{boff}} = 0.32$ ms corresponds to the duration of one backoff period. The value R has to be set by the sensing application, for example, $R = 10$. Satisfying Eq. (16.14) will result in minimal energy consumption. However, we have to note that key exchange overhead will result in an overhead packet rate of $8\tau_0\delta\gamma/t_{\text{boff}}$ packets per second.

16.5.2 Success Probabilities

As we mentioned earlier, we denoted the probabilities that the medium is idle on first and second CCA with α and β, respectively, and the probability that the transmission is successful with γ. Note that the first CCA may fail because a packet transmission from another node is in progress; this particular backoff period may be at any position with respect to that packet. The second CCA, however, will fail only if some other node has *just* started its transmission—that is, the backoff period in which the second CCA is undertaken must be the *first* backoff period of that packet. Note that the first medium access by any node will happen within the first 16 backoff periods of the superframe.

Let the clusters contain n_{bot}, n_{mid}, and n_{top} ordinary sensor nodes, respectively, with the packet arrival rate of λ per node. (References to specific clusters will use the subscripts bot, mid, and top, respectively.) The top cluster coordinator acts as the network sink.

16.5.2.1 Bottom Cluster. We apply the model from Section 16.5 and use expression (16.4) for τ_{bot}. Since τ_{bot} is very small and the number of nodes is large, we may estimate the per-cluster arrival rate of medium access events as

$$\lambda_{c,\text{bot}} = \frac{1}{16}(n_{\text{bot}} - 1)\tau_{\text{bot}}^{(1)} SD \tag{16.15}$$

The probability that the medium is not busy at the first CCA may, then, be approximated with

$$\alpha_{\text{bot}} = \frac{1}{16}\sum_{i=0}^{15} e^{-i\lambda_{c,\text{bot}}} \tag{16.16}$$

The probability that the medium is idle on the second CCA for a given node is, in fact, equal to the probability that neither one of the remaining $n_{\text{bot}}-1$ nodes has started a transmission in that backoff period,

$$\beta_{\text{bot}} = e^{-\lambda_{c,\text{bot}}} \tag{16.17}$$

By the same token, the overall probability of success of a transmission attempt is

$$\gamma_{\text{bot}} = (\beta_{\text{bot}})^{\overline{D_d}} \tag{16.18}$$

16.5.2.2 Middle Cluster. In the middle cluster, besides ordinary nodes, we must account for the presence of the bridge, that is, the coordinator from the bottom cluster. For an ordinary node, we apply the model from Section 16.5 to the environment of middle cluster and use expression (16.4) for τ_{mid}.

The access probability for the bridge coming from the bottom cluster can be modeled as

$$\tau_{\text{bri,mid}} = \frac{1}{16} n_{\text{bot}} \tau_{\text{bot}} SD \tag{16.19}$$

The success probability for bridge transmissions depends on all the nodes in the middle cluster, that is,

$$\gamma_{\text{bri,mid}} = (1 - \tau_{\text{mid}})^{\overline{D_{dn}}\,\text{mid}} \tag{16.20}$$

The medium access event rate for a middle cluster node must also account for both the ordinary nodes and the bridge, hence:

$$\lambda_{c,\text{mid}} = \frac{1}{16}(n_{\text{mid}} - 1)\tau_{\text{mid}} SD + \tau_{\text{bri,mid}} \tag{16.21}$$

Parameters α, β, and γ can, then, be calculated in a similar way as their bottom cluster counterparts, that is,

$$\alpha_{\text{mid}} = \frac{1}{16} \sum_{i=0}^{15} e^{-i\lambda_{c,\text{mid}}} \tag{16.22}$$

$$\beta_{\text{mid}} = e^{-\lambda_{c,\text{mid}}} \tag{16.23}$$

$$\gamma_{\text{mid}} = e^{-\lambda_{c,\text{mid}} \overline{D_d}} \tag{16.24}$$

16.5.3 Sink Cluster

Success probabilities α_{top}, β_{top}, and γ_{top} for the top cluster can be found starting from

$$\tau^{(1)}_{\text{bri,top}} = \frac{1}{16} (n_{\text{bot}} \tau_{\text{bot}} + n_{\text{mid}} \tau_{\text{mid}}) SD \tag{16.25}$$

16.6 MODEL OF ENERGY CONSUMPTION

While the activity management achieves the extension of the lifetime separately for each cluster, individual cluster lifetimes may differ. If this is the case, the network lifetime is determined by the shortest cluster lifetime; it is maximized if all clusters die at approximately the same time. In order to accomplish that, we have looked into the possibility of modifying cluster parameters so as to equalize their respective lifetimes.

The algorithm to calculate node population considers one cluster at a time in an iterative fashion, starting with the cluster that is farthest away from the sink.

As mentioned above, we assume that all transmissions are acknowledged; if the acknowledgment (ACK) packet is not received within the time prescribed by the standard [1], the transmission will be repeated. Let the PGF of the time interval between the data and subsequent ACK packet be $t_{\text{ack}}(z) = z^2$; actually its value is between *aTurnaroundTime* and *aTurnaroundTime* + *aUnitBackoff-Period* [1], but we round the exponent to the next higher integer for simplicity.

According to the standard [1], transmission has to be preceded with the backoff procedure and two CCAs during which the radio part is in the receiving mode. Only after successful CCAs, radio module switches to the transmitting mode. The standard allows m (default value is $m = 5$) backoff attempts during which backoff windows take values of $W_0 = 7$, $W_1 = 15$, $W_2 = W_3 = W_4 = 31$ (if the battery saving mode is not turned on). However, under the sleep management regime, all transmissions will complete in one or two backoff attempts, and battery saving mode is not important. The PGF for the duration of jth backoff time prior to transmission is equal to

$$B_j(z) = \sum_{k=0}^{W_j-1} \frac{1}{W_j} z^k = \frac{z^{W_i} - 1}{W_j(z-1)} \tag{16.26}$$

In order to find energy consumption during the jth backoff attempt, we need to switch to the LST by substitution $z = e^{-s\omega_r}$ (because PGFs don't allow noninteger exponents) and obtain LST:

$$E^*_{B_j}(s) = \frac{e^{-s\omega_r W_i} - 1}{W_j(e^{-s\omega_r} - 1)} \tag{16.27}$$

Let the PGF of the data packet length be $G_p(z) = z^k$, and let $G_a(z) = z$ stand for the PGF of the ACK packet duration. Then the PGF for the total transmission time of the data packet will be denoted with $D_d(z) = z^2 G_p(z) t_{\text{ack}}(z) G_a(z)$; its mean value is $\overline{D_d} = 2 + G'_p(1) + t'_{\text{ack}}(1) + G'_a(1)$. The LST for the energy consumption during pure packet transmission time is $e^{-sk\omega_t}$. The LST for energy consumption during two CCAs is equal to $e^{-s2\omega_r}$. The LST for energy consumption during waiting for and receiving the acknowledgment is $e^{-s3\omega_r}$. The same value has the LST for energy consumption during reception of the beacon frame, which is three backoff periods long.

Then, the PGF for the time needed for one complete transmission attempt including backoffs becomes

$$\mathscr{A}(z) = \frac{\sum_{i=0}^{m} \left[\prod_{j=0}^{i} B_j(z)\right] (1-\alpha\beta)^i z^{2(i+1)} [\alpha\beta G_p(z) t_{\text{ack}}(z) G_a(z)]}{\sum_{i=0}^{m} (1-\alpha\beta)^i \alpha\beta} \tag{16.28}$$

The LST for energy consumption for one transmission attempt then becomes

$$\mathscr{E}^*_{\mathscr{A}}(s) = \frac{\sum_{i=0}^{m} \left[\prod_{j=0}^{i} E^*_{B_j}(z)\right] (1-\alpha\beta)^i e^{-s2\omega_r(i+1)} \left(\alpha\beta e^{-sk\omega_t} e^{-s3\omega_r}\right)}{\sum_{i=0}^{m} (1-\alpha\beta)^i \alpha\beta} \tag{16.29}$$

By taking packet collisions into account, the probability distribution of the packet service time follows the geometric distribution, and its PGF becomes

$$T(z) - \sum_{k=0}^{\infty} [\mathscr{A}(z)(1-\gamma)]^k \mathscr{A}(z)\gamma = \frac{\gamma\mathscr{A}(z)}{1 - \mathscr{A}(z) + \gamma\mathscr{A}(z)} \tag{16.30}$$

In this case, mean packet service time can simply be written as $\overline{T} = T'(1) = \mathscr{A}'(1)/\gamma$.

The LST for the energy spent on a packet service time is then equal to

$$E_T^*(s) = \frac{\gamma \mathscr{E}_A^*(s)}{1 - \mathscr{E}_A^*(s) + \gamma \mathscr{E}_A^*(s)} \tag{16.31}$$

16.6.1 Bottom Cluster

The PGF of the time needed to conduct one transmission attempt is then obtained by substituting α_{bot}, $\beta_{\text{bot}}^{(2)}$, and γ_{bot} in Eq. (16.30). The LST for the energy spent in packet service is obtained by substituting those values in Eq. (16.31). The average value of energy consumed for packet service is obtained as

$$\overline{E_{T,\text{bot}}} = -\frac{d}{ds} E_{T,\text{bot}}^*(s)|_{s=0}$$

The average battery energy consumption per backoff period can be found as

$$u_{\text{bot}} = \frac{\overline{S_1}\omega_r + \overline{S_2}\omega_r + 3\omega_r + \overline{I_{\text{bot}}}\omega_s + \overline{E_{T,\text{bot}}}(1 + 8/n_k)}{\overline{S_1} + \overline{S_2} + 3 + \overline{I_{\text{bot}}} + \overline{T_{\text{bot}}}(1 + 8/n_k)} \tag{16.32}$$

Given the battery budget of b joules, the average number of transmission/sleep cycles in the bottom cluster can be found as

$$n_{c,\text{bot}} = \left\lceil \frac{b}{\overline{S_1}\omega_r + 3\omega_r + \overline{S_2}\omega_r + \overline{E_{T,bot}}(1 + \frac{8}{n_k}) + \overline{I_{\text{bot}}}\omega_s} \right\rceil \tag{16.33}$$

Given the law of large numbers [12], the PGF for total lifetime of the node in bottom cluster becomes

$$L_{\text{bot}}(z) = [S_1(z)S_2(z)T_{\text{bot}}(z)I_{\text{bot}}(z)]^{n_{c,\text{bot}}} \tag{16.34}$$

By differentiating the respective PGFs, we can obtain the standard deviation of the node lifetime as well as the coefficient of skewness, μ, which measures the deviation of a distribution from symmetry [13].

16.6.2 Middle Cluster

By using the appropriate values of α_{mid}, β_{mid}, and γ_{mid} the PGFs for a single transmission attempt and for the overall packet transmission time can be calculated as $\mathscr{A}_{\text{mid}}(z)$ and $T_{\text{mid}}(z)$, respectively. Both PGFs depend on the number of nodes n_{mid} as the parameter. Average battery energy consumption per backoff period is calculated as

$$u_{\text{mid}} = \frac{\overline{S_1}\omega_r + \overline{S_2}\omega_r + 3\omega_r + \overline{E_{T,\text{mid}}}(1 + 8/n_k) + \overline{I_{\text{mid}}}\omega_s}{\overline{S_1} + \overline{S_2} + 3 + \overline{T_{\text{mid}}}(1 + 8/n_k) + \overline{I_{\text{mid}}}} \tag{16.35}$$

Now, if the lifetime of the middle cluster is to be the same as that of the bottom cluster, the average energy that the node consumes per backoff period in both clusters should have equal values:

$$u_{\text{mid}} = u_{\text{bot}} \tag{16.36}$$

from which we can obtain the initial population of the middle cluster n_{mid}. This equation is necessary only if we want to choose n_{mid} in order to equalize the lifetimes of the bottom and middle clusters. Otherwise, n_{mid} can be chosen using some other policy.

Given the battery budget of b backoff periods, the average number of transmission/sleep cycles in bottom cluster can be found as

$$n_{c,\text{mid}} = \left\lceil \frac{b}{\overline{S_1}\omega_r + 3\omega_r + \overline{S_2}\omega_r + \overline{E_{T,\text{mid}}}(1 + 8/n_k) + \overline{I_{\text{mid}}}\omega_s} \right\rceil \tag{16.37}$$

The PGF for total lifetime of the node in the bottom cluster becomes

$$L_{\text{mid}}(z) = [S_1(z)S_2(z)T_{\text{mid}}(z)I_{\text{mid}}(z)]^{n_{c,\text{mid}}} \tag{16.38}$$

The procedure is then repeated for the top cluster, starting from

$$\tau_{\text{bri,top}}^{(1)} = \frac{1}{16}(n_{\text{bot}}\tau_{\text{bot}} + n_{\text{mid}}\tau_{\text{mid}})SD \tag{16.39}$$

This algorithm is scalable since the overall model can be broken in individual cluster models with input from all clusters at lower level. The condition for the correctness for this approximation is that all clusters are not operating in the saturation condition.

16.7 PERFORMANCE EVALUATION

In this section we present numerical results obtained by solving the system of equations that represent the analytical model of the node's MAC with sleep and key exchange, node's queue behavior, and medium behavior. As a solution we obtain system parameters τ_0, τ, P_{sleep}, α, β, γ, and Q_c. We have varied the key exchange threshold between 20 and 100 packets, while the requested event sensing reliability per cluster was kept at $R = 10$ packets per second. Source cluster size was varied between 20 and 100 nodes. We assumed that each node is powered with two AA batteries that supply a voltage between 2.1 and 3.6 V and 500 mA-h as required by tmote_sky [5] operating conditions with total energy $b = 10,260$ J.

We have also assumed that the network operates in the ISM band at 2.45 GHz, with a raw data rate of 250 kbps. The superframe size was controlled with SO, $BO = 0$. The packet size has been fixed at $\overline{G_p} = 12$ backoff periods, while the device buffer had a fixed size of $L = 2$ packets. The packet size includes message authentication code and all physical layer and medium access control protocol sublayer headers and is expressed as the multiple of the backoff period [1]. We also assume that the physical layer header has 6 bytes, and that the medium access control sublayer header and frame check sequence fields have a total of 9 bytes. Other parameters from the medium access control layer were kept at default values. In Fig. 16.9 we present access probabilities in bottom, middle, and top cluster, respectively, and in Fig. 16.10 we present transmission success probabilities.

Figure 16.11 shows a number of nodes in middle and sink cluster with population in source cluster indicated as n_{bot} and period of key exchange indicated as n_k. We notice significant increase of populations as we move

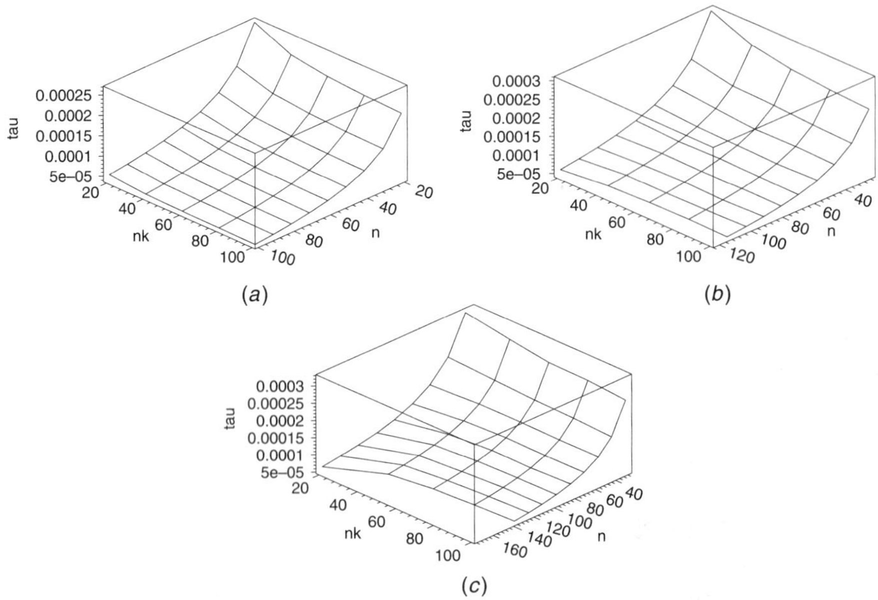

FIGURE 16.9 Access probabilites for node: (*a*) in source cluster, (*b*) in middle cluster, and (*c*) in sink cluster.

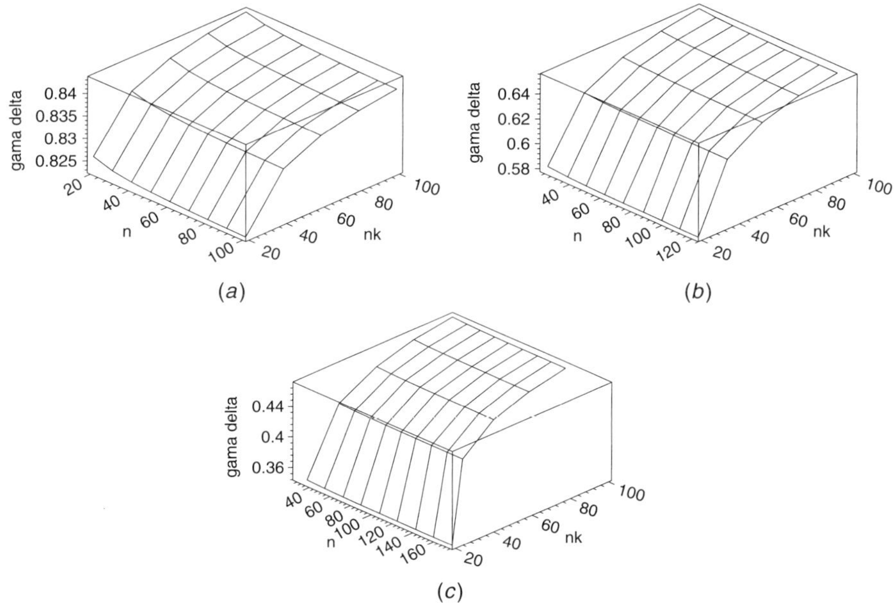

FIGURE 16.10 Medium behavior. Success probability for node: (*a*) in source cluster, (*b*) in middle cluster, and (*c*) in sink cluster.

toward the sink. The task of increase of population in the cluster is mainly to compensate for the drop of transmission success probability and CCA success probabilities, which is caused by the bridges' data and key exchange traffic. Figure 16.12 shows equalized lifetimes of clusters versus recalculated populations and key exchange period.

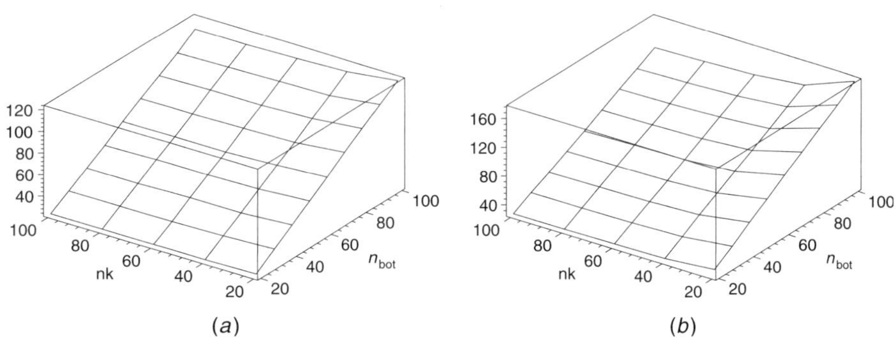

FIGURE 16.11 Node populations: (*a*) in middle cluster; (*b*) in sink cluster.

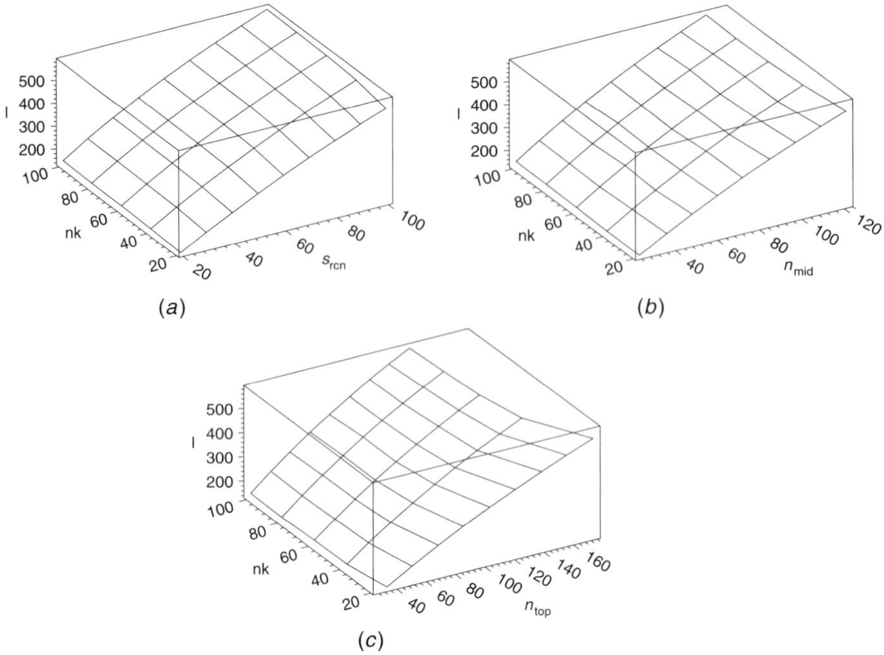

FIGURE 16.12 Cluster lifetimes: (*a*) in source cluster, (*b*) in middle cluster, and (*c*) in sink cluster.

16.8 CONCLUSION AND FUTURE WORK

We have developed an analytical model of the key exchange integrated into the sensing function of the beacon-enabled 802.15.4 cluster. Our results show an important impact of the ratio of the event-sensing reliability and key update threshold on the cluster's energy consumption. We have evaluated the impact of the threshold for key update on the cluster's descriptors. In our future work we plan to model more complex key exchange algorithms.

REFERENCES

1. IEEE 802.15.4-2006 "Wireless MAC and PHY specifications for low rate WPAN," revision of IEEE 802.15.4-2003, IEEE, New York, 2006.
2. ZigBee Alliance, ZigBee Specification, Document 053474r06, Version 1.0, ZigBee Alliance, San Ramon, CA, 2004.
3. J. Mišić, C. J. Fung, and V. B. Mišić, "On node population in a multi-level 802.15.4 sensor network," paper presented at Globecom 2006, San Francico, CA, 2006.
4. J. Mišić, S. Shafi, and V. B. Mišić, "Cross-layer activity management in a 802.15.4 sensor network," *IEEE Commun. Mag.* 44, 131–136 (2006).

5. "tmote sky lowpower wireless sensor module," tmote datasheet 802.15.4, Moteiv, San Francisco, CA, available: www.moteiv.com, 2006.
6. M. Yarvis, N. Kushalnagar, H. Singh, A. Rangarajan, Y. Liu, and S. Singh, "Exploiting heterogeneity in sensor networks," paper presented at INFOCOM05, Vol. 2, Miami, FL, 2005, pp. 878–890.
7. I. Stojmenović (Ed.), *Handbook of Sensor Networks: Algorithms and Architectures*, Wiley, Hoboken, NJ, 2005.
8. Y. Sankarasubramaniam, Ö. B. Akan, and I. F. Akyildiz, "ESRT: Event-to-sink reliable transport in wireless sensor networks," in *Proc. ACM MobiHoc'03*, June 1–3, 2003, Annapolis, Maryland, USA, pp. 177–188.
9. J. Mišić, S. Shafi, and V. B. Mišić, "Maintaining reliability through activity management in an 802.15.4 sensor cluster," *IEEE Trans. Vehic. Technol.* 55, 779–788 (2006).
10. A. Menezes, P. van Oorschot, and S. Vanstone, *Handbook of Applied Cryptography*, CRC Press, Boca Raton, FL, 1997.
11. H. Takagi, *Queueing Analysis*, Vol. 1: *Vacation and Priority Systems*, North-Holland, Amsterdam, The Netherlands, 1991.
12. G. R. Grimmett, and D. R. Stirzaker, *Probability and Random Processes*, 2nd ed., Oxford University Press, Oxford, 1992.
13. P. Z. Pebbles, Jr., Probability, Random Variables, and Random Signal Principles, McGraw-Hill, New York, 1993.

CHAPTER 17

IEEE 802.15.5: RECOMMENDED PRACTICE FOR WPAN MESH NETWORK (LOW DATA RATE)

CHUNHUI ZHU and MYUNG J. LEE

17.1 INTRODUCTION

IEEE 802.15.5 is a recommended practice that provides an architectural framework to allow low rate wireless personal area network (WPAN) devices to form interoperable, stable, and scaleable wireless mesh topologies. Mesh network topologies allow nodes with IEEE 805.15.4 (both 2003 [1] and 2006 [2] versions) compatible medium access control/physical (MAC/PHY) layers to extend the network coverage without increasing the transmit power or the receiver sensitivity. Another key advantage of the mesh network is the enhanced reliability via route redundancy.

Because of the differences in the corresponding MACs and PHYs of IEEE 802.15.3 and 805.15.4, different approaches were used to support the mesh capabilities in high data rate (802.15.3-based) and low data rate (802.15.4-based) WPAN networks. In this chapter, we only discuss the low data rate portion of the WPAN mesh. In addition, due to the limit of the space, among all the supported features of LR-WPAN mesh standard, the addressing scheme, unicast and multicast routing algorithms will be introduced. The power saving mechanism, portability support and trace route functions will be skipped.

Please note when this chapter is being written, the standard is not in its final form. Readers are suggested to refer to the published standard, expected to be released in 2009, for the latest developments and specifications.

17.1.1 Application of WPAN Mesh

The applications of low rate WPAN mesh networks are innumerous, mainly in the area of wireless sensor networks. Typical applications include home and

Emerging Wireless LANs, Wireless PANs, and Wireless MANs. Edited by Y. Xiao and Y. Pan

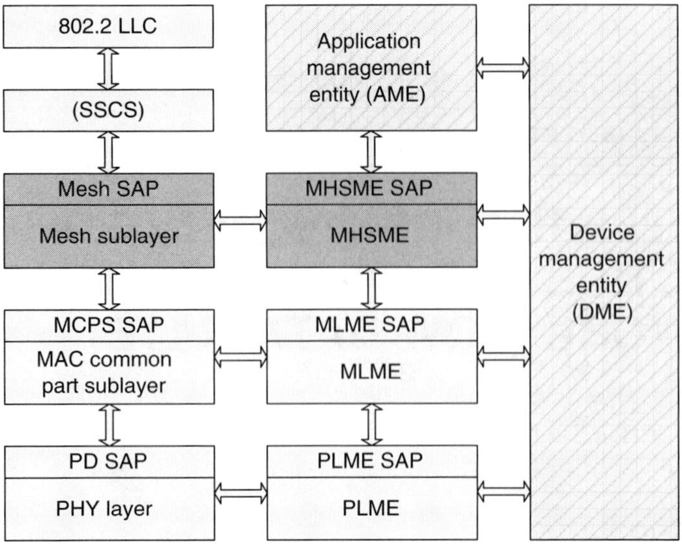

FIGURE 17.1 Reference model used in IEEE 802.15.5.

industrial automation and control, security and environmental monitoring, situational awareness and asset tracking, automatic meter reading, and personal health monitoring.

17.1.2 Architecture of WPAN Mesh

The reference model of the low rate WPAN mesh is illustrated in Fig. 17.1.

In the data plane, the WPAN mesh sublayer resides between the service-specific convergence sublayer (SSCS) of IEEE 802.2 logical link control (LLC) and the IEEE 802.15.4 MAC sublayer. The mesh sublayer provides services to the next higher layer via the mesh service access point (mesh SAP). For implementers to add the mesh function to existing low rate WPAN applications with the least effort, the mesh SAP is made very similar to the MAC common part sublayer SAP (MCPS SAP).

In the management plane, the mesh sublayer management entity (MHSME) resides between the application management entity (AME) and the MAC layer management entity (MLME). The MHSME also interfaces with the mesh sublayer at the same level. As a reference, the device management entity (DME) that has access and control to all layers is also shown in the figure. However,

TABLE 17.1 Mesh SAP Primitives

Name	Request	Confirm	Indication
MESH-DATA	✓	✓	✓
MESH-PURGE	✓	✓	

TABLE 17.2 MHSME SAP Primitives

Name	Request	Indication	Response	Confirm
MHSME-DISCOVER-MESH	✓			✓
MHSME-START-MESH-NETWORK	✓			✓
MHSME-START-MESH-DEVICE	✓			✓
MHSME-JOIN	✓	✓		✓
MHSME-LEAVE	✓	✓		✓
MHSME-RESET	✓			✓
MHSME-GET	✓			✓
MHSME-SET	✓			✓

the specifications of both the AME and the DME are out of the scope of this recommended practice.

17.2 MESH SUBLAYER SERVICE

The mesh sublayer provides two types of services, the mesh data service and the mesh management service, to the next higher layer via two corresponding service access points, the mesh SAP and the MHSME SAP. There is also an internal interface between the mesh sublayer and the MHSME allowing the MHSME to utilize the mesh sublayer data service. On the other hand, the mesh sublayer may also need to get information from the MHSME through this internal interface in order to compose mesh sublayer data frames.

17.2.1 Mesh Data Service

The function of the mesh data service is to support the transport of application protocol data units (APDUs) between peer application entities residing at different nodes which can be multiple hops away from each other. As described in this section, the mesh data service primitives are very similar to the MCPS SAP primitives of the MAC sublayer. The implementers should pay careful attention to the differences between them. Table 17.1 lists the primitives supported by the mesh SAP.

17.2.2 Mesh Management Service

The MHSME SAP allows the transport of management commands between the AME and the MHSME. Table 17.2 summarizes the primitives supported

TABLE 17.3 General Mesh Service Frame Format

Octets: 2	2	8/2	8/2	Variable
Frame Control (FC)	PAN ID	Destination address (DA)	Source address (SA)	Mesh layer payload
Mesh layer header				

TABLE 17.4 Frame Control Field

Bits: 4 bits	1 bit	1 bit	2 bits	8 bits
Protocol version	Frame type	Address mode	Reserved	Transmission options

by the MHSME through the MHSME SAP interface. The primitives are discussed in the sections referenced in the table.

17.3 MESH SUBLAYER FRAME FORMATS

17.3.1 General Mesh Service Frame Format

The general mesh service frame is composed of the following fields: frame control, personal area network (PAN) ID, destination address, source address, and frame payload. The fields appear in a fixed order and the frame should be formatted as illustrated in Table 17.3. Some subfields might be reserved for future use, and they should be set to 0 for all implementations based on this specification version.

17.3.1.1 Frame Control Field. The frame control field consists of five subfields, as illustrated in Table 17.4.

17.3.1.1.1 Protocol Version. The protocol version subfield is 4 bits in length and should be set to 0001 for all implementations based on this specification version.

17.3.1.1.2 Frame Type. Two types of frames are supported, as illustrated in Table 17.5.

17.3.1.1.3 Address Mode. The address mode flag indicates whether the 64-bit extended addresses or the 16-bit short addresses are used in the destination and source address fields. The flag is set to 0 when 64-bit extended addresses are used and to 1 when 16-bit short addresses are used.

17.3.1.1.4 Transmission Options. The transmission options subfield is 8 bits in length and indicates the way the frame should be transmitted (see Table 17.6). To fill in this field at the data frame originator, the TxOptions parameter of the MESH-DATA.request primitive should be referred to. This information will be transmitted with the data payload toward the destination so that all intermediate

TABLE 17.5 Frame Type

Frame Type Value	Frame Type Name
00	Data frame
01	Command frame

TABLE 17.6 Transmission Options

Transmission Options Value	Transmission Options Name
0 × 01	Acknowledged transmission
0 × 02	Guaranteed time slot transmission
0 × 04	Indirect transmission
0 × 08	Security-enabled transmission
0 × 10	Multicast
0 × 20	Broadcast
0 × 40	Reliable broadcast
0 × 80	Reserved

nodes know how to transmit the data frame at both the mesh layer and the MAC layer.

17.3.1.2 Personal Area Network ID. The PAN ID subfield is 16 bits in length. In this version of the specification, only one PAN is supported in one mesh network. Therefore the destination and the source of the data frame should belong to the same PAN. A received frame should be discarded without any process if its PAN ID does not match the PAN ID of the receiving node.

17.3.1.3 Destination Address. The destination address can be a 64-bit extended address or a 16-bit short address, indicated by the address mode subfield of the frame control field.

17.3.1.4 Source Address. The source address can be a 64-bit extended address or a 16-bit short address, indicated by the address mode subfield of the frame control field.

17.3.1.5 Mesh Layer Payload. The mesh layer payload field may contain different subfields and have various lengths depending on the frame type field defined in Table 17.5. The details are described in Section 17.3.2.

17.3.2 Format of Individual Frame Types

17.3.2.1 Data Frame. The data frame contains three subfields in the mesh layer payload, sequence number field, routing control, and data payload (see Table 17.7). The details of the routing control field are shown in Table 17.8.

TABLE 17.7 Data Frame Format

Octets: 2	2	8/2	8/2	1	1	Variable
Frame control	PAN ID	Destination address	Source address	Sequence number	Routing control	Data payload
Mesh layer header				Mesh layer payload		

TABLE 17.8 Routing Control Field

Bits: 7	6–2	1–0
Up–down flag	Hops2Nb	Reserved

The up–down flag is Boolean expression and indicates whether this data frame should be forward up or down the tree. The hops2Nb value is the number of hops from the relaying node to neighbor_found given by the pseudocode in Section 17.4.3 (see Fig. 17.4). The last two bits are reserved for this version.

17.3.2.2 Command Frame. The general command frame format is illustrated in Table 17.9.

The details of the command frame subtype field are illustrated in Table 17.10.

17.4 MESH FUNCTIONS

This chapter describes the functions the mesh layer provided to the next higher layer. These functions include unicast and multicast addressing, unicast routing algorithm, and multicast routing algorithm.

17.4.1 Address Assignment

By binding logical addresses to the network topology, routing can be carried out without going through route discovery. Address assignment is broken down into two stages: association and address assigning.

17.4.1.1 Association. During the association stage, beginning from the root, nodes gradually join the network and a tree is formed. But this tree is not a logical tree yet, since no node has been assigned an address. There is no mesh-level limitation on the number of children a node can have. A node can determine by itself how many nodes (therefore, how many branches) it will accept according to its capability and other factors. Note that in this stage the network is not functional, i.e., no data can be transferred from one node to another.

TABLE 17.9 Command Frame Format

Octets: 2	Variable	1	Variable
Frame control	Routing fields	Command frame sub-type	Command payload
Mesh layer header	Mesh layer payload		

TABLE 17.10 Command Frame Subtypes

Values $B_4b_3b_2b_1b_0$	Command Frame Subtype
00000	Children number report
00001	Address assignment
00010	Hello
00011	Neighbor information request
00100	Neighbor information reply
00101	Link state request
00110	Link state reply
00111	Link state mismatch
01000	Probe
01001–01111	Reserved
10000	Join request
10001	Join reply
10010	Leave request
10011	Leave reply
10100	GC update
10101	Group dismiss
10001–11111	Reserved

17.4.1.2 Address Assigning. After a branch reaches its bottom, that is, there are no more nodes waiting to join the network (a suitable timer can be used for this purpose), a bottom-up procedure is used to calculate the number of nodes along each branch, as shown in Fig. 17.2. The numbers in square brackets indicate the numbers of nodes within branches below a certain node.

To report the number of children, each node proceeds as follows. Whenever a node joins the network, it should start a timer. The duration of the timer is defined by the mesh information base (MeshIB) attribute *meshChildNbReportTime*. A node becomes a leaf node if no other nodes join it before the timer expires. A leaf node should immediately send a children number report frame to its parent, setting the number of the children field to 1 and the number of the requested addresses field to a value equal to 1 or larger than 1 if it wants to reserve some addresses for possible future use. When a nonleaf node receives a children number report frame, it should record the number of children and the number of requested addresses for that branch and then check whether each of its children has reported the number of children. If each of its children has reported the number of children, it should report the number of children to its parent. The number of children field should be set to the number of nodes along its branch, including itself. The number of requested addresses field should be set to the sum of the numbers of requested addresses received from all children plus 1 or some value larger than 1 if the node also wants to reserve some addresses.

Any node can update the number of children and/or the number of requested addresses by sending another children number report frame to its parent if it has not been assigned an address block. For example, if a leaf node

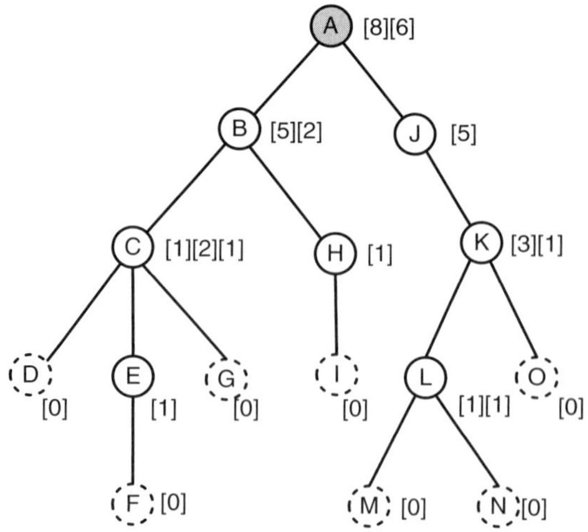

FIGURE 17.2 Calculation of number of nodes along each branch.

becomes a nonleaf node, it should immediately send another children number report frame to update its parent. When a nonleaf node receives a children number report frame and if it has already reported the number of children, it should immediately update its parent. If the nonleaf node has not reported the number of children to its parent, the received children number report frame should be handled normally. If a node receives a children number report frame and it has already been assigned an address block, it should not update its parent by sending another children number report frame. Instead, it should adjust the address assignment as described later.

After the root receives the information from all the branches, it should begin to assign addresses. During the address-assigning stage, a top-down procedure is used. First, the root checks if the total number of nodes in the network is less than the total number of addresses available. If not, address assignment fails. Next, the root assigns a block of consecutive addresses to each branch below it, taking into account the number of children and the number of requested addresses. The address block assigned to each branch is specified by the beginning address field and the ending address field given in the address assignment frame sent to each branch. The actual number of addresses assigned could be less than or more than the number of requested addresses, but no less than the number of children, depending on the availability of addresses. This procedure continues to the bottom of the tree. After address assigning, a logical tree is formed and each node has populated a neighbor table for tracking branches below it. For example, node C in Fig. 17.2 can have a neighbor table as follows (note the begAddr value of each entry represents the one-hop child node's address) (see Table 17.11).

TABLE 17.11 Example of Address Assignment

Beginning Address	Ending Address	Tree Level	Relationship	Number of Hops
6 (node D)	8	3	Child	1
9 (node E)	13	3	Child	1
14 (node G)	14	3	Child	1
...				

The above neighbor indicates that node C has a total of three branches; node D owns address block [6–8] with its own address equal to 6; node E owns address block [9–13] with its own address equal to 9; and node G only owns one address, 14, which is its own address.

17.4.1.3 Adjustment of Address Assignment. More nodes (therefore, more branches) are still allowed to be added at any level of the tree after address assignment if additional addresses (reserved during address assignment) are available. Address assignment can be locally adjusted within a branch if a node runs out of addresses. For instance, a node can request more addresses from its parent. If the parent does not have enough addresses, it can try to either request additional addresses from its parent or adjust address assignment among its children. If there is a substantial change of the node number or network topology, which cannot be handled locally, the network is allowed to go through the address-assigning procedure again. In practice, address assignment is controlled by the application profile. For example, the application profile used for light control can specify that more addresses should be reserved for some special nodes such as those near hallways. This improves the utilization of addresses and reduces the probability of address reassignment as the network evolves. The intelligence of distributing limited addresses among all network nodes is out of the scope of this chapter.

17.4.2 Mesh Topology Discovery and Formation

After a node has been assigned an address block, it should broadcast several hello messages to its neighbors, with the time to live (TTL) field of each hello frame set to meshTTLOfHello. By exchanging hello messages, each node will build a link state table (LST) for all its neighbors within meshTTLOfHello hops. Each neighbor's address block is logged in the LST so that the whole branch below the neighbor is routable. Figure 17.3 shows a two-hop link state view of node J. Note that nodes D, E, F, M, and N are not within two hops of node J, but they are still directly routable since they are the children of those nodes within two hops of node J.

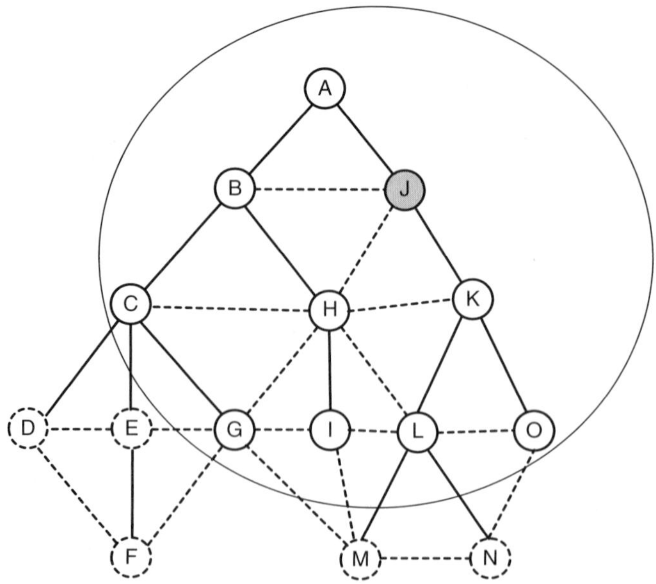

FIGURE 17.3 Example—two-hop link state (view from node J).

17.4.2.1 Link State Generation. The LST of a node, which consists of a meshTTLOfHello hop neighbor list and a connectivity matrix, is updated upon the reception of each hello message.

17.4.2.1.1 Neighbor List. Each node should update its neighbor list, as illustrated in Table 17.12, upon the reception of each hello message. Not only the source of the hello message should be added to the neighbor list, but also the one-hop neighbor of the source should be added to the neighbor list unless the TTL in the incoming hello message is 1. For those entries corresponding to the one-hop neighbors of the source, the endAddr and tree_level fields can not be populated from the incoming hello message and should be temporarily marked

TABLE 17.12 Neighbor List

Beginning Address	Ending Address	Tree Level	Link Quality	Relationship	Number of Hops
begAddr1	endAddr1	tree_level1	LQ1	Parent/child/ sibling	hops1
begAddr2	endAddr2	tree_level2	LQ2	Parent/child/ sibling	hops2
begAddrn	endAddrn	tree_leveln	LQn	Parent/child/ sibling	hopsn

as "unknown". The unknown endAddr and tree_level fields will be replaced with actual values when a hello message is received from the corresponding neighbor. If no hello message is received from some neighbors during the whole hello message exchange procedure, a node can solicit for endAddr and tree_level information by broadcasting a neighbor information request frame to its one-hop neighbors, including all the neighbors whose endAddr and tree_level fields are missing. Each one-hop neighbor that received the message should reply by sending back a neighbor information reply frame if it can provide the endAddr and tree_level information of one or more neighbors included in the neighbor information request frame. The relationship field indicates the relationship between this node and a specific neighbor. The valid relationships include parent, child, and sibling. The field hops in the neighbor list should be calculated according to the connectivity matrix described in next section.

17.4.2.1.2 Connectivity Matrix. From the one-hop neighbor information included in each incoming hello message (except when meshTTLOfHello equals 1), a node can construct a connectivity matrix for neighbors recorded in the neighbor list. Table 17.13 illustrates one example.

The "number of hops" of each node in the neighbor list can be calculated using the connectivity matrix. First the number-of-hops field of each node is set to infinity. Then, all nodes directly connected to the current node (marked as "me" in Table 17.13) are one-hop neighbors (nb_2, nb_{n-1}, ... in the example). Next, all nodes directly connected to one-hop neighbors (and having a hop

TABLE 17.13 An Example of Connectivity Matrix

	me	nb1	nb2	nb3	...	nb$n$−2	nb$n$−1	nb$n$
me	−	−	+	−	...	−	+	−
nb_1		−	+	−	...	+	−	−
nb_2			−	+	...	−	−	−
nb_3				−	...	+	−	−
...					...	...	...	...
nb_{n-2}						−	−	+
nb_{n-1}							−	−
nb$n$								−

Note: (1) The plus or minus sign ("+" or "−") at the cross cell of two nodes indicates they are or are not directly connected (i.e., they are or are not one-hop neighbors).
(2) For bidirectional links, the matrix is symmetric, so only half of the matrix is needed as shown here.
(3) Hop information can be calculated using the connectivity matrix. Here we have:
1-hop neighbors: nb_2, nb_{n-1}, ...
2-hop neighbors: nb_1, nb_3, ...
3-hop neighbors: nb_{n-2}, ...
4-hop neighbors: nb_n, ...

count of infinity) are two-hop neighbors (nb_1, nb_3, ... in the above example). This procedure continues until hop numbers of all neighbors are populated.

17.4.3 Mesh Path Selection and Data Forwarding for Unicast

17.4.3.1 Mesh Path Selection. The pseudocode given in Fig. 17.4 describes how to select the mesh path for data forwarding. When multiple neighbors are available for selection (see lines 11 and 32 in the pseudocode) and there are no other cost metrics indicating one neighbor is preferred over another, a node can randomly select one neighbor for load-balancing purpose. However, to mitigate "out of order" problems, a node should stick to one neighbor for a while once the neighbor is selected (rather than randomly select one neighbor each time).

If no next hop can be found due to route failures, a ring search should be performed. Ring search can be done by exchanging hello messages as in the link state generation stage, but with a larger TTL.

17.4.3.2 Reliability. Data forwarding supported by this recommended practice is not end-to-end reliable, i.e., acknowledgment from destination to source is not supported. When acknowledgment is requested in the TxOptions field, the MAC layer reliable mechanism will be used to achieve hop-by-hop reliability. To ensure the request of acknowledgment to be executed by each of the next hops, the request is carried by the mesh layer frame all the way to the destination header.

17.4.4 Mesh Path Maintenance

In this section, the mechanism of handling link breakage and recovery is described.

17.4.4.1 Sanity and Consistency Checking. To reduce communication overhead and interference, no periodic hello messages should be broadcast after the link state generation stage. After the link state generation stage, hello messages are only broadcast upon the detection of link failures, link recoveries, or new neighbors. If a node misses some hello messages, its link state may not be accurate. An inaccurate link state can result in the selection of detoured routes and, more seriously, routing loops.

To promptly detect an inaccurate link state without using periodic hello messages, a node should include in each data frame a one-bit up–down flag and a hops2Nb value. The up–down flag indicates whether the data frame is forwarded up or down in terms of tree level (TL); the hops2Nb value is the number of hops from the relaying node to the neighbor_found given by the pseudocode in Fig. 17.4. After having received a data frame, a node should calculate the following values:

flag1	Up–down flag calculated using meshTTLOfHello hop link state information and included in incoming data frame
hops2Nb1	The hop2Nb value calculated using meshTTLOfHello hop link state information and included in incoming data frame
flag2	Up–down flag calculated by receiver of data frame using only (meshTTLOfHello-1) hop link state information
hops2Nb2	The hop2Nb value calculated by receiver of data frame using only (meshTTLOfHello-1) hop link state information

Then check if the following equation holds.

$$\left\{\begin{array}{cc} \text{hops2Nb}_1 - \text{hops2Nb}_2 = 1 & (\text{if hops2Nb}_1 > 1) \\ \text{the destination is my descendent} & (\text{if (hops2Nb}_1 = 1) \text{ and (flag}_1 = \text{down)}) \\ \text{myTreeLevel} - \text{preHopTreeLevel} \geq 1 & (\text{if (hops2Nb}_1 = 1) \text{ and (flag}_1 = \text{up)}) \end{array}\right\} \tag{17.1}$$

If Equation (17.1) does not hold, the link state information of the previous hop and/or the receiver is inaccurate. In this case, the receiver should send a link state request frame to the previous hop. Upon the reception of the link state request frame, the previous hop should send back a link state reply frame, in which its complete link state is included. Upon reception of the link state reply frame, the receiver should compare the received connectivity matrix with its own and then broadcast a link state mismatch frame with a TTL of meshTTLOfHello−1. Any node having received a link state mismatch frame should check if its address is included in the addresses of the neighbors field and, if yes, should broadcast several hello messages to update the link state of its neighbors.

17.4.4.2 Link State Maintenance. A node should broadcast several hello messages with a TTL of meshTTLOfHello if it detects its one-hop connectivity has changed due to link failures, link recoveries, or detection of new neighbors.

Transmission failures can be caused by link failures (including node failures), collisions, or background interference. A neighbor to which a transmission has failed should be first put in a probe list. A neighbor in the probe list is either in an unknown or a down state. A neighbor with an unknown state is then probed each meshProbeInterval after the last probe using a timer (timer driven) or probed immediately each time it is selected to be the next hop of a data transmission (data driven). Although the neighbor with an unknown state can still be selected as the next hop, similar to a normal neighbor, it is not used for transmitting data packets. All data packets having this neighbor as the next hop are buffered or dropped if there is not enough memory. The probe should

```
func_nextHop(dst)
neighbor_found = search the neighbor list for the lowest (i.e., with the largest tree level) neighbor who is the ancestor of dst but is not my ancestor;
    if neighbor_found                                             //going down
      next_hop = getOneHopNeighbor(neighbor_found);
      return next_hop;
    else if the destination is not my descendent                  //going up
      found = is there a neighbor who has a tree level less than mine?
      if found
          hops2root = the minimum (hops + tree_level) found among neighbors that have a tree level less than I;
          minHops = the minimum hops found among neighbors that have a (hops + tree_level) of hops2root;
          neighbor_found = select one of the neighbors that have a (hops + tree_level) of hops2root and a hops of minHops;
          next_hop = getOneHopNeighbor(neighbor_found);
          return next_hop;
    else                                                          //should go up, but can't
          return no_next_hop;
    end if
    else                                                          //should go down,but can't
    return no_next_hop;
    end if
end func
 func_getOneHopNeighbor(neighbor_found)
 mark the hop_number of each neighbor as "infinity";
 current_hops = hop number ofthe neighbor_found;
 while current_hops > 1
        for each neighbor nb_i with a hop_number of current_hops
            for each neighbor nb_j directly connected to nbi
                hop_number of nb_j = current_hops – 1;
            end for
        end for
        current_hops = current_hops - 1;
 end while
 return one of the neighbors with hop_number of1;
end func
```

FIGURE 17.4 Mesh path selection and data-forwarding pseudocode.

continue until the link to the neighbor is recovered or the total probe number, including both timer-driven probes and data-driven probes, reaches meshMaxProbeNum.

If a link is recovered, the corresponding neighbor should be removed from the probe list and all packets buffered for this neighbor, if any, should be forwarded to this neighbor. A link is considered recovered if a MAC ACK of a probe is received.

If the probe number reaches meshMaxProbeNum before the link is recovered, the state of the neighbor will be changed to down. The connectivity matrix will be updated accordingly and hello messages will be broadcast with a TTL of meshTTLOfHello. After the broadcast of the first hello message, all packets buffered for the neighbor, if any, will be routed via other routes. Data packets must not be routed via other routes before the original next hop is determined down and at least one hello message has been broadcast to all meshTTLOfHello hop neighbors.

The neighbor remains in the probe list if the link to the neighbor has been determined down, but it should be probed only by timer (it will not be used as the next hop of any data packet) and the probe interval should be increased after each probe, up to a maximum value meshMaxProbeInterval. For example, a neighbor with a state of down can be probed using intervals 2, 4, 6, …, meshMaxProbeInterval, …, meshMaxProbeInterval seconds. This guarantees that, if the link recovers, it will be detected within no more than meshMaxProbeInterval seconds.

17.4.5 Mesh Path Selection and Data Forwarding for Multicast

The multicast routing protocol described in this section utilizes the logical tree built by the unicast routing protocol described in Section 17.4.3. The logical tree is a shared tree rooted at the network coordinator (NC). When the tree is built, the neighbor information as well as its relationship (parent, child, or sibling) to a node is recorded in every node's neighbor list (see Table 17.12).

The goal of multicast routing hence can be defined as finding a minimum subtree of the logical tree which covers all multicast members of a multicast group. Joining and leaving the multicast group are dynamic so the multicast tree is minimal at any time during the multicast session. Due to the use of the tree structure, all control messages are unicast and no multicast routing table is needed. In most cases, the NC is not bothered for transmitting control and data messages and hence the congestion around the NC and the single-point-of-failure problem can be avoided or relieved. Furthermore, multicast data frames do not need to be sent to the NC first. They can be propagated to all other members directly from the data sources to ensure simple and timely data delivery. Nonmembers can also send packets to members but not vice versa.

To better describe the multicast routing protocol, the following entities are defined. They are illustrated in Fig. 17.5. Note tree level (TL) is the hop distance a node is from the root of the logical routing tree.

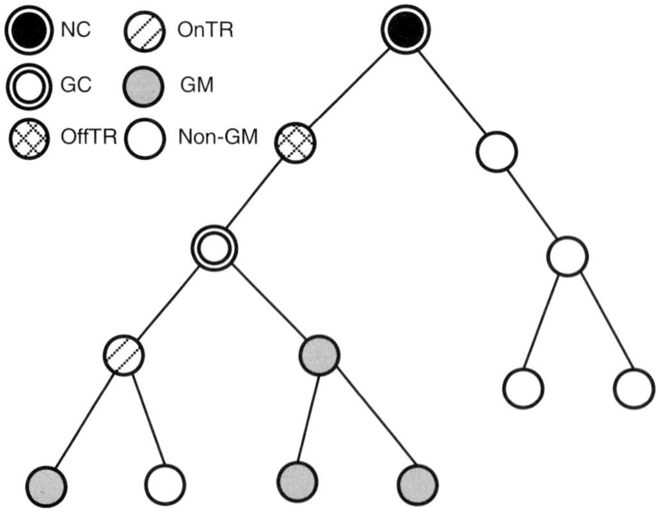

FIGURE 17.5 Entities of multicast protocol.

Network Coordinator (*NC*). The root of the routing tree of a mesh network. It keeps information of all multicast groups in the network so that it always knows from which child(ren) it can reach the multicast tree for a specific group. The NC has a TL equal to zero.

Group Member (*GM*). A node participating in a multicast group. A GM should process any frames sent to its group address and may send frames to its group.

On-Tree Router (*OnTR*). A node on the multicast tree but not a GM. An OnTR relays multicast frames for a multicast group.

Group Coordinator (*GC*). The lowest level GM or OnTR of a specific multicast group. It is the root of the multicast subtree and it sets the lowest TL the multicast frame can propagate in the multicast tree.

Off-Tree Router (*OffTR*). A node which is the GC's ancestor. It always resides between the GC and the NC. An OffTR is not on the multicast tree but it has the routing information of the multicast tree.

17.4.5.1 Group Addressing. The group address is assigned from next higher layer to the mesh layer through the MESH-DATA.request primitive. The selection of the group address for a specific multicast group is out of the scope of this chapter. By setting the transmission options field of the frame control field to multicast, the entire 16-bit short address space can be reused for multicast communication.

TABLE 17.14 Group Communication Table

Group Address	Status (Bitmap)	No. of Links to Group	Link 1	Link 2	Link 3	...
10	GC	2	11	51		
1,000	GM	3	201	456	1087	
11,000	OnTR	2	701	890		

17.4.5.2 Management of Group Membership. This section describes the functions related to membership management of multicast groups, which include joining the group, leaving the group, migration of GC role, and group dismissal. To accomplish these functions, each node involved in multicast communications needs to maintain a group communication table as illustrated in Table 17.14.

The status field describes the function a node plays in a multicast group. The statuses include NC, GC, GM, OnTR, and OffTR. Note that dual status may exist for a node, such as NC and GM, NC but not GM, GC and GM, and GC but not GM. A bitmap can be used to record all applicable statuses a node has for a group (see Table 17.15). The number of links to the group field indicates the number of branches of the multicast tree a node has. The next hops of these branches are listed in the fields such as link 1, link 2, and so on.

17.4.5.2.1 Joining the Multicast Group. Depending on their participating levels in a multicast group, nodes join the multicast group with different processes.

A new node joining the group. A new node is defined as a node which has no information about the multicast group, i.e., it is none of the following: an OnTR, an OffTR, or an NC. A new node should send (by unicast) a join request (JREQ) to its parent node if it wants to join the group. Upon receiving a JREQ from a child node, a parent node should first check its group communication table (see Table 17.14). A parent node with status equal to

TABLE 17.15 Status Field of Group Communication Table

Status (8-bit Bitmap)	Description
0×01	GM
0×02	OnTR
0×04	OffTR
0×08	GC
0×10	NC
0×20	Reserved
0×40	Reserved
0×80	Reserved

GM, OnTR, OffTR, GC, or NC for this multicast group should respond with a join reply (JREP) by unicast. A parent node which is none of the above five entities should forward the JREQ to its own parent. This process will repeat until the JREQ meets a node with status equal to GM, OnTR, OffTR, GC, or NC, which will then reply with a JREP.

Figure 17.6 illustrates different situations when nodes A, B, and C join the multicast group. Node A sends a JREQ to its parent node F. Since node F is an OnTR, it can reply with a JREP back to node A. Node B sends a JREQ to its parent node D, which has no information about the multicast group. So node D forwards the JREQ to its own parent node G, which happens to be a GC. Node G then replies with a JREP back to node B via node D. Node D becomes an OnTR because it receives a JREP with the GC flag unset. Node C joins the multicast group through existing GM H. Node E also becomes an OnTR.

If a JREQ finally meets the NC, it means there is no multicast member in this branch from the NC. The NC should then check the group communication table (see Table 17.14) for this group address. If the NC finds no record of this group address, it means the joining node is the first one for this group. The NC should respond with a JREP with the GC flag set, indicating the joining node to be the GC for this group. If the NC finds the record of this group address and it has GMs of this group in its other branches, then the NC will respond with a regular JREP. The NC should also change its status to GC for this group if it was not GC before.

When an intermediate node, between the joining node and the node which replies with a JREP, receives a JREP, it should check whether the GC flag is set in the JREP frame. If the flag is set, the intermediate node should change its status to OffTR. Otherwise, the intermediate node should change its status to OnTR.

When the joining node receives a JREP, it should check whether the GC flag is set. If it is set, then the node will set itself as the GC for this group. The joining process completes; if it is not set, the route to the multicast tree was

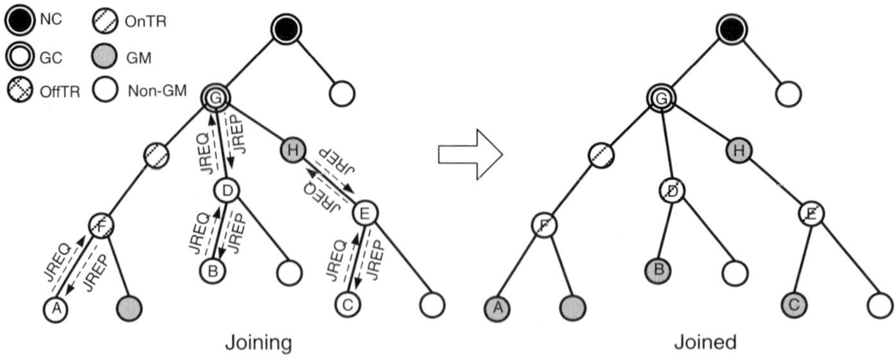

FIGURE 17.6 New nodes A, B, and C joining.

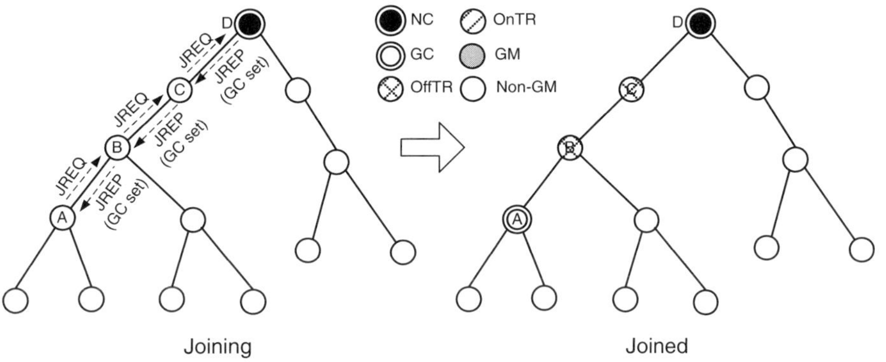

FIGURE 17.7 First GM case.

found—the node should then record this information in its group communication table (see Table 17.14).

Figure 17.7 shows the situation when the joining node is the first member of the group. In this case, the joining node becomes the GC. The intermediate nodes B and C become OffTRs.

A node *may* try to join the group up to MaxMulticastJoinAttempts times if its previous attempts failed. If a node finally receives no JREPs (even no JREP from the NC), it may decide to be the GC for this group. In this case, the node could be in a network partition without the NC.

An OnTR joining the group. An OnTR should simply change its status from OnTR to GM to join the group.

The NC or an OffTR joining the group. The NC or an OffTR, when joining the group, should simply change its status to GC and send a GC UpDate (GCUD) command frame down to the existing multicast tree indicating it will be the new GC for this group. The current GC should give up its GC status upon receiving the GCUD frame and discard the GCUD frame without propagating the packet further.

Figure 17.8 illustrates the process of an OffTR joining the group. In this figure, node C was an OffTR (between the current GC, node B, and the NC, node D) before it decided to join the group. When it wants to join the group, it simply changes its status to GC and sends a GCUD frame to node B. Node B should give up its status as GC because node C is between the NC and itself. Note here we assume node B is not a GM. The NC should follow the same rule when it joins.

17.4.5.2.2 Leave the Multicast Group. This protocol allows any members (including the GC) to drop themselves out from group communications.

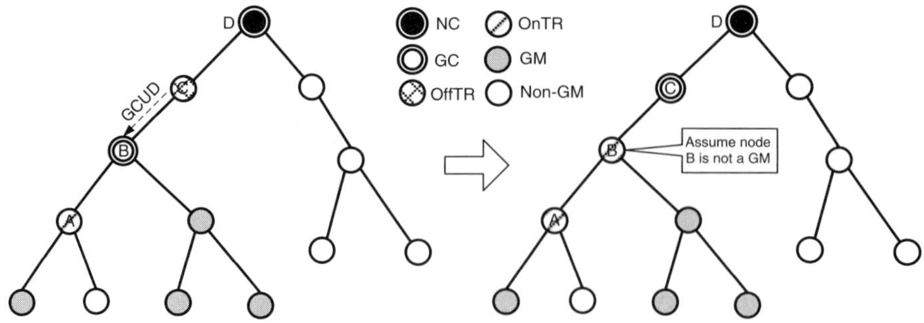

FIGURE 17.8 OffTR joining.

Depending on their statuses of participation, some nodes should still contribute to the communication even after they give up the membership.

To leave a multicast group, a GM should first check whether it is a leaf node. If it is, it should send a leave request (LREQ) to its parent node; otherwise, it can only change its multicast status from GM to OnTR and should not fully leave the tree. Upon receiving a LREQ from a child node, the parent node should respond with a leave reply (LREP) and delete all the multicast information related to this child.

If leaving a child node makes the OnTR parent node a leaf node, then the parent node should also send a LREQ to its parent to prune itself from the multicast tree. If leaving a child makes the GC have only one multicast child for a group, the GC should give up its role as the GC if it is not a GM. If the GC finally finds all of its multicast children have left, it may choose to leave the group too. In this case, the GC should send a LREQ to all OffTRs and NC to delete all the information about this multicast group.

Figure 17.9 illustrates the above leaving rules. Node A is allowed to leave the tree by sending a LREQ to node B since it is a leaf node. The leaving of node A makes node B, an OnTR, a leaf node. Node B will also leave by exchanging

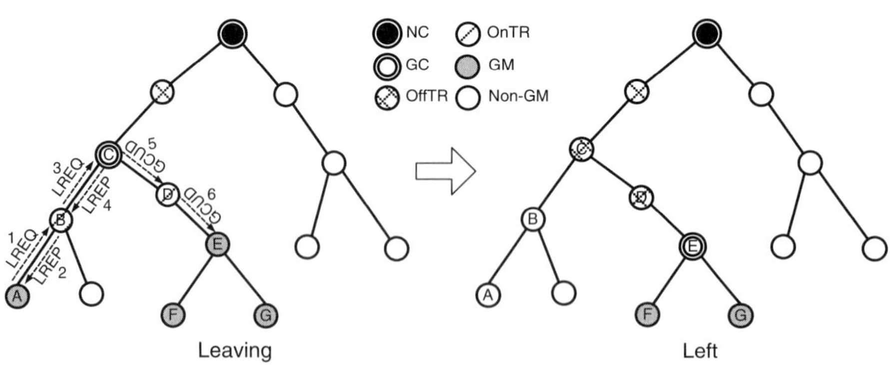

FIGURE 17.9 Leaving multicast group.

LREQ/LREP with its parent node, node C. Assuming group coordinator C is not a GM, it will then find that it now has only one link to the multicast group after B leaves. Node C should give up its role as a GC by sending a GCUD to its only child, becoming an OffTR. Node D changes its status from OnTR to OffTR after receiving the GCUD. The first GM receives the GCUD, node E in this case, and becomes the new GC for the shrunk multicast tree.

17.4.5.2.3 GC Role Migration. The GC's tasks are to set the lowest TL of the multicast tree and to act as the root of the multicast tree so that multicast packets will not propagate outside the multicast tree. However, the lowest TL of the multicast tree is not fixed because this algorithm allows nodes to join and leave the multicast group dynamically. The role of a GC hence needs to be migrated from one node to another in the following two cases: when a new GM joins and when a current GC leaves.

New GM joins. When a new GM outside the existing multicast tree (not in the same branch of the GC) joins the group, the role of the GC may need to migrate from the current GC to the common parent of the newly joined GM and the root of the existing multicast tree. The algorithm is described below.

When an OffTR or the NC receives a JREQ, it means the JREQ is from outside the current multicast tree. The OffTR or the NC should first reply with a JREP to the corresponding child and then send a GCUD packet down to the existing multicast tree, indicating the OffTR/NC itself is now the GC for this multicast group. All intermediate OffTRs should change their status to OnTR upon receiving the GCUD packet. The GCUD will finally reach the current GC. The current GC should give up its role as a GC upon receiving the GCUD command. It should change its status to regular GM or OnTR depending on its participating level for this multicast group.

Figure 17.10 illustrates the GC role switch when the new GM, node A, joined the multicast group. The JREQ from node A finally hits an OffTR, node D. Node D should become the new GC since its TL is smaller than the current GC. Node E changes its status from OffTR to OnTR since it is now on the multicast tree. The current GC, node F, should give up its status as the GC and become either a GM or an MR.

Current GC Leaves. The current GC may leave in the following two cases:

1. The GC is a GM and it wants to leave;
2. The GC is not a GM, but the leaving of a child makes the GC have only one child for a multicast group in its group communication table.

In both cases, the GC should give up its role as the GC for the multicast group. The algorithm is given below.

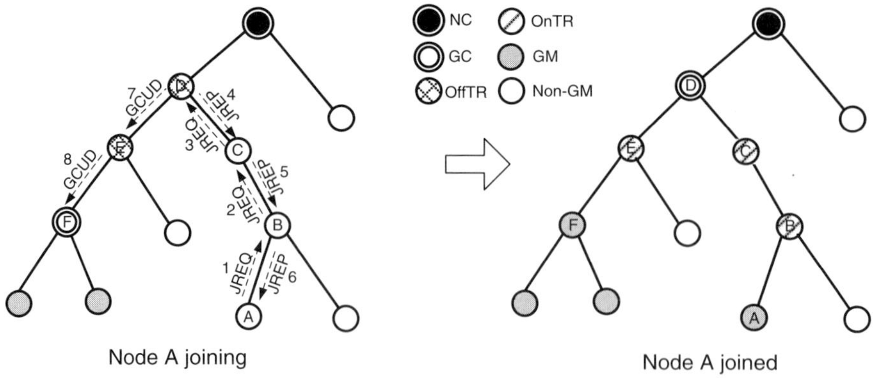

FIGURE 17.10 GC role migration when new node joins.

When the GC detects that it now has only one child for a multicast group in its group communication table and it is not a GM, or it is a member but also wants to leave, it should change its status to OffTR and send a GCUD command down the tree to its child node (illustrated in Fig. 17.9; node C sends a GCUD to node D after both nodes A and B left).

Upon receiving the GCUD, the first GM, or OnTR with more than one child in this group, should become the new GC for this group. The OnTR with only one child in this group should become an OffTR instead of a GC. It is now not on the multicast tree. Note that the GC should still be the OffTR for the group even after it leaves.

17.4.5.2.4 Group Dismiss. When a group finishes its multicast session, one of the members can issue a Group DISmiss (GDIS) packet to all the members to indicate the end of the group communication. The next higher layer should determine which member has the right to issue this GDIS packet. Upon receiving the GDIS packet, all members delete all the information related to this group. The GDIS packets reduce the control traffic led by the GM leaving processes described before.

17.4.5.3 Data Transmission Mechanism

17.4.5.3.1 General Description. Upon receiving the multicast request from the application layer, the mesh layer of the source node puts the multicast group address as the destination address in the mesh layer header and indicates to its MAC layer that the destination address of the MAC frame should be the broadcast address. The source's MAC layer broadcasts the packet by setting both the destination PAN ID and the MAC short address to the broadcast address (0xffff in the case of 802.15.4).

TABLE 17.16 Processes for Neighbors with Different Statuses

Neighbor Status	Process
GM	If the previous hop of the data frame is one of its on-tree neighbors (parent/child), then the GM passes the data frame to the next higher layer; rebroadcasts the data frame if it has on-tree neighbors other than the previous hop (this neighbor check prevents leaf node from rebroadcasting). Otherwise, discard the data frame; in this case, the data frame was received from a sibling node.
OnTR	Same as GM process except not passing the packet to its next higher layer for further process.
OffTR/NC	If the previous hop is not a child who links this OffTR/NC to the multicast tree, the data frame is unicast down to the multicast tree; in this case, a nonmember node is sending the packet to the group and we allow it to be forwarded to the multicast tree. Otherwise, the packet is discarded. In this case, the packet is out of the transmission scope. This happens when the GC broadcasts the packet and its direct parent (immediate OffTR or the NC) receives it.
None of the above	In this case, a nonmember node is sending a data frame to the group and this frame can only follow the tree link toward the NC. So the node should unicast the data frame to its parent. Note most data frames may hit an OffTR and be propagated down to the multicast tree without going all the way up to the NC.

All neighbors that received this broadcast frame (at the MAC layer) will pass it to their mesh layers. By checking its own status of participation in the multicast group, the transmission options field, and the destination address field in the mesh layer header, each neighbor node's mesh layer determines whether it needs to, and how to, process the multicast frame. Table 17.16 describes the process for neighbors with different status.

17.4.5.3.2 Preventing Duplicate Packets. To prevent duplicate data frames when propagating the multicast frames, a multicast transaction table (MTT) is maintained in each node (see Table 17.17). Each entry of the MTT is a multicast transaction record (MTR) which records the following three items of

TABLE 17.17 Multicast Transaction Table

Group Address	Source Address	Sequence Number of Data Frame
grpAddr1	srcAddr1	seqNb1
...		

the last seen multicast frame. A newly received data frame should be silently discarded if the values of these three items are the same as recorded in the MTT.

17.4.5.3.3 Performance Enhancement for Data Propagation. To reduce the processing overhead incurred by the broadcast packet at the MAC layer, a node may choose to broadcast/rebroadcast the multicast data frame only if it has more than two neighbors on the multicast tree. Otherwise, the multicast packet is unicast at the MAC layer (the network layer destination address is still the multicast group address). This can be achieved by utilizing the 802.15.4 MAC layer primitives.

Although the basic algorithm is to broadcast packets along the multicast tree, it is also possible to allow sibling GM/OnTRs to propagate the packets to expedite the packet delivery.

17.4.5.3.4 Nonmember Behaviors. This protocol allows nonmember nodes to participate in the multicast group communication. The following details the behaviors a nonmember node can have.

The non-GM can send data frames to the multicast group but cannot receive data frames from the group. The non-GM can only unicast data frames toward the NC if it does not have a link to the multicast tree. When a non-GM receives a unicast data frame toward a multicast destination and it is neither an OffTR nor the NC, it has to forward the data frame to its parent. The non-GM can unicast data frames toward the multicast tree if it is OffTR or NC because it has the information of the group.

ACKNOWLEDGMENTS

The authors would like to thank all IEEE 802.15.5 task group members for their contributions to the specification.

REFERENCES

1. IEEE 802.15.4-2003, "IEEE standard for information technology—Telecommunications and information exchange between systems—Local and metropolitan area networks—Specific requirements Part 15.4: Wireless medium access control (MAC) and physical layer (PHY) specifications for low-rate wireless personal area networks (LR-WPANs)," IEEE, New York, 2003.
2. IEEE 802.15.4-2006, "IEEE standard for information technology—Telecommunications and information exchange between systems—Local and metropolitan area networks—Specific requirements—Part 15.4: Wireless medium access control (MAC) and physical layer (PHY) specifications for low rate wireless personal area networks (WPANs)," IEEE, New York, 2006.

CHAPTER 18

POWER-SAVING ALGORITHMS ON IEEE 802.15.4 FOR WIRELESS SENSOR NETWORKS

TAE RIM PARK and MYUNG J. LEE

18.1 INTRODUCTION

Wireless sensor networks have received a great deal of attention the last several years. The concept of interconnecting small wireless devices to efficiently share information promises many useful applications ranging from environmental monitoring to home automation. In most of the applications, devices are expected to have low data rates and be battery operated. Since replacing or recharging the battery is difficult in many cases, conserving battery power is one of the most essential challenges for sensor networks. There exists a range of approaches to save battery power at the device level such as voltage scaling, adaptive coding, etc. and the like [1]. In view of communication protocols, however, the power efficiency of the media access control (MAC) protocol plays a critical role because all activities related to wireless communications are controlled in the layer.

IEEE 802.15.4 is the only standard MAC protocol for the low power and low rate wireless networks. Contrary to other MAC protocols [2–6] designed for specific sensor networks, the standard provides interoperable low power solution in conjunction with the reliable physical layer. Compared with the wireless local area network (WLAN) protocols such as IEEE 802.11, low power consumption in IEEE 802.15.4 is attained by following four methods [7, 8]. First, it uses small power levels to transmit a frame. Typically, it targets on the personal operating space of 10 m for direct communication. Wide area can be covered by multihop communication. Second, it has provisions to minimize the frame size. For example, it allows the option of using either a short address or a long address, and the use of only the sequence number

Emerging Wireless LANs, Wireless PANs, and Wireless MANs. Edited by Y. Xiao and Y. Pan

without the address field in its acknowledgment frame. Third, the carrier sense multiple-access/collision avoidance (CSMA/CA) algorithm can operate with minimum energy consumption. Different from the CSMA/CA algorithm in IEEE 802.11, the backoff counter used for random channel access does not freeze when the channel is busy. Thus, when a device performs the backoff for transmission, it can turn off its transceiver to save energy consumption from idle listening if it is not the intended receiver. Last, it provides a mechanism to define an inactive period, for which all devices turn off their transceivers and wait for the next active period for frame transmission. Indirect communication is also supported to exchange a pended frame during the inactive period.

Although the standard supports power saving, the actual power saving does not come about without the proper use of the supported functions. For example, the transceiver cc2420, currently the market leader, drains 17.6 mA when it transmits a frame [9]. The greedier current drainer is, however, idle listening as it consumes 19.6 mA. If a device is operated by two AA batteries of 1600 mAh, the lifetime of the device might be only 3.4 days, even without considering the energy consumptions from other modules in a sensor device such as a microcontroller and sensors,. Similar figures apply to other transceivers, for example, the power consumption ratio of listen to transmit is 42:35 in mc13192 [10].

In order to identify the main sources of the inefficient use of the devices for communication, the power consumption pattern is analyzed in [6]. The authors presented four sources of power consumption, namely collision, overhearing, control packet overhearing, and idle listening. Among these, they reported that useless idle listening is the major power drainer, for which substantial research progress has been made in last few years.

To evaluate energy consumptions in various approaches, we focus on the following sensor network scenarios. We assume a large-scale wireless sensor network. Each device has several neighbors. The data rate of each device is low, a packet per several minutes. One-hop latency is bound within several seconds. The battery should last approximately one year with two AA batteries. Reference to two AA batteries is chosen because prevailing platforms [11, 12] use AA batteries.

The first possible approach to extend the battery life is to add control functions in the MAC layer by modifying the standard. Several algorithms are proposed [13, 14]. In practice, they are acceptable solutions. Unlike the IEEE 802.11 WLAN cards where the MAC is usually included as a part of the chipset, currently, MAC operations in IEEE 802.15.4 are controlled by changeable MAC software in many transceivers [9]. Besides, the internal change such as providing MAC layer status information to the upper layer for cross-layer control is not violating the standard in the view of other devices if it does not affect other devices. However, it may be difficult to implement on some standard complying transceivers if MAC operations are supported by microcode or libraries that are only the manufacturer accessible. Moreover, later on, many functions might be integrated into an unchangeable single chipset. Therefore, the cross-layer solution may lose the benefits from the standard protocol. The second approach is to add control functions above the MAC layer and to use standard interfaces of the protocol. Since it cannot directly access the internal states of the MAC layer other than the ones accessible through the standard interfaces, it may engender additional overhead. The foremost advantage of this approach lies in the fact that it uses standard interfaces without any changes, providing a common solution for all platforms complying with the IEEE 802.15.4 standard. Therefore, we focus on the second approach.

We investigate the following six algorithms: nonbeacon tracking (NBT), beacon tracking (BT), long preamble (LP) emulation, long preamble with acknowledgment (LPA), long preamble with Acknowledgment after local synchronization (LPAS), and global synchronization (GS). Among those, NBT and BT capitalize on the beacon mode of IEEE 802.15.4. The rest of the algorithms are for nonbeacon mode and use command frames designed at above the MAC layer. Each algorithm is analyzed in various scenarios and compared with others to show the best solution in each scenario.

The rest of this chapter is organized as follows. In the next section, an overview of IEEE 802.15.4 and the solutions in the sensor networks are presented. The six algorithms to enable low power on IEEE 802.15.4 are presented in Section 18.3. In Section 18.4, the energy consumption and latency of the algorithms are analyzed and compared. Techniques to improve the energy consumption are also presented. Finally, we conclude with summary and discussion in Section 18.5.

18.2 BACKGROUND

IEEE 802.15.4 supports two modes of operation: a beacon-enabled mode and a non-beacon-enabled mode. In the beacon mode, built-in methods to manage an inactive period are provided. Since the idle listening is the major source of power consumption, the efficient use of the inactive period becomes critical in supporting long battery life in the beacon mode. In the nonbeacon mode, the upper layer should provide equivalent functions. Therefore, the algorithms proposed for proprietary sensor networks can take this approach. In this case, the key issue is figuring out how to merge the two protocols.

In this section, we present the two modes of IEEE 802.15.4 in detail. Then, interfaces of the MAC and the possible issues are presented for the upper layer solution, which would accomplish the power-saving functions in the sensor networks on either mode. For the nonbeacon mode, the applicable sensor network algorithms are presented in the last subsection.

18.2.1 Beacon Mode and Nonbeacon Mode of IEEE 802.15.4

In the beacon mode, devices are classified as either a coordinator or end devices by the role of each in the network. The coordinator provides a synchronization service by periodically broadcasting a beacon frame. Thus, the time line is divided into fixed time intervals, also known as beacon intervals. The beacon interval is divided into two time periods: an active period and an optional inactive period. The beacon interval bounded by two beacon frames is called the superframe structure of IEEE 802.15.4 as shown in Fig. 18.1.

The active period is divided into a contention access period (CAP) and an optional contention free period (CFP). Devices communicate with other devices using slotted CSMA/CA during the CAP, while only a preassigned device by the beacon communicates during guaranteed time slot (GTS) without CSMA/CA in the CFP. In the inactive period, devices and the coordinator turn off their radio circuitry to save energy. In order to keep this period exact, the coordinator transmits a beacon without channel sensing and backoff.

The superframes defined by the coordinator are accessed by devices in two ways: beacon tracking and nonbeacon tracking. First, a device using beacon tracking stores the schedule of the coordinator when it receives a beacon for the first time. Then, the device tries to receive the next beacon by turning on the receiver just before the expected beacon transmission time. By adjusting its own clock to the time of the coordinator whenever it receives the beacon, the device is continuously synchronized with the schedule of the coordinator. The disadvantage of this method is energy consumption. The device has to accept energy consumption for idle listing during every superframe duration after every beacon even though the device seldom transmits and receives a frame.

On the contrary, a device using nonbeacon tracking remains in sleep mode until it has a frame to transmit or when it wants to check a possible pended frame at a coordinator. Since the device does not track the schedule of the coordinator, it has to stay awake until it receives a beacon before transmitting or receiving a frame. Even though this waiting increases latency, it significantly reduces energy consumption if the sleep time is long and the beacon interval is short.

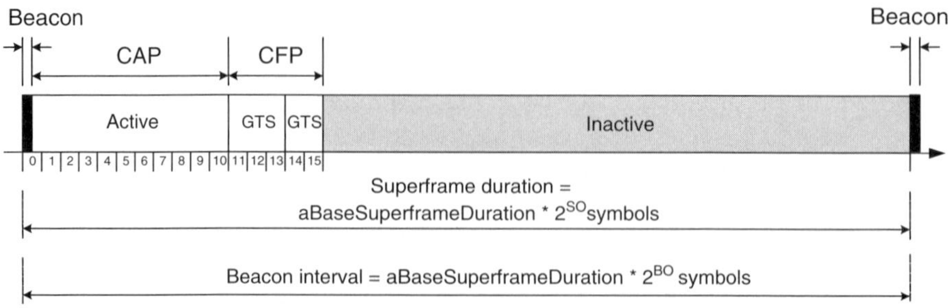

FIGURE 18.1 Superframe structure in beacon mode of IEEE 802.15.4.

The initial version of the IEEE 802.15.4 protocol only considers a one-hop network [7]. Therefore, in order to address the devices outside of the coordinator's transmission range, additional functions are required. A new revision, IEEE 802.15.4-2006, is standardized for this multihop extension [8]. In the standard revision, a coordinator is defined as a device capable of relaying a packet. Among these coordinators, a principal controller of a personal area network (PAN) is named as the PAN coordinator (PNC). The remaining coordinators associate with the PNC or with one of the other coordinators when they join the PAN. The sequence of association builds an association tree. For each coordinator, the beacon interval and the incoming superframe are defined by the beacon frame of the associated coordinator (or parent). Based on the assumption of the very low duty cycle applications, the coordinator transmits its own beacon within the inactive period and can construct an outgoing superframe for other devices (children).

The multihop extension causes a beacon collision problem. Since a beacon is transmitted without carrier sensing and backoff to minimize waiting time for the beacon, it is prone to collision. If a beacon collides with other frames, it is possibly a single event. In the next beacon interval, a beacon may be transmitted successfully without collision. However, when a beacon does collide with other beacons, it becomes a significant problem. Since devices in the network usually keep the same beacon interval, the beacons may collide continuously. Then, other devices that expect the beacon cannot communicate any longer.

To handle the problems, a beacon-scheduling method within the dedicated period for beacons is proposed in [13]. Based on the algorithm proposed in [15], it schedules beacon transmission time. The scheduling algorithm itself is also applicable for Zigbee [16] where superframes are scheduled in the inactive time of the parent's beacon interval. In [14], the authors argued that rescheduling may be required whenever a new device joins in the scheduled network. In addition, they proved that no scheduling method is perfect in removing all geographical areas experiencing continuous beacon collision, named a service hole. Then, they proposed a stochastic transmission algorithm to resolve continuous beacon collision.

Algorithms proposed in [13] and [14] require modifications of the standard. However, offline scheduling may be enough in our scenarios without such modifications. Since the beacon interval is long and the superframe duration is short to enable low power consumption in low rate networks, the probability of beacon collision is very low if the number of neighboring devices is small.

Compared with the beacon mode, the nonbeacon mode of IEEE 802.15.4 is quite simple. The roles of a coordinator and end devices are defined in the same way. However, the coordinator transmits a beacon with CSMA/CA only when needed. Since there is no reference point in the time line, devices use unslotted CSMA/CA to transmit a frame. The inactive period cannot be defined. However, indirect communication is supported for a device with a small battery and low rate traffic in a similar manner as in the beacon mode. Those devices may go to sleep without defining an inactive period. Then, the coordinator has to keep a frame for the device in the pending queue. The device has to transmit a query command to ask the pending frame after a long sleep. Contrary to the beacon mode, the coordinator should keep awake to support those devices. In order to save the energy of coordinators, which are the routers in the multihop sensor networks, a user has to provide the service for periodic wakeup and sleep in the upper layer above the MAC layer.

18.2.2 Upper Layer Solution

Power-saving functions in IEEE 802.15.4 are designed for general-purpose low power networks. For the multihop sensor networks requiring long battery life, an upper layer solution is required for the following two reasons. First, it requires a fine control to maximize the efficiency. For example, the minimum superframe duration in 2.4-GHz channels is 15.36 ms. When a device keeps two superframes (incoming and outgoing) in a beacon interval, the active ratio of the device is more than 6% if the beacon interval is 491.52 ms. With the parameters of the device and two AA batteries discussed in the previous section, the device will last around 2 months. Second, it requires a control to provide reliable communication. As presented in the previous section, communication should be supported without continuous beacon collisions.

Since a user cannot control internal operations of the MAC protocol from the upper layer, the only way of doing so is by using the standard interfaces named MAC primitives. The primitives are interfaces for standard services defined in IEEE 802.15.4. Like a function call in the program language, the primitives convey required information as parameters and then execute appropriate services. There are four types of primitives: *request*, *confirm*, *indication*, and *response*. Request is the primitive called by the upper layer to initiate a service at the lower layer. Confirm is the primitive called by the lower layer to pass the result of the previous request to the upper layer. On the other hand, indication is the primitive called by the lower layer to the upper layer to tell the remote request or an internal event. Response is a primitive to complete the procedure invoked by the previous indication.

IEEE 802.15.4 has two service access points (SAPs). One is the MAC data SAP named MAC common part sublayer (MCPS) data service. The other is the MAC management SAP named MAC sublayer management entity (MLME) service. IEEE 802.15.4 provides two MCPS primitive sets. One is the MCPS-DATA set, which transmits or receives a data frame. The other is the MCPS-PURGE set, which purges out a frame in the transaction queue. MLME provides 15 primitive sets for management functions, including association, scan, and synchronization. The database named MAC PAN Information Base (PIB) for MAC operations is also accessible by two sets of MLME services: MLME-SET and MLME-GET.

A power-saving algorithm in the upper layer uses these primitives. In order to transmit a command frame, MCPS-DATA.request is used. The command frame will invoke MCAP-DATA.indication at the receiver and then the counter part of the algorithm is run. To turn on/off the receiver, MLME-SET.request is used. By setting the macRxOnWhenIdle attribute in MAC PIB to False, the MAC layer turns off the receiver. However, MAC PIB does not have any attribute related to the channel condition. If the algorithm has to make a decision on the channel condition, the only way is by utilizing a short broadcasting frame. At the upper layer, if a device receives the frame, it indicates that channel is occupied by other devices.

Even though the standard interfaces provide substantial control methods, the efficiency of the upper layer solution depends on the cooperation from the MAC layer. For example, if a power-saving algorithm calls MCPS-DATA.request to transmit a command frame, the time starting backoff depends on the MAC implementation. Also, when it calls MLME-SET.request to turn on or off the receiver, the action should be immediate. However, the IEEE 802.15.4 does not specify any time constraint for service primitives. For the complete power saving, we assume that a device supported by prompt operations at the MAC layer because the power saving can be accomplished by the efficient upper layer algorithm and MAC layer cooperation.

18.2.3 Sensor Network MAC

Because the nonbeacon mode does not provide a power-saving method for coordinators in the multihop environment, the power saving should be supported by the upper layer. There has been a lot of research with a similar purpose. Based on the schedules of devices, they are classified into two approaches: asynchronous and synchronous. In any approach, a device has a short active duration for possible reception and a long inactive duration for energy saving. The durations are periodically repeated. In the asynchronous protocols such as BMAC, XMAC, TICER, and WiseMAC [2–5], the schedules of devices are not synchronized. Thus, if a device has a frame in the buffer, it transmits a preamble or a stream of short wakeup requesting frames for longer than the wakeup interval to wake the destination device up. If the destination device recognizes the channel as busy, then it stays awake to receive the data frame after the preamble or a stream of wakeup packets. Usually, a device using asynchronous protocol has a smaller active duration since the duration is only for deciding whether the channel is busy or not.

On the other hand, in the synchronous protocols such as SMAC [6], devices have a common schedule by a time synchronization protocol. Then, the devices wakeup and sleep at the same time. Since the devices share a common active time, frames are easily exchanged during the common active time. However, in order to synchronize the devices, the time synchronization protocol requires a duration for exchanging control frames. Thus, the active duration of the synchronous protocol is usually longer than that of asynchronous protocols. Because of the difference in active duration, it is generally accepted that asynchronous protocols spend less energy compared with synchronous protocols when the traffic is very low.

18.3 POWER-SAVING ALGORITHMS

As introduced earlier, here we present six power-saving algorithms on IEEE 802.15.4. nonbeacon tracking (NBT) and beacon tracking (BT) are based on the beacon mode. The main contribution of these two algorithms is minimizing the active time in the predefined superframe to enhance battery life. Long preamble (LP) emulation, long preamble with acknowledgment (LPA), long preamble with acknowledgment after local synchronization (LPAS), and global synchronization (GS) are based on the nonbeacon mode. These algorithms are implementations of MAC protocols designed for sensor networks. In addition to the effort to minimize the active duration, techniques to implement the MAC functions in the upper layer are presented. The goal of this chapter is to evaluate the energy consumption and to present possible issues. Detailed techniques to overcome each drawback are out of our scope.

18.3.1 Power Saving on Beacon Mode of IEEE 802.15.4

In the beacon mode, all time durations are defined from a constant, aBaseSuperframeDuration. Based on the constant, the beacon interval, t_{BI}, and superframe duration, t_{SD}, are set with two attributes, macBeaconOrder (BO) and macSuperframeOrder (SO), in MAC PIB as follows:

$$t_{\mathrm{BI}} = \text{aBaseSuperframeDuration} \times 2^{\mathrm{BO}} \quad 0 \leq \mathrm{BO} \leq 14 \tag{18.1}$$

$$t_{\mathrm{SD}} = \text{aBaseSuperframeDuration} \times 2^{\mathrm{SO}} \quad 0 \leq \mathrm{SO} \leq \mathrm{BO} \leq 14 \tag{18.2}$$

Since aBaseSuperframeDuration is 15.36 ms in 2.4-GHz channels, the minimum superframe duration is 15.36 ms, and the maximum beacon interval is 251 s. These two attributes should be set when a device starts beacon transmission by calling MLME-START.request. The attributes are also transmitted in the superframe specification field defined in MAC payload of the beacon frame. Based on the assumption of very low traffic, the active time in the superframe duration can be minimized to only allow for the beacon transmission and the maximum size frame transmission.

18.3.1.1 Nonbeacon Tracking. NBT is an asynchronous wakeup algorithm using the nonbeacon tracking method in the beacon mode. A device adopting NBT periodically wakes up and transmits a beacon to notify that it is in the active duration and ready to receive a frame. For transmission, a device wakes up after making a frame internally, and then listens to the beacon of the destination. If the device receives the beacon, it transmits the frame in the queue. Figure 18.2 gives an example of frame transmission from device B to device A.

Energy consumption of NBT is determined by two parts. One is for receiving a frame. For this, beacon interval t_{BI} and the minimum active duration t_{minB}

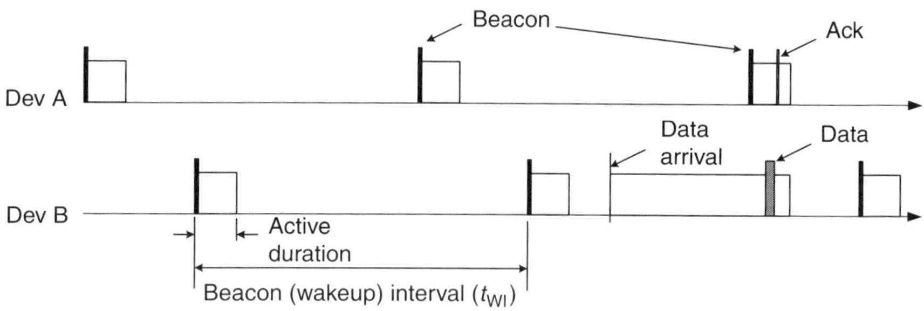

FIGURE 18.2 Time lines of two devices using the nonbeacon tracking algorithm.

comprise the major parameters. The minimum value of t_{BI} is the time duration for a beacon and the time for receiving the maximum size frame. When it comes to energy consumption for frame transmission, t_{BI} is the major parameter in conjunction with the frame arrival rate r. Since the active times of two devices are not synchronized, a device may turn on the receiver and wait for $t_{BI}/2$ on average to receive a beacon of the destination.

In addition to minimizing the active duration, NBT should provide a solution to avoid beacon collision. Because of the clock skew of two oscillators, two superframes may be overlapped after some time, and that results in beacon collision. One possible simple solution is restarting beacon transmission after random delay by calling MLME-START.request if a device receives a beacon from a neighbor in the active duration. However, if the beacon collision is by a device outside of one hop range, the device may not notice the collision. As presented in [14], it is the limitation of the beacon mode.

18.3.1.2 Beacon Tracking. BT is a synchronous wakeup algorithm fully utilizing superfame structures provided by IEEE 802.15.4. By the association procedure, a relation between two devices is decided as a coordinator and a device. Then, the device associated with the coordinator periodically wakes up to receive a beacon frame and share the minimized active duration in the common superframe duration. For its own outgoing superframe, it also periodically transmits its own beacon in the inactive period defined by the beacon of the coordinator. This outgoing superframe enables a new device to associate as a device. Figure 18.3 gives an example of these time lines.

In the figure, device A is at the ith level from the PAN coordinator (PNC) in the association tree. While it tracks the incoming superframe defined by the coordinator it is associated with, it transmits beacons for its children at the $(i + 1)$th level from the PNC. In the same way, device B and C track beacons of device A and transmit their beacons for their children. A data frame is transmitted in the common active duration. Contrary to NBT, a transmitter can wait for the active time without idle listening since it knows the schedule of the coordinator by tracking it. In the figure, device B waits for the next

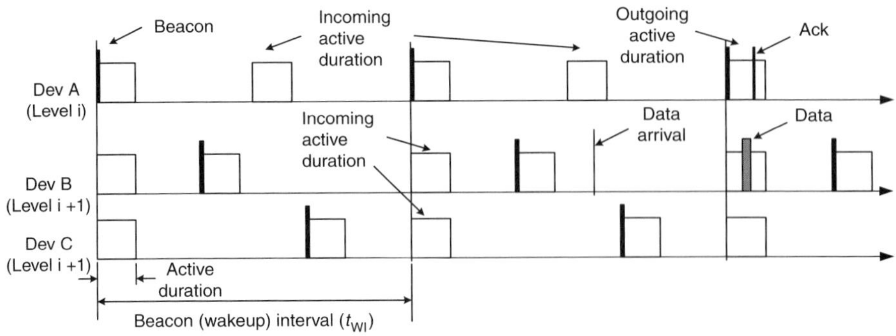

FIGURE 18.3 Time lines of two devices using the beacon tracking algorithm.

superframe without turning on the receiver. It turns on the receiver just before the beacon transmission time of device A, and then it transmits a data frame.

Since transmission activity does not produce any additional overhead as in NBT, energy consumption of BT is decided by t_{minB} and t_{BI}. However, active time in one beacon interval increases to $2t_{\mathrm{minB}}$ since it has two active durations, one incoming and one outgoing active durations.

BT also requires controls to minimize the active time in a superframe duration and to avoid beacon collision. In the case of beacon collision, the time skew is no longer the problem since every time each device receives a beacon, it adjusts its clock. However, the assignment of nonoverlapping active time periods for parent and children nodes remains a problem. In addition, communication in BT is restricted among the parent and children in the association tree. In order to transmit a frame to other neighbors, an upper layer should have their schedules. Otherwise, a frame will be relayed through a coordinator or children.

18.3.2 Power Saving on Nonbeacon Mode of IEEE 802.15.4

The nonbeacon mode of IEEE 802.15.4 provides flexibility to the power-saving algorithms. The wakeup interval, corresponding to the beacon interval of the beacon mode, and the active duration can be assigned freely without considering aBaseSuperframeDuration and beacon transmission time. Therefore, all active times are decided by the algorithm design.

18.3.2.1 Long Preamble Emulation. LP emulates the asynchronous MAC protocol proposed in [2], also known as B-MAC. However, the preamble length defined in IEEE 802.15.4 is 5 bytes, including a start frame delimiter. Thus, making the preamble longer than a wakeup interval is impossible. Similarly to the proposals in [3, 4], also known as X-MAC or TICER, LP uses a short command frame named wakeup request (WR). Contrary to X-MAC, LP does not use an acknowledgment frame to respond to the WR. In addition, the WR

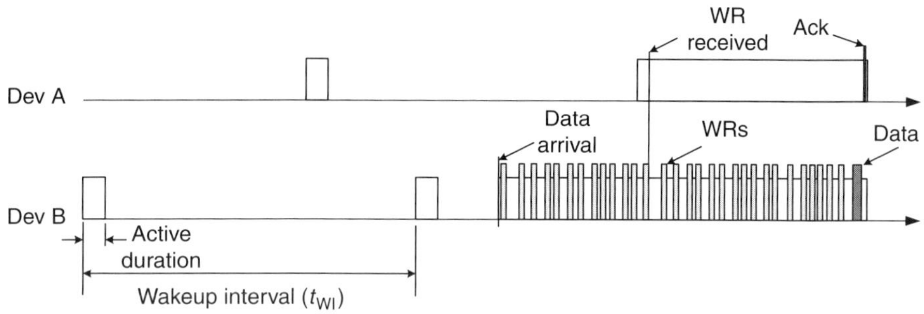

FIGURE 18.4 Time lines of two devices using the long preamble emulation algorithm.

is a broadcasting frame including the destination address in the MAC payload. This enables devices to sense activity of the channel without filtering the MAC layer.

The time line of LP is divided into wakeup intervals. Each wakeup interval is started by a short active duration followed by a long inactive duration. The time line of each device is asynchronous to each other. To transmit a data frame, a device transmits a sequence of WRs longer than a wakeup interval followed by a data frame. While the transmitter transmitts WRs, other devices receive at least one WR in their active duration. If the address is matched, then the device stays awake until it receives the data frame at the end of the WR sequence. If the address is not matched, the device goes to sleep. Figure 18.4 gives an example of the time line of the LP.

Compared to the algorithms on the beacon mode, the wakeup duration lasts only long enough to receive a short WR. Since it does not require time for a beacon and a maximum-sized frame, the active duration can be minimized. However, the transmitter has to transmit WRs longer than the wakeup interval t_{WI} to transmit a data frame. The receiver also has to spend energy for a half of the wakeup interval ($t_{WI}/2$) on average to finally receive a data frame. Therefore, the energy consumption is highly dependent on the frame arrival rate r and t_{WI}.

18.3.2.2 Long Preamble Emulation with Acknowledgment. LPA adopts the asynchronous algorithm X-MAC or TICER proposed in [3, 4]. Based on LP, LPA introduces a new command frame named wakeup acknowledgment (WA). The WR and the WA are broadcasting frames for the same reason given for LP. The transmitter starts with the same manner for LP by transmitting a WR. However, in LPA, in order to reduce the overhead from a long sequence of WRs, the destination device replies with WA, informing the transmitter that the receiver is ready. Then, the transmitter stops transmitting WRs and transmits a data frame. If it does not receive a WA, it keeps transmitting for a wakeup interval. If it does not reply until the end of the sequence, it gives up the data frame transmission and follows the upper layer policy. In addition, the

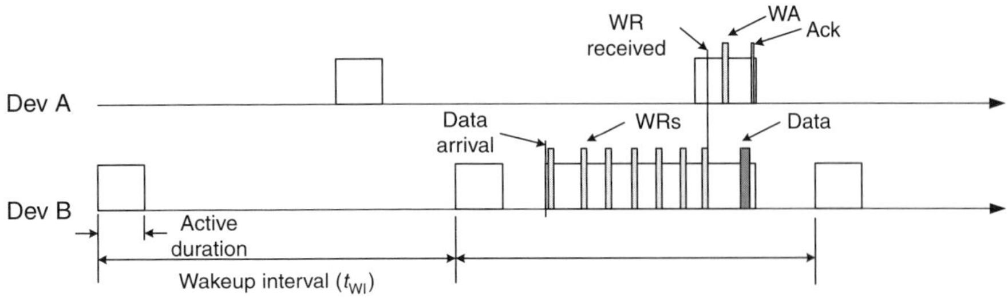

FIGURE 18.5 Time lines of two devices using long preamble emulation with acknowledgment algorithm.

broadcasting frame transmission follows the same procedure as LP since it is not acknowledged. Figure 18.5 gives an example of time lines of LPA.

Compared to LP, the overhead for a frame transmission is reduced by half on average. Moreover, the overhead for reception decreases to only one WR and a data frame. However, broadcasting is not beneficial since it is not acknowledged by a WA. Moreover, the minimum wakeup duration t_{minW} becomes almost twice as long compared with that of LP. It is due to sparse WR transmissions by the transmitter to allow sufficient time for the WA to be received in between two adjacent WRs. Thus, the receiver has to stay awake longer to ensure there is a sequence of WRs or not. Consequently, even though LPA is a solution designed to overcome the disadvantage of LP, LPA spends more energy if r and t_{WI} are small.

18.3.2.3 Long Preamble with Acknowledgment after Local Synchronization. LPAS adopts the local synchronization algorithms proposed in [4, 5]. Even though LPA reduces the overheads of a transmitter and a receiver, it still engenders inevitable overhead at the transmitter because the transmission is started asynchronously. LPAS is a local synchronization algorithm that estimates the schedule of the receiver and transmits a frame on the receiver's schedule.

At first, the frame transmission is the same as LPA. However, when a transmitter receives a WA, it logs the time. Subsequently, when it transmits to the same receiver, it estimates the schedule of the receiver and waits for its next active time. When the time comes, it turns on the transceiver. A WR and data frame transmission procedure is started as in LPA. As described before, it transmits WRs longer than t_{WI} lest it miss a WA by the estimation error. Figure 18.6 gives an example of the time line of LPAS.

Compared to LPA, the overhead for transmitting WRs is reduced. However, the performance of LPAS is highly dependent on the estimation accuracy of the receiver's schedule. Since the active duration of devices is neither synchronized nor adjusted, the accuracy of estimation will suffer if the frame arrival rate is extremely low. In such a case, the energy consumption of LPAS will be the

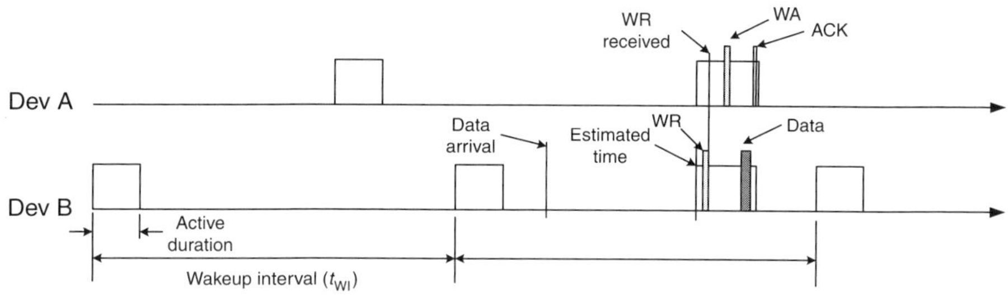

FIGURE 18.6 Time lines of two devices using long preamble with acknowledgment after local synchronization.

same as LPA. In addition, it may also add some errors because of the backoff in the MAC layer and the internal latency between the MAC layer and the upper layer. Depending on the estimation accuracy, additional margin is required to start transmitting WRs. It also requires a table to log the schedule of each neighbor node. If the number of neighbors is large, for example, a dense network, an efficient algorithm is required to handle the table.

18.3.2.4 Global Synchronization. GS is the implementation of the synchronous protocol S-MAC proposed in [6]. It tries to synchronize all devices in the networks by transmitting a SYNC frame. A SYNC is a command frame broadcasted with a schedule information like a beacon. The difference is that the SYNC is not transmitted at every wakeup interval and CSMA/CA is used when it is transmitted. Therefore, it is free from the continuous collision problem. However, a longer time than the time for a beacon should be reserved to receive the SYNC. Figure 18.7 presents an example of time lines of GS.

The time line of GS is divided into wakeup intervals. Each wakeup interval is started from a short active duration and followed by a long inactive duration. The active duration is divided into two time spaces. The first time space is for a SYNC, and the second time space is for a request to send (RTS). When a device transmits a SYNC in the active duration, other devices adjust their clocks.

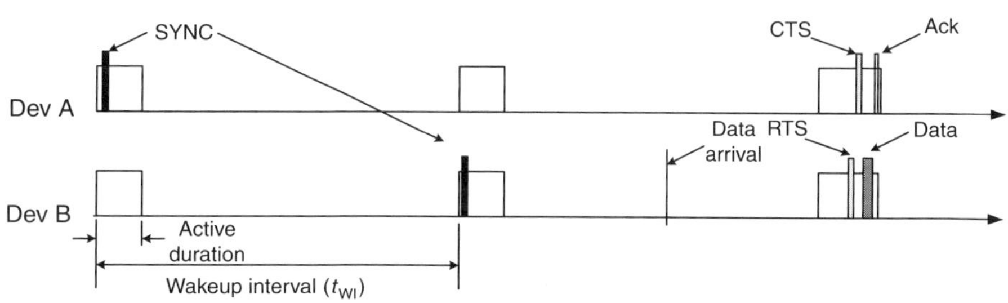

FIGURE 18.7 Time lines of two devices using global synchronization algorithm.

In Fig. 18.7, data frame transmission from device B to device A is presented as an example of frame transmission. When data arrives at device B, it waits for the next common active duration. When the time comes, it turns on the transceiver to receive or transmit a SYNC, and then in the second half of the active duration, it transmits an RTS to announce frame transmission. If device B is replied with a clear to send (CTS) from device A, it transmits the data frame. The rest of the devices that are not related with the RTS and the CTS go to sleep.

The synchronization procedure is started by one of the devices that have the capability of doing so, for example, full function devices. All devices in the network stay awake until a power-saving algorithm is started by the received SYNC. If a device with the capability of initiating the synchronization procedure does not receive a SYNC for some time, it creates and starts transmitting a SYNC. Then, the schedule is propagated by broadcasting SYNCs at the designated time for the SYNC at the beginning of the wakeup interval.

The energy consumption pattern of GS can be compared with that of BT since active times of devices in both protocols are synchronized. The difference is that it requires more active time to receive a SYNC and less time to receive an RTS. Also, it has only one active duration in a wakeup interval. Compared with the asynchronous algorithms presented earlier, GS has the longer active duration since it has to assign two time spaces for SYNC and RTS. However, compared with any other algorithms, it has minimum energy consumption for broadcasting.

One of the biggest problems may be the implementation issue. As presented in the previous section, fine control of packet transmission time depends on the MAC implementation. In practice, local synchronization is acceptable by adding a reasonable margin. However, the global synchronization through multihop links is a difficult problem since clock errors are aggregated over the multiple hops.

18.4 ANALYSIS AND COMPARISON

In order to evaluate the performance of algorithms presented in the previous section, we focus on energy consumption and latency. We note that there exists a trade-off between the battery life and the latency. For example, latency may be sacrificed for long battery life. However, the latency also needs to be bounded to meet the requirements of applications.

As a measure of the energy consumption, we utilize an average active ratio. The ratio is defined as the average active time divided by the total time spent. It is a reasonable measure since the energy consumptions for transmission and reception (including idle listening and channel sensing) are similar in many transceivers compared with the energy consumption of standby [10, 11, 17]. In order to compare the latency, we use the average one-hop latency. We also consider the average two-hop latency to estimate multihop latency. The average one-hop latency is defined as the time from when a transmission activity is started at the upper layer where the power-saving algorithm runs to when a frame is received at the MAC layer of the destination. Based on the arrival rate with uniform distribution, the average one-hop latency of each algorithm is derived. For the two-hop latency, the time from the first hop to the MAC layer of the second hop is added. However, different from at the originator, the starting time is not uniformly distributed due to the intervention of the wakeup interval between the originator and subsequent hops. The internal processing time is assumed to be ignored.

Star topologies consisting of 1–10 devices are used in our analytical study. One of the devices is the common receiver, and the rest are transmitters. In practice, it is expected that the nodes around a sink node or aggregators in a sensor network show a similar energy consumption pattern.

Transmitters transmit frames with the frame arrival rate r to the receiver. We consider different arrival rates r from 0.01 to 0.0003125. Thus, the average interarrival times are from 1.3 to 53 min. We also consider different wakeup intervals from 245 ms to 7.8 s. The exact numbers are calculated from the beacon interval when BI = 4 to BI = 9, while the algorithms on the nonbeacon mode are free from that. Our main focus is when r is 0.0025 (average interarrival time of 6.6 min) and the number of transmitting devices is 3. In all cases, ideal conditions are assumed: no collision, no overhead for turning on the transceiver, and prompt response in the MAC. Although the energy consumption from the processor is also considerable as presented in [18, 19], we leave it out in this analysis for simplicity. Parameters used for the analysis are summarized in Table 18.1. The time duration to transmit any frame is given

TABLE 18.1 Parameters for Analysis

Symbol	Parameter	Values
L_B	Beacon frame length (bytes)	20
L_{WR}	Wakeup request frame length (bytes)	28
L_{WA}	Wakeup ack frame length (bytes)	29
L_{RTS}	RTS frame length (bytes)	28
L_{CTS}	CTS frame length (bytes)	28
L_{SYN}	Sync frame length (bytes)	28
L_D	Data frame length (bytes)	50
L_{Ack}	Acknowledgment frame length (bytes)	11
L_{Max}	Maximum frame length	133
t_b	Transmit/receive one byte (s)	32×10^{-6}
t_{CCA}	Time to perform CCA (s)	128×10^{-6}
t_{TR}	Turnaround time between Tx and Rx (s)	192×10^{-6}
t_{slot}	Backoff slot time (s)	320×10^{-6}
t_{WI}	Wakeup interval (s)	
R	Frame arrival rate (fr/s)	
minBE	Minimum backoff exponent	3

by multiplying the length in byte by t_b if it is not specified. For example, the time duration t_{Ack} for an acknowledgment frame is $L_{\text{Ack}} t_b$.

18.4.1 Energy Consumption

The average energy consumption of each algorithm is decided by the periodic energy consumption for possible frame reception and the energy consumption for frame transmission. When the wakeup interval is the same, the first parameter to compare is the minimum active duration each algorithm can achieve, and the second one is mainly the overhead for frame transmission.

18.4.1.1 Nonbeacon Tracking. The minimum active duration t_{minB} of NBT is the time to receive a beacon and a maximum size frame. Since it operates on the beacon mode using slotted CSMA/CA, t_{minB} is given as follows:

$$t_{\text{minB}} = \left\lceil \frac{L_B}{10} \right\rceil t_{\text{slot}} + (2^{\text{minBE}} - 1) t_{\text{slot}} + 2 t_{\text{slot}} + \left\lceil \frac{L_{\text{Max}}}{10} \right\rceil t_{\text{slot}} \tag{18.3}$$

where $(2^{\text{minBE}} - 1) t_{\text{slot}} + 2 t_{\text{slot}}$ is the time for performing backoff and two CCAs before a maximum size frame. $\lceil L_B/10 \rceil t_{\text{slot}}$ and $\lceil L_{\max}/10 \rceil t_{\text{slot}}$ are the assigned time slots for a beacon and a maximum size frame, respectively. The numbers of occupied slots by frame transmission is the length divided by 10 because L_R and $L_{\max}$ presented by a byte, and the time for one slot is equivalent to the time to transmit 10 bytes. Since a frame is assumed to transmit without experiencing a busy condition of the channel, the time defined in (18.3) is sufficient to sense any frame. Therefore, the average active ratio E_{OB} of a device just observing the network without transmitting or receiving a data frame is

$$E_{\text{OB}} = \frac{t_{\text{minB}}}{t_{\text{WI}}} \tag{18.4}$$

In order to transmit a frame, the first step is to wait for a beacon of the destination. The average time $E[t_{\text{TRK}}]$ for this tracking is simply by the uniform distribution,

$$E[t_{\text{TRK}}] = \tfrac{1}{2} t_{\text{WI}} \tag{18.5}$$

In the slotted MAC, without considering backoff and CCA, the time duration t_{UD} to transmit a unicast frame and to receive an Ack is

$$t_{\text{UD}} = \left\lceil \tfrac{1}{10} L_D + t_{\text{TR}} \right\rceil t_{\text{slot}} + \left\lceil \tfrac{1}{10} L_{\text{Ack}} \right\rceil t_{\text{slot}} \tag{18.6}$$

where t_{TR} is the turnaround time. If the average backoff time $E[t_{BKO}]$ is $(2^{minBE} - 1)t_{slot}/2$, then the average time duration $E[t_{UD}]$ to transmit a unicast frame and receive an Ack is

$$E[t_{TUD}] = E[t_{BKO}] + 2t_{slot} + t_{UD} \tag{18.7}$$

where $2t_{slot}$ is for performing CCA two times defined as IEEE 802.15.4 slotted CSMA/CA. When the time t_B for beacon transmission is $\lceil L_B/10 \rceil t_{slot}$, the average active ratio of the transmitter is derived with the time for periodic energy consumption in (18.4), and the time for the energy consumption to receive a beacon, and the time for transmitting a unicast frame as follows:

$$\begin{aligned} E_{TX} = &\{1 - r(E[t_{TRK}] + t_B + E[t_{TUD}])\}\frac{t_{minB}}{t_{WI}} \\ &+ r(E[t_{TRK}] + t_B + E[t_{TUD}]) \end{aligned} \tag{18.8}$$

where $E[t_{TRK}] + t_B + E[t_{TUD}]$ is the time duration from the time starting to wait for a beacon to the time to receive an acknowledgment. Once the transmission activity is started, it is overlapped with its own wakeup schedule with the probability of $(E[t_{TRK}] + t_B + E[t_{TUD}])/t_{WI}$. Therefore, after multiplying the arrival rate with one wakeup interval, rt_{WI}, the term, $r(E[t_{TRK}] + t_B + E[t_{TUD}])$ t_{minB}/t_{WI} is subtracted from the calculation for the average active ratio as in the first term of (18.8). When n is 3, the results with different t_{WI} and r are presented in Fig. 18.8.

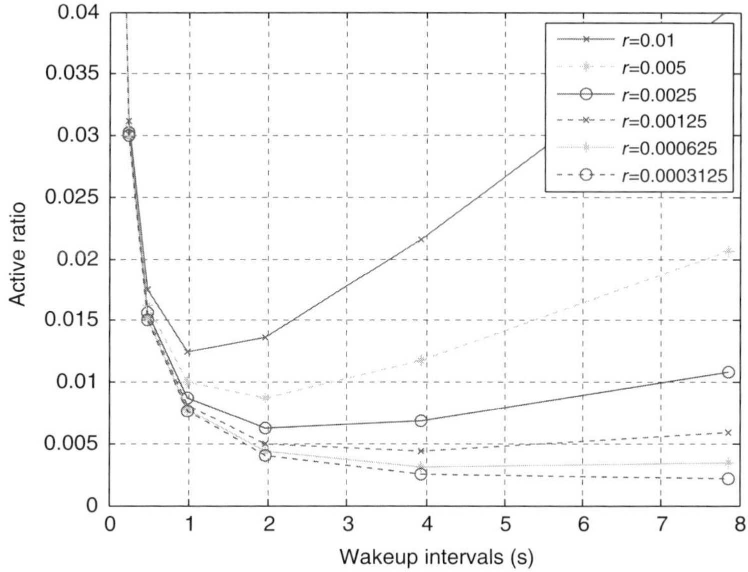

FIGURE 18.8 Comparison of active ratios of transmitters using NBT when $n = 3$.

When t_{WI} is small, the energy consumption of the periodic activity [the first half of (18.8)] is dominant. Therefore, regardless of the arrival rate, all values are over 0.01 until t_{WI} increases to 1 s. On the other hand if t_{WI} increases longer than 2 s, the lines with high arrival rates increase rapidly. In the figure, 0.0025 is the marginal arrival rate to meet the 0.01 active ratio.

The average active ratio of the receiver is derived in the similar manner:

$$E_{\mathrm{RX}} = (1 - nrt_{\mathrm{WI}})\frac{t_{\mathrm{minB}}}{t_{\mathrm{WI}}} + nr\ \max(t_{\mathrm{minB}}, t_{\mathrm{BC}} + E[t_{\mathrm{TUD}}]) \tag{18.9}$$

where n is the number of sources. We assume that the device keeps t_{minB} even if the transmission is finished within the time. In (18.9), the effect of n is small since r has very small values in our scenario. Due to the limited page budget, the numerical results are not presented here. Nonetheless, it is noted that all lines of receivers stay very close to each other and show the similar trend for the case of 0.0003125 in Fig. 18.8.

18.4.1.2 Beacon Tracking. Most values for BT are the same as for NBT since both algorithms operate on the beacon mode. However, as presented in Fig. 18.3, the average active time in a beacon interval is twice because it has two active durations for incoming and outgoing superframes. Therefore, the average active ratio of an observing device is

$$E_{\mathrm{OB}} = 2\,\frac{t_{\mathrm{minB}}}{t_{\mathrm{WI}}} \tag{18.10}$$

Compared with the NBT, the transmitter does not turn on the receiver to wait for a beacon. Thus, the energy consumption for transmitting activity is derived from (18.10) and energy consumption for transmission without tracking overhead is derived from the following:

$$E_{\mathrm{TX}} = (2 - rt_{\mathrm{WI}})\,\frac{t_{\mathrm{minB}}}{t_{\mathrm{WI}}} + r\ \max(t_{\mathrm{minB}}, t_{\mathrm{BC}} + E[t_{\mathrm{TUD}}]) \tag{18.11}$$

The values are compared with those of NBT in Fig. 18.9.

Compared with NBT, BT has no overhead for waiting for a beacon since a transmitter starts transmission activity synchronously based on the schedule information. Thus, the effect from a high arrival rate is small. In the Fig. 18.9, three lines for BT are overlapped. However, when t_{WI} is small, the active ratios are almost twice those of NBT. For the given range of arrival rates, some crossover points are observed. In BT, increasing the wakeup interval to the extent that the latency is acceptable is a good solution to maximize the battery life. The average active ratio E_{RX} of the receiver device is derived in a similar manner:

$$E_{\mathrm{RX}} = (2 - nrt_{\mathrm{WI}})\,\frac{t_{\mathrm{minB}}}{t_{\mathrm{WI}}} + nr\ \max(t_{\mathrm{minB}}, t_{\mathrm{BC}} + E[t_{\mathrm{TUD}}]) \tag{18.12}$$

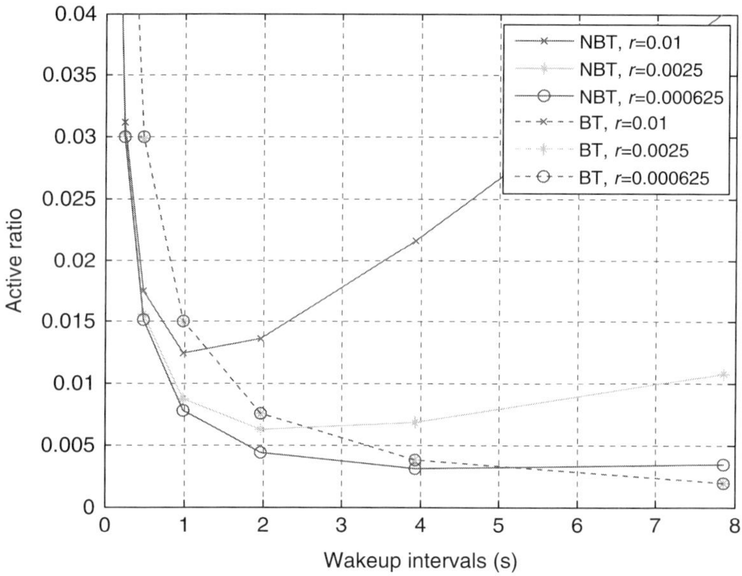

FIGURE 18.9 Comparison of active ratios of transmitters using NBT and NT when $n = 3$.

18.4.1.3 Long Preamble Emulation. The active duration of LP depends on the time to detect the sequence of WRs. If a device wakes up when a WR is just transmitted by another device, it has to pass that WR in transmission and wait for the next WR. This is the worst-case waiting time to receive a WR. Since LP operates on the nonbeacon mode, without considering slotted access, the minimum active duration t_{minW} is decided as

$$t_{\text{minW}} = (2^{\text{minBE}} - 1)t_{\text{slot}} + t_{\text{slot}} + 2t_{\text{WR}} + t_{\text{TR}} \tag{18.13}$$

where $2t_{\text{WR}}$ is the time for two WRs in the worst case, and t_{TR} is the turnaround time to start backoff for the second WR; $(2^{\text{minBE}} - 1)t_{\text{slot}}$ is the maximum bakcoff time, and t_{slot} is the time to perform CCA once as defined in IEEE 802.15.4 for the nonbeacon mode. We define $E[t_{\text{WO}}]$ as the average time for a WR transmission, given by

$$E[t_{\text{WO}}] = E[t_{\text{BKO}}] + t_{\text{slot}} + t_{\text{WR}} + t_{\text{TR}} \tag{18.14}$$

The time to transmit a unicast frame is also defined similar to (18.6) and (18.7), respectively,

$$t_{\text{UD}} = L_{\text{data}}t_b + t_{\text{TR}} + L_{\text{Ack}}t_b \tag{18.15}$$

$$E[t_{\text{TUD}}] = E[t_{\text{BKO}}] + t_{\text{slot}} + t_{\text{UD}} \tag{18.16}$$

The average active ratio E_{OB} of a device just observing without transmitting or receiving a data frame is given similar to (18.4):

$$E_{OB} = \frac{t_{minW}}{t_{WI}} \tag{18.17}$$

Note that t_{minW} is smaller than t_{minB} in (18.3). Therefore, values gathered from (18.17) are smaller than those from (18.4). The average active ratio of a LP transmitter is

$$E_{TX} = (1 - rt_{WI})\frac{t_{minW}}{t_{WI}} + r\left(\left\lceil \frac{t_{WI}}{E[t_{WO}]} \right\rceil E[t_{WO}] + E[t_{UD}]\right) \tag{18.18}$$

where $\lceil t_{WI}/E[t_{WO}] \rceil$ is the number of WRs to cover the time longer than t_{WI}. Once the transmission activity is started, it is always overlapped with its own wakeup. Therefore, rt_{minW} is subtracted in the first term of (18.18). In Fig. 18.10, the active ratios are compared with those of NBT.

Since the active duration and transmission activity of devices are not synchronized, both algorithms are sensitive to t_{WI} and r. Especially, as a transmitter of LP transmits WRs longer than t_{WI}, the average overhead for transmission is twice the averaged overhead to receive a beacon in NBT. However, when the arrival rate is low and the wakeup interval is small, the active ratios are smaller than those of NBT as shown in Fig. 18.10.

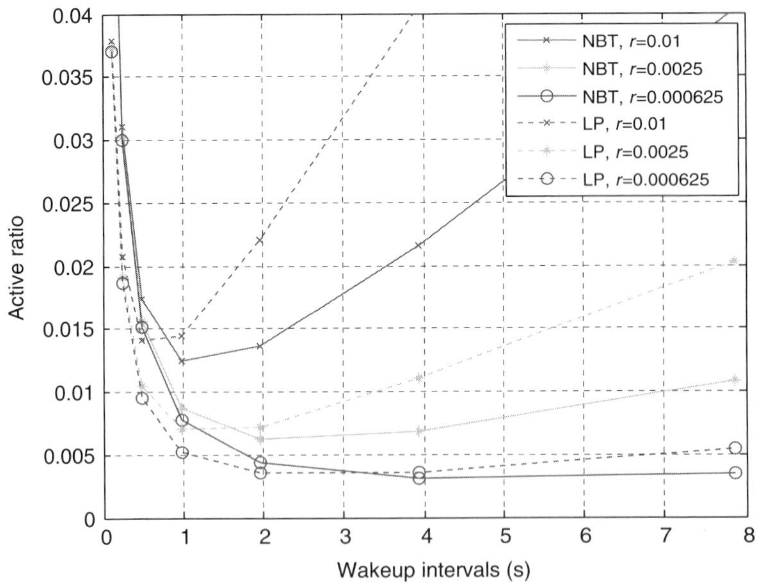

FIGURE 18.10 Comparison of active ratios of transmitters using NBT and LP when $n = 3$.

One possible issue of LP is energy consumption at the receiver. As presented in Fig. 18.4, the receiver also has to wait for the data frame that comes after the sequence of WRs. The average active ratio of the receiver is derived as

$$E_{\mathrm{RX}} = (1 - nrt_{\mathrm{WI}})\frac{t_{\mathrm{minW}}}{t_{\mathrm{WI}}} + nr \left(\left\lceil \frac{t_{\mathrm{WI}}}{E[t_{\mathrm{WO}}]} - 1 \right\rceil \frac{E[t_{\mathrm{WO}}]}{2} + E[t_{\mathrm{WO}}] + E[t_{\mathrm{UD}}] \right) \tag{18.19}$$

Compared to (18.18), the receiver wakes up in the middle of the WR sequence on average by the uniform distribution. Therefore, $\lceil t_{\mathrm{WI}}/E[t_{\mathrm{WO}}] - 1 \rceil E[t_{\mathrm{WO}}]$ is divided by 2. In addition, this energy consumption is required for every frame reception from all transmitters. For n transmitters, the frame reception rate at the receiver is nr. E_{RX} of NBT and LP are compared when n is 3 in Fig. 18.11.

The active ratio of the transmitters, E_{TX}, increases rapidly as t_{WI} and r increase. The slopes of the curves are much steeper than those of E_{TX}. It is because the frame transmission invokes the overhead of the receiver at the rate of nr. If the receiver collects data from a number of transmitters, the energy consumption of the device can be excessive. The effect of varying the number of transmitters n is presented in Fig. 18.12 with r fixed at 0.0025, where $n = 0$ means no transmitter.

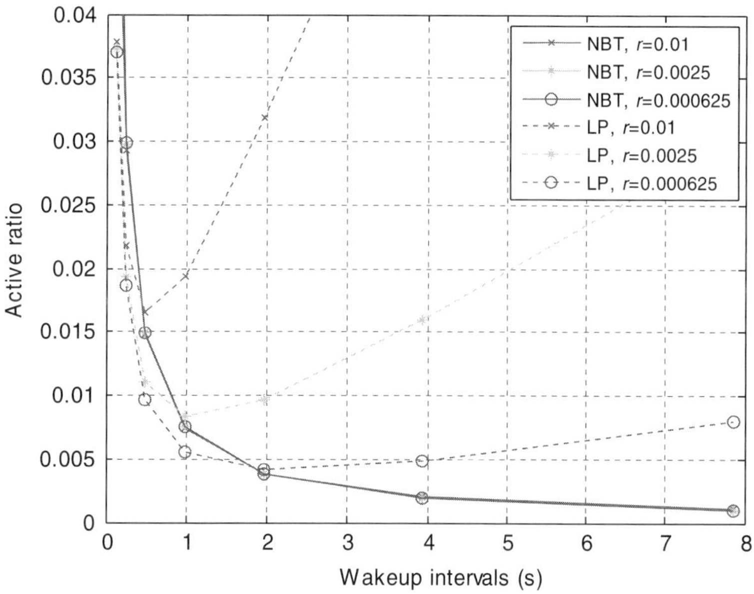

FIGURE 18.11 Comparison of active ratios of receivers using NBT and LP when $n = 3$.

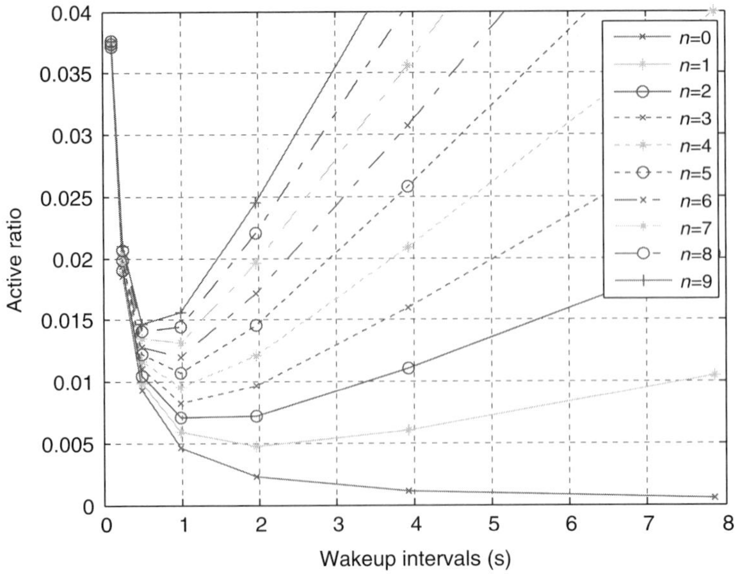

FIGURE 18.12 Comparison of active ratios of receivers using LP when $r = 0.0025$.

18.4.1.4 Long Preamble Emulation with Acknowledgment. LPA evolved from LP to reduce the overhead from the long WR sequence. The WR transmission stops if the transmitter receives a WA. For the WA reception, before transmitting the next WR, a transmitter should wait for at least $(2^{\text{minBE}} - 1)t_{\text{slot}} + t_{\text{slot}} + t_{\text{WA}}$. Therefore, if a device wakes up just after a transmitter transmitted a WR, it has to wait for the time for WA and the next WR to detect the sequence. Based on this, the minimum active duration t_{minW} is defined as

$$t_{\text{minW}} = 2(2^{\text{minBE}} - 1)t_{\text{slot}} + 2t_{\text{slot}} + 2t_{\text{TR}} + 2t_{\text{WR}} + t_{\text{WA}} \tag{18.20}$$

where $2(2^{\text{minBE}} - 1)t_{\text{slot}} + 2t_{\text{slot}}$ are reserved times for performing backoff and CCA for a WR and a WA. The average time $E[t_{\text{WO}}]$ for a WR transmission operation is defined with a manner similar to (18.14).

$$E[t_{\text{WO}}] = E[t_{\text{BKO}}] + (2^{\text{minBE}} - 1)t_{\text{slot}} + 2t_{\text{slot}} + 2t_{\text{TR}} + t_{\text{WR}} + t_{\text{WA}} \tag{18.21}$$

We define the average time $E[t_{\text{ATO}}]$ to transmit a WR, a WA, a data frame, and an Ack as follows:

$$E[t_{\text{ATO}}] = 2t_{\text{TR}} + 2E[t_{\text{BKO}}] + 2t_{\text{slot}} + t_{\text{WR}} + t_{\text{WA}} + E[t_{\text{TUD}}] \tag{18.22}$$

In (18.22), $2t_{\mathrm{TR}}$ is the turnaround time after a WR and a WA to change the modes of the transceiver from transmission to reception. Then, the average active ratio of the LP transmitter is derived as follows:

$$
\begin{aligned}
E_{\mathrm{TX}} = & \left\{1 - r\left(\left\lceil \frac{t_{\mathrm{WI}}}{E[t_{\mathrm{WO}}]} - 1 \right\rceil \frac{E[t_{\mathrm{WO}}]}{2} + E[t_{\mathrm{ATO}}]\right)\right\} \frac{t_{\mathrm{minW}}}{t_{\mathrm{WI}}} \\
& + r\left(\left\lceil \frac{t_{\mathrm{WI}}}{E[t_{\mathrm{WO}}]} - 1 \right\rceil \frac{E[t_{\mathrm{WO}}]}{2} + E[t_{\mathrm{ATO}}]\right)
\end{aligned}
\tag{18.23}
$$

In (18.23), $\lceil t_{\mathrm{WI}}/E[t_{\mathrm{WO}}] - 1\rceil E[t_{\mathrm{WO}}]/2 + E[t_{\mathrm{ATO}}]$ is the average time to transmit one data frame by the uniform distribution. In the best case, a WA is transmitted only after one WR. It is counted in $E[t_{\mathrm{ATO}}]$. In the worst cast, the WR is transmitted $\lceil t_{\mathrm{WI}}/E[t_{\mathrm{WO}}]\rceil$ times. The probability that the transmission activity is overlapped with the periodic active duration is taken into account as in (18.8), considering the transmission time. E_{TX} is compared with those of LP in Fig. 18.13.

When the wakeup interval is long and the arrival rate is high, the active ratios of an LPA transmitter are almost half those of LP. However, when the wakeup interval is short and the arrival rate is small, the ratios are twice those of the LP. Therefore, the two algorithms should be used carefully considering

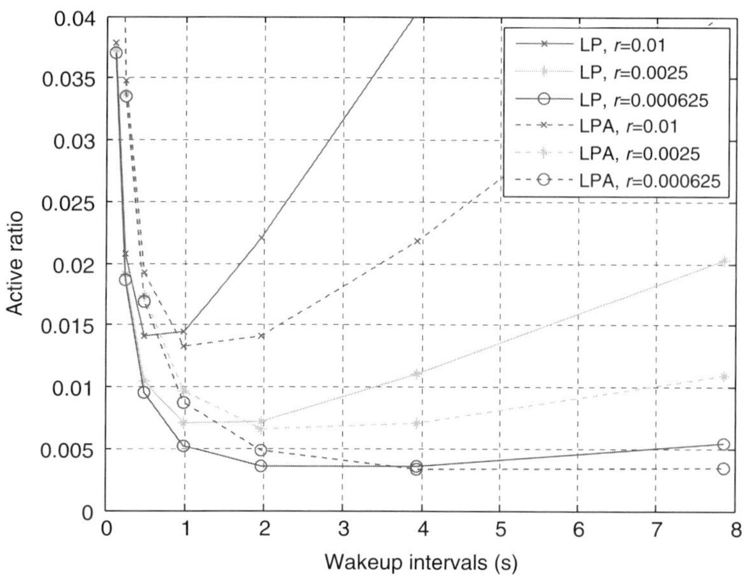

FIGURE 18.13 Comparison of active ratios of transmitters using LP and LPA when $n = 3$.

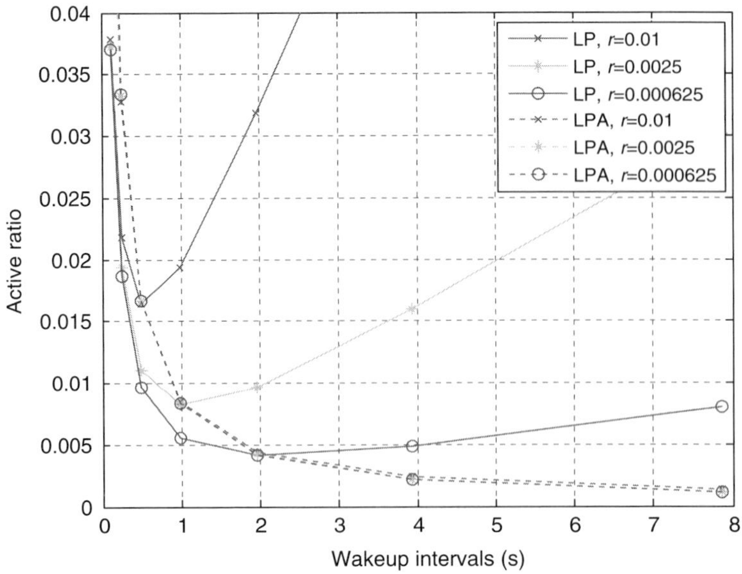

FIGURE 18.14 Comparison of active ratios of receivers using LP and LPA when $n = 3$.

the given environment. The active ratios of the receiver are derived as

$$E_{\mathrm{RX}} = (1 - nr)\frac{t_{\mathrm{minW}}}{t_{\mathrm{WI}}} + nr\left(\frac{E[t_{\mathrm{WO}}]}{2} + E[t_{\mathrm{ATO}}]\right) \tag{18.24}$$

Since the receiver wakes up in the middle of the wakeup operation, it consumes $E[t_{\mathrm{WO}}]/2$ on average before receiving a WR. The values are compared in Fig. 18.14.

Differently from LP, the active ratio of the LPA receiver is not sensitive to the arrival rate since the receiver just replies with a WA and receives a data frame. For the same reason, the effect from the number n of transmitters is small.

18.4.1.5 Long Preamble with Ack after Local Synchronization.

Energy consumption of LPAS highly depends on the accuracy of estimation. If the estimation fails, the average energy consumption will be the same as that of LP. We assume that the estimation is reasonably accurate that the transmitter transmits just one or two WRs on average. Based on that assumption, (18.23) is changed as

$$E_{\mathrm{TX}} = \left\{1 - r\left(\frac{E[t_{\mathrm{WO}}]}{2} + E[t_{\mathrm{ATO}}]\right)\right\}\frac{t_{\mathrm{minW}}}{t_{\mathrm{WI}}} + r\left(\frac{E[t_{\mathrm{WO}}]}{2} + E[t_{\mathrm{ATO}}]\right) \tag{18.25}$$

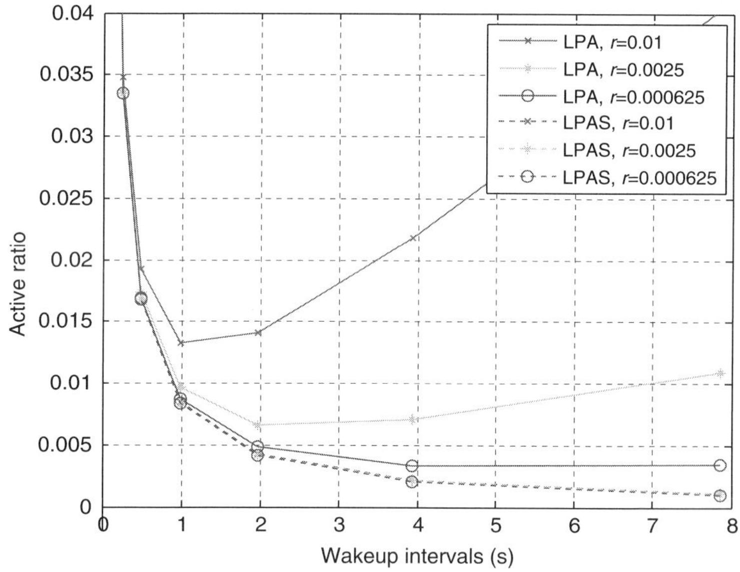

FIGURE 18.15 Comparison of active ratios of transmitters using LP and LPA when $n = 3$.

The results from (18.25) are compared with those of the LPA from (18.23) in Fig. 18.15.

The active ratios of the transmitter do not increase as the wakeup interval increases. The three lines for LPAS are overlapped. When the wakeup interval and the arrival rate are small, the results are close to LPA. The active ratios for the receiver are the same as those of LPA presented in (18.24).

18.4.1.6 Global Synchronization. As explained earlier, GS requires two time spaces for a SYNC packet and an RTS. Since both of them are command frames defined in the upper layer, differently from a beacon, it requires time to perform backoff and CCA. The minimum active duration t_{minS} for GS is defined as

$$t_{\text{minS}} = 2(2^{\text{minBE}} - 1)t_{\text{slot}} + 2t_{\text{slot}} + t_{\text{SYN}} + t_{\text{RTS}} \tag{18.26}$$

where t_{SYN} and t_{RTS} are time for a SYNC and an RTS frame. We define the average time $E[t_{\text{ATO}}]$ to transmit a SYNC, an RTS, a CTS, a data frame, and an Ack similarly to (18.22) as follows:

$$E[t_{\text{ATO}}] = 3t_{\text{TR}} + 3E[t_{\text{BKO}}] + t_{\text{SYN}} + t_{\text{RTS}} + t_{\text{CTS}} + E[t_{\text{TUD}}] \tag{18.27}$$

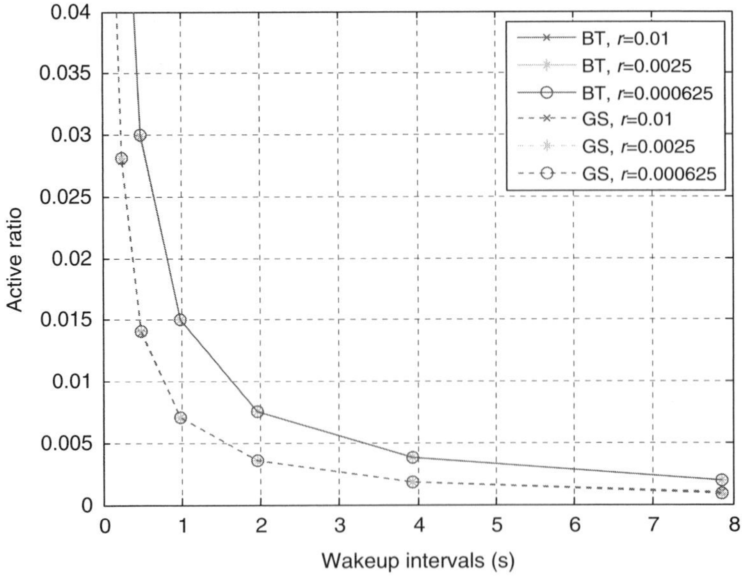

FIGURE 18.16 Comparison of active ratios of transmitters using NBT and GS when $n = 3$.

where $E[t_{\mathrm{TUD}}]$ is defined in (18.16). The average active ratio of the transmitter is derived as

$$E_{\mathrm{TX}} = (1 - rt_{\mathrm{WI}})\frac{t_{\mathrm{minS}}}{t_{\mathrm{WI}}} + rE[t_{\mathrm{ATO}}] \tag{18.28}$$

The E_{TX} is compared with those of BT in Fig. 18.16.

Both algorithms are synchronous. The overhead from the increasing frame arrival rate and the wakeup interval do not increase the active ratios. However, since BT has two active durations, the active ratios of BT are almost twice that of GS. The possible problem with GS is the difficulty of implementation in multihop networks since it has to use functions provided by the MAC layer not directly accessing the channel. The active ratio of a receiver is similarly derived as:

$$E_{\mathrm{RX}} = (1 - nrt_{\mathrm{WI}})\frac{t_{\mathrm{minS}}}{t_{\mathrm{WI}}} + nrE[t_{\mathrm{ATO}}] \tag{18.29}$$

18.4.1.7 Comparison and Optimization. In this section, we present comprehensive comparison of six algorithms and their potential improvements. When n and r are 3 and 0.00025, respectively, the active ratios of transmitters adopting the six algorithms are compared in Fig. 18.17.

The first observation is the fact that there is no clear winner. When the wakeup interval is small, LP claims the edge. On the other hand, when the

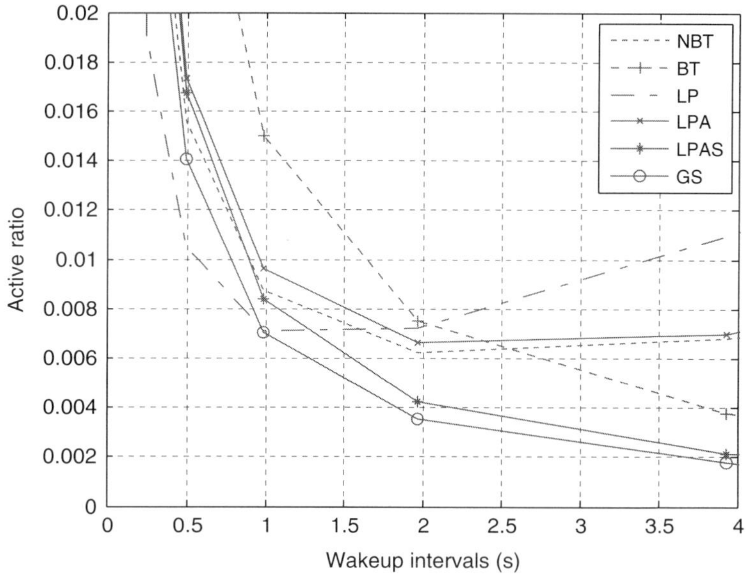

FIGURE 18.17 Comparison of active ratios of transmitters using six algorithms when $n = 3$ and $r = 0.0025$.

wakeup interval is large, GS is the winner. Therefore, in the given scenario, in order to maximize the battery life for small wakeup intervals, LP is the best option. If, however, we can choose a long wakeup interval without considering the latency, LP is not a good choice. Another fact seen in the figure is that the algorithms that include NBT, LP, and LPA are sensitive to the wakeup interval and the arrival rates. This is because the transmission activity is started asynchronously.

In the given scenarios, the average active ratios of all algorithms are over 0.01 when the wakeup interval is smaller than 0.5 s. It means that the lifetimes of the devices are less than one year. A device using any power-saving algorithm wakes up periodically for possible reception. Therefore, the active ratio is decided by the active duration and the wakeup interval when the frame arrival rate is extremely low. If the wakeup interval is fixed, the key contributing factor is the minimum active duration. The value of the minimum active duration of each algorithm evaluated using the parameters in Table 18.1 is presented in Table 18.2. When the wakeup interval is 0.5 s, the only way to support a one-year lifetime with the batteries is to reduce the active duration further. In every algorithm, there are some possibilities to reduce the active duration.

In NBT and BT, a device assigns the active duration for the maximum size frame. Since the frame arrival rate is low, the time duration wastes energy considerably. If a short command frame such as a RTS in GS is transmitted

TABLE 18.2 Minimum Active Duration

Algorithms	Values (s)
NBT	0.0074
BT	0.0147
LP	0.0045
LPA	0.0082
LPAS	0.0082
GS	0.0069

ahead of the data frame, the time duration defined in (18.3) can be changed as

$$t_{\text{minB}} = \left\lceil \tfrac{1}{10} L_B \right\rceil t_{\text{slot}} + (2^{\text{minBE}} - 1)t_{\text{slot}} + 2t_{\text{slot}} + \left\lceil \tfrac{1}{10} L_{\text{RTS}} \right\rceil t_{\text{slot}} \tag{18.30}$$

The minimum wakeup time of LP can be reduced by decreasing initial values of the backoff counter. In LP, a transmitter performs backoff before transmitting every WR. However, if a sequence of WRs is started, the channel should be occupied by the sequence, and it does not need to contend anymore. If the initial selection of backoff counter is minimized to $(2^1-1)t_{\text{slot}}$, the active duration defined in (18.13) can be optimized as follows:

$$t_{\text{minW}} = 2t_{\text{slot}} + 2t_{\text{WR}} + t_{\text{TR}} \tag{18.31}$$

The active duration of LPA and LPAS defined in (18.20) can also be optimized in the same way:

$$t_{\text{minW}} = 4t_{\text{slot}} + 2t_{\text{TR}} + 2t_{\text{WR}} + t_{\text{WA}} \tag{18.32}$$

It means that a transmitter waits for only $t_{\text{TR}} + 2t_{\text{slot}} + t_{\text{WA}}$ after transmitting a WR. Even though we do not analyze the effect, if a receiver replies the WA a little late, the transmitter may find the channel busy. In this case, the CSMA/CA mechanism at the MAC layer will retry after selecting a backoff counter value among 0 to 2^2-1.

The active duration of GS can be similarly reduced by minimizing the initial backoff counter, but doing so may cause SYNC collision as discussed above. However, if the wakeup interval is long and the number of devices within one hop range is small, the probability of SYNC collision is very low. If we select a backoff counter between 0 and 1, then the minimum active duration (18.26) is changed as

$$t_{\text{minS}} = (2^{\text{minBE}} - 1)t_{\text{slot}} + 3t_{\text{slot}} + t_{\text{SYN}} + t_{\text{RTS}} \tag{18.33}$$

where $(2^{\text{minBE}}-1)t_{\text{slot}}$ is for the backoff before an RTS. Even with the very low arrival rate, if the wakeup interval becomes very large, the collision probability

TABLE 18.3 Optimized Minimum Active Durations

Algorithms	Values (s)
ONBT	0.0038
OBT	0.0077
OLP	0.0026
OLPA	0.0044
OLPAS	0.0044
OGS	0.0050

in the active duration after a SYNC can be considerable because the frame arrival rate in the active time becomes rt_{WI}. Thus, reducing the backoff time can invoke addition overhead. It is the reason $(2^{\mathrm{minBE}}-1)t_{\mathrm{slot}}$ is kept in (18.30). The optimized values are shown in Table 18.3.

With the values in Table 18.3 and the same condition as in Fig. 18.17, the active ratios of optimized algorithms are presented in Fig. 18.18. The trend of lines is similar to those in Fig. 18.17. However, notice in the figure that all algorithms but the optimized BT (OBT) have values under 0.01 when the wakeup interval is 0.5 s. When the wakeup interval gets long, values of the optimized NBT (ONBT), the optimized LP (OLP), and the optimized LPA (OLPA) become close to those in Fig. 18.17. As the overhead from the

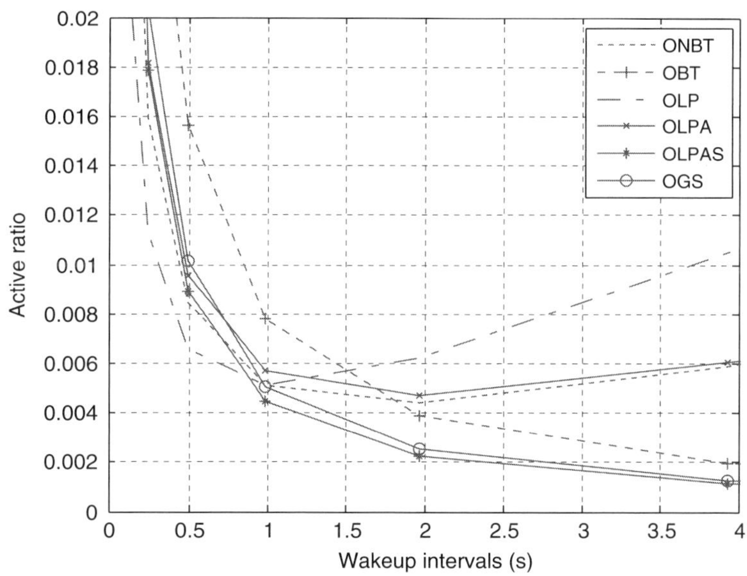

FIGURE 18.18 Comparison of active ratios of transmitters using six algorithms when $n = 3$ and $r = 0.0025$.

transmission becomes dominant when the wakeup interval is large, the optimization efforts in those algorithms do not affect much. The active ratios of the optimized LPAS (OLPAS) and the optimized GS (OGS) show good results. However, when it comes to real implementation, additional margin should be considered for those synchronous algorithms.

18.4.2 Latency

Latency in the power-saving MAC has two meanings. First, latency itself is a fundamental performance measure for any applications. Especially, the long inactive period may cause excessive latency not seen in mains-powered networks. Second, it can provide the guideline to select a power-saving algorithm and parameters to use to satisfy the latency required by applications. In the previous section, we compared the active ratios based on the same wakeup interval. However, if a certain algorithm has small latency compared with others, it may be compared with the active ratio calculated with longer wakeup intervals. As presented above, the average one-hop latency $E[D]$ is defined as the time when a transmission activity is started at the upper layer of the source to when a frame is received at the MAC layer of the destination. Therefore, it consists of times to wait for the schedule of the destination and to exchange command frames and the data frame.

In NBT, a device waits for the beacon of the destination device for $t_{\mathrm{WI}}/2$. Thus, the average latency $E[D]$ of NBT is derived as

$$E[D] = \tfrac{1}{2} t_{\mathrm{BI}} + \left\lceil \tfrac{1}{10} L_B \right\rceil t_{\mathrm{slot}} + E[t_{\mathrm{TUD}}] \tag{18.34}$$

where $E[t_{\mathrm{TUD}}]$ is defined in (18.7) and L_B is the length of a beacon. The device of BT transmitting a frame gives the same result since it has to wait for the same time even though it does not turn on the receiver while waiting. In LP, a device always transmits WRs for more than t_{WI}; therefore the latency is derived as

$$E[D] = \left\lceil \frac{t_{\mathrm{WI}}}{E[t_{\mathrm{WO}}]} \right\rceil E[t_{\mathrm{WO}}] + E[t_{\mathrm{UD}}] \tag{18.35}$$

where $E[t_{\mathrm{WO}}]$ and $E[t_{\mathrm{TUD}}]$ are given in (18.14) and (18.16), respectively. By introducing the WA, the time for WRs of LPA reduces to the half of LP. Thus, the average latency of LPA is

$$E[D] = \left\lceil \frac{t_{\mathrm{WI}}}{E[t_{\mathrm{WO}}]} - 1 \right\rceil \frac{E[t_{\mathrm{WO}}]}{2} + E[t_{\mathrm{ATO}}] \tag{18.36}$$

where $E[t_{\mathrm{WO}}]$, $E[t_{\mathrm{TUD}}]$, and $E[t_{\mathrm{ATO}}]$ are given in (18.21), (18.16), and (18.22), respectively. In LPAS, a transmitter waits for $t_{\mathrm{WI}}/2$ on average, then it transmits one or two WRs. Therefore, the latency is estimated as

$$E[D] = \tfrac{1}{2} t_{\mathrm{WI}} + \tfrac{1}{2} E[t_{\mathrm{WO}}] + E[t_{\mathrm{ATO}}] \quad (18.37)$$

where $E[t_{\mathrm{WO}}]$ and $E[t_{\mathrm{ATO}}]$ are given in (18.21) and (18.22), respectively. Finally, the latency of GS is derived as

$$E[D] = \tfrac{1}{2} t_{\mathrm{WI}} + E[t_{\mathrm{ATO}}] \quad (18.38)$$

where $E[t_{\mathrm{ATO}}]$ is defined in (18.27). The average latency for each algorithm is compared in Fig. 18.19.

Since the wakeup interval is very large compared to any other latency components, the waiting time for the active duration of the destination device becomes the dominant term for the overall latency. If a data frame is generated according to the uniform distribution, the average waiting time for the next active duration of the receiver is around $t_{\mathrm{WI}}/2$. Thus, except for LP, all algorithms have almost the same latency. In the case of LP, the latency is more than t_{WI} all the time since the destination cannot stop the transmitter from transmitting WRs.

However, the latency in the multihop path takes a little different pattern. Since the active time is minimized for one data frame transmission, at each hop

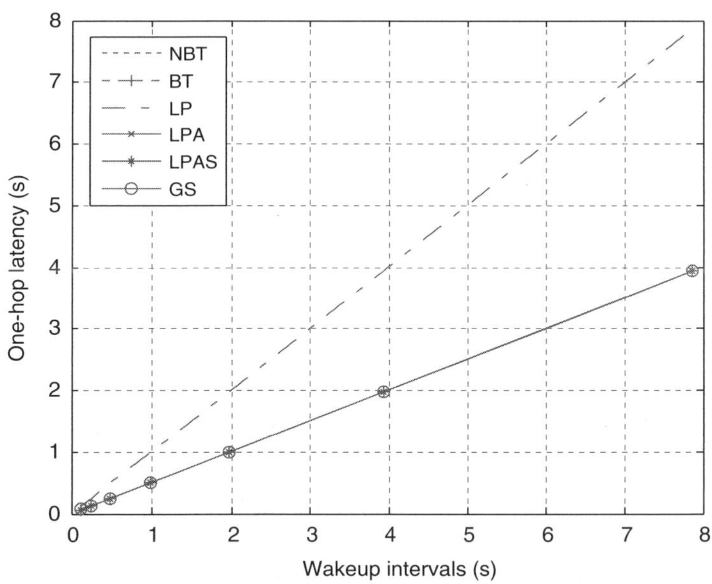

FIGURE 18.19 Comparison of latencies of transmitters using six algorithms when $n = 3$ and $r = 0.0025$.

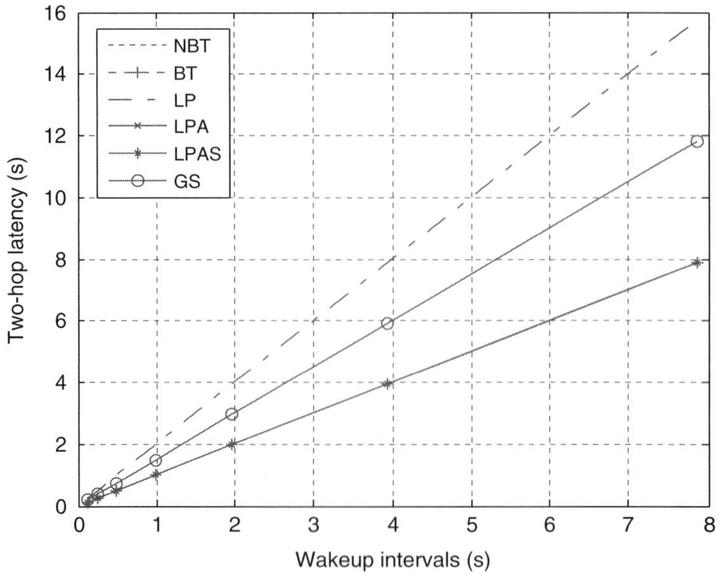

FIGURE 18.20 Comparison of latencies of transmitters using six algorithms when $n = 3$ and $r = 0.0025$.

a packet has to experience additional delay to wait for the next active duration of the next hop. If the wakeup schedules of two consecutive devices on the path are synchronized, a packet will be delayed for t_{WI} between two devices. Thus, the latency of GS for the second-hop will be similar to those of LP. The average two-hop latencies of the algorithms are presented in Fig. 18.20. The line for GS is in the middle of LP and other algorithms. If the number of hops increases, the latency will be closer to LP since at each hop t_{WI} is added.

18.5 CONCLUSIONS

In this chapter, we presented and analyzed six power-saving algorithms on IEEE 802.15.4. In order to minimize the power consumption from idle listening, all algorithms repeat short wakeup and long sleep periodically. In order to support communication in the active duration, each algorithm behaves differently. In order to evaluate the algorithms, the average active ratios are analyzed with the measure of energy consumption in various scenarios. The average latency to transmit a frame is also derived to consider effects from each power-saving algorithm.

In the active ratio comparison, LP shows the best results when the frame arrival rate and the wakeup interval are small since it has the minimum active duration among the algorithms compared. However, when those parameters increase, LPAS and GS show the best results. Since transmission activity of LPAS and GS are synchronized to the schedule of the destination, these algorithms can decrease active ratios without overhead on increasing wakeup intervals. However, one of the major concerns for these synchronous algorithms is the overhead from synchronization in real implementation. Since the algorithm is operated above IEEE 802.15.4, real implementation may require additional margin for estimating schedules and transmitting frames.

In the case of latency, the wakeup interval is the dominant factor. Therefore, the latencies of all algorithms increase linearly with the increasing wakeup intervals. In single-hop communication, only LP requires the full wakeup interval before transmitting a frame. If a packet is transmitted through multiple hops, GS has the similar characteristics to LP. Since the schedules of all devices are synchronized at each hop, it has to wait for the next wakeup interval to transmit a packet. Given the same latency for the multihop sensor networks, the active ratio of LPAS is the best since the longer wakeup interval than those of LP and GS can be used.

When the scenario is fixed with 0.5-s wakeup interval and 0.0025-frame/s arrival rate, all algorithms have the active ratios over 0.01. With the ratio, it is difficult to operate more than one year with most platforms today. Methods to extend battery life have also been presented and evaluated. With the extension, the active ratios of all optimized algorithms except BT are improved to under 0.01.

Currently, significant senor network research on IEEE 802.15.4 is underway. However, it is difficult to find real applications that are battery powered. The power-saving algorithms may be the first step to enable such applications. In practice, an adaptive algorithm would be helpful since the performance of algorithms highly depends on scenarios. Finally, finding suitable parameters by extensive experiments with real devices will be required.

REFERENCES

1. H. Karl and A. Will, *Protocols and Architectures for Wireless Sensor*, John Wiley, Hoboken, NJ, 2005.
2. J. Polastre, J. Hill, and D. Culler, "Versatile low power media access for wireless sensor networks," in *Proceedings of the 2nd International Conference on Embedded Networked Sensor Systems*, ACM Press, 2004, pp. 95–107.
3. M. Buettner, G. V. Yee, E. Anderson, and R. Han, "X-MAC: A short preamble MAC protocol for duty-cycled wireless sensor networks," in *Proceedings of the 4th International Conference on Embedded Networked Sensor Systems*, ACM Press, 2006, pp. 307–320.
4. E.-Y. A. Lin, J. M. Rabaey, and A. Wolisz, "Power-efficient rendez-vous schemes for dense wireless sensor networks." in *Proceedings of IEEE International Conference on Communications (ICC'04)*, Paris, France, June 2004.
5. C. C. Enz et al., "WiseNET: An ultralow-power wireless sensor network solution," *IEEE Comp*. 37(8) (2004).

6. W. Ye, J. Heidemann, and D. Estrin, "An energy-efficient MAC protocol for wireless sensor networks," paper presented at IEEE INFOCOMM, 2001, pp. 1567–1576.
7. IEEE 802.15.4-2003, "Part 15.4: Wireless LAN medium access control (MAC) and physical layer (PHY) specifications for low-rate wireless personal area networks (LR-WPANs)," IEEE, New York, 2003.
8. IEEE 802.15.4-2006, "Part 15.4: Wireless LAN medium access control (MAC) and physical layer (PHY) specifications for low-rate wireless personal area networks (LR-WPANs)," IEEE, New York, 2006.
9. Chipcon, 2.4GHz IEEE802.15.4/ZigBee-ready RF transceiver datasheet (rev. 1.2), Chipcon AS, Oslo, Norway, 2004.
10. Freescale, Technical data: MC13192/MC13193 2.4 GHz low power transceiver for the IEEE 802.15.4 standard (rev. 2.9), Freescale Semiconductor, 2005.
11. Crossbow Micaz motes, http://www.xbow.com.
12. Crossbow Telos motes, http://www.xbow.com.
13. M. Lee, J. Zheng, Y. Liu, H.-R. Shao, H. Dai, J. Zhang, and H. Jeon, "Combined proposal of enhancements to IEEE 802.15.4," available: ftp://ftp.802wirelessworld.com/15/04/15-04-0536-00-004b-combinedbeacon-scheduling-15-4b.ppt, 2004.
14. T. R. Park and M. J. Lee, "Stochastic beacon transmission in wireless sensor networks: IEEE 802.15.4 Case," paper presented at IEEE CCNC'07, Jan. 2007.
15. I. Chlamtac and S. Kutten, "Tree-based broadcasting in multihop radio networks," *IEEE Trans. Comput* 36(10), 1209–1223 (1987).
16. ZigBee Newtorking Group, "Network specification version 1.0," Zigbee document 02130r10, ZigBee Alliance, San Ramon, CA, Dec 2004.
17. Texas Instruments, cc2430 preliminary datasheet (rev. 2.01), Texas Instruments, Dallas, TX, 2006.
18. ATmega 128(L) Preliminary Complete, ATmel product documentation, Atmel, 2004.
19. MSP430 × 1xx Family User's Guide, Texas Instruments product documentation, Texas Instruments, 2004.

PART V

IEEE 802.16 WIRELESS MANs

CHAPTER 19

IEEE 802.16 MEDIUM ACCESS CONTROL AND PHYSICAL LAYERS

YANG XIAO, MICHAEL J. PLYLER, TIANJI LI, and FEI HU

19.1 INTRODUCTION OF IEEE 802.16

The IEEE 802.16 and IEEE 802.16a standards are the topic of this chapter. One of the most compelling aspects of broadband wireless access (BWA) technology is that networks can be created in just weeks by deploying a small number of base stations (BSs) on buildings or poles to create high-capacity wireless access systems. This type of standard is important for the developing world where wired infrastructures are limited.

The Institute of Electrical and Electronics Engineers (IEEE) decided to make BWA more available and standardized by developing the IEEE 802.16 standard. This standard addresses the connections in wireless metropolitan area networks (WMANs). It focuses on the efficient use of bandwidth and defines the medium access control (MAC) layer protocols that support multiple physical (PHY) layer specifications. These can easily be customized for the frequency band of use. This will allow them to be used in many countries around the world.

In recent years there has been increasing interest shown in wireless technologies for subscriber access as an alternative to THE traditional twisted-pair local loop.

These approaches are generally referred to as wireless local loop (WLL) or fixed-wireless access. To provide a standardized approach to WLL, the IEEE 802 committee set up the 802.16 working group (WG) in 1999 to develop broadband wireless standards.

IEEE 802.16 standardizes the air interface and related functions associated with WLL. Three WGs have been chartered to produce standards: IEEE 802.16.1—air interface for 10–66 GHz, IEEE 802.16.2—coexistence of BWA

Emerging Wireless LANs, Wireless PANs, and Wireless MANs. Edited by Y. Xiao and Y. Pan

systems, and IEEE 802.16.3—air interface for licensed frequencies, 2–11 GHz [1].

The work of 802.16.1 is the farthest along, and it is likely that it will generate the most interest in the industry as it targets at available frequency bands.

The 10–66-GHz part of the standard supports various traffic levels at many different frequencies for bidirectional communication. These bands are both licensed and unlicensed.

The IEEE 802.16 systems will allow thousands of users to share capacity for data, voice, and video. These networks are also easily scalable. As the demand grows for subscriber channels or cells, the carriers can expand these network implementations.

There are mechanisms in the WMAN MAC layer that provide different levels of quality of service (QoS) to support different application needs. Examples of these could be a network that requires voice and video, which require low latency but would allow for some transmission errors. The way that these different implementations for applications are done is by layering control over the MAC layer.

There have been systems in the past with fixed modulation. This standard supports an adaptive modulation and is more robust. These also support lower data rates. This is good in the sense that it would be compatible with more lower end devices (DEVs). Adaptive modulation also allows for more efficient use of bandwidth, which would allow for more customers.

The IEEE 802.16 standard supports frequency division duplexing (FDD) and time division duplexing (TDD). For the most part, FDD has been used in cell phone technology. The reason that TDD is also used is that it can dynamically allocate upstream and downstream bandwidth depending on traffic requirements.

An 802.16 WMAN provides communication between the subscribers and the core network. An example of a core network would be the Internet. Another example would be a phone company. As stated before, 802.16 is concerned with the air interface between the subscriber and a base transceiver station.

There are several protocols that have been addressed to handle these issues. There are essentially three layers that deal with this architecture.

The first, or lowest, layer is the PHY layer. This layer specifies the frequency band, the modulation scheme, error correction techniques, synchronization between transmitter (TX) and receiver (RX), data rate, and time division multiplexing (TDM) structure.

Just above this layer, the functions are associated with providing service to customers. Some of these functions are transmitting the data in frames with controlled access to the shared medium. These are grouped into a MAC layer. These define how and when a transmission on a channel is accessed and initiated. The MAC protocol must be able to allocate channels correctly in order to maintain QoS standards for the network.

The third layer is a convergence layer that provides functions that are specific for the service being currently provided.

IEEE 802.16 defines how wireless traffic will move between subscribers and core networks. First, a subscriber sends wireless traffic at speeds of 2–155 Mbps from a fixed antenna or building. Second, the BS receives transmissions from multiple sites and sends traffic over wireless or wired links to a switching center using the 802.16 protocol. The switching center sends traffic to an Internet service provider (ISP) or to a public switched telephone network (PSTN).

Figure 19.1 shows the IEEE 802.16 standard in reference to other standards that are similar to it. The first circle represents a smaller area of service. The smallest area is shown to be covered by the IEEE 802.15 wireless personal area network (WPAN). The second largest coverage area is the IEEE 802.11 wireless local area network (WLAN). The 802.16 network is the third largest coverage area according to the figure, leaving the wide area network, 802.20, as the largest coverage area of the wireless technologies currently available [1].

Figure 19.2 is an example of the layout of an 802.16 network. The parts of this network will be explained fully later. Notice that we are dealing with buildings and residents, not small DEVs. This network concentrates on filling in the gaps left by sparse local area networks (LANs) [1].

Established in 1999, the IEEE 802.16 WG produced the IEEE 802.16 specification for broadband WMANs, specifically called BWA, defining the PHY layer and MAC protocol, to provide wireless last mile broadband access as an alternative way for fixed broadband access such as cable modem and digital subscriber line (DSL). The specification was first approved in 2001 [1] and enhanced and approved again in 2004 for the IEEE 802.16-2004 version.

Established in 2001, the Worldwide Interoperability for Microwave Access (WiMAX) Forum aims to promote conformance and interoperability of the IEEE 802.16 specification and issues WiMAX Forum certification to vendors that pass conformance and interoperability testing. WiMAX for IEEE 802.16 is similar to WiFi for IEEE 802.11.

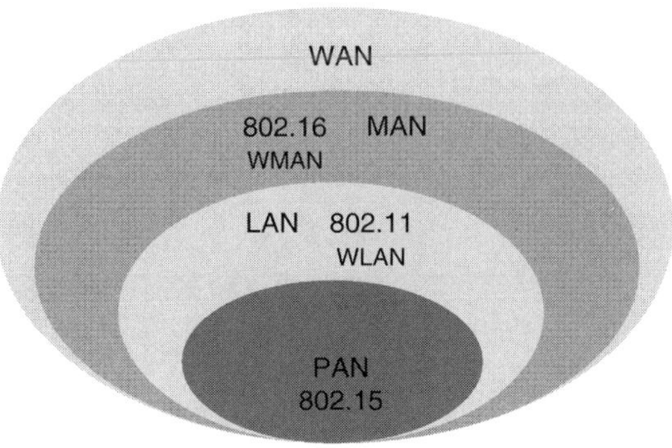

FIGURE 19.1 IEEE 802.16 standard in relation to other wireless standards.

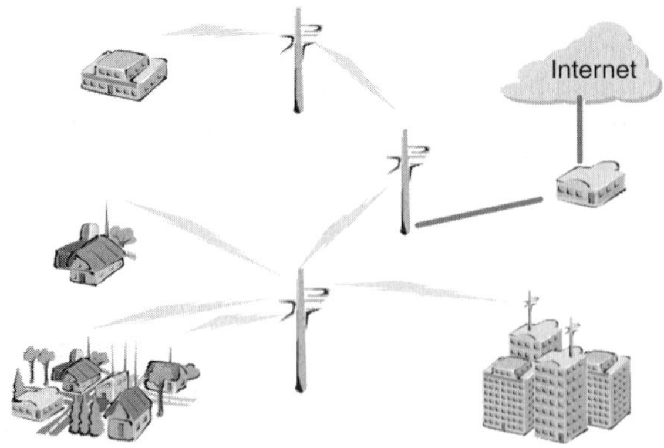

FIGURE 19.2 Example of IEEE 802.16 WMAN.

19.2 802 FAMILY

This standard defines an air interface for fixed BWA systems. This standard deals with metropolitan area networks (MANs), which are capable of supporting multiple PHY layer specifications for the frequency bands of the current application being implemented. The 10–66-GHz air interface is based on a single-carrier modulation known as the wireless MAN-SC air interface. 802.16a is an amendment that supports 2–11 GHz using an enhanced version of the same basic MAC layer along with PHY layer specifications.

Figure 19.3 shows that this standard is part of a family for LANs and MANs. Figure 19.3 shows the 802 family, which consists of the 802.10 security standard; 802.1 management standard; 802.2 logical link standard; 802.1 bridging standard; 802.3, 802.4, 802.5, 802.6, 802.11, and 802.12 MAC and PHY layer standards; etc.

The 10–66 GHz for this network, the IEEE 802.16 standard, provides a PHY layer environment where, due to the short wavelength, line of sight is required between the subscriber and transmitter. In this case, there are not multiple

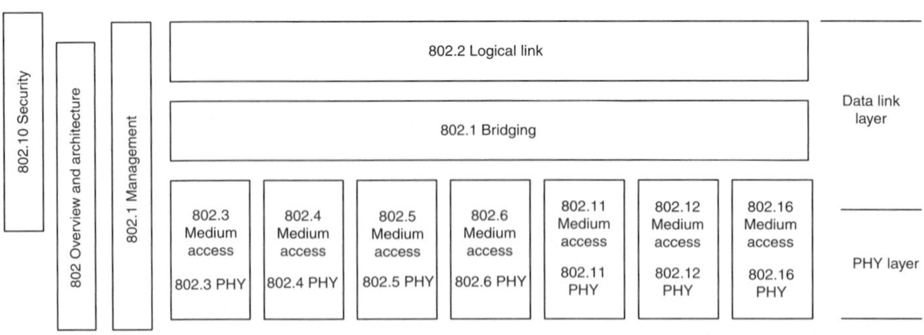

FIGURE 19.3 The 802 family.

paths possible for this network. Because of this, the channels are typically large. This means that the channels are 25–28 MHz wide. This is a typical physical configuration.

19.3 802.16 FAMILY

The IEEE 802.16 standardizes the air interface and related functions associated with the WLL. There have been three groups that have been characterized for this standard. The 802.16.1 is the air interface for the 10–66 GHz. The 802.16.2 is the standard for the coexistence of BWA systems. The 802.16.3 standard is the air interface for the licensed frequencies of 2–11 GHz.

Of the standards listed above, 802.16.1 has had the most work completed thus far. This standard will probably generate the most interest in industry since it targets available frequency bands.

An 802.16 wireless service provides a communications path between a subscriber site and a core network. A few examples of core networks are the public telephone network and the Internet.

Some protocols are defined specifically for wireless transmission of blocks of data over a network. The standards are organized into a three-layer architecture.

The lowest layer is the PHY layer. This specifies the frequency bands, modulation schemes, error correction techniques, synchronization between TX and RX, data rate, and TDM structure.

The layer above the PHY layer, the one associated with the functions that provide service to subscribers, has functions associated with transmitting data on frames and controlling access to the shared wireless medium. These are all grouped into the MAC layer. The protocol in this layer defines how and when a BS or subscriber station (SS) may initiate transmission on the channel. The MAC layer must also allocate the medium resources to satisfy the QoS demands.

The highest of the three layers of the architecture described in this section is above the MAC layer. This layer is a convergence layer that provides functionality to the service that is currently being provided to the subscriber. For IEEE 802.16.1, this includes digital, audio/video multicasting, digital telephony, asynchronous transfer mode (ATM), Internet access, frame relay, and wireless trunks in telephone networks and frame relay.

The 802.16 networks work in the following way: A subscriber sends wireless traffic at speeds ranging from 2 to 155 Mbps from a fixed antenna on a building. The BS receives transmissions from multiple sites and sends traffic over wireless or wired links to a switching center using the 802.16 protocols. Then, the switching center sends traffic to an ISP or the PSTN.

19.4 SUBLAYERS FOR MAC

The raw data rates seen in the 802.16 standard can be seen with an excess of 128 Mbps, and there have been several implementations that have reached

155 Mbps. This is suitable for point-to-multipoint configurations. An example of a configuration like this would be a small office/home office (SOHO). This configuration of the medium could be used to implement large office applications.

Figure 19.4 shows the reference model for the 802.16 standard. This shows the protocol layering and service access points (SAPs). The MAC layer has three sublayers, according to this figure. The first layer is the service specific convergence sublayer (CS). This provides any transformation or mapping of external network data received through CS SAP into MAC service data units (SDUs) received by the MAC common part sublayer (CPS) through the MAC SAP.

This first layer includes classifying external network SDUs and associating them to the proper MAC service flow and connection identifier (CID). This data packet could also include some degree of header suppression.

The second layer is where the MAC CPS provides the core MAC functionality of system access, bandwidth allocation, connection establishment, and connection maintenance. It receives data from the various CSs through the MAC SAP classified to a particular MAC connection. QoS is applied to the transmissions and receptions and scheduling data over the PHY layer.

The third layer is the PHY layer, which can include multiple specifications, each appropriate to a particular frequency range and application.

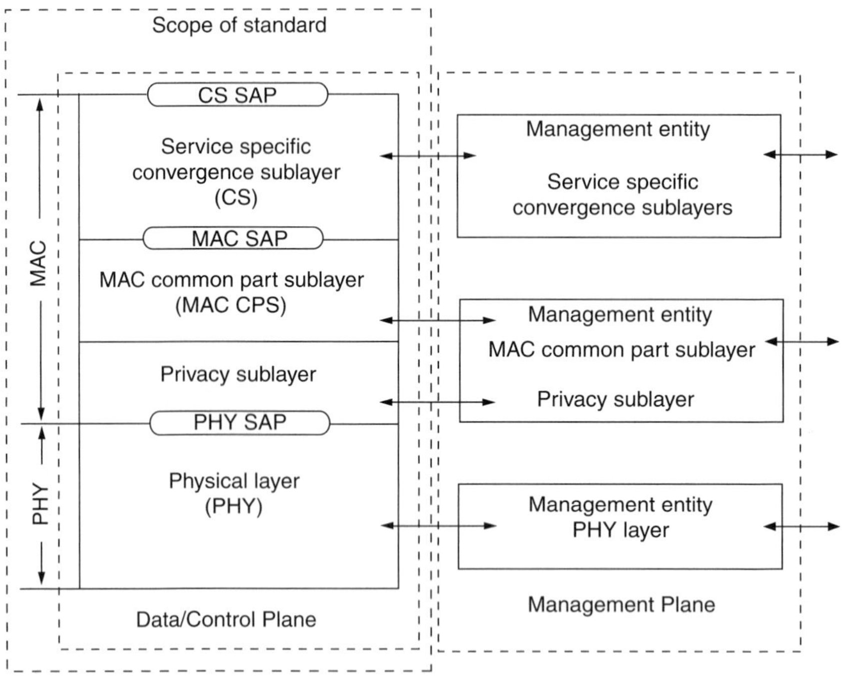

FIGURE 19.4 Protocol layering with SAPs.

19.4.1 Service Specification CS

The service specification CS resides on top of the MAC CPS. By way of the MAC SAP, it utilizes the CS. It performs several different functions. Some of these are accepting high layer packet data units (PDUs) from the higher layer, performing classification of higher layer PDUs, processing (if it is required) the higher layer PDUs based on the classification, and delivering CS PDUs to the appropriate MAC SAP, where it receives CS PDUs from the peer entity in a peer-to-multipoint application.

There are two different specifications provided for the CS in the IEEE 802.16 standard: the *ATM and CS* and the *pocket CS*. These are the only two existing specifications; others may be specified in the future. Currently, work is being done to this effect.

19.4.2 ATM CS

The ATM CS is a logical interface that associates different ATM services with the MAC CPS SAP. The ATM CS accepts ATM calls from the ATM layer, performs classification and, if provisional, payload header suppression (PHS), and delivers CS PDUs to the appropriate MAC SAP.

ATM connections may be either virtual circuit (VC) switched or virtual path (VP) switched. Additionally, there are a variety of ATM adaptation layers (AALs), such as AAL-1, AAL-2, and AAL-5. The ATM CS must be able to efficiently support all of these ATM services without burdening the MAC with knowledge of ATM. To this end, the ATM CS allows connections to be handled in either of two modes. Which mode is used for a particular connection is part of the initial connection provisioning [1].

19.4.3 Packet CS

The packet CS resides on the top of the MAC CPS layer. It performs the following functions: classification of the higher layer protocol PDU into the appropriate connector, suppression of payload header information, delivery of the resulting CS PDU to the MAC SAP associated with the service flow for transport to the peer MAC SAP, receipt of the CS PDU from the peer MAC SAP, and the rebuilding of any suppressed payload header information. The suppression of payload header information and the rebuilding of any suppressed payload header information are optional.

The sending CS is responsible for delivering the MAC SDU to the MAC SAP. The MAC is responsible for the delivery of the MAC SDU to peer MAC SAP in accordance with the QoS, fragmentation, concatenation, and other transport connections. These connections have to do with the connections of the service flow and their characteristics. The receiving CS is responsible for accepting the MAC SDU from the peer MAC SAP and delivering it to a higher layer entity.

The packet CS is used for transporting. This is true of all packet-based protocols such as Internet protocol (IP), point-to-point protocol (PPP), and the IEEE 802.3 standard, i.e., the ethernet standard.

For the downstream channel, the BS decides what to put in which subframe. We have stated that the upstream channel is more complicated. The allocation is tied to satisfy the QoS [1].

There are basically four classes of service. The classes are the constant-bit-rate service, real-time variable-bit-rate service, non-real-time variable-bit-rate service, and best-efforts service [1].

19.4.4 Constant-Bit-Rate Service

The constant-bit-rate service is intended for transmitting uncompressed voice, such as on a T1 channel. With this service, there is a predetermined amount of data sent at a predetermined time interval. This is achieved by dedicating certain time slots to each connection that is of the same type. Once the bandwidth has been allocated, the time slots are available automatically. This means that there is no need to send requests to the BS for the time slots [1].

19.4.5 Real-Time Variable-Rate Service

The real-time variable-rate service is intended for compressed multimedia and other soft, real-time applications. In this type of application, the bandwidth needed may differ at each instance. This is accommodated by the BS polling the subscriber at a fixed interval to ask how much bandwidth is needed at that interval time [1].

19.4.6 Non-real-time Variable-Bit-Rate Service

The non-real-time variable-bit-rate service is intended for transmissions that involve things such as file transfers. This is accomplished by the BS polling the subscriber frequently. This, however, does not have to be at a regular interval. There is a bit that can be set by the subscriber to make the BS poll in order to vary the bit traffic rate [1].

If a station does not respond to a poll for a certain number of times, the BS puts the station into a multicast group. The SS's poll is taken away [1]. When the multicast group is polled, any of the stations in the group can respond and contend for the service. This process helps to prevent stations with little or no traffic from wasting valuable polling and bandwidth time [1].

19.4.7 Best-Effort Service

The best-effort service is intended for all other types of transmission services. In this case, there is no polling. The subscriber must contend for bandwidth with all of the other subscribers grouped into this category. A request (REQ) for bandwidth is made in the time slots marked in the upstream map as available for contention. If a REQ is not successful, the subscriber has to try again at a later time. You might think at this point that there will be collisions, but this is avoided by using the ethernet's binary exponential backoff algorithm. This is where the contention window doubles when a backoff period of time still does not result in a channel access [1].

19.5 MAC SDU FORMAT

Figure 19.5 shows the MAC SDU format. The higher layer PDUs should be encapsulated in the MAC SDU format. For some payload protocols, each payload consists of an 8-bit PHS index (PHSI) field followed by the actual payload. Other protocols map the higher layer PDU directly to the MAC SDU. A value of zero in the PHSI indicates no PHS has been applied to the PDU.

19.5.1 Classification

Classification is the process by which a MAC SDU is mapped into a particular connection for transmission between MAC peers. This mapping process associates a MAC SDU with a connection, which also creates an association with the service flow characteristics of that connection. This process facilitates the delivery of the MAC SDUs with the correct QoS level [1].

The classifier itself is a set of matching criteria applied to each packet entering the network. These sets of criteria consist of some of the protocol-specific packet-matching criteria. An example of this could be a destination IP address. The sets also consist of a classifier priority and a reference to a CID. If a packet matches the specified packet-matching criteria, it is then delivered to the SAP for delivery on the connection defined by the CID. The QoS is provided by the service flow characteristics of the connection.

It is possible for a packet to fail to match the set of defined classifiers. The CS may either associate the packet with a default CID or discard the packet. This action, as well as others, is vendor specific. Most IEEE standards deal more with MAC and PHY layer specifications. They leave room for manufacturer

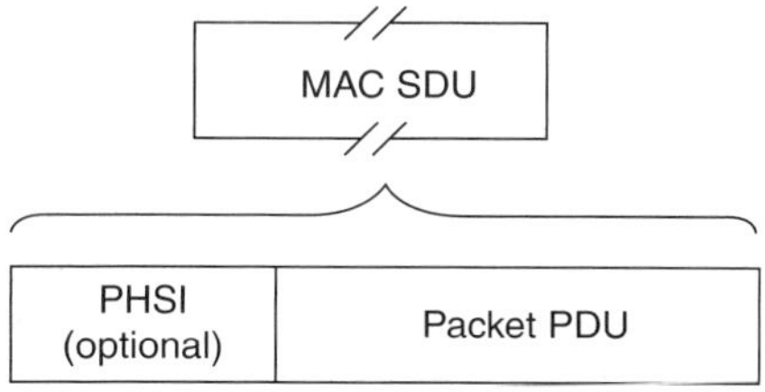

FIGURE 19.5 MAC SDU format.

changes and specifications. The main part of the standard is still intact, but some of the other implementation details that do not have to do with the MAC and PHY specifications are left for open interpretation or implementation with other things that are already proven to work well [1].

19.5.2 Payload Header Suppression

In PHS, a repetitive portion of the payload headers of the higher layer is suppressed in the MAC SDU by the sending entity and restored by the receiving entity. On the uplink, the sending entity is the SS and the receiving entity is the BS. On the downlink, the sending entity is the BS and the receiving entity is the SS. Each MAC SDU is prefixed with a PHSI, which references the payload header suppression field (PHSF) [2].

The sending entity uses classifiers to map packets into a service flow. The classifier uniquely maps packets to its associated PHS rule. The receiving entity uses the CID and the PHSI to restore the PHSF. Once a PHSF has been assigned to a PHSI, it should not be changed. To change the value of a PHSF on a service flow, a new PHS rule should be defined, the old rule removed from the service flow, and the new rule added. When a classifier is deleted, any associated PHS rule should also be deleted. PHS has a payload header suppression valid (PHSV) option to verify or not verify the payload header before suppressing it. PHS also has a payload header suppression mask (PHSM) option to allow select bytes not to be suppressed. This is used for sending bytes that change, such as IP sequence numbers, while still suppressing bytes that do not change.

The BS should assign all PHSI values just as it assigns all CID values. Either the sending or the receiving entity should specify the PHSF and the payload header suppression size (PHSS). This provision allows for preconfigured headers or for higher level signaling protocols outside the scope of this specification to establish cache entries. PHS is intended for unicast service and is not defined for multicast service [1].

The sending entity uses classifiers to map packets into a service flow. The classifier uniquely maps packets to its associated PHS rule. The receiving entity uses the CID and the PHSI to restore the PHSF. Once a PHSF has been assigned to a PHSI, it will not be changed. To change this value of a PHSF on a service flow, a new PHS rule has to be defined and added; then the old rule is removed from the service flow and the new rule is added. Once a classifier is deleted, any associated PHS rule should also be deleted.

It is the responsibility of the high layer service entity to generate a PHS rule which uniquely identifies the suppressed header within the service flow. It is also the responsibility of the higher layer service entity to guarantee that the byte strings being suppressed are constant from packet to packet for the duration of the active service flow.

19.5.3 PHS Signaling

The PHS signaling section of the IEEE 802.16 standard has the PHS requiring three different objects: the service flow, the classifier, and the PHS rule. The three objects may be created either simultaneously or in separate message flows.

PHS rules are created in one of two ways: by dynamic service addition (DSA) or by dynamic service change (DSC) messages. The BS should define the PHSI when the PHS rule is created. PHS rules are deleted with DSC or dynamic service deletion (DSD) messages. The SS and the BS may define the PHSS and PHSF. To change the value of a PHSF on a service flow, a new PHS rule should be defined, the old rule removed from the service flow, and then the new rule added.

19.6 MAC CPS

A network that utilizes a shared medium requires a mechanism to efficiently share it. This is important in any type of network situation. Shared resources save time and space, which equals money. A two-way peer-to-multipoint wireless network is a good example of a shared medium. The medium is the space through which the radio waves move. Another term for this is propagation.

In a layered protocol system, the information flow across the boundaries between the layers can be defined in terms of primitives that represent different items of information and cause actions to take place. These primitives do not appear as such on the medium (the air interface) but serve to define more clearly the relations of the different layers. The semantics are expressed in the parameters that are conveyed with the primitives [1].

The downlink on a peer-to-multipoint basis is the BS to the user. This is within a given frequency channel and antenna sector, where all stations receive the same transmission or parts thereof. The BS is the only transmitter operating in this direction, so it transmits without having to coordinate with other stations, except for the overall response during downlink transmission periods. It broadcasts to all stations in TDD that may divide into uplink and the sector and frequency. The stations check the address in the received messages and return only those addressed to them.

The user stations share the uplink to the BS on a demand basis. Depending on the class of service utilized, the SS may be issued confirming the right to transmit or the right to transmit may be granted by the BS after receipt of a request from the user [1].

Connections are identified by a 16-bit CID. At SS initialization, three management connections in each direction (uplink and downlink) should be established between the SS and the BS. These CIDs should be assigned in the ranging response (RNG-RSP) and registration response (REG-RSP) messages and should reflect the fact that there are inherently three different QoSs of

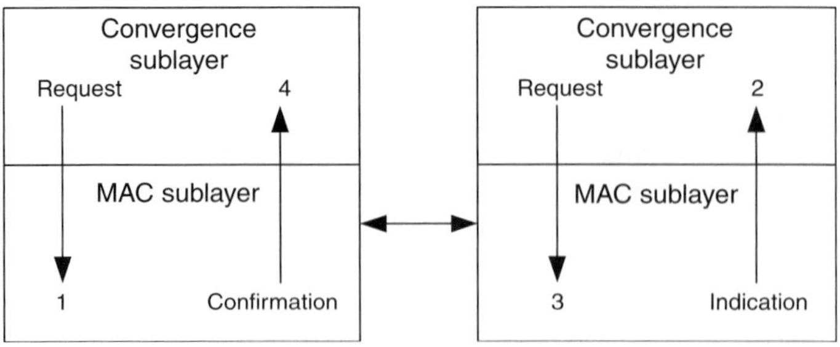

FIGURE 19.6 Use of primitives to request service of MAC sublayer.

management traffic between an SS and the BS. The basic connection is used by the BS MAC and SS MAC to exchange short, time-urgent MAC management messages. The primary management connection is used by the BS MAC and SS MAC to exchange longer, more delay tolerant MAC management messages.

19.6.1 MAC Service Definition

The MAC service definition is a logical interface. The purpose of the primitives is to describe the information that must necessarily be exchanged between the MAC and the CSs to enable each to perform its requirements.

The IEEE 802.16 standard is a layered protocol system. This means that the information flows across the boundaries between layers. They can be defined as primitives that represent different items of information and cause actions to take place. Those primitives define more clearly the relations of the difficult layers.

These primitives can be seen in Fig. 19.6. The initial request for service from a lower layer is provided by a request. Once this request is send, it generates an indicator. It is handled with a response. Then there is confirmation of the transmission and data reception.

19.6.2 MAC PDU Formats

The MAC PDUs are shows in Fig. 19.7. Each has a fixed-length generic MAC header, a dynamic/optional payload, and an optional cyclic redundancy check (CRC).

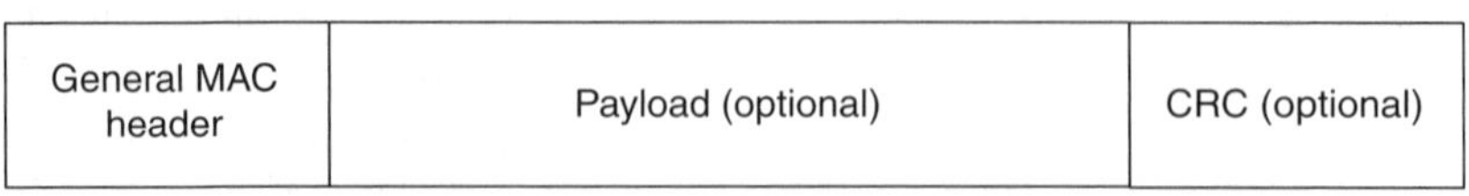

FIGURE 19.7 Generic MAC PDU format.

The payload information can vary in length. Therefore, a MAC PDU can be represented with a variable number of bytes.

Two MAC header formats are defined. The first is the generic MAC header that begins each MAC PDU containing either MAC management messages or CS data. The second is the bandwidth request header used to request additional bandwidth. The single-bit header type (HT) field distinguishes the generic and bandwidth request header formats. The HT field should be set to zero for the generic header and to unity for a bandwidth request header [1].

Requests for transmission are based on these CIDs, since the allowable bandwidth may differ for different connections, even within the same service type. For example, an SS unit serving multiple tenants in an office building would make requests on behalf of all of them, although the contractual service limits and other connection parameters may be different for each of them. Many higher layer sessions may operate over the same wireless CID. For example, many users within a company may be communicating with the transmission control protocol (TCP)/IP to different destinations, but since they all operate within the same overall service parameters, all of their traffic is pooled for request/grant purposes. Since the original LAN source and destination addresses are encapsulated in the payload portion of the transmission, there is no problem in identifying different user sessions [1].

19.7 MAC FRAMES

Figure 19.8 shows the layout of the generic MAC frame. The frame contains an encryption control (EC) bit, type field, CI field, EK field, length field, connection ID, and header CRC field. The EC bit tells whether the payload is encrypted or not. The type field identifies the frame type. The CI field indicates the absence or presence of a check sum field. The EK field tells which encryption keys are being used. The length field gives a complete length (size) of the frame. This also includes the header. The CID indicates to which connection the frame belongs. The header CRC field contains information for the checksum for the header [1].

Two MAC header formats are defined. The first is the generic MAC header that begins each MAC PDU containing either MAC management messages or CS data. The second is the bandwidth request header used to request additional bandwidth. The single-bit HT field distinguishes the generic and bandwidth request header formats. The HT field should be set to zero for the Generic Header and to unity for a bandwidth request header [1].

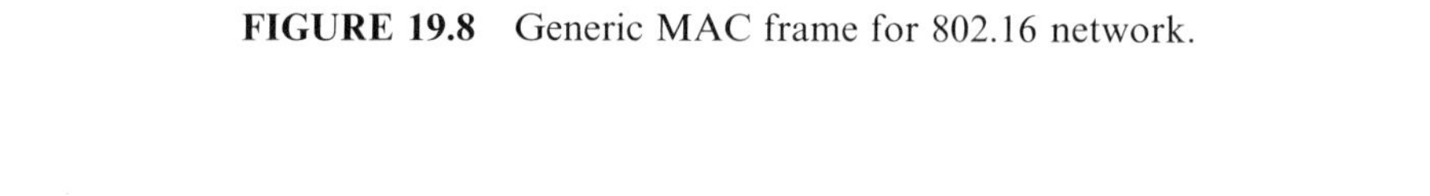

FIGURE 19.8 Generic MAC frame for 802.16 network.

Requests for transmission are based on these CIDs, since the allowable bandwidth may differ for different connections, even within the same service type.

For example, an SS unit serving multiple tenants in an office building would make requests on behalf of all of them, though the contractual service limits and other connection parameters may be different for each of them.

Many higher layer sessions may operate over the same wireless CID. For example, many users within a company may be communicating with TCP/IP to different destinations, but since they all operate within the same overall service parameters, all of their traffic is pooled for request/grant purposes. Since the original LAN source and destination addresses are encapsulated in the payload portion of the transmission, there is no problem in identifying different user sessions [1].

19.8 SIGNALING PROCEDURES

To establish a soft permanent virtual circuit (PVC), the network management system provisions one end of the soft PVC with the address identifying the egress ATM interface of the ATM network. The calling end has the responsibility for establishing and releasing the connection. It is also has the responsibility of the calling party (if necessary) to reestablish the connection in case of switching system or link failure. It should be the responsibility of the implementation of the BS to map ATM signaling messages to corresponding MAC CPS service primitives.

On the downlink direction, the signaling starts at an "end user" of the ATM backhaul network that implements an ATM UNI and terminates at the BS that should implement either an ATM user-to-network interface (UNI) or an ATM network-to-network interface (NNI). The signaling may be mapped by an interworking function (IWF) and extended to some user network on the SS side. On the uplink direction, the signaling starts at the ATM interface of the BS and ends at the ATM UNI of an end user. In addition, the signaling may be originated by an end user of some user network and mapped by the IWF. Note that mapping of data units carried by the air link should be limited to only cell-level convergence. If required by a user network, other levels of mappings should be handled by the user network's IWF exclusively [1].

19.9 BANDWIDTH REQUESTS

There are several items that must be addressed with bandwidth requests: The length of the header should always be 6 bytes; the EC field should be set to zero, indicating no encryption; the CID should indicate the service flow for which uplink bandwidth is requested; the bandwidth request (BR) field should indicate the number of bytes requested; and the allowed types for bandwidth requests are 000000 for incremental and 000001 for aggregate.

If transmit diversity is used, a portion of the downlink (DL) frame (called a zone) can be designated to be a transmit diversity zone. All data bursts within the transmit diversity zone are transmitted using space time coding (STC). Finally, if an adaptive antenna system (AAS) is used, a portion of the DL subframe can be designated as the AAS zone. Within this part of the subframe, the AAS is used to communicate to AAS-capable SSs. The AAS is also supported in the uplink (UL) [2].

There are two forms of bandwidth allocation: per station and per connection. Per-station allocation allows the SS to aggregate the needs of all the users in the building and make a collective request for them. Once the bandwidth is given, the SS gives out as it sees the fit to the rest of the subscribers. Per-connection allocation allows the BS to manage each connection directly [1].

19.10 MAC SUBHEADERS

There are three types of MAC headers: per-PDU subheaders, fragmentation, and grant management. The grant management header comes first when indicated. This is the 2 bytes in length and is used by the SS to convey bandwidth management needs to the BS. It is encoded differently based on the type of uplink scheduling service for the connection. This is given by the CID [1].

The main difference between the point-to-multipoint (PMP) and optional mesh modes is that in the PMP mode traffic only occurs between the BS and SSs, while in the mesh mode traffic can be routed through other SSs and can occur directly between SSs. Depending on the transmission protocol algorithm used, this can be done on the basis of equality using distributed scheduling, on the basis of superiority of the mesh BS, which effectively results in centralized scheduling, or on a combination of both [1].

The CID is the connection ID in the mesh mode as conveyed in the generic MAC header. The length parameter specifies the length of the MAC SDU in bytes. The data parameter specifies the MAC SDU as received by the local MAC entity. The priority/class parameter embedded in the CID specifies the priority class of the MAC SDU. The reliability parameter embedded in the CID specifies the maximum number of transmission attempts at each link [2].

The drop precedence parameter embedded indicates the relative MSDU dropping likelihood. The encryption flag specifies that the data sent over this link are to be encrypted if on. If this is set to an off position, then no encryption is used for this MAC frame [2].

19.11 MAC MANAGEMENT MESSAGES

There are a set of MAC management issues in IEEE 802.16. These messages are carried in the payload of the MAC PDU. All MAC management messages begin with a management message type field and could contain other additional fields.

Management message type	Management message payload

FIGURE 19.9 MAC management message format.

The messages for basic, broadcast, and initial ranging connections do not get packed. These are not fragmented either. The format for the MAC management messages can be seen in Fig. 19.9. The encoding has several connection types, including broadcast, initial range, primary management, and basic. MAC management messages are not carried on the transport connections.

19.12 PHY LAYER

The PHY layer service is provided to the MAC entity at both the BS and SS through the PHY layer SAP, and PHY layer service is described using a set of primitives. The primitives associated with communication between the MAC and PHY layers fall into three basic categories: (a) Service primitives that support the data transfer, thus participating as intermediate signals in MAC peer-to-peer interactions. These are the PHY_MACPDU primitives. (b) Service primitives that have local significance and support sublayer-to-sublayer interactions related to layer control. These include the PHY_TX_START primitives. (c) Service primitives that support management functions, such as the PHY_DCD primitives. Primitives with names of the form PHY*.request are generated by the MAC layer and addressed to the PHY layer to invoke some PHY layer function(s).

As we can see in Fig. 19.10, the further away from a BS a SS is, the slower the baud rate of transmission. The closer the BS, the higher the baud rate. The closest subscribers to the BS have a rate of 6 bits/baud. A little further away from that, the rate is 4 bits/baud. Even further away from that, but still within transmission range, the rate is 2 bits/baud [1].

For transmission from the SS to the BS, demand assignment multiple access-time division multiple access (DAMA-TDMA) is used. DAMA is a capacity assignment technique that adapts as needed to respond to demand change among multiple stations. TDMA is the technique of dividing time as a channel into a sequence of frames. Each one of the frames has slots, allocating one or more slots per frame to form a logical channel [1].

Figure 19.11 shows the typical frame for the TDD. As we can see, the frames are divided into three basic parts: the downstream time slots, the guard time, and the upstream time slots. The downstream traffic is mapped onto the time slots by the BS. However, the upstream traffic is more complex and depends on the QoS-level requirement of the network implementation [1].

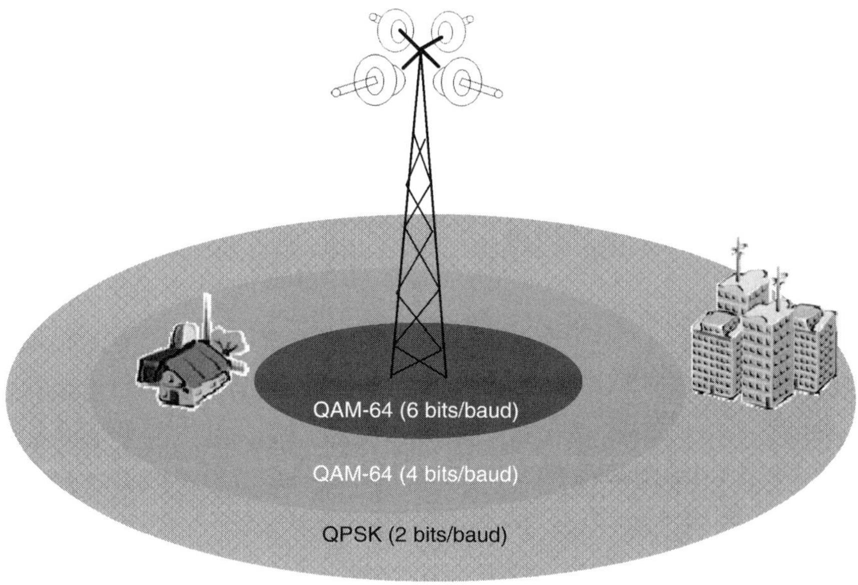

FIGURE 19.10 Levels of throughput with respect to distance from BS.

Using the DAMA-TDMA technology, the slot assignments to the channels conduct dynamically. There are essentially two modes of operation specified: Mode A is targeted to support a continuous transmission stream (audio or video transmissions) [1]; mode B is used for burst transmission streams. These transmissions would be IP-based network traffic [1].

The management information specific to the PHY layer is represented as a management information base (MIB) for this layer. The PHY layer management entity is viewed as "containing" the MIB for this layer. The generic model of MIB-rclated management is to allow the management system (which is out of the scope of this standard) to either *get* the value of an MIB attribute or *set* the value of an MIB attribute. The setting act may require that the layer entity perform certain defined actions [1].

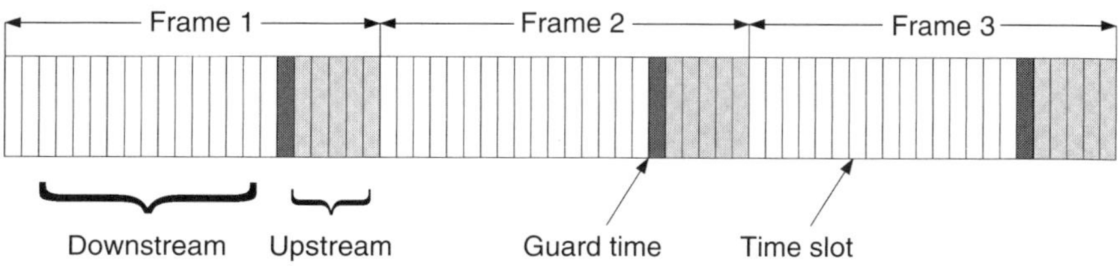

FIGURE 19.11 Example of TDD frame.

19.12.1 Parameter Sets

Several service primitives use parameter vectors. Such a vector is a list of values that may be certain MIB parameters or may be derived by the PHY layer from MIB parameters or from measurement of the airlink characteristics.

Connections are identified by a 16-bit CID. At SS initialization, three management connections in each direction (uplink and downlink) should be established between the SS and the BS. These CIDs should be assigned in the RNG-RSP and REG-RSP messages and should reflect the fact that there are inherently three different QoSs of management traffic between an SS and the BS. The basic connection is used by the BS MAC and SS MAC to exchange short, time-urgent MAC management messages. The primary management connection is used by the BS MAC and SS MAC to exchange longer, more delay tolerant MAC management messages.

The list itself and the set of possible parameter values may vary depending on the PHY layer details [1].

19.12.2 802.16 Specifications

This PHY layer specification, targeted for operation in the 10–66-GHz frequency band, is designed with a high degree of flexibility in order to allow service providers the ability to optimize system deployments with respect to cell planning, cost, radio capabilities, services, and capacity. In order to allow for flexible spectrum usage, both TDD and FDD configurations are supported. Both cases use a burst transmission format whose framing mechanism supports adaptive burst profiling in which transmission parameters, including the modulation and coding schemes, may be adjusted individually to each SS on a frame-by-frame basis. The FDD case supports full-duplex SSs as well as half-duplex SSs, which do not transmit and receive simultaneously.

The uplink PHY layer is based on a combination of TDMA and DAMA. In particular, the uplink channel is divided into a number of time slots. The number of slots assigned for various uses (registration, contention, guard, or user traffic) is controlled by the MAC layer in the BS and may vary over time for optimal performance. The downlink channel is time division multiplexed, with the information for each SS multiplexed onto a single stream of data and received by all SSs within the same sector. To support half-duplex FDD SSs, provision is also made for a TDMA portion of the downlink [1].

Figure 19.12 gives an illustration as to how the communication in the IEEE network happens. In the first step, a subscriber sends wireless traffic and speeds ranging from 2 to 155 Mbps from a fixed antenna or building.

In the second step, the BS receives transmission from multiple sites and sends traffic over wireless or wired links to a switching center using the 802.16 protocol.

The last step of this process is where the switching center sends traffic to an ISP or the PSTN.

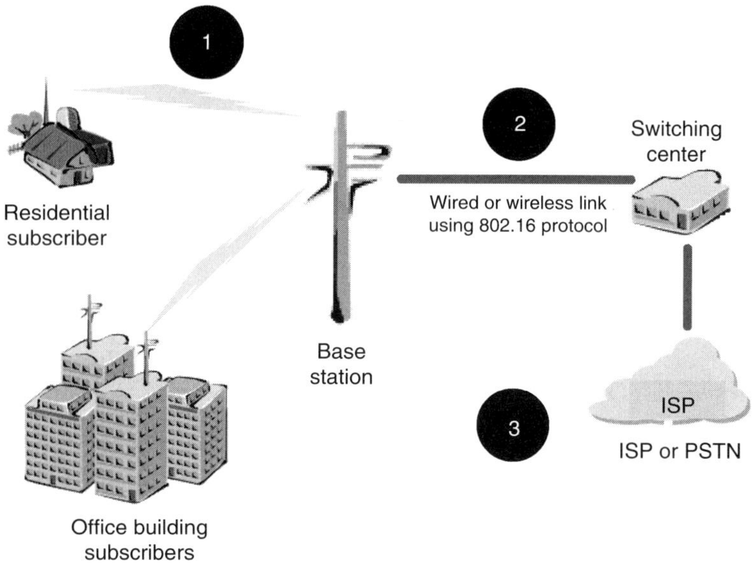

FIGURE 19.12 Example of traffic flow of IEEE 802.16 network.

19.12.3 PHY Layer Frames

Within each frame are a downlink subframe and an uplink subframe. The downlink subframe begins with information necessary for frame synchronization and control. In the TDD case, the downlink subframe comes first, followed by the uplink subframe. In the FDD case, uplink transmissions occur concurrently with the downlink frame.

A network that utilizes a shared medium should provide an efficient sharing mechanism. Two-way point-to-multipoint and mesh topology wireless networks are examples for sharing wireless media. Here the medium is the space through which the radio waves propagate [2].

Though the MAC specification invokes the Internet protocols, they are required only as a standard basis for element management rather than MAC operation, since, in all practicality, element management is necessary in this type of network [2].

Each SS should attempt to receive all portions of the downlink except for those bursts whose burst profile either is not implemented by the SS or is less robust than the SS's current operational downlink burst profile. Half-duplex SSs should not attempt to listen to portions of the downlink coincident with their allocated uplink transmission, if any, adjusted by their TX time advance [1].

The receipt of the primitive causes the MAC entity to process the MAC SDU through the MAC sublayer and pass the appropriately formatted PDUs to the PHY layer for transfer to the peer MAC sublayer entity using the node ID specified [2].

Elements within this PHY layer include TDD and FDD support, TDMA UL, TDM DL, block adaptive modulation and FEC coding for both UL and DL, framing elements that enable improved equalization and channel estimation performance over the non-line-of-sight (NLOS) and extended-delay spread environments, symbol unit granularity in packet sizes, concatenated forward error correction (FEC) using Reed–Solomon and pragmatic TCM with optional interleaving, FEC options using block turbo code (BTC) and convolutional turbo code (CTC), no-FEC options using automatic repeat request (ARQ) for error control, STC transmit diversity option, and parameter settings and MAC/PHY layer messages that facilitate optional AAS implementations [2].

This layer also specifies that there must be frequency band, a modulation scheme, error correction techniques, synchronization bandwidth TX and RX data rates, and TDM architecture [1].

19.13 802.16 DESIGN AND FUNCTIONALITY

This section takes a look at Fig. 19.13 and takes a last look at the implementation and reasoning behind the IEEE 802.16 network. Some of the design factors of the network are that the configuration consists of a BS mounted on a building or tower that communicates on a PMP basis with the SS [1].

Typically the transmission range for this network would be in the range of 30 miles or so. The typical cell would be in the radius of about 4–6 miles [1]. Within one of the cells the NLOS performance and throughputs are optimal. This is due to the use of backhaul technology that helps to connect 802.11 WLANs with the Internet. This enables the flexibility of being able to deploy more 802.11 hotspots in areas where they are nonexistent [1].

These networks have typically shared data rates of around 75 Mbps. These networks also have robust security implementation, which is outside the scope of this chapter, and QoS.

Essentially these networks fill in the gaps of coverage for larger areas [1].

In Fig. 19.13 we have T1 service subscribers, DSL subscribers (which also include small home and office users), residential subscribers, and business subscribers which contain multiple users on WiFi access. All of these connections go to a central switching center, as seen in the figure, using either a wired or wireless implementation of the 802.16 protocol [1].

Throughput. The IEEE 802.16 network delivers high throughput at long ranges with a high level of spectral efficiency. This is also tolerant of signal reflections that could occur with interference [1]. The protocol uses an adaptive modulation that allows the BS to trade off throughput for range. This way, the BS can make decisions about the signal strength based on the location of the SS [1].

Scalability. This network protocol supports flexible channel bandwidths. For example, if an operator is assigned 20 MHz of the bandwidth

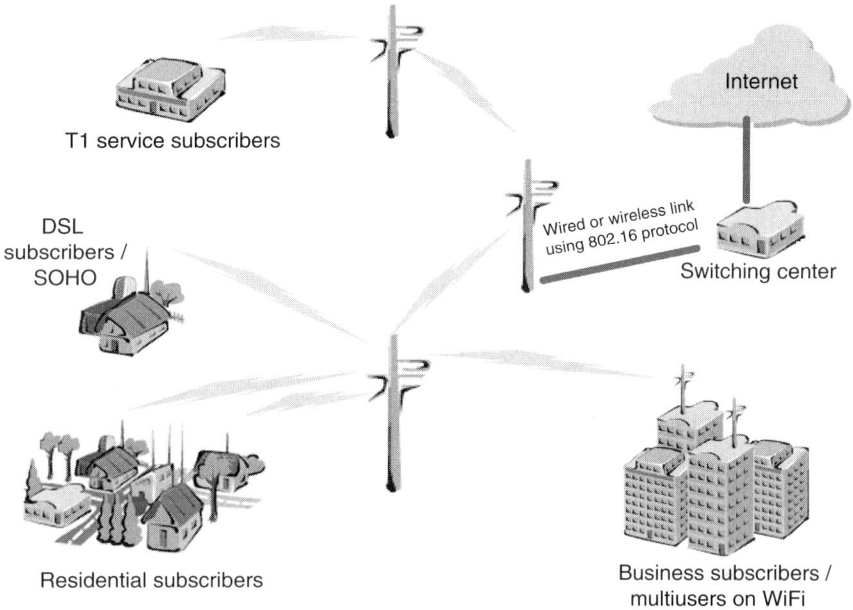

FIGURE 19.13 Example of complete IEEE 802.16 WMAN implementation.

spectrum, the operator could divide it into two sectors of 10 MHz or four sectors of 5 MHz.By focusing power on increasingly narrow sectors, the operator can increase the number of users while maintaining range and throughout [1].

19.14 CONCLUSIONS

This chapter has reviewed the IEEE 802.16 and IEEE 802.16a standards. These standards have WMAN MAC and PHY layer specifications. They mainly deal with the air interface for fixed BWA systems. The main focus of this chapter was the IEEE 802.16 standard.

The IEEE decided to make BWA more available and standardized by developing the IEEE 802.16 standard. This standard addresses the connections in WMANs. It focuses on the efficient use of bandwidth and defines the MAC layer protocols that support multiple PHY layer specifications. These can easily be customized for the frequency band of use. This will allow them to be used in many countries around the world. The 10–66-GHz part of the standard supports various traffic levels at many different frequencies for bidirectional communication. These bands are both licensed and unlicensed.

The IEEE 802.16 systems will allow thousands of users to share capacity for data, voice, and video. These networks are also easily scalable. As the demand grows for subscriber channels or cells, the carriers can expand these network implementations. There are mechanisms in the WMAN MAC that provide different levels of QoS to support different application needs. An example is a network that requires voice and video, which require low latency but would allow for some error rate. The way that these different applications are implemented is by layering control over the MAC. There have been systems in the past with fixed modulation. This standard supports an adaptive modulation and is more robust. These also support lower data rates. This is good in the sense that it would be compatible with more lower end DEVs. Adaptive modulation also allows for more efficient use of bandwidth, which would allow for more customers.

An 802.16 WMAN provides communications between the subscribers and the core network. An example of a core network is the Internet. Another example is a phone company. As stated before, the 802.16 is concerned with the air interface between the subscriber and a base transceiver station.

IEEE 802.16 defines how wireless traffic will move between subscribers and core networks. First, a subscriber sends wireless traffic at speeds of 2–155 Mbps from a fixed antenna or building. Second, the BS receives transmissions from multiple sites and sends traffic over wireless or wired links to a switching center using the 802.16 protocols. The switching center sends traffic to an ISP or to a PSTN.

The broadband wireless architecture is being standardized by the IEEE 802.16 WG and the WiMAX Forum. The 802.16 WG is developing standards for the PHY and MAC layers as well as for the security and higher layer network model. In this chapter we concentrated on the MAC layer and the QoS support provided by the IEEE 802.16 standard.

REFERENCES

1. LAN/MAN Standards Committee of the IEEE Computer Society, 802.16, "IEEE standard for local and metropolitan area networks, Part 16: Air interface for fixed broadband wireless access systems," IEEE Computer Society, Apr. 8, 2002.
2. LAN/MAN Standards Committee of the IEEE Computer Society, "802.16a IEEE standard for local and metropolitan area networks, Part 16: Air interface for fixed broadband wireless access systems—Amendment 2: Medium access control modifications and additional physical layer specifications for 2–11 GHz," IEEE Computer Society, Apr. 1, 2003.

CHAPTER 20

QoS SUPPORT FOR WiMAX

USMAN A. ALI, QIANG NI, YANG XIAO, WENBING YAO,
and DIONYSIOS SKORDOULIS

20.1 INTRODUCTION

There is a high demand for broadband access which provides multimedia Internet services such as voice-over-IP (VoIP) and video streaming. These services require high bandwidth (BW) along with quality-of-service (QoS) support. In particular, broadband wireless access (BWA) has recently received very high interest due to the merits of lower infrastructure cost, easier deployment, and higher flexibility than its wired counterpart. The IEEE 802.16 worldwide interoperability of microwave access (WiMax) has been one of the most promising BWA standards and technologies. However, it is challenging to maintain various QoS requirements (e.g., guaranteed BW allocation, bounded delay, jitter, and packet loss ratio) due to time-varying wireless channel characteristics. Figure 20.1 shows the QoS architecture specified in the WiMax standard to support various multimedia applications with different QoS requirements.

The chapter consists of five sections. Section 20.2 deals with the issues related with QoS. Section 20.3 provides a description of the four scheduling mechanisms including their flow charts. Section 20.4 describes BW request and allocation mechanisms specified in the WiMax standard. Section 20.5 explains the QoS support mechanisms in WiMax. Section 20.6 concludes the chapter with simulations results.

20.2 QoS PARAMETERS IN WiMAX

In the WiMax standard, the medium access control (MAC) layer is connection oriented where different application flows are categorized as QoS and non-QoS

Emerging Wireless LANs, Wireless PANs, and Wireless MANs. Edited by Y. Xiao and Y. Pan

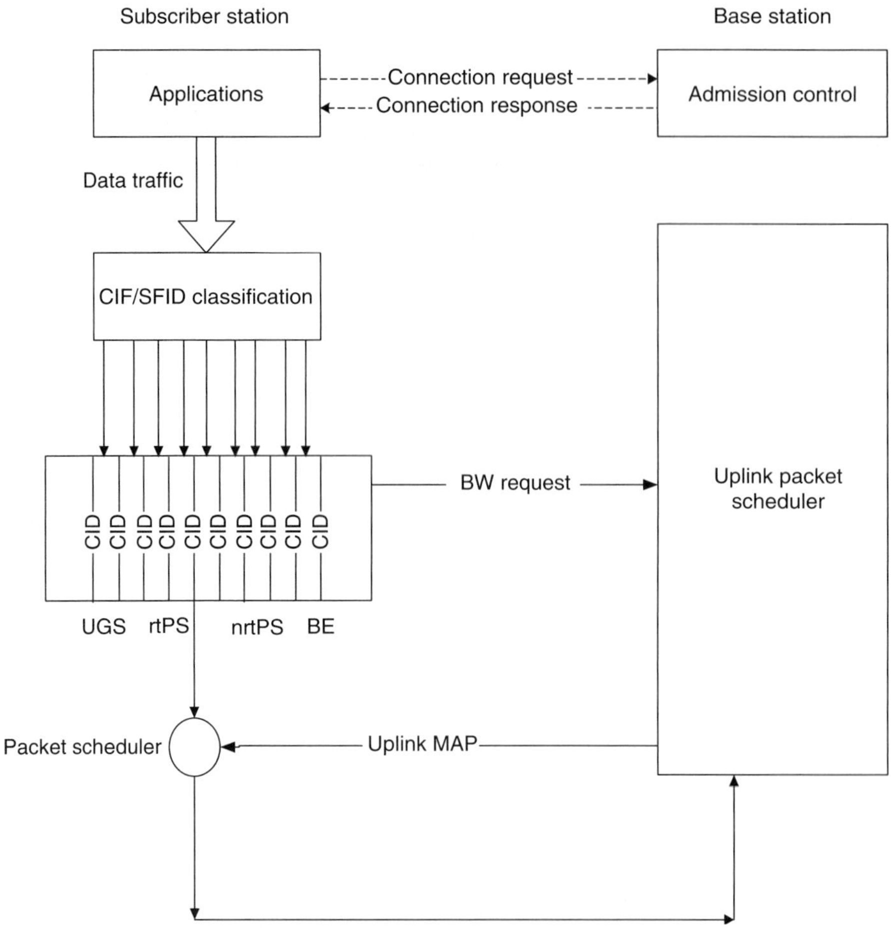

FIGURE 20.1 QoS Architecture in 802.16.

services and are mapped to connections with different scheduling services, facilitating equally assured handling and traffic enforcements.

There are a number of ways to classify QoS support in communication networks. The two main categories are *parameterized QoS* and *prioritized QoS*. Parameterized QoS characterize strict QoS requirements which are expressed in terms of quantitative values, for example, data rate, delay, and jitter bound. On the other hand, prioritized QoS is not very strict and deals with relative delivery priority.

In WiMax, a base station (BS) requires configuration and registration functions in order to configure and register the service flows that will be used mutually by the BS and a subscriber station (SS). Likewise a signaling function is essential for the BS and SS to correspond with each other for activation, creation, admission, modification, and deletion of the service flows. For service flow, the QoS parameter set is vital and introduction of different service classes are also needed but they are discretionary. The following parameters are defined in the QoS parameter set in the WiMax standard:

- *Maximum Sustained Traffic Rate* (*MSR*). The MSR defines the peak information rate of the service in bits per second. This constraint does not contain MAC overhead such as cyclic redundancy check (CRC) or MAC headers. This constraint does not limit the instantaneous rate of service and is managed by the physical attributes of the ingress port.
- *Minimum Reserved Traffic Rate* (*MRR*). The MRR defines the minimum rate reserved for this service to flow. The rate is represented in bits per second and illustrates the smallest amount of data to be broadcasted for the service flow when it is averaged over time.

TABLE 20.1 Request/Transmission Policy

Type	Length	Value	Scope
[146/147.12]	4	Bit 0: Broadcast bandwidth request opportunities should not be used by service flow.	Dynamic service addition request (DSA-REQ)[a]
		Bit 1: It is reserved and should be set to 0.	Dynamic service addition response (DSA-RSP)[b]
		Bit 2: The service flow is not allowed to use piggyback requests.	Dynamic service addition acknowledgment (DSA-ACK)[c]
		Bit 3: The service flow is not permitted to fragment any data.	
		Bit 4: The payload headers will not be suppressed by the service flow, i.e., CS parameter.	
		Bit 5: The packing of multiple service data units into one MAC PDU is not prohibited for the service flow.	
		Bit 6: CRC may not be incorporated in MAC PDU.	
		Bit 7: This bit is reserved and set to zero.	

[a]This request is sent by the SS to create a new service flow.
[b]The response is sent by the BS in response to a DSA-REQ.
[c]The SS sends an acknowledgment back to the BS.

- *Maximum Latency*. Maximum latency identifies the peak latency between the reception of a packet by the BS or SS present on its network interface and dispatching of the packet to its RF interface.
- *Maximum Traffic Burst*. Maximum traffic burst describes the maximum burst size in bytes of service flow.
- *Tolerated Jitter*. Tolerated jitter describes the maximum delay variation for the connection.
- *Traffic Priority*. Traffic priority defines the peak information rate of the service in bits per second. This constraint does not contain MAC overhead such as CRC or MAC headers.
- *Request/Transmission Policy*. The value of this parameter illustrates the capability to specify certain attributes for the connected service to flow, as shown in Table 20.1. These attributes include certain options for the following parameters: configuration of protocol data unit (PDU, uplink services to flow, and restrictions on the type of BW request that may be used.

20.3 SCHEDULING MECHANISMS

Scheduling services are designed to improve the efficiency of the polling and BW grant processes. They represent the data-handling mechanisms supported by the MAC scheduler for the purpose of data transportation on a connection. Every connection is associated with a single data service and each data service is linked with a set of QoS parameters which enumerate aspects of its performance. These constraints are handled using the two message dialogs dynamic service addition (DSA) and dynamic service change (DSC).

There are five kinds of scheduling mechanisms suggested in the WiMax standard to support different types of applications.

20.3.1 Unsolicited Grant Services

Unsolicited grant services (UGSs) are designed to support constant-bit-rate (CBR) services, e.g., T1/E1 emulation and VoIP without silence suppression. These kinds of applications require firm scheduling/guarantee on throughput, latency, and jitter. They are designed to guarantee the requirements of real-time service flows that are capable of producing fixed-sized packets on periodic basis and to reduce overhead and latency introduced by the BW request mechanisms.

A BS periodically awards data grant burst information elements (IEs) to a UGS service flow based on its MSR. The MSR describes the peak information rate of the service, which is represented in bits per second and belongs to the service data unit (SDU) present at the input of the system.

The size of these grants is satisfactory enough to grasp the fixed-length data related to the service flow. This service works correctly only when request/

transmission policy forbids the SS from employing a contention request opportunity for this connection.

QoS service parameters used by the UGS are MSR, tolerated jitter, and request/transmission policy. If the MRR is present, then this value will be equal to the MSR.

20.3.2 Real-Time Polling Services

The real-time polling service (rtPS) is designed for the creation of variable-size data packets on a periodic basis, e.g., VoIP with silence suppression, or video streaming. They also guarantee throughput but do not emphasize latency. The rtPS maintains variable grant sizes for the best possible data transport effectiveness. However, this service has more request overhead than the UGS as it provides real-time, periodic, unicast request opportunities and also matches the flow's real-time needs and allows the SS to indicate the size of the required grant.

This service works well when a BS allows the unicast request opportunities and the request/transmission policy does not allow the SS to use any contention request opportunity for that connection.

In order to issue a unicast request, the BS should only grant when previous requests are satisfied. This allows an SS to make use of only unicast request opportunities with the intention of obtaining uplink transmission opportunities.

QoS service parameters used by the rtPS are MSR, MRR, request/transmission policy, and maximum latency.

20.3.3 Non-Real-Time Polling Services

The non-real-time polling service (nrtPS) is in charge of providing non-real-time services that regularly create variable-size data grant burst. This service only guarantees throughput. Timely unicast opportunities are provided by the BS and this service works well when the request/transmission policy allows the SS the access to exercise contention request opportunities because the nrtPS gives unicast polls regularly and also promises the service flow to accept request opportunities even when there is congestion in the network.

QoS service parameters used by the nrtPS are MSR, MRR, request/transmission policy, and traffic priority.

20.3.4 Extended Real-Time Polling Service

The extended real-time polling service (ErtPS) is responsible for providing VoIP services with fixed-size packets using silence suppression. An activity algorithm is used by the BS to examine the flow state. When the ErtPS changes its state from active to inactive, the BS degenerates unicast request polling.

QoS service parameters used by the ErtPS are MSR, maximum latency, and request/transmission policy.

20.3.5 Best-Effort Services

The main purpose of the best-effort (BE) services is to provide an efficient mechanism for the BE traffic. This service does not provide any guarantee and the user can use the maximum available data rate. This service works only when the request/transmission policy allows the SS full access to the contention request opportunities. This permits the SS to use contention request opportunities and also utilizes unicast request opportunities and unsolicited data grant burst types.

QoS service parameters used by the BE are MSR, request/transmission policy, and traffic priority.

20.4 BW REQUEST AND ALLOCATION MECHANISMS

BW request and allocation mechanisms are responsible for providing the uplink BWs requested by different SSs from the BS. These requests are always connection based.

Every SS is allocated three dedicated connection IDs (CIDs) such that it can send and receive control messages while entering a network and during initialization. These pairs of connections are allowed to be used in different levels of QoS so that they can be applied in different connections that carry MAC management traffic. Demand assigned multiple access (DAMA) services are provided to resources on an on-demand basis. QoS is looked up by the BS and is established at the creation of connection.

The following BW request/allocation mechanisms are specified in the WiMax standard and we classify them into different classes according to different criteria.

20.4.1 BW Request

In WiMax, an SS can request uplink BW from the BS in different ways. A BW request can be considered as either a *stand-alone request* versus a *piggyback request* or an *incremental request* versus an *aggregate request*:

- *Stand-Alone Request.* A standard-alone request is transmitted by each SS to the BS independently from the actual data packet transmission. In WiMax, random access and polling-based access are two main means for an SS to issue a standard-alone request.
- *Piggyback Request.* A piggyback request is transmitted along with an uplink data packet transmission, which introduces less overhead. It is a 16-bit entity, which corresponds to the number of uplink bytes of BW requested for the connection.

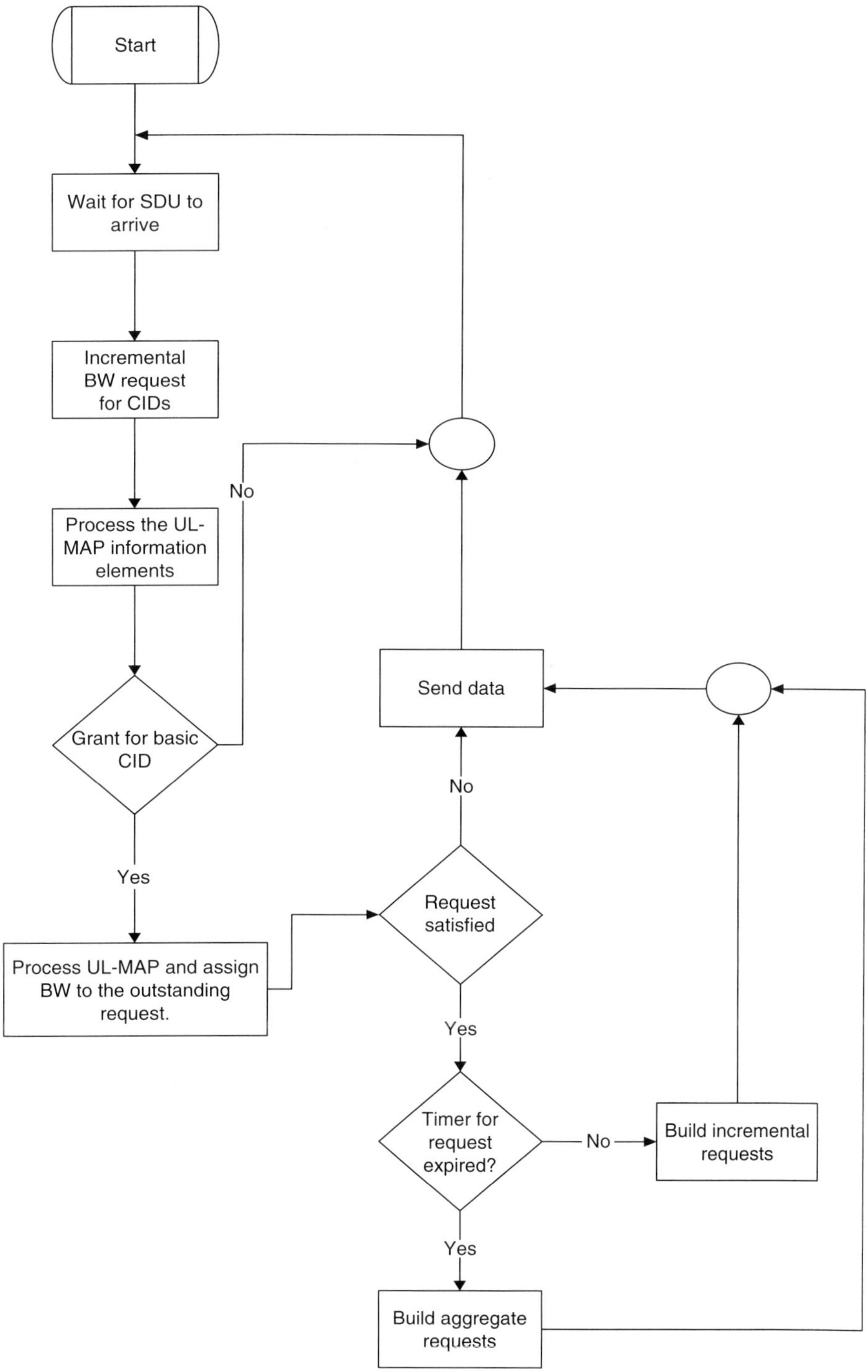

FIGURE 20.2 WiMax grant mechanism.

- *Incremental BW Request*. When a BS receives an incremental BW request, the BS will add the quantity of BW requested to its current BW that is required for connection.
- *Aggregate BW Request*. When a BS receives an aggregate BW request, the BS will replace the current BW and will allocate the amount of BW requested.

The type field present in the header of the BW request can be aggregate or incremental. Piggyback requests are always incremental, as they do not have a type field. Its protocol is defined in such a way that the incremental BW request is always used and is of a self-correcting nature.

20.4.2 BW Grant (Allocation)

A BW request is issued per connection while the BW grant is addressed to the SS as a whole, not to an individual connection. The flow chart of the grant mechanism is shown in Fig. 20.2. Normally a grant is awarded to an SS by polling. The following section describes polling mechanisms in WiMax.

20.4.3 Polling

Polling is a process by which the BS polls SSs to either issue BW requests or grant the BW allocations to SSs. These allocations can either be individual or be grouped from the SSs.

Polling can be unicast or multicast as follows:

- *Unicast Polling*. When an SS is polled individually, no explicit message is transmitted to poll the SS. Rather, in the uplink map (UL-MAP), the SS is allocated BW sufficient to respond with a BW request. The flow chart of unicast polling is shown in Fig. 20.3.
- *Multicast Polling*. When insufficient polling is available for unicast, multicast polling is used so that the SS may be polled in multicast groups and broadcast polls are issued. The flow chart of multicast polling is shown in Fig. 20.4.

20.5 QoS SUPPORT IN WiMAX

In a point-to-multipoint mode, uplink and downlink data transmission occurs in separate time frames. The BS transmits a large amount of MAC PDUs in the downlink subframe. The SS uses time division multiplexing (TDD) to send MAC PDUs to the BS. The phenomenon is explained in Fig. 20.5.

The uplink and downlink subframes occur simultaneously on different frequencies in frequency division multiplexing (FDD) but on the same

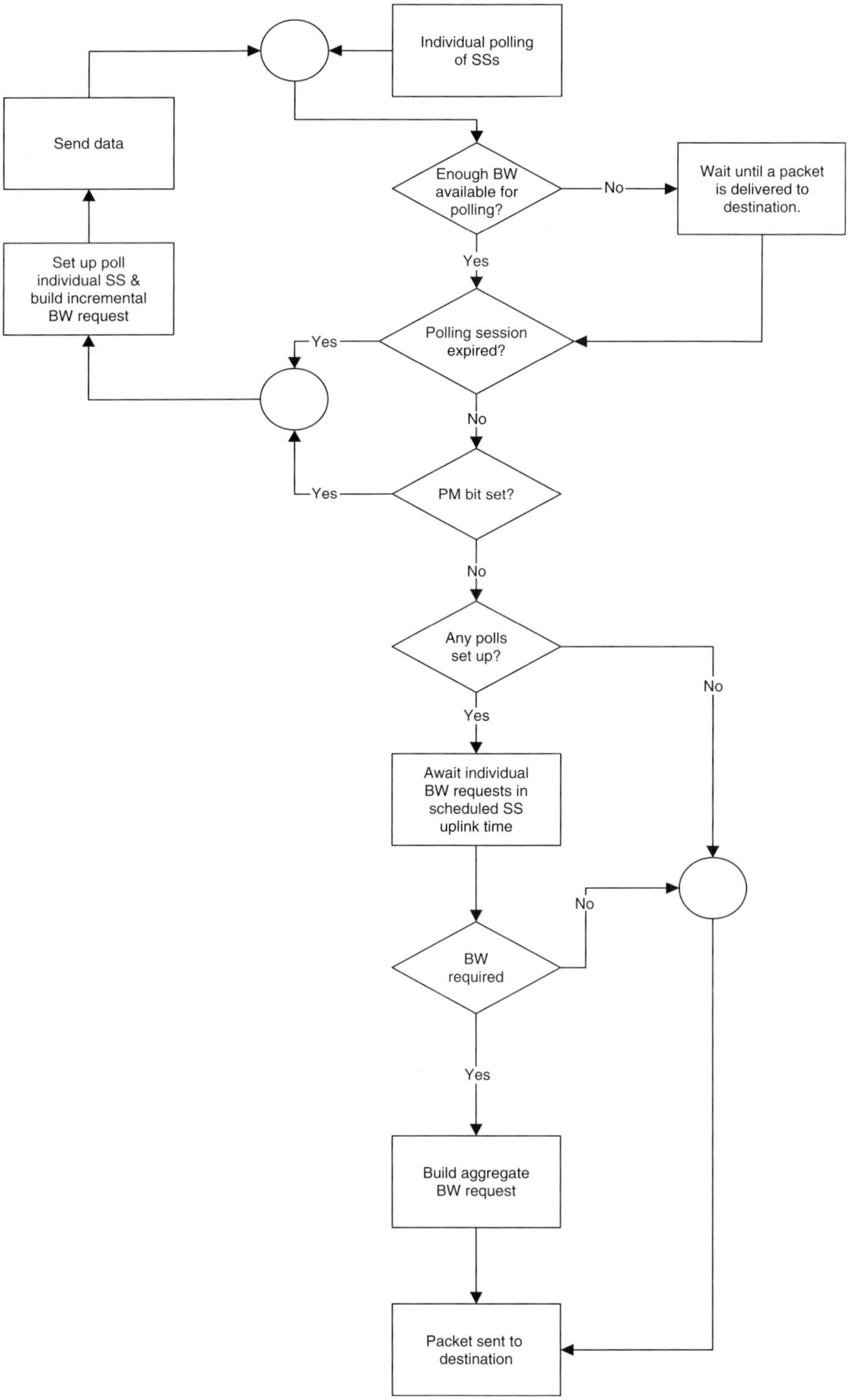

FIGURE 20.3 Flow chart of unicast polling.

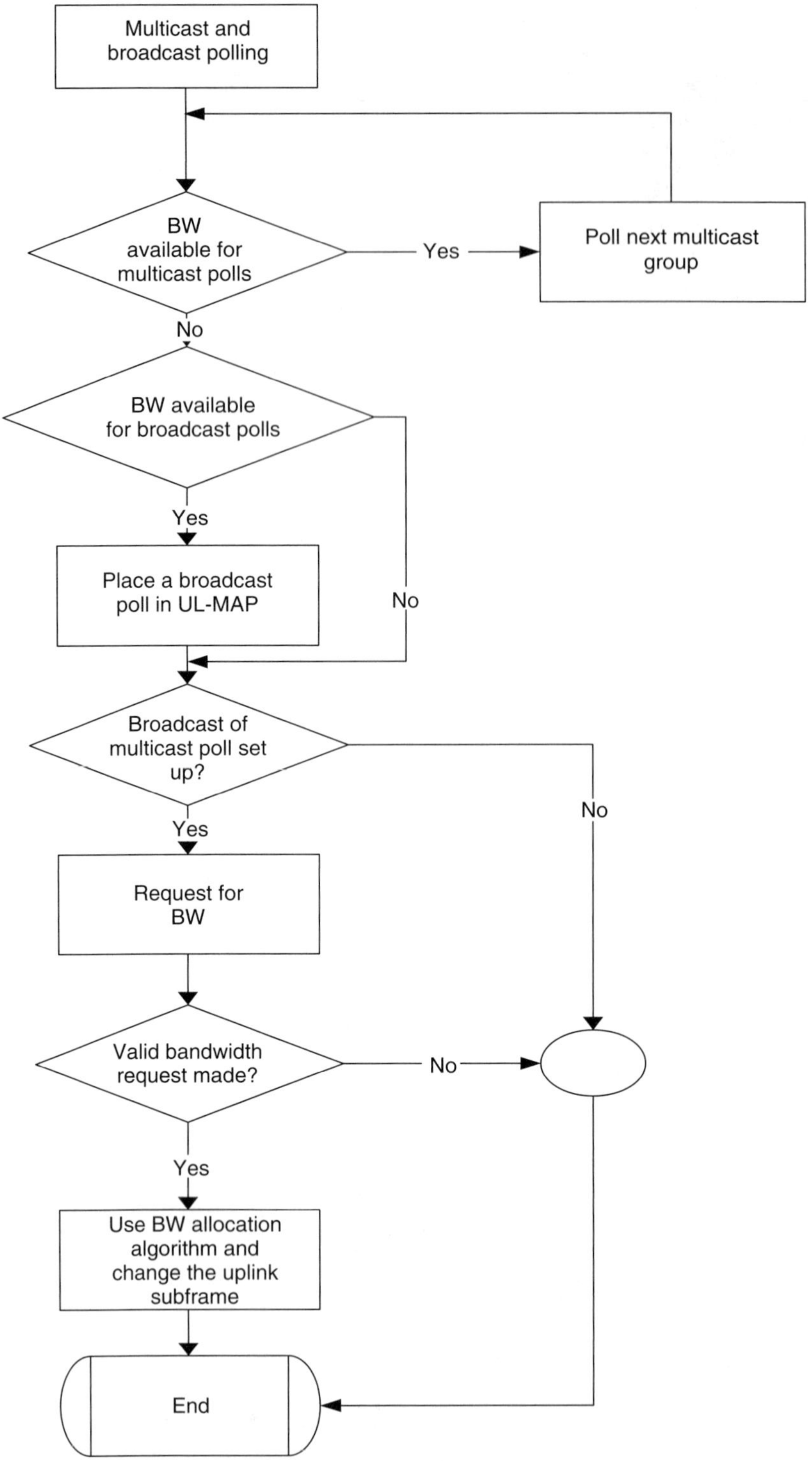

FIGURE 20.4 Flow chart of multicast polling.

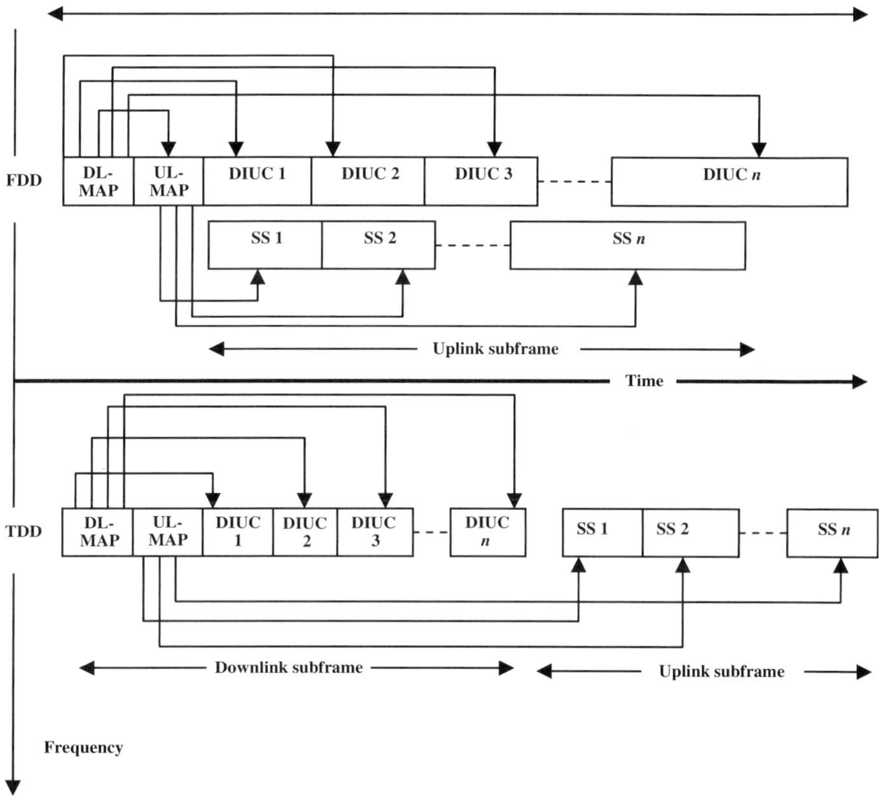

FIGURE 20.5 WiMax frame structure including FDD and TDD.

frequency in TDD. The MAC protocol is designed to be connection oriented for controlling and transportatiing data. The BS schedules the downlink and uplink grants in order to deal with QoS requirements at the beginning of every frame. The SS then decodes its UL-MAP messages and extracts the limit of allocation. However, the timetable for downlink grants in the forthcoming downlink subframes are contained in the downlink map (DL-MAP).

The downlink grants that are directed to the SS with the same downlink interval message code (DIUC, an interval usage code used for downlink) are transmitted by the DL-MAP in a single burst. Afterward both the uplink and downlink maps are transmitted by the BS at the start of each downlink subframe, as shown in Fig. 20.5.

20.5.1 QoS Architecture

Figure 20.6 explains the QoS architecture with the help of service flows. A unidirectional logical link is created between the peer MACs of the BS and the

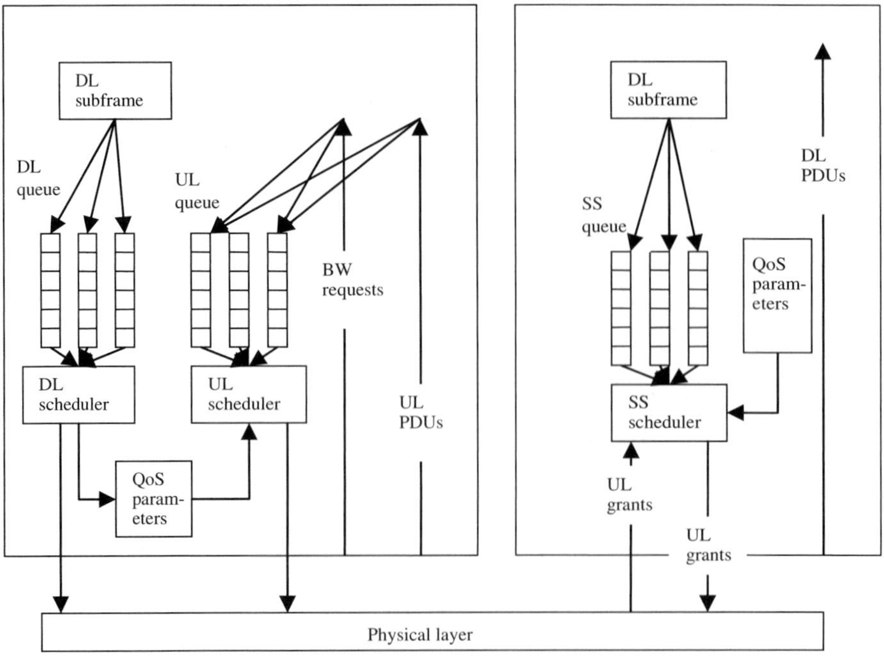

FIGURE 20.6 QoS architecture.

SS before providing any QoS. The outbound MAC then links the packets passing through the MAC interface into a service flow so that they can be transported over the connection already established. The QoS parameter set associated with the service flows identifies the order of transmission and scheduling mechanism on the air interface. As a result, a connection-oriented QoS is capable of offering precise control over the air interface. As the air interface is generally a bottleneck, the connection-oriented QoS can efficiently allow end-to-end control. The service flow parameters are controlled via MAC messages in order to provide the dynamic service demand. This service flow is applied to both the DL and UL, thereby providing a bidirectional QoS.

As WiMax MAC protocol is connection oriented, the application creates a link first with a BS and then with its related service flows (UGS, rtPS, nrtPS, ErtPS, BE). Upon connection, the BS provides a unique CID. This connection can correspond to a single application or a number of applications. All the PDUs present in the application layer of the BS and the SS are categorized by the connection classifier, which is based on the CID, and are then forwarded to the appropriate queue. Afterward, the scheduler of the BS/SS determines which traffic will be mapped first into a queued frame. Afterward a burst is generated along with the appropriate UL-MAP/DL-MAP information element and all service data are scheduled according to respective services classes (UGS, rtPS, nrtPS, ErtPS, BE). The UL-MAP and DL-MAP have all the information for transmission (to/from) all SSs for each frame as well as their size, coding type, modulation, and position of allocation.

BW requests are sent to SSs, which then calculate the approximate value of the residual backlog of UL connections. The BS calculates this residual backlog by estimating the amount of BW granted and requested up to now and then assigns UL grants according to the particular QoS parameter set. The mechanism is shown in Fig. 20.6. In the WiMax standard, the BW grant is approved on a per-connection basis and sent to the SS altogether. Hence, when an SS receives a UL grant, its internal scheduler is required to reallocate the granted capacity to the entire links that it possesses. This mechanism of connection per subscriber station provides a low granting load and also allows more sophisticated response to QoS requirements.

20.5.2 Connection Admission Control

Connection admission control (CAC) is an important QoS process. It determines how latency and BW are granted to different streams that have a variety of service requirements. For this reason, it is necessary to implement the CAC scheme between core and network edges in order to control the traffic coming into a network. Even though the CAC is allowed in the WiMax standard, its detail, efficient BW reservation, and allocation are left open for vendors. Also the biphase BW reservations for activating and admitting service flows are not

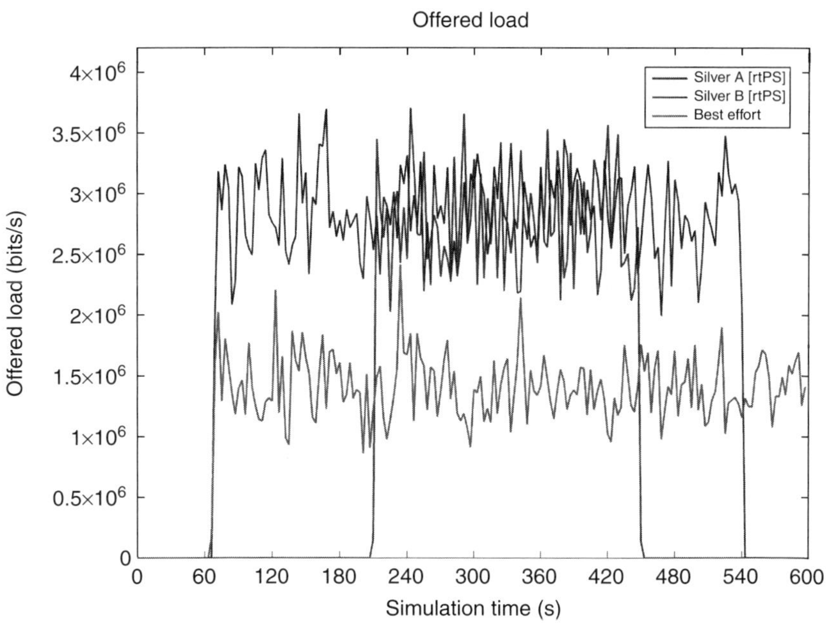

FIGURE 20.7 Offered loads.

appropriate for all real-time multimedia services since it does not guarantee real-time services to attain the required BW in time.

20.6 SIMULATION EVALUATIONS

This section presents a simulation analysis of different QoS classes supported in WiMax. We use OPNET Modeler version 11.0 to carry out the simulations.

The scenario contains two rtPS SSs, namely silver A SS and silver B SSs with different BW requests, one BE SS, one UGS SS, and one BS. The QoS classes defined for silver A SS and silver B SS are Silver_A and Silver_B and both of them use the rtPS as the scheduling type. The UGS uses gold as its QoS class, while the BE SS has the lowest priority of service class.

The UGS SS can request up to 16.8 Mbps. While there is no traffic activity defined in the UGS SS, it is introduced to reserve some BW resources from rtPS and BE services. The silver A SS and silver B SS have limits of 2 and 0.5 Mbps, respectively. The BS scheduler provides all the services to the two silver SSs based on the ratio of 4:1. The BE SS uses a default system class, so there are no service guarantees as it only uses the resources left by other higher priority services.

Two video traffic flows are sent with two different types of service classes (Silver_A and Silver_B). The interarrival time of both video flows is set at 0.05 s

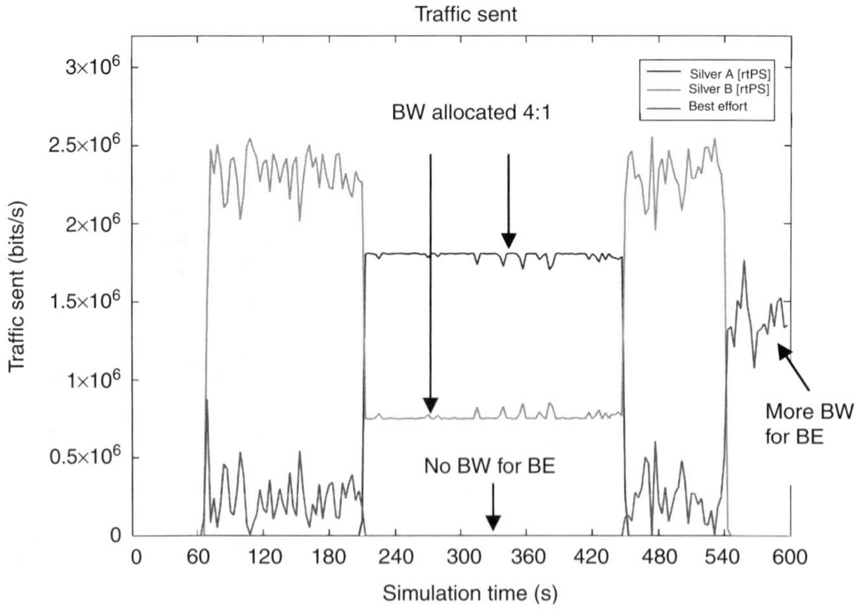

FIGURE 20.8 Traffic sent.

and the outgoing frame size is set at exponential 17,280 bytes in both service classes. So the traffic-sending rate is approximately 2.8 Mbps. The sending rate of the BE SS is set at 1.4 Mbps.

Figure 20.7 shows the offered load at each of the three SSs. The silver B class sends traffic at 60 s, its offered load is approximately 2.8 Mbps, and it stops at 540 s. The BE also starts sending traffic at 60 s with the rate of 1.4 Mbps and stops at 600 s. The silver A class sends traffic in the time interval of 200–460 s, and its offered load is also 2.8 Mbps.

Figure 20.8 shows how QoS is supported when the rtPS (silver A SS and silver B SS) and BE traffic flows are transmitted over a WiMax network. Silver B starts first and the remaining BW is used by the BE SS. After 200 s, silver A starts sending traffic. No BE packets can be transmitted because all the available BW is used by the two silver SSs. The BW is allocated by the BS to the two silver SSs according to the predefined ratio 4:1. After 450 silver A stops sending traffic. Silver B can then send traffic at 2.8 Mbps again and stops at 540 s. After that the BE flow uses all the available resources until the end of simulation.

The offered load and throughput are shown in Figure 20.9. The offered load remains zero from 0 to 60 s as no node sends traffic. After 60 s silver B and the BE send traffic so the offered load increases and it doubles up between 200 and 450 s because both the silver SSs send traffic. There is no BE traffic in this

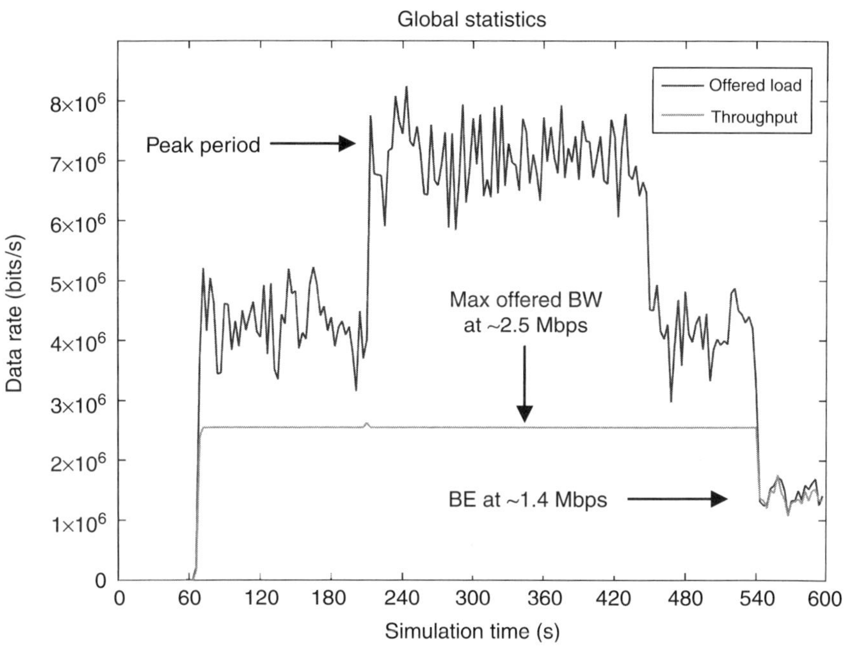

FIGURE 20.9 Offered load and throughput.

interval because there are no resources available for BE to send any traffic. The two silver SSs stop at 450 and 540 s, respectively. After 540 s all the channel BW is available to the BE flow.

Abbreviations and Acronyms

ACK	Acknowledgment
ADSL	Asymmetric digital subscriber line
ARQ	Automatic repeat request
BE	Best effort
BER	Bit error rate
BS	Base station
BSID	Base station identification
BW	Bandwidth
BWA	Broadband wireless access
CBR	Constant bit rate
CDMA	Code division multiple access
CID	Connection identification
DAMA	Demand assigned multiple access
DIUC	Downlink interval usage code
DL-MAP	Downlink map
DSA	Dynamic service addition
DSC	Dynamic service change
FTP	File transfer protocol
HT	Header type
IEEE	Institute of Electrical and Electronics Engineers
IP	Internet protocol
MAC	Medium access control
MAC CPS	MAC common part convergence sublayer
MRR	Minimum reserved traffic rate
MSR	Maximum sustained traffic rate
NLOS	Non–line of sight
nrtPS	Non-real-time polling services
PDU	Protocol data unit
PMP	Point to multipoint
QoS	Quality of service
REQ	Request
RSP	Response
rtPS	Real-time polling services
SAP	Service access point
SS	Subscriber station
TC	Transmission convergence sublayer
TCP/IP	Transmission control protocol/internet protocol

TDD	Time division duplex
TDM	Time division multiplexing
TDMA	Time division multiple access
UGS	Unsolicited grant service
UL-MAP	Uplink map
UMTS	Universal mobile technology system
VoIP	Voice over internet protocol
WiMax	Worldwide interoperability of microwave access
WirelessMAN	Wireless metropolitan area network

REFERENCES

1. "Part16: air interface for fixed broadband wireless access systems," IEEE standard for local and metropolitan area networks, IEEE, New York, May 2004.
2. WiMAX Forum, available: http://www.wimaxforum.org/.
3. Q. Ni, A. Vinel, Y. Xiao, et al., "Investigation of bandwidth request mechanisms under point-to-multipoint mode of WiMAX networks," *IEEE Commun. Mag.* 45(5), 132–138 (2007).
4. Y. Xiao, "Energy saving mechanism in the IEEE 802.16e wireless MAN," *IEEE Commun. Lett.* 9(7), 595–598 (2005).
5. C. Eklund, R. B. Marks, and K. L. Stanwood, "IEEE standard 802.16: A technical overview of the wireless MAN air interface for broadband wireless access," *IEEE Commun. Mag.* 98–107 (2002).
6. OPNET Modeler 11.5, online documentation and models, OPNET Technologies, available: 2005.
7. A. Ghsosh, D. R. Wolter, J. G. Andrews, and Runhua, "Broadband wireless access with WiMax/802.16: Current performance benchmarks and future potentia," *IEEE Commun. Mag.* 23(2), 129–136 (2005).
8. G. Chu, D. Wang, and S. Mei, "A QoS architecture for the MAC protocol of IEEE 802.16 BWA system," paper presented at the IEEE 2002 International Conference, Vol. 1, July 2002, pp. 435–439.
9. I. Koffman and V. Roman, "Broadband wireless access solutions based on OFDM access in IEEE 802.16," *IEEE Commun. Mag.* 40(4), 96–103 (2002).

CHAPTER 21

SUBCHANNEL ALLOCATION AND CONNECTION ADMISSION CONTROL IN OFDMA-BASED IEEE 802.16/ WiMAX-COMPLIANT INFRASTRUCTURE WIRELESS MESH NETWORKS

DUSIT NIYATO and EKRAM HOSSAIN

21.1 INTRODUCTION

Also known as the WiMAX (worldwide interoperability for microwave access), IEEE 802.16 (e.g., 802.16a, 802.16-2004, 802.16e, WiBro)–based technology can potentially deliver fixed, portable, and mobile wireless solutions enabling high bandwidth services with an array of multimedia features [1]. Specifically designed for outdoor non-line-of-sight communication environment, IEEE 802.16a air interface operating in the 2- to 11-GHz band provides high speed transmission rate (in the range of 32–130 Mbps) based on orthogonal frequency division multiplexing (OFDM) along with adaptive modulation and coding. One of the radio interfaces for 802.16a, namely, wirelessMAN-OFDM, is based on the orthogonal frequency division multiple access (OFDMA) scheme in which the entire transmission bandwidth is divided into subchannels that are dynamically allocated among the different connections.

IEEE 802.16/WiMAX supports point-to-multipoint architecture as well as mesh architecture among WiMAX subscriber stations (i.e., client meshing in IEEE 802.16j). Wireless mesh networks are basically multihop relay networks in which the wireless routers perform the relay functionality [2]. The mesh networking mode of operation extends the service coverage area of a base station (BS) with quality of service (QoS) support. Although meshing/relaying

Emerging Wireless LANs, Wireless PANs, and Wireless MANs. Edited by Y. Xiao and Y. Pan

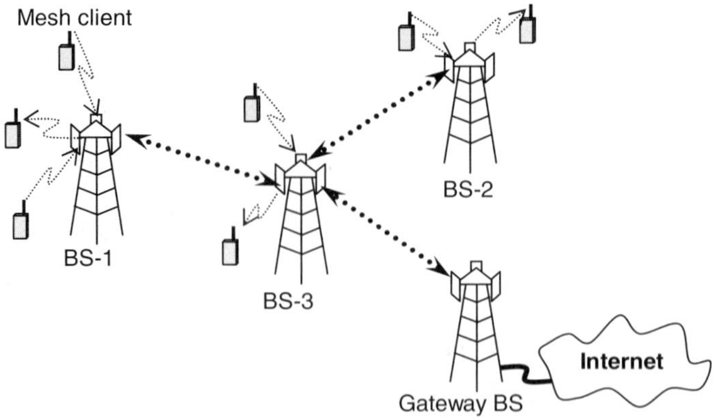

FIGURE 21.1 Mesh infrastructure with tree topology.

among the BSs has not been supported in the WiMAX standard yet, this is a viable option that would be useful for many applications such as backhauling wireless traffic to the core of the Internet. For such an infrastructure mesh network, efficient radio resource management and admission control mechanisms would be necessary to provide effective service to the users as well as to achieve efficient usage of the network resources.

In this chapter, we review the major radio resource management issues in OFDMA-based wireless infrastructure mesh networks. The related works in the literature are then summarized. Then we present a radio resource management framework for subchannel allocation and connection admission control in IEEE 802.16/WiMAX–compliant OFDMA-based wireless infrastructure mesh networks where the WiMAX BSs work as the mesh routers. The WiMAX BSs form a backhaul network to relay traffic from WiMAX clients to Internet gateways (as shown in Fig. 21.1).

The rest of this chapter is organized as follows. Background information is presented in Section 21.2. Section 21.3 summarizes the related works in the literature. The major components of the radio resource management model are described in Section 21.4. Section 21.5 presents the subchannel allocation and the admission control schemes. The queueing analytical model used for performance analysis of the subchannel allocation and the admission control schemes is presented in Section 21.6. Section 21.7 presents the performance evaluation results. Conclusion and future work are stated in Section 21.8.

21.2 BACKGROUND

21.2.1 Wireless Mesh Networks

Wireless mesh networks are based on multihop wireless communication where some of the nodes referred to as mesh nodes/mesh routers form a backbone

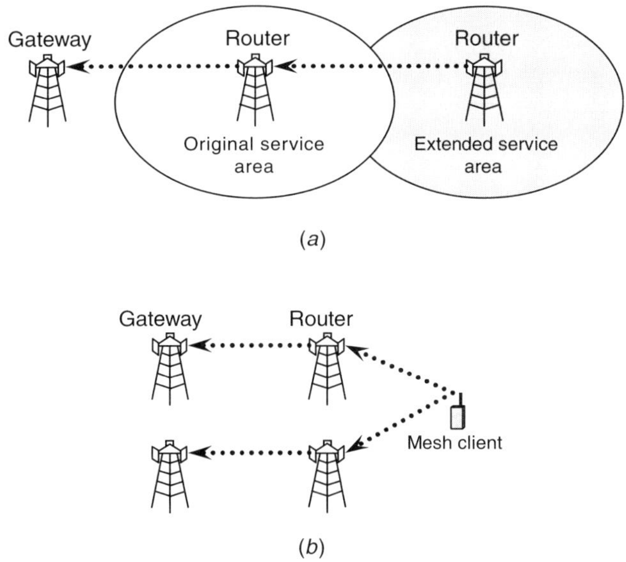

FIGURE 21.2 (*a*) Extension of coverage area and (*b*) enhancement of throughput.

network to communicate among each other in a peer-to-peer manner to serve traffic from the mesh clients (Fig. 21.1). Multihop communication among the mesh routers can provide service coverage extension (Fig. 21.2*a*) and throughput enhancement (Fig. 21.2*b*).

In the mesh network shown in Fig. 21.1, there are two major types of router nodes, namely mesh router (e.g., BS-3) and mesh gateway (e.g., gateway BS). At each of the routers, there are two types of connections, that is, local and relay connections.[1] While traffic from a local connection is transmitted to a mesh client served by the same router, traffic from a relay connection is transmitted and relayed to a gateway router. Since these two types of connections share the same pool of radio resource [e.g., time slot in time division multiple access (TDMA), transmit power in code division multiple access (CDMA), and subchannel in OFDMA] at a mesh router, resource allocation methods at a mesh router ensuring efficiency and fairness are required.

21.2.2 Buffer Management for Relay Traffic in Wireless Mesh Routers

There are two approaches for buffering relay traffic in a mesh router, namely buffering on a per-flow basis and buffering on an aggregated basis. As shown in

[1] We refer to the connections involving multihop transmissions to reach an Internet gateway as the *relay* connections. The connections involving a single-hop transmission are referred to as *local* connections.

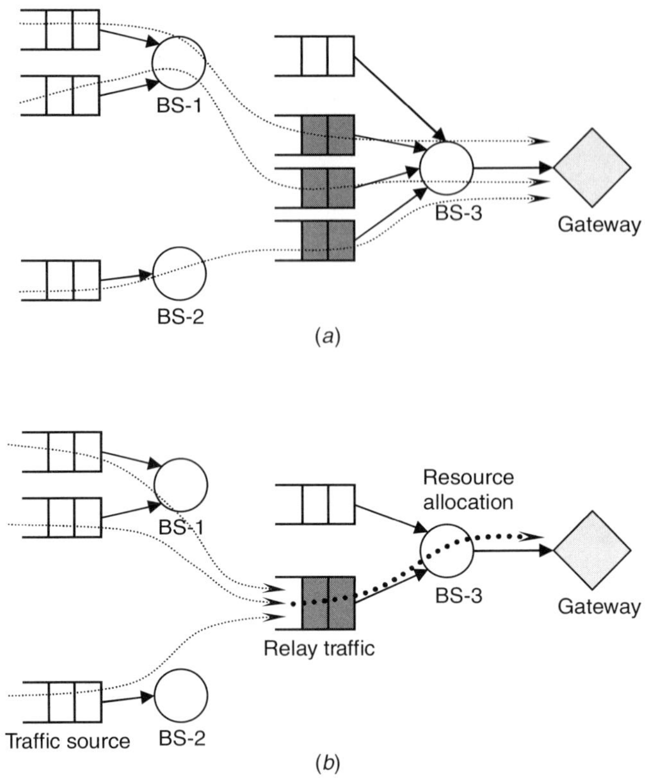

FIGURE 21.3 Queueing on an (*a*) per-flow basis and on a (*b*) aggregated basis.

Fig. 21.3, with the per-flow buffering approach at BS-3, traffic from each connection from the upstream routers (e.g., BS-1 and BS-2) are buffered into a separate queue. The per-flow queueing approach provides more flexibility for resource allocation among the different flows to satisfy their QoS requirements. However, this per-flow buffering approach suffers from the scalability problem. On the other hand, with the aggregated buffering approach, traffic from an upstream router is aggregated into a single queue. Despite simplicity in implementation, satisfying the different QoS requirements for the different traffic flows may not be possible with this aggregated approach.

The amount of buffer allocated to relay traffic impacts the QoS performance (i.e., loss, delay, and throughput performance). Upon arrival of a data packet, if the buffer is full, the packet will be lost. The queueing delay for a data packet measures the waiting time of the packet in the queue. The queue throughput measures the amount of traffic successfully transmitted from the queue per unit time. Since the end-to-end QoS performance depends on the QoS performance at each of the mesh routers along the routing path, resource allocation and route selection should be performed such that the QoS performance can be maintained at the target level.

21.2.3 Radio Resource Management in OFDMA-Based Wireless Infrastructure Mesh Networks

The major challenges in designing radio resource management schemes in a wireless mesh network arise due to the following reasons:

- *Lack of Global Information.* Due to the distributed architecture, radio resource management needs to be designed differently from that in a centralized model where all of the network information is available to the central controller.
- *Shared Resource.* Since mesh routers may operate within the range of each other, the available frequency bands need to be carefully allocated among them to avoid interference and at the same time maximize spectrum utilization.
- *Lack of Absolute Control over the Network.* In an infrastructure mesh network, the routers may operate independently in a very dynamic environment. Therefore, the radio resource management mechanism at each router has to observe, learn, and adapt the allocation (based on the actions from other routers) to achieve the desired objective.
- *Requirement for Diverse QoS Support.* Radio resource management in a wireless mesh network becomes more challenging due to the requirement of QoS support for different types of traffic (voice, video, data traffic, local traffic, and relay traffic).

The major components in radio resource management in an OFDMA-based wireless infrastructure mesh network are the following: subchannel allocation, route selection, and admission control. The interaction among these components is shown in Fig. 21.4. In particular, subchannel allocation should be performed first to obtain the amount of resource allocated to the connections. Then, the performance due to allocated resource is estimated and used for route

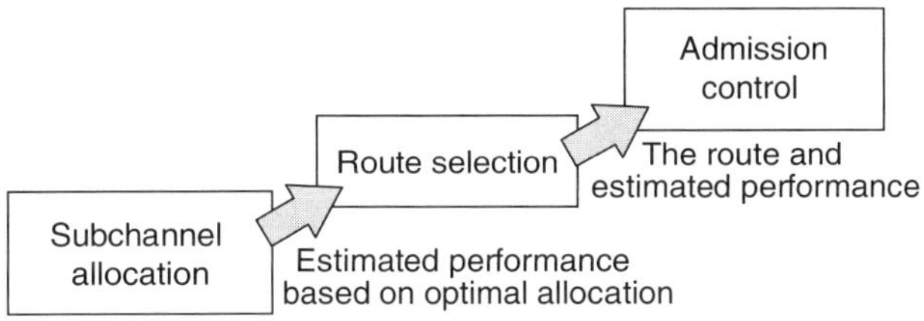

FIGURE 21.4 Major components of radio resource management in OFDMA-based wireless mesh networks.

selection. Then, the decision on admission control is made based on the assigned route and estimated QoS performances.

21.2.3.1 Subchannel Allocation in an OFDMA Wireless Mesh Router.

In OFDMA, one or more of the subchannels are allocated to a user/connection. For this, an optimization problem can be formulated to obtain the assignment of subchannels among the users. However, in a distributed scenario, if frequency bands are shared among multiple mesh routers, subchannel allocation in each mesh router must consider the allocations in the neighboring routers as well.

The performance of subchannel allocation in an OFDMA system can be enhanced by exploiting multiuser diversity. In a multiuser OFDMA system, different users may have different channel quality on different subchannels at different points in time. Therefore, the resource allocation to multiple users can be performed so that the performance (e.g., capacity) of the system is maximized. This can be illustrated by the example shown in Fig. 21.5. There are seven users (i.e., connections) and nine subchannels in the system. The average values of spectral efficiency (in bits/hertz) for the set of users and the subchannels are shown (each row corresponds to user and each column corresponds to subchannel). Due to wireless channel fading, the spectral efficiency of different users are different on different subchannels. To exploit this diversity, the subchannel allocation problem can be formulated as an optimization problem with an objective to maximizing system capacity (and/or maintaining fairness among the users). One common constraint is that one subchannel must be allocated to only one user. If the system has to provide QoS guarantee, another constraint would be to maintain the throughput of each user at least as high as the minimum requirement. Standard techniques (e.g., Hungarian method [3]) can be used to solve this assignment problem formulation.

When the wireless channel experiences fast fading (e.g., due to high mobility of the user), the channel quality and spectral efficiency (e.g., obtained from channel estimation) can vary in each transmission frame. In such a case, a subchannel allocation algorithm can be invoked in every frame to optimize the system capacity (Fig. 21.6). Note that even though the allocation in each frame

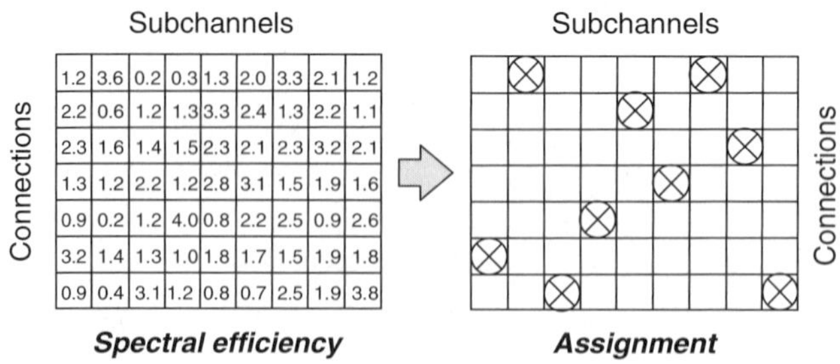

FIGURE 21.5 Subchannel allocation in OFDMA system.

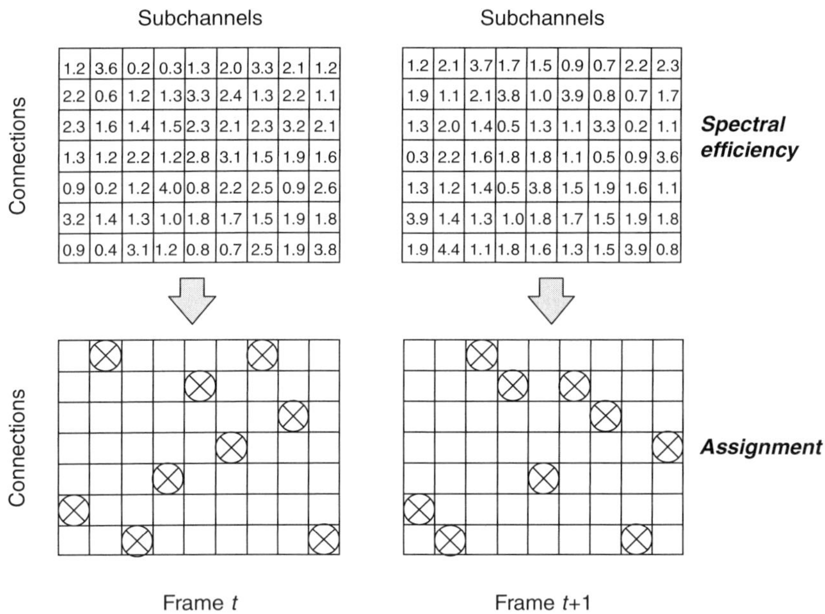

FIGURE 21.6 Subchannel allocation in OFDMA system of fast fading wireless channel.

is optimal (e.g., maximizes the system capacity), long-term performances of all users need to be guaranteed as well [4]. Since some of the users can experience deep fading, subchannel allocation must be able to maintain fairness in the multiple transmission frames.

21.2.3.2 Route Selection in the Mesh Backbone. This determines the route of data transfer from the source router to the destination router. The routing metric should consider the available radio resources at each of the mesh routers as well as the channel quality. Therefore, the routing algorithms for multihop ad hoc/sensor networks proposed in the literature need to be modified to consider subchannel allocation as well as the QoS requirements of the different types of flows (e.g., local, relay).

21.2.3.3 Admission Control in a Mesh Router. Admission control is crucial for QoS support in an infrastructure mesh network. The objective of an admission control scheme is to limit the number of ongoing connections in a mesh router so that radio resources can be efficiently utilized while satisfying the QoS requirements of the ongoing connections. To satisfy the end-to-end QoS requirement, the admission control mechanism should consider the decision from route selection and subchannel allocation mechanisms.

21.3 RELATED WORK

21.3.1 Wireless Mesh Networks

A comprehensive survey on wireless mesh networks was provided in [5]. Major research challenges in the design of architecture and protocols for wireless mesh network were discussed in [2]. An overview on the multihop relay networking and the related issues was presented in [6]. Several physical layer techniques to enhance the transmission rate were discussed. Also, the routing and the radio resource management issues for a prototype cellular relay network, namely, the *wireless media system* (WMS) were presented. An analysis of multihop diversity in the physical layer of wireless relay networks was presented in [7] where four different channel models were considered and various performance measures (e.g., outage and error probability and optimal power allocation) were obtained. In the medium access control (MAC) layer, a cooperative medium access scheme (i.e., MACA-P) for wireless mesh networks was proposed and analyzed in [8]. This MAC protocol was designed to solve the problem of simultaneous transmission in IEEE 802.11 distributed coordination function (DCF) MAC. A fair subcarrier and power allocation scheme for OFDM-based wireless mesh networks was proposed in [9]. The proposed distributed algorithm has the objective to maximize fairness by performing the resource allocation in two steps—allocating the subcarrier first and then allocating power. This allocation problem was formulated as a nonlinear integer and nonlinear mixed integer programming problem. A similar problem was investigated in [10] where graph theory was used to obtain the solution.

21.3.2 IEEE 802.16–Based Wireless Mesh Networks

An optimal solution for designing IEEE 802.16–based backhaul topology was presented in [11]. With this solution, the number of WiMAX links in the backhaul network can be reduced significantly compared to that for a ring topology. Frequency allocation is an important issue to maximize the radio resource utilization. A scheme for spatial frequency reuse for IEEE 802.16 mesh networks was proposed in [12]. An interference-aware routing mechanism for IEEE 802.16/WiMAX mesh networks was proposed in [13]. The objective of this routing scheme is to minimize interference among the links while the throughput is maximized. Performance of real-time traffic in IEEE 802.16 mesh network was studied in [14]. A QoS management framework was proposed in [15]. Performances of different scheduling schemes in IEEE 802.16 mesh networks were investigated in [16]. Routing as well as centralized scheduling algorithms were presented in [17]. A directional antenna is one of the techniques that can be used to enhance the efficiency of IEEE 802.16 mesh networks. However, the bandwidth allocation needs to be redesigned to achieve the optimal performance in the presence of a directional antenna [18]. Note that the problem of admission control based on QoS requirement and resource reservation for local and relay traffic was not considered in any of these works.

21.3.3 Resource Management in WiMAX and OFDM Networks

Resource allocation and admission control in WiMAX and OFDM wireless networks, respectively, were studied extensively in the literature [19–34]. Truncated generalized processor sharing (TGPS) scheduling scheme, derived from the well-known generalized processor sharing (GPS) scheduling discipline, was proposed for power and subcarrier allocation in the downlink of an OFDM system [19]. In OFDM, the subcarrier and power allocations are crucial to achieve the highest performance. Optimization-based approaches were presented in [20, 21] where the objective is to maximize user utility, which is a nondecreasing and a concave function of the performance and fairness measures.

Since WiMAX was designed to support multiple types of traffic (e.g., voice, video, data), QoS is one of the important issues in WiMAX-based broadband wireless access. Resource allocation needs to be designed within the predefined QoS framework in the IEEE 802.16 standard to ensure the QoS guarantee. A queue-aware bandwidth allocation and admission control scheme was proposed in [24] for WiMAX subscriber stations. A bandwidth allocation scheme for a WiMAX base station was presented in [25] in which queueing performance was used to determine the resource requirement of a connection. Also, an admission control scheme was proposed. Performance of IEEE 802.16 MAC for QoS support was evaluated and investigated in [26]. A framework to provide efficient resource management for QoS support was considered in [27]. Note that all of these works considered single-cell WiMAX and OFDM systems with centralized control at the base station. However, similar issues in multihop/mesh networks need to be investigated.

21.3.4 Radio Resource Management/Scheduling in Multihop Wireless Networks

A number of works in the literature considered the scheduling problem in multihop wireless networks. An incentive-based scheduling for cooperative relay in wireless wide area network and local area network (WWAN/WLAN) two-hop-relay network was proposed in [35]. Game theory (e.g., *Nash equilibrium* as a solution) was used to determine the optimal strategy for traffic scheduling in a relay node. A multihop-scheduling scheme, namely *coordinated multihop scheduling* (*CMS*), was proposed in [36] with a view to achieving the desired end-to-end performances. Also, the admission control problem was studied to guarantee QoS of ongoing connections. The end-to-end schedulability condition for several classes of coordinated schedulers was derived in [37]. The efficiency of the coordinated scheduling algorithm was illustrated. An analytical model to study the cooperation in wireless ad hoc networks was presented in [38] and a distributed admission control method was proposed. However, these works did not consider the radio link level queueing dynamics. Also, adaptive modulation and coding (AMC) technique (to achieve multirate transmission at the physical layer as in the IEEE 802.16 standard) was not taken into account.

21.4 RADIO RESOURCE MANAGEMENT FRAMEWORK FOR IEEE 802.16/WIMAX–COMPLIANT WIRELESS INFRASTRUCTURE MESH NETWORK

21.4.1 Network Topology

We consider an OFDMA-based infrastructure wireless mesh network (e.g., as shown in Fig. 21.1) in which the physical and the MAC layers are compliant with the IEEE 802.16a standard [1]. In this case, a subscriber station (SS) connects to the corresponding parent BS, and this parent BS is connected to other BSs for relaying traffic to the destination BS or to an Internet gateway. In Fig. 21.1, the BSs serve both the *local* and the *relay* connections. For the relay connections, data packets are transmitted to the Internet gateway through multihop transmissions. The BSs connected directly to Internet gateways are called *gateway* BS. The neighboring BSs in the network use different frequency bands. For example, with 10 MHz of bandwidth allocated to each BS, BS-1 in Fig. 21.1 is allocated with the 2.000- to 2.010-GHz band while BS-2 is allocated with the 2.010- to 2.020-GHz band and so on.

21.4.2 MAC and Physical Layer Transmission Model

The frame structure for communication between a BS (i.e., mesh node/mesh router) and an SS/mesh client is composed of downlink and uplink subframes. Each subframe consists of multiple bursts and each burst is used for transmission of protocol data units (PDUs) corresponding to one connection. Adaptive modulation and coding is used to adjust the transmission rate in each subchannel dynamically according to the channel quality.

The AMC modes and the required signal-to-noise ratios (SNRs) for the different modes in the IEEE 802.16 standard are listed in Table 21.1. With basic

TABLE 21.1 Adaptive Modulation and Coding Defined in the IEEE 802.16 Standard

Rate ID	Modulation Level (coding)	Information Bits/Symbol	Required SNR (dB)
0	BPSK (1/2)	0.5	6.4
1	QPSK (1/2)	1	9.4
2	QPSK (3/4)	1.5	11.2
3	16QAM (1/2)	2	16.4
4	16QAM (3/4)	3	18.2
5	64QAM (2/3)	4	22.7
6	64QAM (3/4)	4.5	24.4

modulation and coding scheme (i.e., *rate ID* = 0), one subchannel can transmit one PDU, and the total PDU transmission rate for a connection depends on the number of allocated subchannels and the rate ID used in each subchannel. We assume that the subchannel condition remains stationary over a frame interval ($\leq$2 ms), and all the PDUs transmitted in the same subchannel during one frame interval use the same rate ID.

21.4.3 Wireless Channel Model

We consider a Nakagami-m channel model for each subchannel c in which the channel quality is determined by the instantaneous SNR at the receiver $\gamma_{b,j,c}$ for connection j at BS b. With adaptive modulation, the SNR at the receiver is divided into $R+1$ nonoverlapping intervals (i.e., $R=7$ in IEEE 802.16) by thresholds Γ_r ($r \in \{0,1,\ldots,R\}$), where $\Gamma_0 < \Gamma_1 < \cdots < \Gamma_{R+1} = \infty$. The subchannel is said to be in state r (i.e., rate ID = r will be used) if $\Gamma_r \leq \gamma_{b,j,c} < \Gamma_{r+1}$. To avoid possible transmission error, no PDU is transmitted when $\gamma_{b,j,c} < \Gamma_0$. Note that these thresholds correspond to the required SNR specified in the IEEE 802.16 standard, that is, $\Gamma_0 = 6.4, \Gamma_1 = 9.4, \ldots, \Gamma_R = 24.4$. With Nakagami-$m$ fading, the probability of using rate ID = r [i.e., Pr(r)] can be obtained as follows [39]:

$$\Pr(r) = \frac{\Gamma(m, m\Gamma_r/\bar{\gamma}_{b,j,c}) - \Gamma(m, m\Gamma_{r+1}/\bar{\gamma}_{b,j,c})}{\Gamma(m)} \qquad (21.1)$$

where $\bar{\gamma}_{b,j,c}$ is the average SNR, m is the Nakagami fading parameter ($m \geq 0.5$), $\Gamma(m)$ is the gamma function, and $\Gamma(m, \gamma)$ is the complementary incomplete gamma function.

21.4.4 Components of the Radio Resource Management Framework

The proposed radio resource management framework consists of three components (Fig. 21.7): subchannel allocation algorithm, route selection algorithm, and connection admission control algorithm. While the subchannel allocation (among connections served by a particular BS) is performed by a BS in a distributed fashion, route selection and admission control are performed by the parent BS associated with an arriving new connection.

We propose that the resource allocation process works as follows. First, a new connection submits the PDU arrival rate (i.e., the traffic description) and the delay requirements (i.e., the QoS constraints) to the parent BS. Then, the BS sends a connection initiation request to other BSs along the possible routes to an Internet gateway. Upon receiving this request, each BS executes the subchannel allocation algorithm (to be presented later in this chapter) and responds with the information on the allocated subchannels to other BSs. Based on the information in this message, the parent BS calculates the average

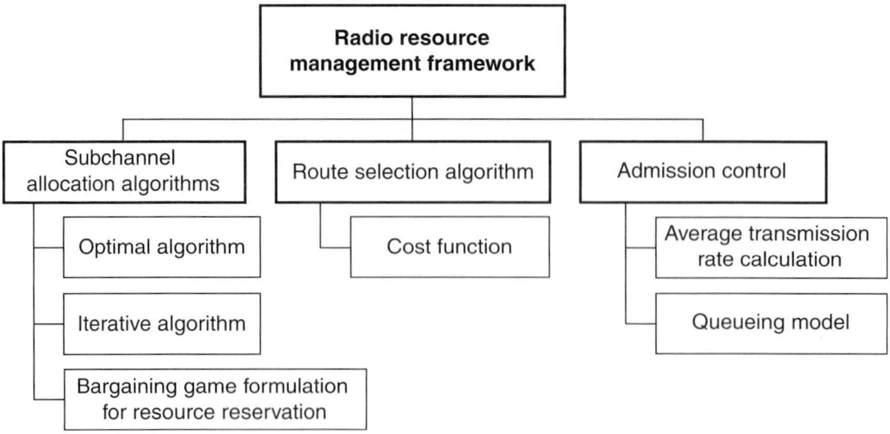

FIGURE 21.7 Major components of proposed radio resource management framework.

end-to-end queueing delay and the average PDU dropping probability. Route selection chooses the routing path with the best performance. Then, the admission controller (at the parent BS of the new connection) decides whether the delay and the PDU dropping rate requirements for the new connection can be satisfied or not. If so, the new connection is accepted (and the route is selected accordingly); otherwise, the new connection is rejected. Note that the admission control scheme applies to the relay connections only, although the subchannel allocation algorithm can be used at a mesh node for both the admitted local connections and the relay connections.

We present two subchannel allocation algorithms, namely the *optimal* algorithm and the heuristic (iterative) algorithm. The iterative algorithm takes advantage of multiuser diversity on the OFDM subchannels to maximize the throughput at a mesh router (i.e., WiMAX BS). While the *optimal*[2] algorithm can achieve maximum throughput under predefined constraints, it incurs large computational complexity. In contrast, the iterative algorithm is able to achieve similar throughput performance to that of the optimal algorithm under a heavy traffic load scenario with much less computational complexity. Both of these algorithms are implemented in a distributed manner in a mesh router where the subchannels are completely partitioned among the local and the relay connections. The value of the partitioning threshold is optimized by using the concept of the bargaining game. The objective of this bargaining game is to allocate the transmission rate among the local and the relay connections in a fair manner.

Route selection is used to select the best route for a new connection in the mesh backbone. The connection admission control mechanism is used to ensure that upon admission of a new connection, the QoS performances of all the

[2] The *optimality* here is achieved in a per-node basis rather than an end-to-end basis.

connections in the network can be maintained at the desired level. For this, the average transmission rate for a connection is obtained based on channel quality (i.e., average SNR) under Nagakami fading. Then, we formulate a tandem queueing model to obtain the QoS performance measures such as PDU dropping probability, throughput, and average queueing delay for relay connections in an end-to-end basis. The end-to-end average queueing delay is used by the route selection and admission controller to decide whether the delay requirement for a relay connection can be satisfied or not.

The performances of the proposed subchannel allocation and the admission control schemes are evaluated by extensive simulations. Also, the performance of the distributed subchannel allocation scheme is compared to that of a globally optimal subchannel allocation scheme from an end-to-end QoS perspective. Even though in this chapter we consider the QoS on a per-connection basis, the proposed framework can be easily extended for the case when the QoS needs to be maintained in an aggregate basis [such as in a *Differentiated Services* (*DiffServ*) IP network].

21.5 SUBCHANNEL ALLOCATION, ROUTE SELECTION, AND CONNECTION ADMISSION CONTROL

21.5.1 Objectives of Subchannel Allocation Algorithm

In the system model under consideration for the OFDMA-based infrastructure mesh network, the requirements for a subchannel allocation algorithm can be summarized as follows:

1. The algorithm must be performed in a distributed manner to reduce communication overhead.
2. The total transmission rate at a mesh router (i.e., WiMAX BS) needs to be maximized.
3. The subchannels need to be allocated such that the transmission rate and/or the delay requirements for the connections are satisfied.
4. Optimal resource reservation among different types of connections (i.e., local traffic and relayed traffic) can be performed.

For the third requirement, a queueing analysis will be used to determine whether the end-to-end delay requirements can be satisfied or not and the subchannel allocation algorithm must ensure that all the queues at the BSs remain stable. In particular, the allocated transmission rate for a connection must be higher than the PDU arrival rate. For the fourth requirement, a complete partitioning of the subchannels to be allocated among the local and the relay connections is assumed. In this case, from the total number of subchannels N, a threshold T_r is used to limit the number of subchannels to be allocated to the relay connections; therefore, $T_l = N - T_r$ subchannels are allocated to local connections. The value of this threshold is optimized based on a bargaining game formulation between two players (i.e., groups of local and relay connections) so that fair resource allocation among the local and the relay connections can be achieved at a relay BS.

With a view to achieving the above requirements, we present two distributed subchannel allocation algorithms, namely the optimal and the iterative algorithms.

21.5.2 Optimal Subchannel Allocation

Since the transmission rate ($\alpha_{b,j,c}$) for a connection j in subchannel c at BS b depends on the average SNR $\bar{\gamma}_{b,j,c}$, an optimization problem (more specifically, an assignment problem) can be formulated to satisfy the aforementioned requirements. Let $a_{b,j,c}$ be the assignment variable, that is, $a_{b,j,c} = 1$ if subchannel c is allocated to connection j at BS b, otherwise $a_{b,j,c} = 0$. The optimization problem can be formulated to maximize the transmission rate at a mesh router as follows:

$$\text{Maximize:} \quad \sum_{\forall j} \sum_{\forall c} \alpha_{b,j,c} a_{b,j,c} \tag{21.2}$$

$$\text{Subject to:} \quad \sum_{\forall j} a_{b,j,c} = 1 \tag{21.3}$$

$$\sum_{\forall c} \alpha_{b,j,c} a_{b,j,c} \geq \lambda_j \tag{21.4}$$

$$\sum_{c \in \mathscr{R}_b} a_{b,j,c} = T_r \tag{21.5}$$

$$\sum_{c \in L_b} a_{b,j,c} = T_l \tag{21.6}$$

where λ_j denotes the PDU arrival rate for connection j, $\mathscr{R}_b$ and $\mathscr{L}_b$ denote the sets of relay and local connections, respectively, at base station b. Note that the constraint in (21.3) is due to the one-to-one relationship between a connection and an allocated subchannel, the constraint in (21.4) indicates that the transmission rate requirements must be satisfied for connection j, and the constraints in (21.5) and (21.6) ensure that the total number of subchannels allocated to local and relay connections is limited to $N = T_r + T_l$.

This optimization problem can be solved by using binary integer linear programming in which the solution is obtained by a linear programming-based branch-and-bound algorithm [40]. Note that the gross upper bound time complexity of this algorithm is $O(2^{C^2(N)^2})$ [41] where C and N refer to the total number of connections and the number of available subchannels, respectively.

21.5.3 Iterative Subchannel Allocation

The iterative subchannel allocation is based on a water-filling algorithm. First, the algorithm tries to allocate subchannels among the connections such that the transmission rate requirements are met. In this case, the number of subchannels allocated to relay and local connections must be limited to the corresponding threshold (i.e., T_r and $T_l = N - T_r$, respectively). If the requirements for all the connections are satisfied, the algorithm tries to allocate the subchannels among the different connections so that the total transmission rate at the BS is maximized. This procedure is shown in Algorithm 21.5.1. Note that the complexity of the iterative subchannel allocation algorithm is $O(N \times C)$, which is much smaller than that of the optimal algorithm.

21.5.4 Optimization of the Resource Reservation Threshold

We formulate a bargaining game [42] to obtain the optimal threshold T_r to reserve the number of subchannels for relay connections during subchannel allocation. The motivation of using this bargaining game formulation is to ensure that both these groups of connections are assigned with the fair number of subchannels.

In general, for a two-person bargaining game, two players try to make an agreement on trading a limited amount of resource. These two individuals have a choice to *bargain* with each other so that both of them can gain benefit higher than they could have obtained by playing the game without cooperation.

At each relay BS, the two players refer to the groups of relay and local connections. Both groups of connections share the common pool of resources of N subchannels. The solution (i.e., equilibrium point) of this bargaining game seeks to satisfy both groups of connections. In this case, the amount of resource that relay and local connections receive is T_r and $T_l = N - T_r$, respectively. The payoff received by both players is defined by using the sigmoid utility function [43]. This utility function represents user satisfaction on the perceived transmission rate $\alpha_{b,j}$ (i.e., $\alpha_{b,j} = \sum_{j \in \mathscr{C}_{b,j}} \alpha_{b,j,c}$, where $\mathscr{C}_{b,j}$ is the set of subchannels allocated to connection j) of connection j at BS b. This function is defined as follows:

$$W(\alpha_{b,j}) = \frac{1}{1 + \exp[g(\alpha_{b,j} - h)]} \tag{21.7}$$

where g and h are the parameters of the sigmoid function. Specifically, g determines the steepness (i.e., sensitivity of the utility function to the transmission rate) and h represents the center of the utility function.

Algorithm 21.5.1. Iterative Algorithm for Subchannel Allocation (*input:* transmission rate matrix for the connections and the subchannels)

comment: allocate subchannels such that the QoS requirements for the connections are satisfied

repeat

find subchannel j_{max} and connection c_{max} which correspond to the highest transmission rate

if rate requirement of connection c_{max} is not satisfied

then

assign subchannel j_{max} to connection c_{max}

remove subchannel j_{max} from the set of unallocated subchannels

if rate requirement of connection c_{max} is satisfied

then remove connection c_{max} from the set of unsatisfied connections

if the number of allocated subchannels to the relay connections reaches threshold T_r

then remove relay connections from the set of candidate connections

if the number of allocated subchannels to the local connections reaches threshold $N - T_r$

then remove local connections from the set of candidate connections

until requirements of all connections are satisfied or all subchannels are allocated

comment: allocate subchannel such that total transmission rate is maximized

if there are some subchannels which are not allocated

then

repeat

find subchannel j_{max} and connection c_{max}, which correspond to the highest transmission rate

assign subchannel j_{max} to connection c_{max}

remove subchannel j_{max} from the set of candidate subchannels

if the number of allocated subchannels to the relay connections reaches threshold T_r

then remove relay connections from the set of candidate connections

if the number of allocated subchannels to the local connections reaches threshold $N - T_r$

then remove local connections from the set of candidate connections

until all subchannels are allocated

Without loss of generality, the payoffs for the players are given by $\{(\Psi(T_r), \Omega(T_l)) : 0 \leq \Psi(T_r), 0 \leq \Omega(T_l)\}$ (i.e., feasible set). These payoffs can be obtained from $\Psi(T_r) = \sum_{j \in \mathcal{R}} W(\alpha_{b,j})$ and $\Omega(T_l) = \sum_{j \in \mathcal{L}} W(\alpha_{b,j})$, where T_r and T_l denote the number of subchannels allocated to relay and local connections, respectively. If an agreement between both players in the game cannot be reached, the utility that the players will receive is given by the threat point $(\Psi'(.), \Omega'(.))$. In particular, $(\Psi'(.), \Omega'(.)) = (0, 0)$ is the threat point for this game.

The strategic form of this bargaining game can be expressed as in (21.8) for $T = N/2$ where the rows and the columns represent the strategies corresponding to the relay and the local connections, respectively.

	$\cdots$	$T-1$	T	$T+1$	$\cdots$
$\vdots$	$\ddots$	$\vdots$	$\vdots$	$\vdots$	$\iddots$
$T-1$	$\cdots$	$(\Psi(T-1), \Omega(T-1))$	$(\Psi(T-1), \Omega(T))$	$(\tilde{\Psi}(T-1), \tilde{\Omega}(T+1))$	$\cdots$
T	$\cdots$	$(\Psi(T), \Omega(T-1))$	$(\tilde{\Psi}(T), \tilde{\Omega}(T))$	$(0, 0)$	$\cdots$
$T+1$	$\cdots$	$(\tilde{\Psi}(T+1), \tilde{\Omega}(T-1))$	$(0, 0)$	$(0, 0)$	$\cdots$
$\vdots$	$\iddots$	$\vdots$	$\vdots$	$\vdots$	$\ddots$

(21.8)

We can divide the set of strategies into three categories. The first category corresponds to the case $T_r + T_l < N$ in which some of the subchannels are allocated to neither relay nor local connections. Therefore, the payoffs of the players are not maximized. The second category corresponds to the case $T_r + T_l = N$ in which the payoff of at least one of the players is maximized. The set of strategies in this category establish Pareto optimality. This Pareto optimality is defined as a set of strategies for which one player cannot increase its payoff without decreasing the payoff of the other player. The Pareto optimality in the above example is denoted by the set of payoffs $(\tilde{\Psi}(.), \tilde{\Omega}(.))$. The third category corresponds to the case $T_r + T_l > N$ in which the bargaining among the players fails to achieve a feasible solution, and the payoff received for any player is given by the threat point, which is (0,0) in our case.

We define the equilibrium as follows:

$$(\Psi^*(T_r), \Omega^*(T_l)) = \arg \max_{(T_r, T_l)} [\Psi(T_r)^{w_r} \times \Omega(T_l)^{w_l}] \tag{21.9}$$

where w_r and w_l denote the weights corresponding to the utilities for relay and local connections, respectively.

In particular, the equilibrium lies on the Pareto optimality to ensure that the payoff is maximized for either group of relay or local connections. In particular, $(\Psi^*(T_r), \Omega^*(T_l)) = (\Psi^*(T_r), \Omega^*(N - T_r))$. Note that this threshold T_r can be adjusted adaptively based on the traffic load of relay and local connections.

21.5.5 Route Selection and Connection Admission Control

For route selection, let $\hat{\alpha}_{b,j}$ denote the transmission rate allocated to connection j at base station b (i.e., $\hat{\alpha}_{b,j} = \sum_{\forall c} \alpha_{b,j,c} a_{b,j,c}$). Then, the path $i_j = \max_i \left(\min\left(\hat{\alpha}_{b,j}\right)\right)$, $\forall b \in \mathscr{B}_i$ will be selected for connection j where $\mathscr{B}_i$ is the set of BSs along path i.

Connection admission control is performed based on the outcome of the execution of the subchannel allocation algorithm, resource reservation optimization, and route selection. A new connection is admitted if end-to-end average queueing delay is less than the requirement (i.e., $\bar{d}_j \leq d_{\text{req}}$). Also, to ensure reliable transmission, end-to-end PDU dropping probability is required to be less than the target level (i.e., $P_j^{\text{drop}} \leq P_{\text{req}}^{\text{drop}}$). Note that d_{req} and $P_{\text{req}}^{\text{drop}}$ denote target average end-to-end delay and PDU dropping probability, respectively (e.g., $d_{\text{req}} = 50\,\text{ms}$ and $P_{\text{req}}^{\text{drop}} = 0.05$); $\bar{d}_j$ and P_j^{drop} denote end-to-end average PDU delay and PDU dropping probability, respectively. These performance measures can be computed from the queueing model presented in the next section. The new connection j is admitted if $\bar{d}_j(i) \geq \bar{d}_j'$ and rejected otherwise, where $\bar{d}_j(i)$ is the average end-to-end delay for connection j when using route i and $\bar{d}_j'$ is the delay requirement for connection j.

21.6 QUEUEING ANALYTICAL MODEL

We assume that in each BS, a dedicated queue is used to store the PDUs of each connection. In Fig. 21.8, data traffic corresponding to a connection originating at BS-1 needs to be relayed via BS-2, BS-3, BS-4, and BS-5 before it is routed to BS-6, which interfaces with an Internet gateway. For a particular connection, the PDUs transmitted from one BS will be stored in the corresponding queue in the next BS to be relayed along the route to the gateway BS. Since the gateway

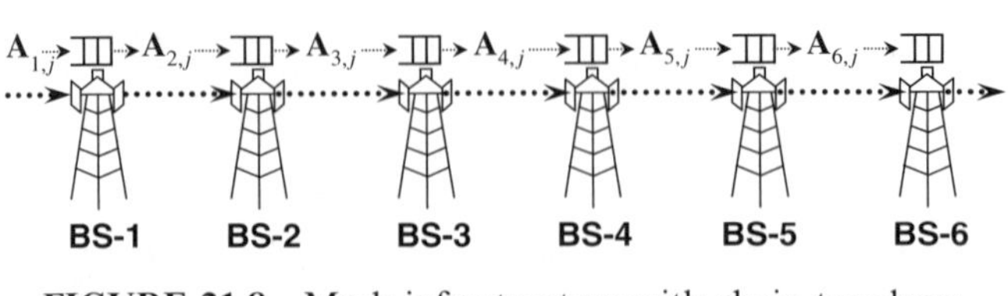

FIGURE 21.8 Mesh infrastructure with chain topology.

BS interfaces directly with the Internet gateway, the PDU transmission rate from this queue is assumed to be much higher (e.g., due to a high speed wired connection) than that from a relay BS. Therefore, we ignore the queueing delay at the queue in the gateway BS while calculating the average end-to-end delay[3] for a PDU.

Note that the presented framework can be used also to analyze QoS performance measures such as PDU dropping probability and connection blocking probability for the admitted local connections at each mesh node (i.e., IEEE 802.16 BS).

21.6.1 Transmission Rate Distribution for a Connection at a Base Station

Let $\mathscr{C}_{b,j}$ denote the set of allocated subchannels to connection j at base station b. We can define a row matrix $\mathbf{D}'_{b,j,c}$ whose elements d_k correspond to the probability of transmitting k PDUs in one frame on one subchannel c ($c \in \mathscr{C}_{b,j}$) as follows:

$$\mathbf{D}'_{b,j,c} = [d_0 \cdots d_k \cdots d_9] \tag{21.10}$$

where $d_{(I_r \times 2)} = \Pr(r)$ is the probability that $I_r \times 2$ PDUs are transmitted in one frame, and I_r is the number of transmitted bits per symbol corresponding to rate ID $= r$, and $d_0 = 1 - \sum_{k=1}^{9} d_k$. Note that, in one frame one PDU can be transmitted with rate ID 0 over one subchannel.

The matrix for probability mass function (pmf) of total PDU transmission rate can be obtained by convoluting matrices $\mathbf{D}'_{b,j,c}$ as follows:

$$\mathbf{D}_{b,j} = \bigodot_{\forall c \in \mathscr{C}_{b,j}} \mathbf{D}'_{b,j,c} \tag{21.11}$$

where $\mathbf{a} \odot \mathbf{b}$ denotes discrete convolution [44] between matrices $\mathbf{a} = [a_0\, a_1 \cdots a_{u-1}]$ and $\mathbf{b} = [b_0\, b_1 \cdots b_{v-1}]$ and can be expressed as $[\mathbf{a} \odot \mathbf{b}]_{i+1} = \sum_{j=0}^{i} a_j b_{i-j}$ for $i = 0, 1, 2, \ldots, u + v - 1$, and $[\mathbf{a}]_i$ denotes the element at column i of row matrix $\mathbf{a}$. Note that matrix $\mathbf{D}_{b,j}$ has size $1 \times U + 1$, where $U = (9 \times |\mathscr{C}_{b,j}|)$ indicates the maximum number of PDUs that can be transmitted during one frame interval, and $|\mathscr{C}_{b,j}|$ is the number of elements in the set $\mathscr{C}_{b,j}$ indicating the number of subchannels allocated to connection j in BS b.

The PDU transmission rate in subchannel c for connection j can be obtained as follows:

$$\alpha_{b,j,c} = \sum_{k=1}^{9} k \times [\mathbf{D}'_{b,j,c}]_{k+1} \tag{21.12}$$

[3] This is measured as the total delay that a PDU incurs (since it is generated) while waiting in the queues along the route to a gateway BS.

and the total transmission rate for connection j at BS b can be calculated from

$$\alpha_{b,j} = \sum_{k=1}^{U} k \times [\mathbf{D}_{b,j}]_{k+1}. \tag{21.13}$$

21.6.2 Traffic Source and Arrival Probability Matrix

We assume that the traffic arrival process for connection j is the truncated Poisson with mean rate λ_j. Therefore, at the first queue (e.g., at BS-1 in Fig. 21.8) the probability of arrival of a PDUs with mean rate λ_j during time interval t (i.e., one frame interval) is given by

$$f_a(\lambda_j) = \frac{e^{-\lambda_j t}(\lambda_j t)^a}{a!} \tag{21.14}$$

where $a \in \{0, 1, \ldots, A\}$ and A is the maximum batch size for PDU arrival (i.e., $1 \leq A \leq \infty$). The complementary cumulative probability mass function for this arrival process is given by

$$F_a(\lambda_j) = \sum_{i=a}^{\infty} f_i(\lambda_j). \tag{21.15}$$

Note that A can be obtained such that $F_A(\lambda_j) < \epsilon$ where ϵ is a small number (e.g., $\epsilon = 10^{-5}$). The PDU arrival probability $\mathbf{A}_{1,j}$ for connection j at the first hop is obtained as follows:

$$\mathbf{A}_{1,j} = [f_0(\lambda_j) \quad \cdots \quad f_a(\lambda_j) \quad \cdots \quad F_A(\lambda_j)]. \tag{21.16}$$

For a relay connection, since the PDUs transmitted from BS b will arrive at BS $b+1$ as the relay traffic, the distribution of PDU arrival at BS $b+1$ (denoted by arrival probability matrix $\mathbf{A}_{b+1,j}$, for connection j) is the same as the distribution of PDU transmission rate at BS b (as shown in Fig. 21.8). This matrix can be obtained by considering the transmission probability matrix and the steady-state probability of queue length as follows:

$$[\mathbf{A}_{b+1,j}]_{i+1} = \sum_{x=0}^{X} [\mathbf{D}_{b,j}^{(x)}]_{i+1} \pi_{b,j}(x) \tag{21.17}$$

for $i = 0,1,\ldots U$, and $\pi_{b,j}(x)$ denotes the steady-state probability for the number of PDUs in the queue j at base station b. This probability will be obtained later in this chapter.

21.6.3 State Space and Transition Matrix

At each BS, a separate queue with size X PDUs is used for buffering the PDUs from each connection. The state of a queue is observed at the beginning of each frame interval. A PDU arriving during frame interval f will not be transmitted until the next frame interval (i.e., frame $f + 1$) at the earliest. The state space of a queue can be defined as follows:

$$\Phi = \{(\mathscr{X}), 0 \leq \mathscr{X} \leq X\} \tag{21.18}$$

where $\mathscr{X}$ indicates the number of PDUs in the queue. The transition matrix $\mathbf{P}_{b,j}$ of the queue for connection j at base station b can be expressed as in (21.19). The rows of matrix $\mathbf{P}_{b,j}$ represent the number of PDUs in the queue, and element $p_{x,x'}$ inside this matrix denotes the probability matrix for the case when the number of PDUs in the queue changes from x in the current frame to x' in the next frame.

$$\mathbf{P}_{b,j} = \left[\begin{array}{cccccc}
p_{0,0} & \cdots & \cdots & p_{0,A} & & \\
\vdots & \ddots & \ddots & \ddots & & \\ \hline
p_{U,0} & \cdots & p_{U,U} & \cdots & p_{U,U+A} & \\
 & \ddots & \ddots & \ddots & \ddots & \\
 & p_{x,x-U} \;\cdots & & p_{x,x} & \cdots & p_{x,x+A} \\ \hline
 & & \ddots & \ddots & \ddots & \ddots \\
 & & & p_{X,X-U} & \cdots & p_{X,X}
\end{array}\right]. \tag{21.19}$$

Then, following a similar procedure as in [25], different QoS measures can be obtained.

21.6.4 Quality of Service Measures

21.6.4.1 PDU Dropping Probability. Denoted by $P_{b,j}^{\text{drop}}$, it refers to the probability that an incoming PDU will be dropped due to insufficient buffer space at the queue for connection j at BS b. Since we do not consider any error recovery mechanism at the hop level (i.e., the dropped PDUs are not retransmitted at a BS), the end-to-end PDU dropping probability of connection j along route i is calculated from $P_j^{\text{drop}} = 1 - \Pi_{b \in \mathscr{B}_i}(1 - P_{b,j}^{\text{drop}})$, where $\mathscr{B}_i$ is the set of BSs along route i.

21.6.4.2 Queue Throughput. It measures the number of PDUs transmitted during one frame interval and can be obtained from $\eta_{b,j} = \lambda_j(1 - P_{b,j}^{\text{drop}})$. The end-to-end throughput for connection j can be obtained from $\eta_j = \lambda_j(1 - P_j^{\text{drop}})$.

21.6.4.3 Average Delay. It is defined as the number of frame intervals that a PDU waits in the queue since its arrival until it is transmitted. We use Little's law to obtain the average delay as follows: $\bar{d}_{b,j} = \bar{x}_{b,j}/\eta_{b,j}$, where $\eta_{b,j}$ is the throughput (same as the effective arrival rate at the queue) and $\bar{x}_{b,j}$ is the average number of PDUs in the queue. The average number of PDUs in the queue corresponding to connection j at base station b is obtained as follows: $\bar{x}_{b,j} = \sum_{x=1}^{X} x \times \pi_{b,j}(x)$.

Since the queues along the route are in tandem, the average end-to-end delay for connection j can be simply obtained from $\bar{d}_j = \sum_{b \in \mathcal{B}_j} \bar{d}_{b,j}$.

21.7 PERFORMANCE EVALUATION

21.7.1 Parameter Setting and Simulation Environment

We consider both a tree and a chain topology for an OFDMA-based infrastructure wireless mesh network as shown in Figs. 21.1 and 21.9, respectively, in which the frequency bands of operation of the adjacent BSs are nonoverlapping. For example, the frequency band of BS-1 is nonoverlapping with that of BS-2, but it can be the same as that of either BS-3 or BS-4.

Each BS has 48 subchannels (i.e., $N = 48$) and each subchannel has a bandwidth of 160 kHz. The total bandwidth required (including the guard bands) is 10 MHz. The length of a subframe for uplink transmission is 1 ms, and therefore, the transmission rate in one subchannel with rate ID = 0 [i.e., binary phase shift keying (BPSK) modulation and coding rate of $\frac{1}{2}$] is 80 kbps.

Adaptive modulation and coding is performed independently in each subchannel to increase the transmission rate if the channel quality (i.e., average SNR) permits. Although we assume that the average SNR is the same for all the allocated subchannels to a particular connection, the instantaneous SNR, and consequently, the selected rate ID in the different subchannels can be different during the same subframe.

The PDU arrivals for a connection follow a truncated Poisson process and the maximum batch size of arrival is 100 (i.e., $A = 100$). We assume that the queue size is 400 PDUs (i.e., $X = 400$) and it is the same for all BSs. To investigate the impacts of traffic load on the network performance, we vary the average connection arrival rate at each BS while the connection holding time is exponentially distributed with mean 10 minute. Note that the total traffic arrival rate at any BS consists of traffic due to the new connections at that BS and traffic due to the connections relayed from the upstream BSs.

The connection blocking probability for a relay connection with a given number of hops is calculated by dividing the number of blocked connections (due to the admission controller at the first hop BS for that connection) by the total number of connections initiated at that BS.

The transmission radius for a BS transmitter is 800 m. We assume that communication between adjacent BSs is line-of-sight while that between an SS and a BS can be non-line-of-sight. For line-of-sight communication, we assume a Nakagami-m fading channel with $m = 2.0$ (which corresponds to the Rice factor $K = 4$ dB for a Ricean fading channel) and a relatively high value for average SNR (e.g., 15–25 dB). For non-line-of-sight communication, we use $m = 1.1$ (corresponding to the Rayleigh fading channel) and assume that the SNR at the receiver varies in the range of 7–24 dB depending on the location of the SS. Note that this parameter setting for non-line-of-sight communication scenario is similar to that used in the literature [45]. For simulations, we use an event-driven simulator developed in MATLAB.

21.7.2 Numerical and Simulation Results

21.7.2.1 Queueing Performances. We investigate the queueing performance (average end-to-end delay and cumulative distribution of PDU dropping probability with the number of hops) for a chain topology with 6 hops as shown in Fig. 21.8. We assume that the average SNR for each of the allocated subchannels is 20 dB and the average PDU arrival rate is 15,000 PDUs per second.

The queueing delay and the PDU dropping probability for a tagged connection in each BS (obtained using the analytical model and the simulations) are shown in Figs. 21.9*a* and 21.9*b* respectively. The numbers inside the brackets indicate the number of allocated subchannels in each BS. For example, [3,3,2,2,3,3] refers to the case in which three subchannels are allocated to the tagged connection at each of the BSs 1, 2, 5, and 6 while two subchannels are allocated to the connection at each of the BSs 3 and 4. As expected, the average queueing delay for a connection increases as the connection traverses an increasing number of hops to reach an Internet gateway.

The number of subchannels allocated to the connections in each BS has a significant impact on both average delay and PDU dropping probability. In Fig. 21.9*a*, we observe that, with three subchannels allocated to the tagged connection in each hop, the average end-to-end delay is less than 20 ms. With two subchannels allocated to the connection at each BS (i.e., for [2,2,2,2,2,2]), the average end-to-end delay becomes very high. In fact, the transmission rate of the allocated subchannels for a connection should be higher than the PDU arrival rate corresponding to that connection in order to minimize average end-to-end delay and PDU dropping probability.

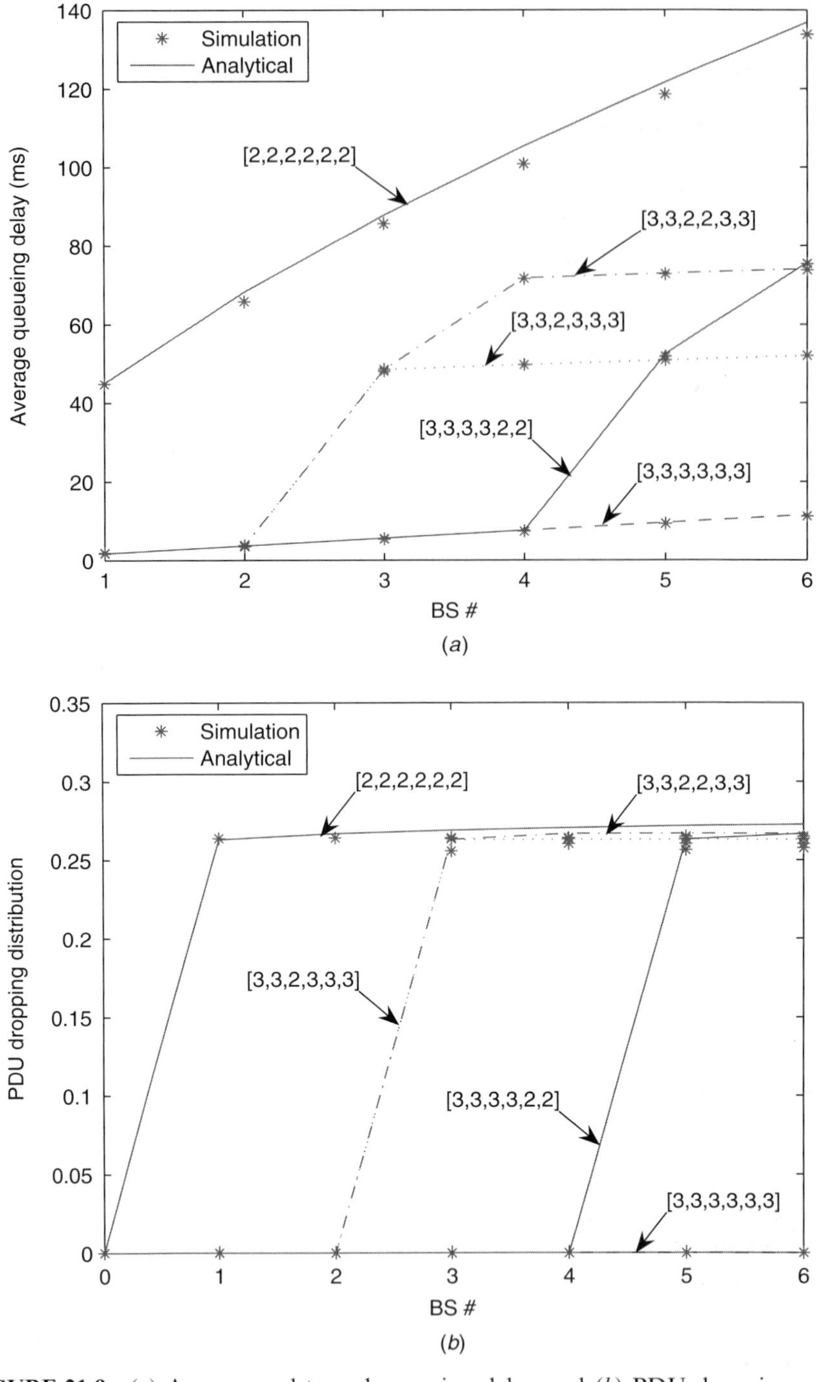

FIGURE 21.9 (*a*) Average end-to-end queueing delay and (*b*) PDU dropping probability under different number of allocated subchannels to a tagged connection.

We observe that the numerical results on queueing delay and PDU dropping probability obtained from the analytical model are close to those obtained from simulations. This validates the accuracy of the analytical model.

21.7.2.2 Relative Performances of the Different Subchannel Allocation Algorithms. Figure 21.10 shows typical variations in average transmission rate in the subchannels allocated to the tagged connection for both the optimal and the iterative algorithms. In this case, when the average SNR for the relay connections is 20 dB, we set $T_r = 30$, and the number of relay connections is one third of the total number of connections. As expected, for both the algorithms the transmission rate decreases as the number of connections increases. The optimal algorithm performs slightly better when the number of connections is relatively small (e.g., 10–15 connections). However, when the number of connections is relatively large (e.g., 16–22 connections), both the algorithms achieve similar transmission rate. In addition, the computational time[4] for the optimal algorithm is much larger than that for the iterative

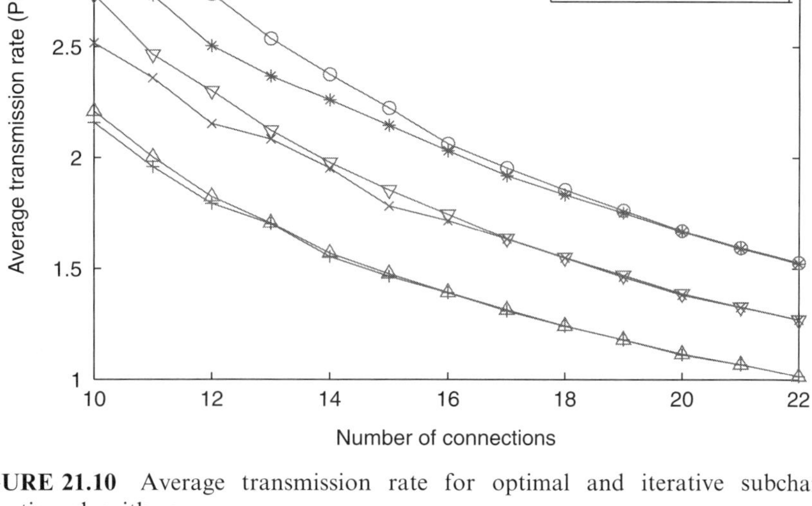

FIGURE 21.10 Average transmission rate for optimal and iterative subchannel allocation algorithms.

[4] Using MATLAB on a Pentium III 2.0-GHz PC with 512-MB RAM.

algorithm. In particular, for 20 connections, the iterative algorithm requires 0.01 s, while the optimal algorithm requires 1.5 s.

21.7.2.3 End-to-End Performance. We evaluate the end-to-end performance (i.e., transmission rate) of a connection for the proposed distributed subchannel allocation algorithm. Also, we compare the performance of the proposed distributed scheme with that of a globally optimal subchannel allocation scheme. For this, we formulate an optimization problem to maximize the end-to-end performance based on (21.2) as follows:

$$\text{Maximize:} \quad \sum_{\forall c} \min_{b} \left(\alpha_{b,j,c} a_{b,j,c}\right) \tag{21.20}$$

$$\text{Subject to:} \quad \sum_{\forall j} a_{b,j,c} = 1 \tag{21.21}$$

$$\sum_{\forall c} a_{b,j,c} a_{b,j,c} \geq \lambda_j \tag{21.22}$$

where the objective is to maximize the total end-to-end transmission rate. This end-to-end transmission rate is a minimum of transmission rate among all hops. We solve this optimization problem by enumeration. Note that the computational complexity of solving this optimization is substantially large and also the subchannel information for all the mesh routers along the route to the destination would be required. Therefore, a distributed implementation of this global optimization might not be possible.

For numerical results, we consider a chain topology with three hops (i.e., BS-4 to BS-6 in Fig. 21.8). The end-to-end transmission rate per connection is shown in Fig. 21.11. As expected, with the objective defined in (21.19) the end-to-end transmission rate achieved via global optimization is only slightly larger (i.e., about 10–18%) than that for the proposed subchannel allocation algorithm, which is performed in a distributed manner. Since the objective in global optimization is to maximize the transmission rate at the bottleneck BS (i.e., BS with minimum transmission rate), with the global information of all BSs, end-to-end transmission rate for a connection obtained from this optimization formulation is larger than that from the proposed distributed algorithm. However, note that the proposed distributed subchannel allocation approach does not require global information of all the mesh routers along the route to the destination, and it incurs much lower computational complexity.

21.7.2.4 Optimization of the Resource Reservation Threshold—Performance of the Bargaining Game. To evaluate the performance of the bargaining game formulation for optimizing the resource reservation threshold for subchannel allocation, we consider 15 connections (including both local and relay connections) at a particular BS. While the average SNR for

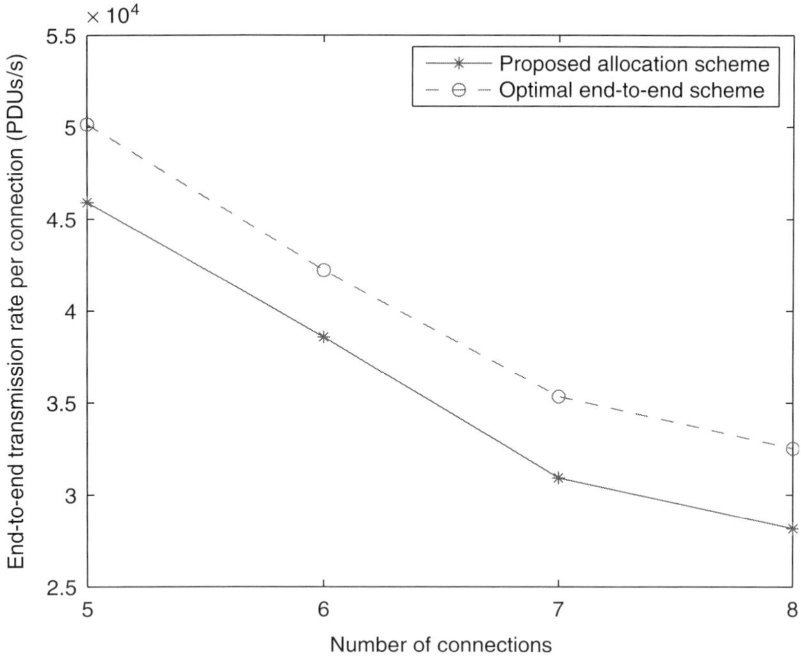

FIGURE 21.11 End-to-end performance for proposed distributed subchannel allocation scheme compared to that for globally optimal scheme.

a local connection is 20 dB, that for a relay connection (i.e., between current BS to the next relay BS) is 15 or 20 dB, depending on the evaluation scenario. Since the PDU arrival rate is 15,000 PDUs/second, we set $g = 7.5$ (per frame) and $h = 1$ so that $W(\alpha_{b,j}) = 1$ for $\alpha_{b,j} \geq 15$. For the bargaining solution, we assume $w_r = w_l = 1$.

First, in Fig. 21.12, we show the equilibrium point for threshold optimization. In this figure, we plot total utility of relay connections against that of local connections. The Pareto optimality for two cases, that is, when the number of local and relay connections are 5 and 10 for case 1 ($|\mathscr{R}| = 5, |\mathscr{L}| = 10$) and 10 and 5 for case 2 ($|\mathscr{R}| = 10, |\mathscr{L}| = 5$) are shown. We observe that the Pareto optimality is rectangular in shape. This is due to the fact that when we increase the number of subchannels allocated to either the group of relay or local connections, the total utility of that group increases while that of another group remains the same. However, after reaching the equilibrium (i.e., at the top right corner of Pareto optimality), increasing the number of allocated subchannels results in constant total utility (i.e., transmission rate requirements for all connections in that group are satisfied) while the total utility for the other group of connections decreases (i.e., requirements for some of the connections in this group are not satisfied). From the Pareto optimality, the equilibrium can be determined easily.

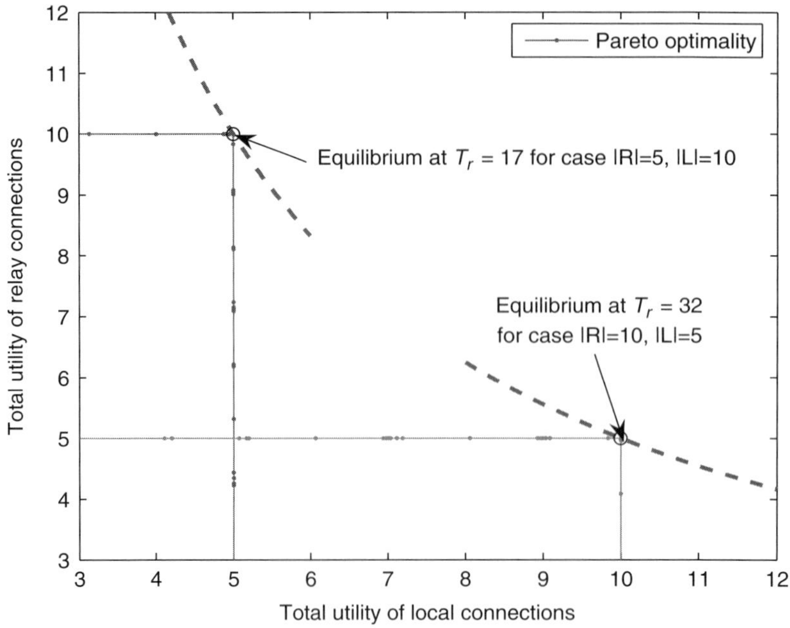

FIGURE 21.12 Illustration of equilibrium for bargaining game between relay and local connections.

Next, we change the number of local connections while the total number of ongoing connections is fixed at 15. The average transmission rates for both relay and local connections are shown in Fig. 21.13. We observe that under different average SNR for relay connections, the equilibrium threshold changes accordingly. As expected, when the number of local connections increases, the number of subchannels required for this group of connections increases. Also, with better channel quality for relay connections, the number of subchannels allocated to relay connections increases linearly. Since higher average SNR results in enough transmission rate for both relay and local connections, the equilibrium threshold changes smoothly. In contrast, when the relay connections experience low SNR, the transmission rate requirements for some of these connections cannot be satisfied. Consequently, depending on the variation of channel quality of local connections, the number of subchannels allocated to relay and local connections may fluctuate significantly.

For the average transmission rate, when the channel qualities of local and relay connections are identical (i.e., 20 dB), the equilibrium strategy of the bargaining game results in a fair transmission rate allocation. In particular, the transmission rate for local connections/relay connections increases/decreases linearly as the number of local connections increases. Similarly, the transmission rate for relay connections decreases linearly. However, when the average SNR is different for local and relay connections (i.e., 20 and 15 dB,

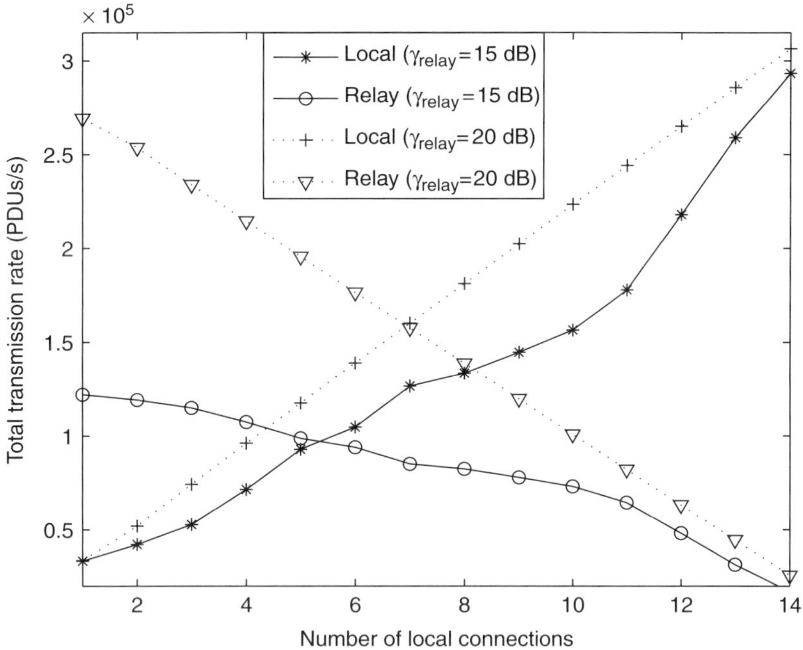

FIGURE 21.13 Transmission rate under different number of local connections.

respectively), the transmission rate of local connections increases proportionally to the number of local connections. However, the transmission rate of relay connections reduces significantly compared with the case of identical channel quality. Since the channel quality of relay connections is worse than that of local connections, a BS tries to allocate less transmission rate (i.e., a smaller number of subchannels) to the relay connections and allocate higher transmission rate to local connections, which have better channel quality. Therefore, optimized resource reservation based on the bargaining game can adapt the subchannel allocation threshold by taking the channel quality into account.

The total utility of a BS (i.e., sum of total utility of local and relay connections) obtained at each equilibrium strategy is shown in Fig. 21.14. This variation in total utility suggests that there is an optimal point for the number of local and relay connections at a BS. As expected, with identical channel qualities for both the relay and the local connections, the maximum total utility is achieved when the number of local and relay connections is 8 and 7, respectively. However, in case the channel quality of local connections is better than that of relay connections, the maximum total utility is achieved when the number of local and relay connections is 10 and 5, respectively. Since the average SNR for local connections is higher, a BS prefers to accept more numbers of local connections.

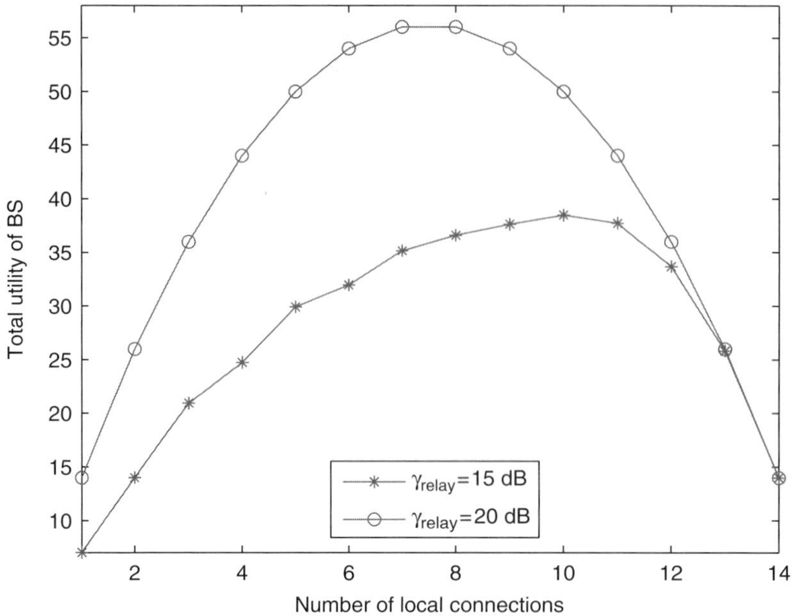

FIGURE 21.14 Total utility of BS under different number of local connections.

21.7.2.5 Performance of Admission Control. We first consider a network having chain topology as shown in Fig. 21.8. The arrival rate of local and relay connections at each BS is 0.25 connection/minute and average connection holding time is 10 minute. We compare the performances between the cases when the resource reservation threshold is fixed (i.e., the number of subchannels for relay connections is 36) and when the resource reservation threshold is optimized based on the bargaining game.

The connection blocking probability and the average transmission rate for a relay connection at each BS are shown in Figs. 21.15*a* and 21.15*b*, respectively. As expected, the connection blocking probabilities for local and relay connections are identical for all BSs in case of a fixed threshold. Since BS-6 becomes a bottleneck, admission of relay connections is limited by transmission rate at BS-6 (i.e., transmission rate per connection is the smallest at BS-6 in Fig. 21.15*b*). In addition, as T_r increases/decreases, the blocking probability of relay connections decreases/increases accordingly.

However, with an optimized resource reservation, even though BS-6 is a bottleneck, the reservation thresholds at the other BSs are adjusted dynamically. In particular, BS-6 allocates an equal number of subchannels to local and relay connections which results in a similar connection blocking probability. However, BS-1–BS-5 allocate a larger number of subchannels to local connections. Since relay connections experience a bottleneck at BS-6, it is better for BS-1–BS-5 to allocate more subchannels to local connections. Consequently,

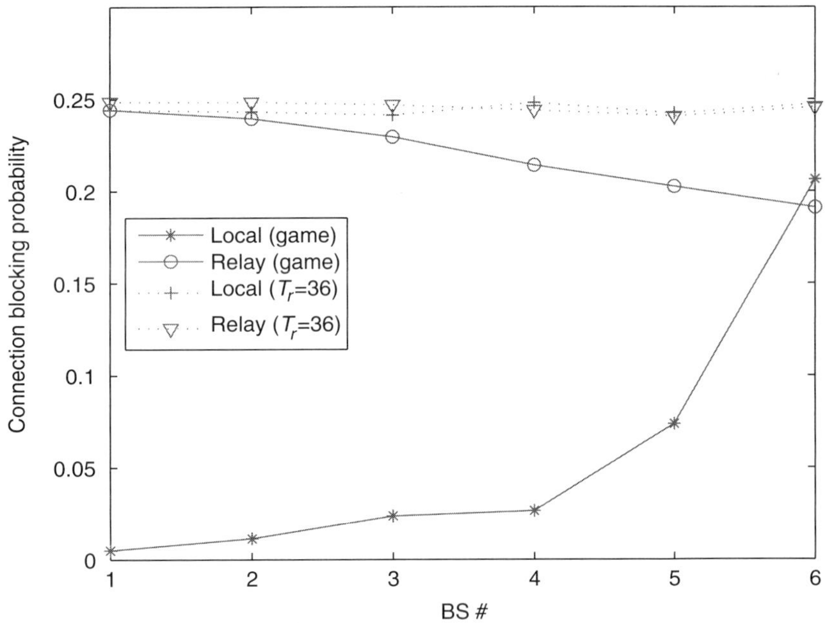

(*a*)

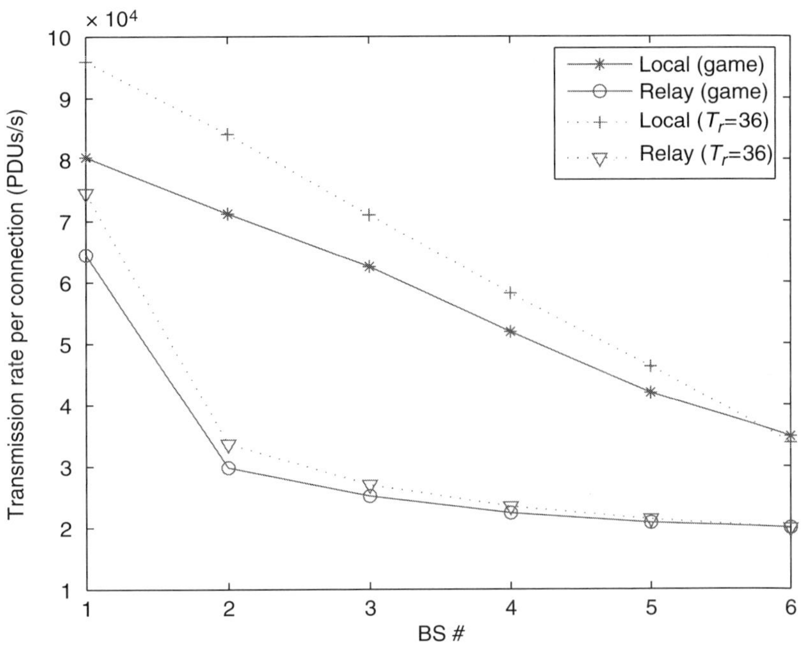

(*b*)

FIGURE 21.15 (*a*) Connection blocking probability and (*b*) average transmission rate per connection in chain topology.

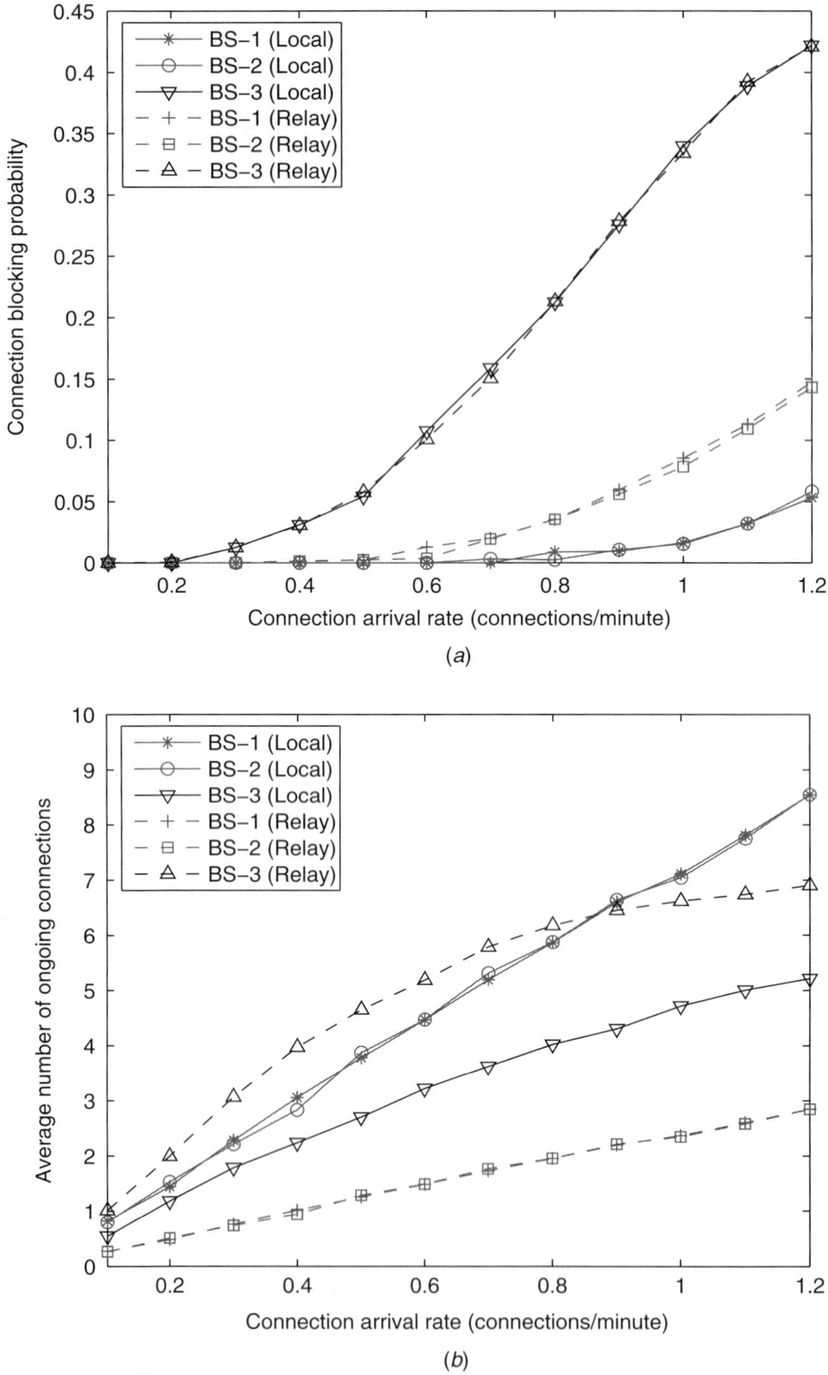

FIGURE 21.16 (*a*) Connection blocking probability and (*b*) average number of ongoing connections in tree topology.

while the connection blocking probabilities for relay connections are slightly different in each BS, those of local connections decrease significantly from BS-5 to BS-1. This result shows that higher resource utilization can be achieved by minimizing connection blocking probability through optimizing the resource reservation threshold.

Then, we vary traffic load (i.e., connection arrival rate) and observe the performance measures. For the tree topology (Fig. 21.1), variations in the connection blocking probability and the average number of ongoing connections are shown in Figs. 21.16*a* and 21.16*b*, respectively. As expected, both the connection blocking probability and the average number of ongoing connections increase as traffic load increases. In this case, BS-3 becomes a bottleneck as indicated by a higher blocking probability. Since the proposed reservation threshold optimization allocates a fair transmission rate to both local and relay connections, at BS-3 the number of relay connections is larger than that of local connections. Since BS-3 is the bottleneck, blocking probabilities for local and relay connections are identical while those at BS-1 and BS-2 are significantly different. For inconnection level performances, we observe that average end-to-end PDU delay and PDU dropping probability are maintained below 50 ms and 0.05, respectively. Therefore, when a bottleneck situation occurs, effective admission control and fair resource reservation schemes are required to guarantee QoS and achieve high system utilization in the mesh infrastructure.

21.7.2.6 Performance of the Route Selection Mechanism for a Tree Topology. We evaluate the performance of the proposed route selection assuming a tree topology for the mesh infrastructure topology shown in Fig. 21.17. In this topology, there are four mesh routers (i.e., BS-1, BS-2, BS-3, and BS-4) and two gateway routers (i.e., BS-5 and BS-6). While BS-1 does not serve any relay connection, BS-2, BS-3, and BS-4 serve both relay and local connections. At BS-2, Internet traffic from the mesh clients can split through

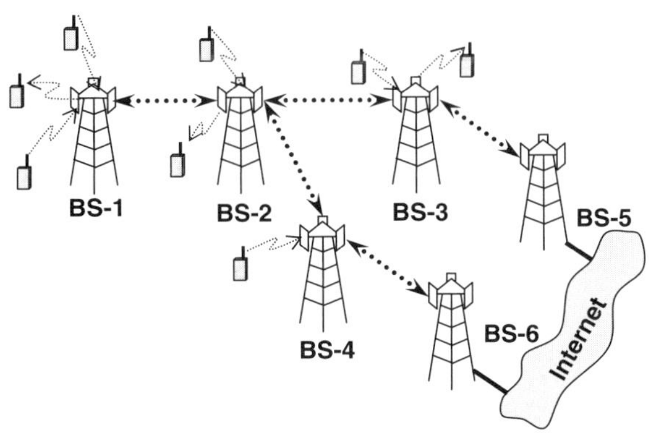

FIGURE 21.17 Tree topology for infrastructure mesh.

either of the two routes via BS-3 or BS-4. In this case, the route selection is used to determine the best route in terms of the QoS performances.

The PDU arrival rate for a connection is 15,000 PDUs per second and the average end-to-end delay requirement is 10 ms, which could be achieved when the queue is stable. We assume that half of the ongoing connections in each of the BSs are relay connections. Variations in the connection blocking probability and the average number of relay and local connections with traffic load are shown in Figs. 21.18, 21.19*a*, and 21.19*b*, respectively.

When traffic load increases, the number of ongoing local and relay connections increases. However, at a certain point a feasible subchannel allocation cannot be found such that the delay requirements for all the connections can be met. Therefore, the connection blocking probability starts increasing. Since BS-1 does not serve any relay connection, the connection blocking probability at BS-1 is significantly smaller than that in each of BS-2, BS-3, and BS-4. We observe that the connection blocking probability in BS-2, BS-3, and BS-4 are almost the same. This is due to the fact that the relay traffic load at BS-2 comes only from BS-1, and the total relay traffic load from BS-1 and BS-2 is divided equally to BS-3 and BS-4. Therefore, the total traffic load at these three BSs are more or less the same, which results in similar connection blocking probability performance.

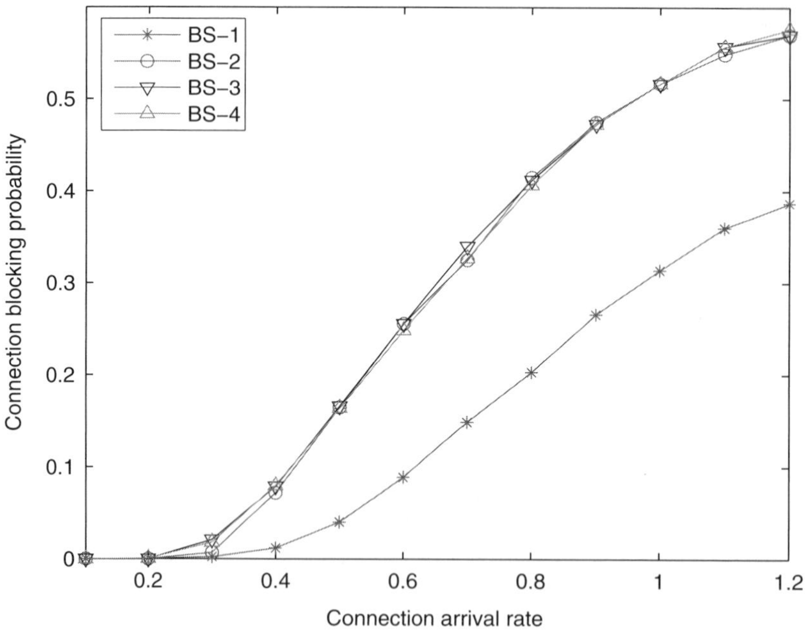

FIGURE 21.18 Variation in connection blocking probability with connection arrival rate.

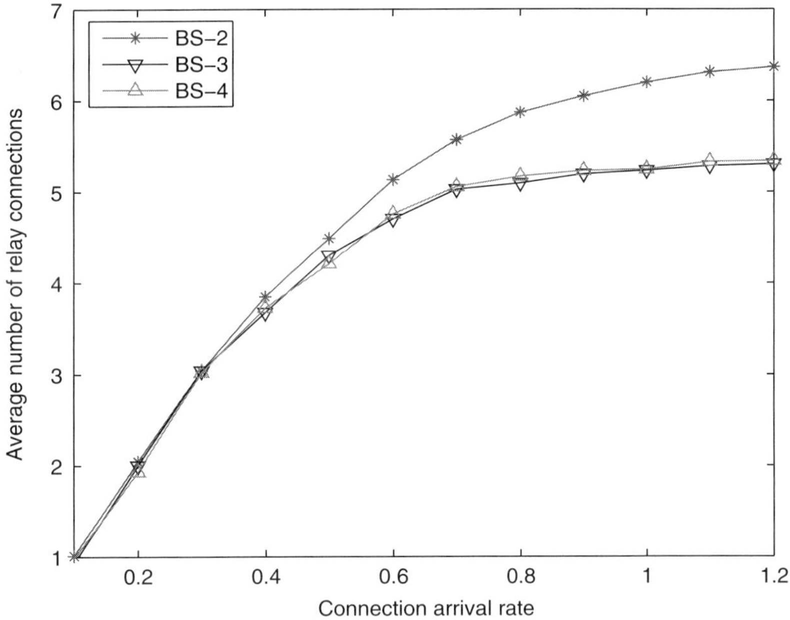

(a)

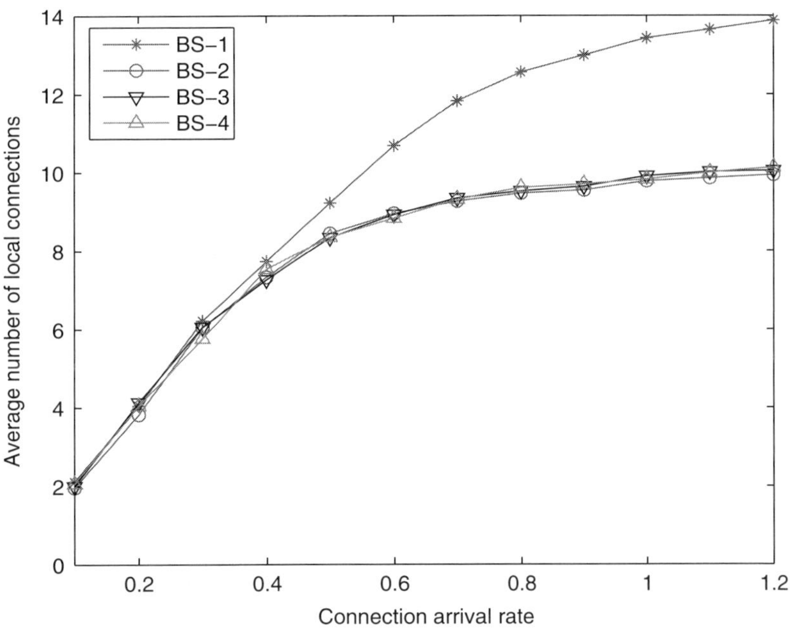

(b)

FIGURE 21.19 Variationin average number of (*a*) relay connections and (*b*) local connections with connection arrival rate.

21.8 CONCLUSION

We have presented a radio resource management framework for subchannel allocation and admission control in IEEE 802.16–compliant OFDMA wireless mesh networks. While the optimal subchannel allocation algorithm can guarantee that the throughput at a BS is maximized, the iterative algorithm achieves slightly inferior throughput performance with much smaller computational complexity. The problem of optimization of resource reservation for local and relay connections in these allocation algorithms has been formulated as a bargaining game. A tandem queueing model has been used to investigate the QoS performances (e.g., average queueing delay and PDU dropping probability) on an end-to-end basis and to decide whether the QoS requirements for a new connection can be guaranteed or not. Both the connection level and the in-connection level performance measures for the relay traffic have been studied extensively and the analytical results have been verified through simulations.

REFERENCES

1. IEEE 802.16a-2003, "Local and metropolitan area networks—Part 16," IEEE, New York, 2003.
2. R. Bruno, M. Conti, and E. Gregori, "Mesh networks: Commodity multihop ad hoc networks," *IEEE Commun. Mag.* 43(3), 123–131 (2005).
3. H. W. Kuhn, "The Hungarian method for the assignment problem," *Naval Res. Logist. Q.* 2, 83–97 (1955).
4. D. Niyato and E. Hossain, "Adaptive fair subcarrier/rate allocation in multirate OFDMA networks: Radio link level queuing performance analysis," *IEEE Trans. Vehic. Technol.* 55(6), 1897–1907 (2006).
5. I. F. Akyildiz and X. Wang, "A survey on wireless mesh networks," *IEEE Commun. Mag.* 43(9), S23–S30 (2005).
6. R. Pabst, B. H. Walke, D. C. Schultz, P. Herhold, H. Yanikomeroglu, S. Mukherjee, H. Viswanathan, M. Lott, W. Zirwas, M. Dohler, H. Aghvami, D. D. Falconer, and G. P. Fettweis, "Relay-based deployment concepts for wireless and mobile broadband radio," *IEEE Commun. Mag.* 42(9), 80–89 (2004).
7. J. Boyer, D. D. Falconer, and H. Yanikomeroglu, "Multihop diversity in wireless relaying channels," *IEEE Trans. Commun.* 52(10), 1820–1830 (2004).
8. A. Acharya, A. Misra, and S. Bansal, "Design and analysis of a cooperative medium access scheme for wireless mesh networks," paper presented at IEEE BroadNets'04, 25–29 October 2004, pp. 621–631.
9. K.-D. Lee and V. C. M. Leung, "Fair allocation of subcarrier and power in an OFDMA wireless mesh network," *IEEE J. Sel. Areas Commun.* 24(11), 2051–2060 (2006).
10. G. Li and H. Liu, "Resource allocation for OFDMA relay networks with fairness constraints," *IEEE J. Sel. Areas Commun.* 24(11), 2061–2069 (2006).

11. T. Bu, M. C. Chan, and R. Ramjee, "Designing wireless radio access networks for third generation cellular networks," paper presented at IEEE INFOCOM'05, Vol. 1, Mar. 2005, pp. 68–78.
12. L. Fu, Z. Cao, and P. Fan, "Spatial reuse in IEEE 802.16 based wireless mesh networks," paper presented at IEEE ISCIT'05, Vol. 2, Oct. 2005, pp. 1358–1361.
13. H. Y. Wei, S. Ganguly, R. Izmailov, and Z. J. Haas, "Interference-aware IEEE 802.16 WiMAX mesh networks," paper presented at IEEE VTC'05 (Spring).
14. C. Schwingenschlogl, V. Dastis, P. S. Mogre, M. Hollick, and R. Steinmetz, "Performance analysis of the real-time capabilities of coordinated centralized scheduling in 802.16 mesh mode," paper presented at IEEE VTC'06 (Spring), Vol. 3, 2006, pp. 1241–1245.
15. M. S. Kuran, B. Yilmaz, F. Alagoz, and T. Tugcu, "Quality of service in mesh mode IEEE 802.16 networks," paper presented at IEEE SoftCOM'06, Sept. 2006, pp. 107–111.
16. B. Han, F. P. Tso, L. Ling, and W. Jia, "Performance evaluation of scheduling in IEEE 802.16 based wireless mesh networks," paper presented at IEEE MASS'06, Oct. 2006, pp. 789–794.
17. H. Shetiya and V. Sharma, "Algorithms for routing and centralized scheduling in IEEE 802.16 mesh networks," paper presented at IEEE WCNC'06, Vol. 1, Apr. 2006, pp. 147–152.
18. Y. Lu and G. Zhang, "Optimum fair bandwidth allocation scheme for IEEE 802.16 mesh mode with directional antenna," paper presented at IEEE VTC'06 Fall, Sept. 2006, pp. 1–5.
19. J. Cai, X. Shen, and J. W. Mark, "Downlink resource management for packet transmission in OFDM wireless communication systems," *IEEE Trans. Wireless Commun.* 4(4), 1688–1703 (2005).
20. G. Song and Y. Li, "Cross-layer optimization for OFDM wireless networks—part I: Theoretical framework," *IEEE Trans. Wireless Commun.* 4(2), 614–624 (2005).
21. G. Song and Y. Li, "Cross-layer optimization for OFDM wireless networks—part II: Algorithm development," *IEEE Trans. Wireless Commun.* 4(2), 625–634 (2005).
22. G. Kulkarni, S. Adlakha, and M. Srivastava, "Subcarrier allocation and bit loading algorithms for OFDMA-based wireless networks," *IEEE Trans. Mobile Comput.* 4(6), 652–662 (2005).
23. Z. Han, Z. Ji, and K. J. R. Liu, "Fair multiuser channel allocation for OFDMA networks using Nash bargaining solutions and coalitions," *IEEE Trans. Commun.* 53(8) 1366–1376 (2005).
24. D. Niyato and E. Hossain, "Queue-aware uplink bandwidth allocation and rate control for polling service in IEEE 802.16 broadband wireless networks," *IEEE Trans. Mobile Comput.* 5(6), 668–679 (2006).
25. D. Niyato and E. Hossain, "A queuing-theoretic and optimization-based model for radio resource management in IEEE 802.16 broadband wireless networks," *IEEE Trans. Comput.* 55(11), 1473–1488 (2006).
26. C. Cicconetti, A. Erta, L. Lenzini, and E. Mingozzi, "Performance evaluation of the IEEE 802.16 MAC for QoS support," *IEEE Trans. Mobile Comput.* 6(1), 26–38 (2007).

27. D. Tarchi, R. Fantacci, and M. Bardazzi, "Quality of service management in IEEE 802.16 wireless metropolitan area networks," paper presented at IEEE ICC'06, Vol. 4, No. 1789–1794, June 2006.

28. D. Niyato and E. Hossain, "Delay-based admission control using fuzzy logic for OFDMA broadband wireless networks," paper presented at IEEE International conference on communications, 12, 5511–5516 (2006).

29. R. Iyengar, K. Kar, and B. Sikdar, "Scheduling algorithms for point-to-multipoint operation in IEEE 802.16 networks," paper presented at WiOpt'06, Apr. 2006, pp. 1–7.

30. J. Chen, W. Jiao, and H. Wang, "A service flow management strategy for IEEE 802.16 broadband wireless access systems in TDD mode," paper presented at IEEE ICC'05, Vol. 5, May 2005, pp. 3422–3426.

31. H. Wang, B. He, and D. P. Agrawal, "Admission control and bandwidth allocation above packet level for IEEE 802.16 wireless MAN," paper presented at IEEE ICPADS'06, Vol. 1, July 2006.

32. Y. Ge and G.-S. Kuo, "An efficient admission control scheme for adaptive multimedia services in IEEE 802.16e networks," paper presented at IEEE VTC'06 Fall, Sept. 2006, pp. 1–5.

33. B. Rong, Y. Qian, and H.-H. Chen, "Adaptive power allocation and call admission control in multiservice WiMAX access networks," *IEEE Wireless Commun.* 14(1), 14–19 (2007).

34. S. H. Ali, K.-D. Lee, and V. C. M. Leung, "Dynamic resource allocation in OFDMA wireless metropolitan area networks," *IEEE Wireless Commun.* 14(1), 6–13 (2007).

35. H. Y. Wei and R. D. Gitlin, "Incentive scheduling for cooperative relay in WWAN/WLAN two-hop-relay network," paper presented at IEEE WCNC'05, Vol. 3, Mar. 2005, pp. 1696–1701.

36. C. Li and E. W. Knightly, "Coordinated multihop scheduling: A framework for end-to-end services," *IEEE/ACM Trans. Networking* 10(6), 776–789 (2002).

37. C. Li and E. W. Knightly, "Schedulability criterion and performance analysis of coordinated schedulers," *IEEE/ACM Trans. Networking* 13(2), 276–287 (2005).

38. V. Srinivasan, P. Nuggehalli, C. F. Chiasserini, and R. R. Rao, "An analytical approach to the study of cooperation in wireless ad hoc networks," *IEEE Trans. Wireless Commun.* 4(2), 722–733 (2005).

39. Q. Liu, S. Zhou, and G. B. Giannakis, "Queuing with adaptive modulation and coding over wireless links: Cross-layer analysis and design," *IEEE Trans. Wireless Commun.* 4(3), 1142–1153 (2005).

40. L. A. Wolsey, *Integer Programming*, Wiley, New York, 1998.

41. C. Bastarrica, A. A. Shvartsman, and S. Demurjian, "A binary integer programming model for optimal object distribution," in *Proceedings of the International Conference on Principles of Distributed Computing*, Amiens, France, December 1998, pp. 91–105.

42. M. J. Osborne, *An Introduction to Game Theory*, Oxford University Press, New York, 2003.

43. M. Xiao, N. B. Shroff, and E. K. P. Chong, "Utility-based power control in cellular wireless systems," paper presented at IEEE INFOCOM'01, Vol. 1, 2001, pp. 412–421.

44. C. D. Meyer, Matrix Analysis and Applied Linear Algebra, Siam, Philadelphia, 2000.

45. I. Koffman and V. Roman, "Broadband wireless access solutions based on OFDM access in IEEE 802.16," *IEEE Commun. Mag.* 40(4), 96–103 (2002).

CHAPTER 22

UNIVERSAL AUTHENTICATION AND BILLING ARCHITECTURE FOR WIRELESS MANs

XIAODONG LIN, HAOJIN ZHU, MINGHUI SHI, RONGXING LU, PIN-HAN HO, and XUEMIN (SHERMAN) SHEN

22.1 INTRODUCTION

With the advance of wireless technology and the increasing worldwide wireless network deployment, *wireless metropolitan-area networks* (wireless MAN) have been on an upswing to enable ubiquitous Internet access. Developing and employing the wireless mesh networking technology for creating wireless MAN defined in IEEE 802.11s has been well recognized as a promising solution. Recently, higher speed IEEE 802.16a WiMax serves as the technology of choice to connect metropolitan-area *wireless mesh networks* (WMNs) to the Internet, which extends IEEE 802.11–based WMNs to cover a large geographical area, as shown in Fig. 22.1. As a promising alternative yet complement to the state-of-the-art last-mile technologies such as *digital subscriber lines* (DSL) and cable modem, WMNs have been proved with the ability of reducing network operational cost and extending service coverage by taking advantage of the infrastructure-free and low maintenance characteristics for broadband wireless access.

Different from the wired Internet, a metropolitan-area WMN is expected to be operated locally, which is mainly composed of a cloud of distributed *mesh access points* (MAPs) as the backhaul. A wireless mesh link exists between two MAPs if one is located within the coverage area of the other and vice versa. A number of physically adjacent MAPs are grouped to form an *extended service set* (ESS) corresponding to a *mesh gateway* (MGW), which interfaces the ESS with IEEE 802.16a *base stations* (BSs), and then connects to the public Internet. Within the WMN backhaul, numerous independent *wireless Internet service*

Emerging Wireless LANs, Wireless PANs, and Wireless MANs. Edited by Y. Xiao and Y. Pan

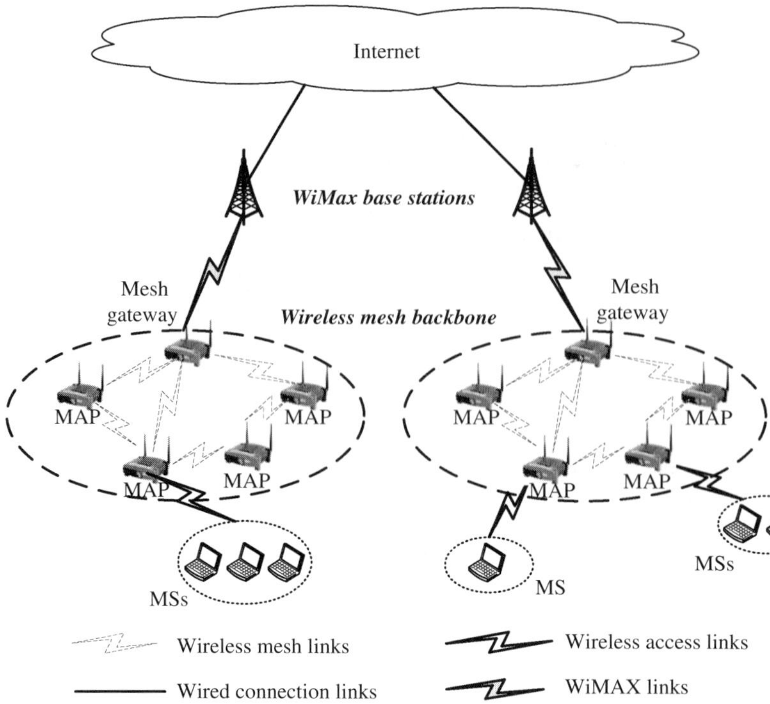

FIGURE 22.1 Wireless metropolitan area network architecture.

providers (WISPs) are supported, where each WISP can register a number of MAPs to run its business.

Due to the intrinsically autonomous and distributed characteristics in wireless networks, privacy and security issues are expected to be among the most important issues in pushing the success of wireless MANs for supporting service-oriented applications. In particular, the wide-open communication media and intrinsically multihop connections make the operation of wireless MANs subject to many security threats, which may hinder wireless MANs from being practically launched. Thus, it is desired to have a security architecture for wireless MANs where user authentication, privacy, and billing are essential and crucial for the ongoing success of wireless MANs. However, the issues become more critical in the wireless MAN application scenario due to small coverage of each MAP and the stringent demand for many emerging online multimedia services such as voice over Internet protocol (VoIP), video phone, and gaming, where user handover and roaming along with authentication requests will be issued much more frequently than that in the traditional GSM (Global System for Mobile Communications) or code division multiple access (CDMA) cellular networks.

The current widely accepted security solution is based on *authentication, authorization, and accounting* (AAA) architecture [1], where the authentication request is issued by the *mobile user* (MU) and is sent through the *serving MAP* (sMAP) and the MGW, until reaching the AAA server of the MU's home network. The AAA server of the home network then authenticates the MU based on the received authentication credentials and sends the authentication decision back to the sMAP [2]. If successful, the MU will be granted the access right to the network. Such a long signaling path, however, could take up to one or a few seconds of propagation and might cause fatal impairment on the emerging real-time services. Recently, many fast authentication schemes such as predictive authentication [3], localized authentication [4], prekey distributions [5], and enhanced Interaccess point protocol (IAPP) [6], have been reported to support seamless handover when an MU roams among adjacent MAPs under a common WISP domain (also referred to as intradomain handoff). On the other hand, the existing fast authentication techniques cannot be directly applied to interdomain handoff since it requires a bilateral *service-level agreement* (SLA) established between each pair of WISPs. Such a peer-to-peer approach may lead to a scalability problem in the presence of numerous wireless Internet service providers (WISPs) in the wireless MANs [7].

The best practice for establishing a trust relationship among different WISPs so far is by way of a centralized *roaming broker* (RB) trusted by all the WISPs [8]. Under this framework, when an MU roams into a foreign network domain, the foreign WISP simply forwards the corresponding AAA session of the MU to the home WISP of the MU for authorization via the RB. A more elaborated approach can be devised on the top of the centralized RB architecture by taking advantage of the *public key infrastructure* (PKI), where the RB serves as not only a *trusted third party* (TTP) but also a *certificate authority* (CA) that issues public key certificates to each WISP and MU. The trust relationship among WISPs, or between a WISP and MUs, can be easily established by validating the public key certificates issued by the RB [5, 8]. In both cases, the foreign WISP reports the accounting information of the roaming MU to its home WISP at the completion of the session, by which the home WISP will pay the bill and then charge the MU in terms of the MU's spending. The RB architecture can effectively solve the interdomain roaming and billing problem; unfortunately, the RB will also become the performance bottleneck for the interdomain handoff authentication and billing. In addition, the long signaling propagation latency of every transaction may not be tolerable to the real-time services in the interdomain roaming events.

Further, the issues on user privacy are subject to more concerns due to the fast booming wireless communication markets. In addition to keeping communication content private, users are also concerned with the commercial misuse of their personal data such as personal traveling preference and whereabouts. Concern about the more heterogeneous and complicated environments of wireless MANs, users would usually prefer to travel incognito [9].

In addition, there has been consideration of privacy regulations that enforce WISPs to adopt appropriate administrative, technical, and physical security measures to protect user privacy [10]. However, none of the aforementioned solutions provide strong privacy preserving for mobile users while roaming across different WISPs.

In order to cope with the above-mentioned challenges, a universal, efficient, and simplex authentication and billing architecture for wireless MAN is highly demanded and is essential to its prevalence. This architecture should be capable of efficiently authenticating users roaming from numerous WISP domains, and charging users in a single bill and providing strong privacy protection functionality so that mobile users can have complete control of their location privacy. Most importantly, the aforementioned features should be achieved without a sacrifice of system performance. To design such a scalable, secure, and efficient authentication and billing architecture, the following three observations are made [11–14].

- Partially blind signature-based e-cash, denoted U-coin, can be used by the MUs as authentication credentials to gain access to the wireless Internet service provided by the WISPs, where U-coin is issued by a TTP, such as bank. There are several advantages to using U-coin as authentication credentials over the conventional user authentication mechanisms. First, it provides better user flexibility and convenience. With U-coin, there is no need for an MU to subscribe to any WISP. Second, it not only provides user privacy but also solves the issue of billing based on the network usage of the MUs. Third, the MUs can roam without preagreements among WISPs. The WISPs in the proposed architecture do not require roaming service agreements with each other. Instead, each WISP establishes the trust relationship with the TTP. Finally, U-coin doesn't have the issue of an unlimitedly growing central database, which is used to prevent double spending of e-cash. In an e-cash system based on a partially blind signature scheme, the bank (or signer, respectively) assures that each issued e-cash (or signature, respectively) contains some desired information, such as the face value and expiration date. By embedding the information of expiration date into each e-cash issued by the bank, all the corresponding records of the expired e-cashes in the bank database can thus be removed. In other words, the database of the bank only needs to keep the unexpired e-cashes held by a user to prevent double spending.
- Since each WISP not only serves as a vendor providing services to the MUs but also as a buyer, which purchases services from other WISPs for the MUs, a multi-WISP wireless MAN can be taken as both a business-to-business (B2B) system (WISP–WISP) and a business-to-consumer (B2C) system (WISP–MU). Therefore, from the WISPs' point of view, an interdomain handoff can be taken as an inter-WISP payment; while from the MUs' point of view, an MU can roam into another WISP

domain if and only if it has enough remaining credits. Thus, a WISP can issue a digital signature based on PKI, which serves as the digital currency for electronically performing interdomain payment with another WISP without the intervention of the RB. In addition, this digital signature can also be taken as an authentication credential of the corresponding MU, which is authenticated every time when the MU requests for interdomain roaming. Such a digital signature is referred to as "D-coin."

- By preloading each MAP with some cryptographic information, some required security capability can be achieved such that the roaming/handoff authentication and billing can be performed in a localized manner with much better scalability. Such a localized authentication and billing scheme is expected to effectively solve the scalability problem due to the centralized RB and dramatically reduce the interdomain roaming latency by avoiding any intervention of the RB.

The advantages gained in localizing the authentication and billing, however, are at the expense of reduced security level of the system due to the compromise-prone MAPs, which are most likely low cost devices without expensive and wholesome protection [15]. In a compromise event, the cryptographic secrets, such as the public/secret key pairs, could be deprived by an attacker, which may launch some serious attacks by manipulating the secret. For example, the attacker can manipulate a compromised MAP to arbitrarily issue D-coins to an illegal MU or accept a D-coin without granting services to the MU (or referred to as the *coin fraud attack*), or overcharge an MU by holding the connection even when the MU has disconnected from the MAP (or referred to as the *overcharge attack*).

This chapter proposes a universal authentication and billing architecture, called UAB, based on partially blind signature, *local voting strategy*, and the threshold digital signature scheme. MUs purchase U-coin from the TTP. A WISP has a mutual agreement with the TTP to allow the MUs with U-coin to gain wireless network access. Furthermore, with the local voting strategy, the D-coin is issued under the endorsement of the *serving MAP* (sMAP) and its *neighbor MAPs* (nMAPs) instead of by any single MAP. To perform billing in a single stage, a *local user accounting profile* (LUAP) is maintained at both sMAP and nMAPs to locally record an MU's spending information. With UAB, an interdomain handoff authentication and billing can be performed in a peer-to-peer manner from sMAP to the *target MAP* (tMAP) of different WISPs without the intervention of the RB. The RB, on the other hand, only needs to be involved during the clearance phase, which can be performed offline, in which every WISP submits its collected D-coins issued by the other WISPs to the RB for payment. To further reduce the workload of the RB in the clearance phase, we take advantage of the short and aggregate digital signature technique [16] to effectively reduce the computational and storage cost on the RB due to the D-coin verification and storing.

The remainder of this chapter is organized as follows. Section 22.2 gives the background knowledge. In Section 22.3, the details of our universal authentication and billing architecture are described. Sections 22.4 and 22.5 present discussions on the security and efficiency of the proposed architecture, respectively. Section 22.7 contains some concluding remarks.

22.2 PRELIMINARIES AND BACKGROUND

22.2.1 Partially Blind Signature

The notion of blind signature was first introduced by Chaum [17]. As a variant of digital signature, blind signature allows a user to get a signature without giving the signer any information about the actual message. Just due to the unlinkability between the user and the signature, blind signature is very useful in privacy-oriented e-services such as e-cash and e-voting system [18]. However, when the blind signature technique is applied in the practical e-cash system, the blindness property also incurs two obvious issues. For example, to prevent a customer's double-spending his e-cash, the bank has to keep a database that stores all spent e-cash to check whether a specific e-cash has been spent or not. Clearly, such a database kept by the bank may unlimitedly grow. The other issue is that the bank isn't able to inscribe the value on the blindly issued e-cash. To believe the face value of e-cash, there are two conventional solutions:

1. The bank uses different public keys to link with such common information. In this case, the shops and the customers must always carry a list of those public keys in their electronic wallet, which is typically a smart card whose memory is very limited.
2. The bank uses the cut-and-choose algorithm in the withdrawal phase. But this solution is not efficient.

To address such issues in practice, the notion of partially blind signature was introduced by Abe and Fujisaki [19]. A partially blind signature allows the signer to explicitly include some agreed-upon information in the blind signature. Thus, using the partially blind signature in the e-cash system, we can prevent the bank's spent database from growing unlimitedly. Because the bank assures that each e-cash issued by itself contains the information it desires, such as the date information. By embedding an expiration date into each e-cash issued by the bank, all expired e-cash recorded in the bank's database can be removed. At the same time, each e-cash can be embedded with the face value, and the bank can know the value on the blindly issued e-cash. In addition, many partially blind signature schemes have appeared recently [19, 20]. In this study, the first provably secure partially blind signature due to Abe and Okamoto [20] is adopted to realize the proposed universal authentication and billing architecture.

22.2.2 Short Digital Signature and Aggregate Signature

The proposed universal authentication and billing architecture is based on a short and aggregate digital signature technique. The short digital signature has been considered as an effective tool in the classic cryptographic research area for reducing digital signature overhead. A number of short signature schemes have been reported in the literature. Boneh et al. used Weil pairing to build the shortest digital signature [16]. Compared with the RSA signature sized at 1024 bits and ECDSA signature sized at 320 bits, a short digital signature is only 160 bits. In the wireless communication scenario, adopting a signature with an extremely small size can save the precious wireless communication resources and device transmission power. Furthermore, in case a lot of D-coins are submitted to the RB for verification and clearance, it is desirable to aggregate the D-coins into a single short signature by applying the aggregate signature technique [16] in order to save both transmission cost and computation cost. The short and aggregate signature can be achieved by bilinear pairing, which is briefly introduced below. Let $\mathbb{G}_1$, $\mathbb{G}_1'$ be two cyclic additive groups and G_2 be a cyclic multiplicative group of the same prime order q, that is, $|\mathbb{G}_1| = |\mathbb{G}_1'| = |\mathbb{G}_2| = q$. Let P be a generator of $\mathbb{G}_1$, P' be a generator of $\mathbb{G}_1'$, and ψ be an isomorphism from $\mathbb{G}_1'$ to $\mathbb{G}_1$, with $\psi(P') = P$. An efficient admissible bilinear map $\hat{e} : \mathbb{G}_1 \times \mathbb{G}_1' \rightarrow \mathbb{G}_2$ with the following properties:

- Bilinear: For all $P_1 \in \mathbb{G}_1$, $Q_1 \in \mathbb{G}_1'$, and $a, b \in \mathbb{Z}_q^*$, $\hat{e}(aP_1, bQ_1) = \hat{e}(P_1, Q_1)^{ab}$.
- Nondegenerate: There exist $P_1 \in \mathbb{G}_1$ and $Q_1 \in \mathbb{G}_1'$ such that $\hat{e}(P_1, Q_1) \neq 1$.
- Computable: There is an efficient algorithm to compute $\hat{e}(P_1, Q_1)$ for any $P_1 \in \mathbb{G}_1$, $Q_1 \in \mathbb{G}_1'$.

Such an admissible bilinear map $\hat{e}$ can be constructed by Weil or Tate pairings on the elliptic curves. By software and hardware acceleration, pairing operations can be efficiently accomplished within 1.3 ms [21].

22.3 DESCRIPTION OF THE UNIVERSAL AUTHENTICATION AND BILLING ARCHITECTURE

22.3.1 System Architecture

For the considered system architecture, there are five types of entities: the MUs, the TTP, the WISPs, MGW, and the MAPs, and their relationship is shown in Fig. 22.2. A WISP may operate multiple MAPs, which may or may not be adjacent to each other. The MUs can request for wireless Internet access by subscribing a U-coin, which is purchased from the TTP. In addition to the role of issuing U-coin, the TTP also serves as an RB. Without loss of generality, we use the terms RB and TTP interchangeably in the following context.

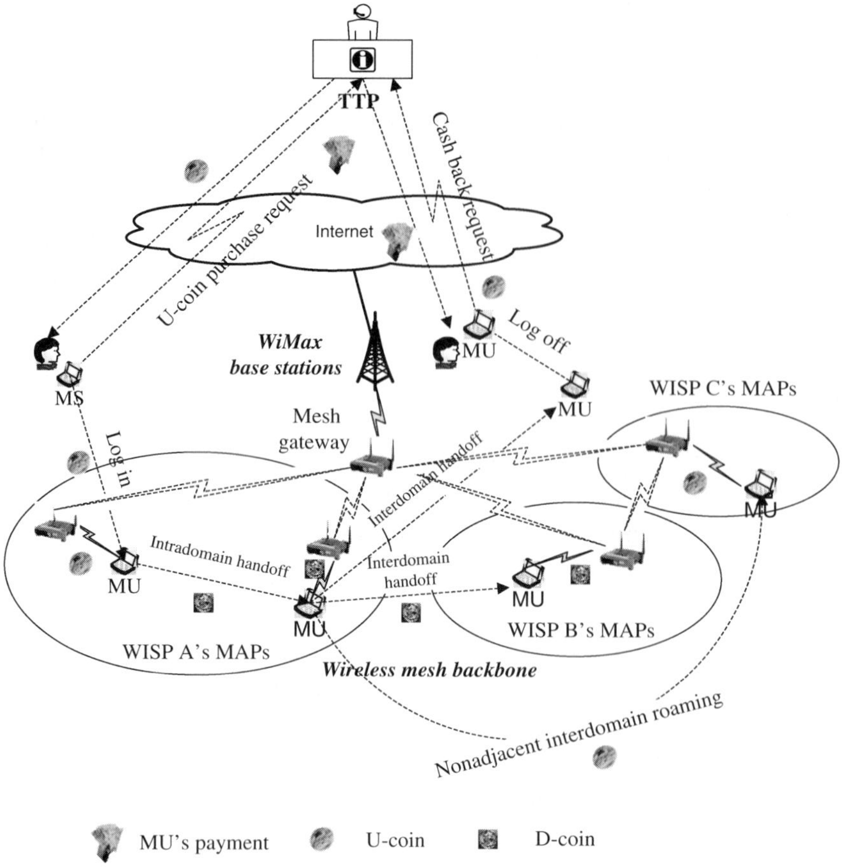

FIGURE 22.2 System architecture.

The proposed UAB takes advantage of the traditional PKI architecture to build the trust relationship among different WISPs and between WISPs and MUs by way of a RB. Similar to [11], the RB can serve as a *certificate authority* (CA) and issue every legitimate WISP with its corresponding certificate such that each WISP can check the validity of another. We assume that a legitimate WISP does not intentionally misbehave, which is reasonable since the attacks on its MUs will decrease the satisfactory MUs on the WISP, and will lead to reduction of its long-term revenue. On the other hand, the attacks launched by a WISP can be easily detected by the RB, and the malicious WISP will be deprived of its WISP qualification with subsequent penalties. Further, the RB also issues certificates to MAPs. The certificate issued to an MAP is a digital signature signed by the RB on its public key as well as the linkage between the public key and the MAPs identity. In this case, a mutual public key authenticated key agreement can be derived based on MU's U-coin and

MAP's public key certificate. Finally, the limited number of revoked WISPs make real-time updating and distributing *certificate revocation list* (CRL) of WISPs feasible. The trust relationship also exists between MUs and WISPs, where an MU can check the validity of a WISP by verifying the WISP's certificate issued by the RB.

We assume that a hierarchical public key system is established in every WISP domain, which includes a domain public/private key preloaded at the MGW along with a number of MAP-level public/private keys corresponding to every MAP. Since the MGW is difficult to be compromised, it can thus be fully trusted to serve as the security administrator in any metropolitan-area WMN domain. However, since the proposed UAB is a localized and distributed security scheme, MGW gets involved only during the key distribution procedure in the system setup phase or when some attacks or disputes take place. For simplicity, the communication among MAPs within a common WISP can be transmitted in a secure channel since it is easy for different MAPs to make an authenticated key agreement with their corresponding public/private key. In this chapter, we focus on the authentication and billing-related attacks.

The proposed UAB consists of the following seven phases: (1) system setup phase, (2) U-coin purchasing phase, (3) login and mutual authentication phase, (4) *local user accounting profile* (LUAP) generation and maintenance phase, (5) localized LUAP transfer during intradomain handoff phase, (6) D-coin issuing and interdomain handoff authentication phase, and (7) clearance phase. Only phases 3, 5, and 6 are performed online during MU handoffs, while the others are conducted for the maintenance or preparation of the future handoff events.

22.3.2 System Setup Phase

22.3.2.1 TTP Initialization. Let p, q be two large prime numbers such that $|p| = 1024$, $|q| = 160$, and $q|p-1$. Let g be a generator of order q in $\mathbb{Z}_p$, and $\langle g \rangle$ denote a subgroup in $\mathbb{Z}_p^*$ generated by g. Let F_1, F_2 be two public hash functions, where $F_1 : \{0,1\}^* \rightarrow \mathbb{Z}_q^*$ and $F_2 : \{0,1\}^* \rightarrow \langle g \rangle$.

The signer $\mathscr{S}$ chooses a random number $x \in \mathbb{Z}_q^*$ as his private key and computes the corresponding public key $y = g^x$.

22.3.2.2 WISP Initialization. Let WISP A represents the currently serving WISP of the MU. It can generate the system parameters $(q^A, \mathbb{G}_1^A, \mathbb{G}_1'^A, \mathbb{G}_2^A, \hat{e}^A, P^A)$ and then choose a random number $s^A \in Z_{q^A}^*$ as its private key which corresponds to the public key expressed as $Y^A = s^A P^A$. In addition, two hash functions are formed: $H : \{0,1\}^* \rightarrow \{0,1\}^l$ and $H_2 : \{0,1\}^* \rightarrow G_1^A$, where l is a predefined security parameter. The public key and system parameters $(q^A, \mathbb{G}_1^A, \mathbb{G}_1'^A, \mathbb{G}_2^A, \hat{e}^A, P^A, Y^A, H, H_2)$ along with a public key certificate issued by the trusted RB will be periodically broadcasted to each MU and MAP within the WISP A domain. In addition, MGW will serve as the security administrator to generate the D-coin signing key for every MAP inside the WISP A domain. It can be described as follows. Let $\overline{P^A}$ be a generator of $G_1{}^A$ such that $\overline{P^A} = \alpha P^A$ for some $\alpha \in Z^*_{q^A}$, while it is infeasible to derive α given P^A and $\overline{P^A}$. The MGW randomly picks up two polynomials $f(x) = s + a_1x + \cdots + a_{k-1}x^{k-1}$ and $f'(x) = b_0 + b_1x + \cdots + b_{k-1}x^{k-1}$ of degree $k-1$ such that $f(0) = a_0 = s$ and $f'(0) = b_0$. Then, MGW computes and broadcasts $C_i = a_iP^A + b_i\overline{P^A}$ for $i = 0, 1, \ldots, k-1$ to all the MAPs. Furthermore, MGW computes $f(j)$ and $f'(j)$ secretly and sends them to MAP_j, where $j = 1, \ldots n$. Any MAP_j can verify the received share by checking whether $f(j)P^A + f'(j)\overline{P^A} = \sum_{i=0}^{k-1} j^i \times C_i$ holds. If the verification holds, $s_j = f(j)$ will be stored by MAP_j as its secret share.

22.3.3 U-coin Purchasing Phase

An MU has to purchase U-coin from the TTP in order to gain wireless Internet access at any MAP run by a WISP. The U-coin purchasing process is detailed as follows [13], as shown in Fig. 22.3:

Step 1. The MU sends a U-coin purchase request to the TTP. This request includes (ID_{MU}, d), where ID_{MU} is a random identity, which is an MU-controlled unique representation, called *virtual identity* (VID), and d is the denomination of the request. The VID is only used for the purpose of billing. In this case, user privacy can be preserved by not disclosing the real identity of the MU.

Step 2. The TTP, upon receiving the purchasing ordering message, returns a "payment request" message to the MU. The payment request message contains the following information:

payment request = transaction ID, purchase information, validity period, CERT_{TTP}, *purchase agreement, Sig_{TTP} (transaction ID, amount, purchase, validity period, CERT_{TTP}, *purchase agreement),

where transaction ID is generated by the TTP and used by the TTP to keep track of all the transactions. The purchase information is the same as that provided by the MU in step 1. The validity period specifies the time during which the payment must be confirmed. The TTP's certificate, CERT_{TTP}, can be used by the mobile user to verify the TTP's signature. The purchase agreement is an optional field, which contains information such as refund policy, product quality, warranty, and so on. A digital signature is included as part of the payment request message.

Step 3. When the payment request is received, the MU verifies the TTP's signature. After verifying the TTP's signature, the MU proceeds to sending a "payment" message to the TTP by using various payment methods such as a credit card. The following is an example based on credit card payment.

	TTP		MU
Step 1		$\xleftarrow{\text{ID}_{\text{MU}}, d}$	
Step 2		Payment Request $\longrightarrow$	
Step 3		$\xleftarrow{\text{Payment}}$	
Step 4	$u, s, d \xleftarrow{R} \mathbb{Z}_q^*$		
	$z = F_2(\text{ExpDate})$		
	$a = g^u,\ b = g^s \cdot z^d$	$\xrightarrow{a, b, \text{ExpDate}}$	
Step 5			$t_1, t_2, t_3, t_4 \xleftarrow{R} \mathbb{Z}_q^*$
			$z = F_2(\text{ExpDate})$
			$\alpha = ag^{t_1}y^{t_2},\ \beta = bg^{t_3}z^{t_4}$
			$\varepsilon = F_1(\alpha \Vert \beta \Vert z \Vert d)$
			$e = \varepsilon - t_2 - t_4 \bmod q$
		$\xleftarrow{e}$	
Step 6	$c = e - d \bmod q$		
	$r = u - cx \bmod q$	$\xrightarrow{r, s, c}$	
Step 4			$\rho = r + t_1 \bmod q$
			$\omega = c + t_2 \bmod q$
			$\sigma = s + t_3 \bmod q$
			$\delta = e - c + t_4 \bmod q$
			$\omega + \delta \stackrel{?}{=} F_1(g^\rho y^\omega \Vert g^\sigma z^\delta \Vert z \Vert d)$
		U-coin =	$(\text{ExpDate}, \rho, \omega, \sigma, \delta, d, TTP)$

FIGURE 22.3 U-coin purchasing protocol.

The "payment" message contains the following information: payment = E_{SSL} (payment information, amount, TTP, transaction ID, *timestamp, *$CERT_{MU}$), *Sig_{MU} (payment information,amount, TTP, transaction ID, *timestamp).

The payment information field contains information such as the number, the holder's name, and the expiration date of the credit card. The transaction ID is the same as that provided by the TTP. In addition, a digital signature is included as part of the payment message if the MU has its own certificate. The MU's certificate can be used by the TTP to verify the MU's signature on the payment message. Again, an optional timestamp could be included to defend a replay attack. This message has to be encrypted by using either the TTP's public key or any other existing security protocol such as secure sockets layer (SSL) to ensure confidentiality and integrity of the payment message.

Step 4. After receiving the payment information and checking MU's corresponding payment, the TTP first chooses three random numbers $u, s, d \xrightarrow{R} \mathbb{Z}_q^*$. Then the TTP computes $z = F_2(\text{ExpDate})$, $a = g^u$ and $b = g^s \cdot z^d$, where ExpDate refers to the date when the issued U-coin is invalid. The TTP sends a, b, ExpDate to the MU in the end.

Step 5. Upon receiving a, b, ExpDate, the MU first chooses four random numbers $t_1, t_2, t_3, t_4 \in \mathbb{Z}_q^*$, computes $z = F_2(\text{ExpDate})$, $\alpha = ag^{t_1}y^{t_2}$, $\beta = bg^{t_3}z^{t_4}$, $\epsilon = F_1(\alpha||\beta||z||d)$, and $e = \varepsilon - t_2 - t_4 \bmod q$. Finally, the MU sends e back to the TTP.

Step 6. When the TTP receives e, he computes $c = e - d \bmod q$ and $r = u - cx \bmod q$, and sends (r, s, c) to the MU.

Step 7. The MU computes $\rho = r + t_1 \bmod q$, $\omega = c + t_2 \bmod q$, $\sigma = s + t_3 \bmod q$, and $\delta = e - c + t_4 \bmod q$. In the end, the MU issues $(\rho, \omega, \sigma, \delta)$ as the signature of d.

The validation of (ExpDate, $\rho, \omega, \sigma, \delta, d$, TTP) can be examined by observing the following congruence:

$$\omega + \delta = F_1(g^\rho y^\omega||g^\sigma z^\delta||z||d), \quad \text{where } z = F_2(\text{ExpDate})$$

If it holds, the signature can be accepted, otherwise it is rejected. Since

$$\omega + \delta = c + t_2 + e - c + t_4 = e + t_2 + t_4 = \epsilon = F_1(\alpha||\beta||z||d)$$

and

$$g^\rho y^\omega = g^{r+t_1} y^{c+t_2} = g^{r+xc} g^{t_1} y^{t_2} = g^u g^{t_1} y^{t_2} = ag^{t_1} y^{t_2} = \alpha$$

$$g^\sigma y^\delta = g^{s+t_3} z^{e-c+t_4} = g^s z^{e-c} g^{t_3} z^{t_4} = g^s z^d g^{t_3} z^{t_4} = bg^{t_3} z^{t_4} = \beta$$

For detailed security analysis of the proposed U-coin issuing mechanism, please refer to [20].

22.3.4 Login and Mutual Authentication Phase

In this phase, an MU authenticates itself to a MAP and the MAP authenticates itself to the MU in such a way that both parties are assured of the others' legitimacy whenever the MU wishes to gain wireless Internet access. However, the proposed authentication protocol is different from previously reported work in the sense that the authentication functionalities at the MU and MAP sides are taken asymmetric. At the MU side, the MU needs to ensure the MAP is who it claims it is. On the other hand, a MAP just needs to make sure that the MU possesses enough U-coin or D-coin to gain wireless access.

First, the MAP broadcasts its public parameters containing its public key certificate, periodically to all the associated MUs. In this case, the MU can easily ensure the security of the MAP's public key after validating the TTP's signature on it. If the validation fails, the MU aborts the login process since the MU could be subject to impersonation attack by the MAP.

Second, the MU randomly chooses a nonce n and encrypts n with the MAP's public key. Then, the MU sends the result to the MAP. Upon receiving encrypted n, the MAP decrypts it and sends the result back to the MU.

Finally, the MU checks if its received number matches its chosen nonce. If not, the MU aborts the login process since the MU could be subject to impersonation attack by the MAP as well. Otherwise, the MU continues by subscribing his purchased U-coin to the MAP. Upon the receipt of the U-coin, the MAP ensures the validity of the U-coin by doing the following steps:

1. Verify the signature with the TTP's public key.
2. Check that this U-coin doesn't expire.
3. Search the TTP's database to make sure that this U-coin has not been spent before.

If these three conditions are satisfied, the MAP accepts the MU as a legitimate user having access to the network.

22.3.5 Local User Accounting Profile Generation and Maintenance Phase

With UAB, the sMAP of an MU has to collaboratively generate and maintain the *local user accounting profile* (LUAP) of the MU with some of the nMAPs in order to timely reflect the spending information of the MU. However, updating the LUAP relies on a secure accounting protocol; otherwise, a compromised MAP may arbitrarily change the accounting information of an MU. Therefore, to achieve incontestable payment and authenticity, the idea of a micropayment [22] is exercised to maintain a secure communication session and the LUAP, by which the MUs are forced to periodically submit a nonrepudiation proof of the previous spending information to maintain the session consistency.

Every MU has a predefined maximum consuming credit $\mathscr{B}$. If the MU is gaining access with its home WISP, this balance is the remaining credit in its account. If the MU is gaining access with a foreign WISP using D-coin, $\mathscr{B}$ is taken as the face value defined in the D-coin. Based on $\mathscr{B}$, an MU selects a random integer M and computes a one-way hash chain $H^m(M) = H(H(\cdots(H(M)\cdots)))$ by applying the one-way function $H()$ to M for m times, where every hash token $H^i(M), i \in [1, \ldots, m]$ stands for a monetary value τ such that $\mathscr{B} = m \times \tau$. At the beginning, the MU sends $H^m(M)$ to the MGW in a full IEEE 802.1x authentication scheme or by embedding $H^m(M)$ in a D-coin, which will be further discussed in Section 22.3.7. Thus, MGW can distribute $H^m(M)$ to the nMAPs and sMAP with the neighbor graphs technique introduced in [5]. The sMAP and nMAPs will initialize a local user accounting profile for this new user and store $H^m(M)$ as the commitment. The initial LUAP can therefore be defined as $(\mathrm{ID_{MU}}, \mathscr{B}, H^m(M), \mathrm{PMK})$, where PMK is the pairwise master key between sMAP and the MU, and can be used to establish a secure channel between MU and sMAP. It is important to point out that, since UAB requires prestoring PMK at nMAPs, any existing key predistribution technique [5] can be employed to realize the fast intradomain handoff.

When the spending of the MU equals to τ money units, it triggers the submission of the first spending proof $\mathrm{SP}_1 = H^{m-1}(M)$, which is sent by the MU to its sMAP. The sMAP can check the validity of this proof by simply verifying if $H(\mathrm{SP}_1) = H^m(M)$ holds. If valid, the SP_1 will be forwarded to the nMAPs. If the verification passes, the ith nMAP will send back the receiving acknowledgment $\mathrm{AK}^1_{\mathrm{nMAP}_i} = H(\mathrm{SP}_1 || K_{\mathrm{nMAP}_i,\mathrm{MGW}})$ to the sMAP, where $K_{\mathrm{nMAP}_i,\mathrm{MGW}}$ refers to the shared key between nMAP_i and MGW. After receiving the first $k-1$ acknowledgments $\mathrm{AK}^i_{\mathrm{nMAP}_i} \in \{1, \ldots, k\}$ from its $k-1$ one-hop neighbors, the sMAP can also compute its acknowledgment $\mathrm{AK}^1_{\mathrm{sMAP}} = H(\mathrm{SP}_1 || K_{\mathrm{sMAP,MGW}})$ and the aggregate acknowledgment $\mathrm{AK}^1 = H(\mathrm{AK}^1_{\mathrm{sMAP}} || \mathrm{AK}^1_{\mathrm{nMAP}_1} || \cdots || \mathrm{AK}^1_{\mathrm{nMAP}_{k-1}})$. Then, the first spending proof as well as the LUAP updating acknowledgment $(\mathrm{SP}_1, \mathrm{AK}^1)$ are submitted to MGW. Meanwhile, sMAP and nMAPs update their stored LUAP to $(\mathrm{ID_{MU}}, \mathscr{B} - \tau, \mathrm{SP}_1, \mathrm{PMK})$. After receiving $(\mathrm{SP}_1, \mathrm{AK}^1)$, MGW can check the validity of the received message by checking if the following two conditions hold:

$$
\begin{aligned}
H(\mathrm{SP}_1) &= H^m(M) \\
\mathrm{AK}^1 &= H(\mathrm{AK}^1_{\mathrm{sMAP}} || \mathrm{AK}^1_{\mathrm{nMAP}_1} || \cdots || \mathrm{AK}^1_{\mathrm{nMAP}_{k-1}})
\end{aligned}
\tag{22.1}
$$

If these two conditions hold, the MGW will accept SP_1 as the first spending proof of the MU, and the LUAPs stored at the sMAP and nMAPs can be updated to $(\mathrm{ID_{MU}}, \mathscr{B} - \tau, \mathrm{SP}_i, \mathrm{PMK})$. In this way, the MU could reveal $\mathrm{SP}_2 = H^{m-2}(M), \ldots, \mathrm{SP}_i = H^{m-i}(M), \ldots, \mathrm{SP}_m = H^0(M) = M$ one after another to prove the spending for m times. In case the hash chain runs out or the MU cannot submit a valid chain token on time, the MGW can detect it and terminate the ongoing communication session immediately.

22.3.6 Localized LUAP Transfer During Intradomain Handoff

The user mobility may result in a switch of the sMAP and the corresponding set of nMAPs. Thus, the LUAP transfer is performed to ensure that every new nMAP of this MU can obtain a copy of the MU's authentic LUAP. To reduce the multihop signaling, we propose a localized LUAP transfer algorithm based on localized voting strategy, which is defined as that the LUAP can be accepted as a valid one if and only if there are more than k valid LUAP copies from the nMAPs being consistent. Let Neighbor(MAP_i) denote the set of nMAPs of MAP_i, Local(MAP_i) = Neighbor(Map_i) $\bigcup$ MAP_i denotes the nMAPs and itself, LUAP(MU) denotes the LUAP of the MU, and Cache(MAP_i) denotes the caches maintained at MAP_i. Let Obtain_LUAP($\text{MAP}_{\text{Source}}$, MU, $\text{MAP}_{\text{Destination}}$) be the function invoked by $\text{MAP}_{\text{Destination}}$ for obtaining an LUAP copy of an MU. Let Check_LUAP(MAP_i, MU) be the function invoked by MAP_i for checking the LUAPs in hand and decide if they are consistent. This function will return the maximum number of consistent LUAPs. Let Insert_Cache(MAP_i, LUAP(MU)) and Remove_Cache(MAP_i, LUAP(MU)) be the functions that insert and remove the LUAP of the MU to and from the cache of MAP_i, respectively. The LUAP transfer algorithm is presented as follows:

Algorithm 22.1. Localized LUAP Transfer

```
Input (sMAP, MU, tMAP)
Output valid or invalid
for MAP_i ∈ Neighbor(tMAP) ∧ LUAP(MU) ∉ cache(MAP_i) do
   for MAP_j ∉ Local(sMAP) do
      Obtain_LUAP(MAP_j, MU, MAP_i)
   end for
   if Check_LUAP(MAP_i, MU) ≥ k then
   Insert_Cache(MAP_i, LUAP(MU))
   else
      return invalid
   end if
end for
for MAP_i ∈ Neighbor(sMAP) ∧ ∉ Local(tMAP) do
   Remove_Cache(MAP_i, LUAP(MU))
end for
return valid
```

Suppose that an MU is currently associated to MAP $\{G\}$ and will handoff to MAP $\{H\}$, as shown in Fig. 22.4 this process, the sMAP and nMAP set of the MU will switch from Local(G) = $\{D, E, G, H\}$ to Local(H) = $\{D, E, F, G, H, I\}$. Therefore, two new nMAPs $\{F, I\}$ need to obtain the LUAP of MU from Local(G). $\{F, I\}$ can contact any MAPs in Local(G) to obtain their stored LUAP(MU). If more than k LUAPs among the obtained LUAP(MU) are

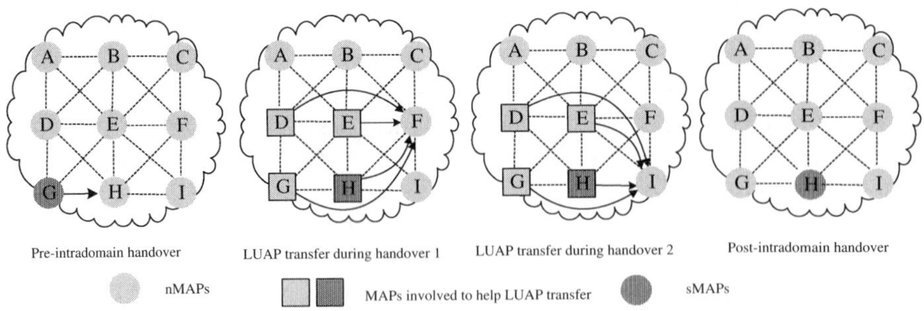

FIGURE 22.4 LUAP transfer during intradomain handoff.

consistent, $\{F, I\}$ will store the consistent LUAP in its caches. Otherwise, it will return a fault alert to the MGW. Because the LUAP transfer algorithm can be proceeded in a peer-to-peer fashion among MAPs without intervention of the MGW, the LUAP transfer can be performed in a localized manner.

22.3.7 D-coin Issuing and Interdomain Handover Phase

Once an MU is going to handoff to an adjacent MAP of the other WISP domain, the MU will be collaboratively issued a D-coin generated by the sMAP and the nMAPs.

A D-coin is composed of seven components, as shown in Fig. 22.5, where Issuer is the current serving WISP, Receiver is the tMAP of the handoff target WISP, $\mathscr{B}$ represents the face value of this D-coin, Enc(PMK) refers to a pairwise master key (PMK) encrypted with the receiver's public key, SP is a hash chain newly generated by the MU as the spending proof described in Section 22.3.5, Exp is the expiration date, and Sig is the issuer's signature on the above six components. Among the seven components, a digital signature issued by the issuer plays a critical role in building up the trust relationship among WISPs. In the following, we will show how a sMAP collaborates with its nMAPs to locally issue a D-coin.

Assume that every neighboring MAP periodically broadcasts its public key certificate along with *service set identifier* (SSID). After deciding the handoff target MAP_{tMAP}@$WISP_B$, the MU can easily ensure the security of tMAP's public key by (1) validating the public key certificate by checking the RB's signature on it and (2) ensuring that this public key certificate has not been revoked by checking the CRL stored at the cMAP. After that, the D-coin issuing protocol can be performed as follows.

Issuer	Receiver	$\mathscr{B}$	Enc(PMK)	SP	Exp	Sig

FIGURE 22.5 Components of D-coin.

The MU generates a new hash chain $H^m(M')$ and derives a new PMK, which will be used to establish a secure channel with $tMAP_B$. MU encrypts the new PMK with $tMAP_B$'s public key and obtains Enc(PMK). The encryption method can adopt any existing paring based encryption such as [23]. After that, MU sends a handoff request hREQ = ($tMAP@WISP_B$, $H^m(M')$, Enc(PMK)) to $sMAP_A$. $sMAP_A$ broadcasts this message to its one-hop neighbors and initializes a D-coin issuing algorithm as shown in Algorithm 22.2.

Alogarithm 22.2. Localized D-Coin Issuing

Input (hREQ)

Output a valid D-coin

for each $MAP_A[i] \in \{sMAP_A, nMAP_As\}$ **do**

  Based on LUAP, summarize the MU's remaining credits $\mathscr{B}$, generate a partial D-coin by computing

$$Psig_i = s_i H_2(WISP_A||tMAP@WISP_B||\mathscr{B}||H^m(M')||Enc(PMK)||Exp);$$

  Send $Psig_i$ to $sMAP_A$;

end for

for $sMAP_A$ **do**

  Successfully collects k valid partial signatures (including the one generated by itself), denoted as $Psig_i$, $1 \le i \le k$; generate a full signature (D-coin) by computing

$$Sig = \sum_{i=1}^{k} \prod_{j=1, j \neq 1}^{k} \frac{0-i}{j-i} \cdot Psig_i;$$

  Send Sig to the MU;

end for

for MU **do**

  Obtain the Sig, check the validity by computing

$$\hat{e}(Sig, P^A) = \hat{e}(H_2(WISP_A||tMAP@WISP_B||\mathscr{B}||H^m(M')||Enc(PMK)||Exp), Y^A);$$

  Get a valid D-coin

$$(WISP_A, tMAP@WISP_B, \mathscr{B}, H^m(M'), Enc(PMK), Exp, Sig)$$

end for

With a valid D-coin, the MU can successfully handoff to the tMAP operated by WISP B, where tMAP only needs to verify the validity of the D-coin to ensure that this D-coin has not been spent before and decrypt the future pairwise master key PMK from Enc(PMK). Note that a double-spending check is critical to ensure the security of a signature-based authentication scheme since a digital signature can be used more than once without such a check, which immediately leads to a service fraud. In addition, the double-spending check is normally performed by the RB, which will certainly cause extra delay. To avoid the centralized double-spending check, the proposed D-coin is localized by containing the name of tMAP and its WISP. The localization of the D-coin can effectively avoid the double-spending fraud by restricting the validity of the D-coin only within a specific MAP of the WISP domain. Therefore, the tMAP only needs to maintain a local cache to check double spending without going through the RB.

22.3.8 Clearance Phase

With the proposed UAB, the RB also serves as an *automated clearing house* (ACH) to enable the inter-WISP payment to be handled and processed in an efficient way. The UAB clearance procedure is based on an event-driven model with batching, where every D-coin is regarded as an event while the D-coin can only be submitted to the RB when a batch of a given size of D-coin is gathered or after a minimum time period has elapsed. By dealing with a batch of clearance requests at a time, the centralized RB can be relieved from involving every interdomain handoff and transaction. In addition, when the RB verifies the gathered D-coin, the aggregate signature [16] is performed for reducing the transmission and verification cost. The following are the detailed clearance action steps:

Step 1. D-coin aggregation and submission: Let WISP B collect n D-coins from the same WISP A in the clearance phase: $\text{D}-\text{coin}_i = (\text{WISP}_A, \text{T}_i, \text{Sig}_i)$, where $i = 1, \ldots, n$ and $T_i = \{\text{tMAP@WISP}_B, \mathscr{B}, H^m(M'), \text{Enc(PMK)}, \text{Exp}\}$. Then, we can take advantage of the aggregate signature technique to merge the n D-coin into one single D-coin by computing $\overline{\text{Sig}} = \prod \text{Sig}_i$. The aggregated D-coin can be represented as $(\text{WISP}_A, T_1, \ldots, T_n, \overline{\text{Sig}})$. Then WISP B submits the aggregated D-coin as a clearance request to the RB.

Step 2. D-coin batch verification: After receiving the clearing request, the RB needs to verify the aggregate D-coin as follows:

a. Ensure that all the T_i are different and have not expired.
b. Batch the D-coin by further computing

$$\overline{T} = \prod_{i=1}^{n} H_2(\text{WISP}_A || \text{tMAP}_i\text{@WISP}_B || \mathscr{B}_i || H^m(M'_i) || \text{Enc}(\text{PMK}_i) || \text{Exp}_i)$$

c. Check the validity of the set of D-coins using the following equation:

$$\hat{e}(\overline{\text{Sig}}, P^A) = \hat{e}(\overline{T}, Y^A)$$

Step 3. Payments deposit: After ensuring the validity of the D-coin, the RB evaluates the amount of the D-coin by computing $\overline{\mathscr{B}} = \sum_{i=1}^{n} \mathscr{B}_i$. A specific amount of money $\overline{\mathscr{B}}$ will be transferred from WISP A's account to WISP B's account.

22.4 SECURITY ANALYSIS

22.4.1 Overall Security Improvement

With UAB, a D-coin is issued under the endorsement of k or more nMAPs. Let the number of MAPs in a WISP domain be n. In a normal case, a WISP is considered to be compromised as long as any registered MAP is compromised. On the other hand, with UAB, a WISP is considered to be compromised only if k or more MAPs are simultaneously compromised, where the compromise resilience of the authentication and billing functionality can be further improved.

22.4.2 Prevention of D-coin Fraud Attack

A compromised MAP may launch a D-coin fraud attack by denying having accepted a piece of D-coin and refuse to offer services to the MU, even if the MAP did accept the D-coin. Furthermore, the compromised MAPs can sell the D-coin to other unauthorized MUs, which will lead to an immediate loss to the MUs. With UAB, since an MU needs to submit the spending proofs (hash-chain tokens) on time to maintain a consistent session, even if a compromised MAP can perpetrate a fraud on a piece of D-coin and transfer it to another unauthorized MU, the unauthorized D-coin holder cannot take advantage of this D-coin to gain access to the WMN without a valid usage proof on time. Since a session without submitting a spending proof will be terminated by the MGW, even if the compromised MAP defrauds the D-coin at the authentication phase, the compromised MAPs cannot transfer it to the other MUs, by which the D-coin fraud attack can be thwarted.

22.4.3 Prevention of Overcharge Attack

The overcharge attack can be performed by a compromised MAP in such a way that the compromised MAP maliciously fails to inform the accounting server when the MU has disconnected from the MAP. The UAB can successfully resist the overcharge attack due to the intrinsic nonrepudiation feature. When an MU disconnects from a MAP, the MU will receive a D-coin indicating its remaining credits. Therefore, the D-coin can be utilized to resolve the possible dispute between MUs and WISPs resulting from an overcharge attack.

22.4.4 Other Security Properties

22.4.4.1 Location Privacy Protection. Location privacy is another important issue related to roaming. In [24], the risks associated with the unauthorized disclosure, collection, retention, and usage of location data is discussed. A secure roaming scheme should be able to keep the MU's identity unknown to the foreign networks. In the proposed UAB, the MU's privacy can be well guaranteed through the employment of the D-coin since the identity information of its holder is not included.

22.4.4.2 Impersonation Attack. A malicious attacker may impersonate a legitimate MAP and broadcast bogus beacons to attract the MUs. Therefore, mutual authentication is necessary. This can be achieved in the proposed UAB. In specific, when an MU sends a piece of D-coin to an MAP for authentication, the D-coin will include a PMK, which is encrypted with the public key of the MAP. The encrypted PMK, denoted as Enc(PMK), can serve as a challenge by the MU to the MAP. Only a real MAP with the corresponding secret key can obtain the PMK by computing decryption operations so as to perform the subsequent reassociation operation with the MU.

22.5 PERFORMANCE EVALUATION

The applicability of the UAB (denoted as a short and aggregate signature–based UAB, or SAS-UAB) is evaluated through extensive simulation in terms of the resultant interdomain handoff delay and the workload at the RB. To further demonstrate the superiority of UAB, we also evaluate a number of other existing interdomain authentication solutions for comparison, including the IEEE 802.1x authentication scheme and public key certificate–based localized authentication (PKC-LA) [7].[1] In addition, since UAB can adopt different encryption and signature schemes as the building block, we introduce two UAB variations called RSA-UAB and ECC-UAB for comparision, which are based on RSA and ECC encryption and signature, respectively.

22.5.1 Average Interdomain Authentication Latency

In [25], the authentication latency is defined as the time from the instant when the MU sends an authentication request to the instant when the MU receives

[1] In [7], there is no particular encryption or signature schemes defined. Without loss of generality, two classic encryption [23] and signature [26] are adopted as an example to evaluate the performance of PKC-LA.

TABLE 22.1 Explanation of Authentication Latency

Notation	Explanation
T_{TR}	Message transmission time on one hop
$T_{\mathrm{PK\_RSA}}$	RSA-based public key operation time
$T_{\mathrm{PK\_ECC}}$	ECC-based public key operation time
$T_{\mathrm{PK\_SAS}}$	Short-signature-based public key operation time
$T_{\mathrm{PK\_PKC}}$	ID-based public key operation time in scheme PKC-LA
T_{CRL}	Certificate revocation list online checking time

the authentication reply. Since interdomain roaming is focused, the authentication latency can be expressed as

$$T_{\mathrm{AD}}(i) = \mathbf{d} \cdot \mathbf{t} \tag{22.2}$$

where i is a specific interdomain authentication scheme, $\mathbf{t}$ is a vector representing the authentication operations that may contribute to the authentication latency and is defined as $\mathbf{t} = [T_{\mathrm{TR}}, T_{\mathrm{PK\_RSA}}, T_{\mathrm{PK\_ECC}}, T_{\mathrm{PK\_SAS}}, T_{\mathrm{PK\_PKC}}, T_{\mathrm{CRL}}]$, where all the time components are defined as in Table. 22.1.

Also, $\mathbf{d}$ is a vector denoting the amount of time for each time component. In the SAS-, RSA-, and ECC-UAB, the localized authentication is supported, and thus the authentication message can traverse directly from the sMAP through the nMAP, the sMAP, and the MU to the tMAP, which takes only four hops. Furthermore, SAS-, RSA-, and ECC-UAB also require one public key encryption, one decryption, one signature generation, and two verification operations based on pairing, RSA, and ECC, respectively. According to [7], the interdomain authentication in PKC-LA is a three-way handshake protocol. At each step, one signature or encrypted message is transmitted. Further, two signature generation/verification operations and one encryption/decryption need to be performed. In addition, one certificate revocation list checking operation is inevitable to defend the service abuse attack. Finally, in the IEEE 802.1x authentication scheme, the authentication request should be transmitted to the home network via N hops, and the home network will send the authentication result back via another N hops, where N refers to the distance between the sMAP and MU's home WISP. Notice that we do not take the symmetric key processing time into consideration since the running time of symmetric key is negligible compared with the other operations. As the summary of above discussion, we can define d by introducing Eq. (22.3) as follows:

$$\begin{pmatrix} d_1 \\ d_2 \\ d_3 \\ d_4 \\ d_5 \end{pmatrix} = \begin{pmatrix} 4 & 0 & 0 & 1 & 0 & 0 \\ 2N & 0 & 0 & 0 & 0 & 0 \\ 3 & 0 & 0 & 0 & 1 & 1 \\ 4 & 1 & 0 & 0 & 0 & 0 \\ 4 & 0 & 1 & 0 & 0 & 0 \end{pmatrix} \tag{22.3}$$

TABLE 22.2 Summary of Transmission Delay per Hop in Different Schemes

	SAS-UAB	IEEE 802.1x	PKC-LA	RSA-UAB	ECC-UAB
Delay/hop (ms)	0.5	20	0.2	1.5	0.6

To investigate the average interdomain authentication delay, the mobility model in [27] is adopted. The probability density function (PDF) of the number of inter-domain handoff j can be written as follows:

$$\alpha(j) = \begin{cases} 1 - 1/\rho_{\text{WISP}}[1 - f^{*}_{\text{WISP}}(t)] & \text{if } j = 0 \\ 1/\rho_{\text{WISP}}[1 - f^{*}_{\text{WISP}}(t)]^{2} * [f^{*}_{\text{WISP}}(t)]^{i-1} & \text{if } j > 0 \end{cases} \tag{22.4}$$

where the residence time of the MU follows a general distribution of $(1/\mu_{\text{WISP}})$, and its probability density function (pdf) is $f_{\text{WISP}}(t)$ and Laplace transform is $f^{*}_{\text{WISP}}(t)$. Let the interarrival time of each MU entering a network domain follow an exponential distribution with a mean of $1/\lambda$ and $\rho_{\text{WISP}} = \lambda/\mu_{\text{WISP}}$. The average authentication delay for any specific authentication scheme i can be defined as

$$T_{\text{inter}}(i) = \sum_{j} \mathbf{d}_i \cdot \mathbf{t} \cdot j \cdot \alpha(j) \quad \forall i = 1, 2, 3, 4, 5 \tag{22.5}$$

22.5.2 Parameter Setting

The parameter settings of the simulation are as follows. In IEEE 802.1x, the maximum authentication message is 4096 bytes, the transmission delay per hop is about 20 ms provided with 2 Mbps link capacity [25]. The transmission delay per hop with different message sizes for each scheme is listed in Table 22.2. We evaluate the delay of cryptographic operations on a 3.0-GHz Intel Pentium machine with 1-GB RAM running Fedora Core 4 based on the cryptographic library MIRACL [28] except the pairing operations. As reported in [29], by most efficient software optimization and hardware acceleration, the pairing calculation can be accomplished within 1.3 ms. Therefore, we can estimate all of the pairing-based public key operations shown in Table 22.3. Note that since

TABLE 22.3 Summary of Various Public Key Operations Computational Cost

	RSA	SAS	ECC	PKC-LA
Encrypt (ms)	0.03	1.3	1.55 (ECDH)	1.3
Decrypt (ms)	4.49	1.3	1.55 (ECDH)	1.3
Sign (ms)	4.49	1	1.55 (ECDSA)	1.3
Verify (ms)	0.03	1.3	1.95 (ECDSA)	2.6

TABLE 22.4 Summary of Simulation Parameters

	$T_{\text{PK\_RSA}}$	$T_{\text{PK\_ECC}}$	$T_{\text{PK\_SAS}}$	$T_{\text{PK\_PKC}}$	T_{CRL}
Delay (ms)	9.07	8.55	6.2	9.1	500

the computation cost of the pairing-based operations dominates the pairing-based public key process, we mainly consider the pairing calculation time.

According to [21], the latency incurred by the certificate revocation list checking is about 0.5 s and the majority of it is from network latency. Then, all of the parameters to evaluate the authentication delay are summarized in Table 22.4.

We evaluate the effect of user mobility and the average hop count between each MAP and the MGW in terms of the average authentication latency. Figure 22.6 shows the impact by varying the WISP domain residential time of each MU upon the average authentication latency, where the distance between an MAP and the MGW is 4. For the certificate revocation list-checking operation, PKC-LA is subject to the longest authentication latency, followed by the IEEE 802.1x full

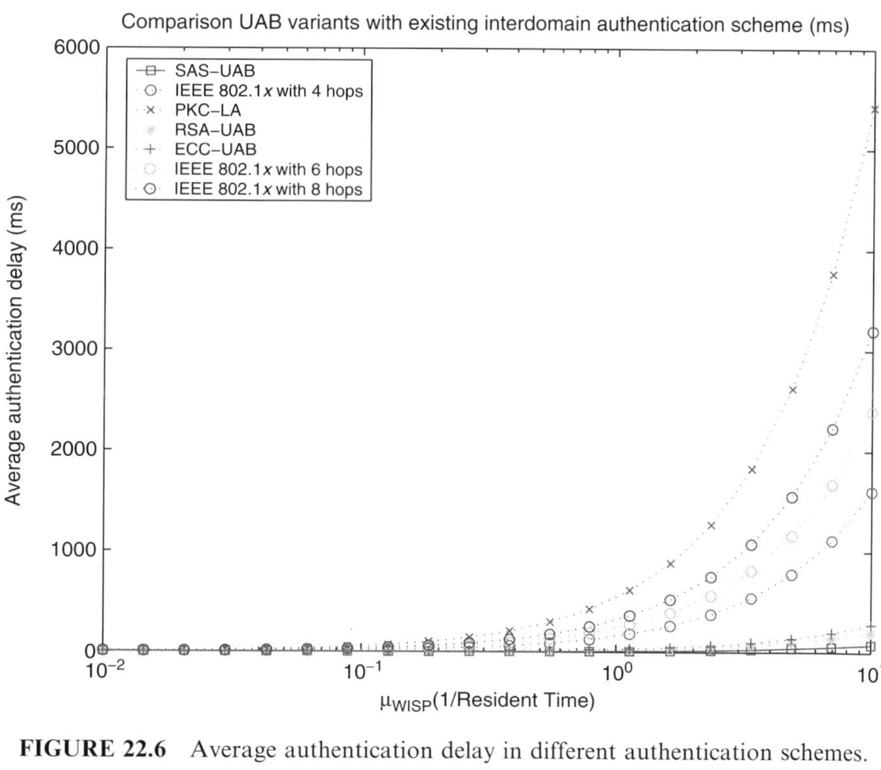

FIGURE 22.6 Average authentication delay in different authentication schemes.

authentication scheme. It can be seen that the proposed SAS-UAB yields the shortest interdomain handoff latency, while the authentication delay caused by two other UAB variants are very close to each other. In addition, the average hop count between the MAPs and the MGW plays an important role in the average authentication delay when a centralized authentication method is in place. The delay of the full IEEE 802.1x authentication scheme increases significantly with the hop count from an MAP to the WMG. On the other hand, the hop count is seen with very little impact on the authentication cost of the proposed UAB schemes. This further demonstrates that achieving localized authentication could be very critical to seamless mobility support.

22.5.3 Workload on Roaming Broker

The advantage of the UAB variants against IEEE 802.1x and PKC-LA is straightforward. IEEE 802.1x and PKC-LA require the RB to be online during an interdomain roaming event; while in UAB, the interdomain authentication and billing can be proceeded in a peer-to-peer fashion. However, it is not so obvious to see the advantages of the proposed SAS-UAB over RSA-UAB and ECC-UAB in terms of load on RB. Therefore, in this section, we will examine the efficiency of the proposed SAS-UAB scheme in terms of storage consumption and computation workload on RB. The following analysis will focus on SAS-, RSA-, and ECC-UAB schemes.

22.5.3.1 Space Analysis. The approximate length of components of the D-coin in SAS-UAB is shown in Table 22.5.

It is important to point out that by adopting the short signature technique [16], the signature field is only 20 bytes, which is much shorter than the length of an RSA and ECDSA signature that is 128 and 40 bytes, respectively. By considering the public key encryption size, the size of a single D-coin in SAS-UAB, RSA-UAB, and ECC-UAB is 100, 296, and 120 bytes, respectively.

Furthermore, we evaluate the overall storage consumptions for different schemes by considering the batch verification. Let the total number of D-coin be N and assume that a clearance action is automatically triggered when m D-coin is collected. Then the total storage consumption of the SAS-UAB scheme can be computed as follows:

$$S_{\text{SAS-UAB}} = 80N + \frac{20N}{m} \quad 1 < m < N \tag{22.6}$$

TABLE 22.5 Size of Each Component of Authentication Message in SAS-UAB

Components	Issuer	Receiver	$\mathcal{B}$	Enc(PMK)	SP	Exp	Sig
Size	8	8	8	40	8	20	8

According to [21], RSA also supports aggregation mode. Therefore, the total storage consumption of RSA-UAB can be computed as

$$S_{\text{RSA-UAB}} = 168N + \frac{128N}{m} \quad 1<m<N \tag{22.7}$$

and the storage consumption in the ECC-UAB can also be calculated as

$$S_{\text{ECC-UAB}} = 120N \quad 1<m<N \tag{22.8}$$

For a large-scale wireless MAN containing numerous independent MUs with frequent interdomain transaction events, the storage saving achievement will yield great benefit. Suppose $N = 50{,}000$, we can manipulate the clearance speed in order to achieve desired storage consumption. The numerical results are shown in Fig. 22.7. It can be seen that once the clearance speed is set large enough (e.g., $m > 10$), SAS-UAB can save about 33 and 50% storage consumption compared with that by ECC-UAB and RSA-UAB, respectively.

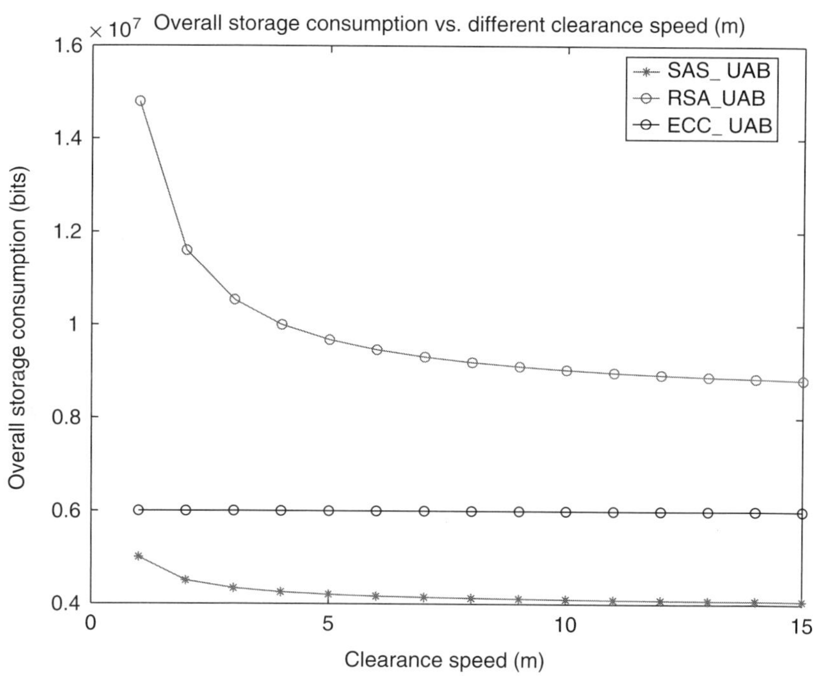

FIGURE 22.7 Comparison of storage consumption on RB under different UAB schemes.

TABLE 22.6 Summary of Computational Cost of Various Signature

	RSA (1024 bits)	SAS-UAB (160 bits)	ECDSA (160 bits)
Verification cost for one signature	0.03	1.3	1.95
Aggregate signing cost	$0.015(k + 1)$	$1.3(k + 1)$	×

22.5.3.2 Computation Workload for the RB. Similar to storage consumption, the aggregate technique may be employed to decrease the computation load of the RB. Assume that a WISP submits a piece of aggregate D-coin composed of N pieces of single D-coin. Before evaluating the computation workload on the RB, we need to break down the computation load of each algorithm and obtain the running time of each step. For the RSA scheme, when the public key e chooses a small prime such as 3, the aggregation cost of two signature operations is almost half of the RSA verification cost. Therefore, the overall RSA-based aggregate verification on k distinct messages takes about $(k + 1)/2$ times of verification for a single D-coin. For a short and aggregate

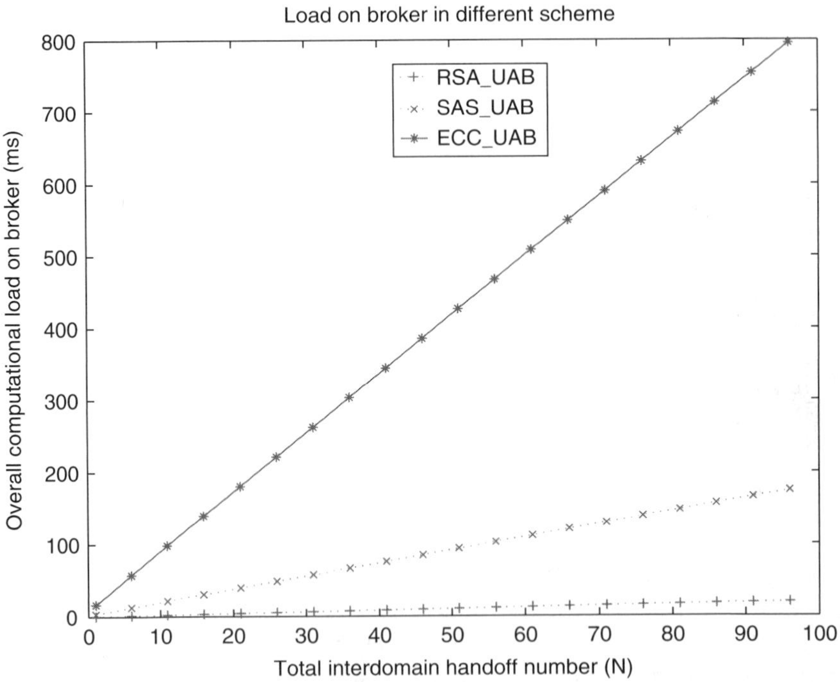

FIGURE 22.8 Comparison of computational overhead on RB in different schemes.

signature, a single verification takes two times of pairing computation. Thus, an aggregated verification on k distinct messages requires $k + 1$ times of pairing. We list the primitive computation cost for all the UAB variants in Table 22.6.

Given a specific D-coin number N, we can obtain the computational cost on RB in different UAB variants as shown in Fig. Fig. 22.8. It is observed that the performance of SAS-UAB is close to RSA-UAB even when RSA-UAB also supports aggregation operations.

22.6 DISCUSSIONS

22.6.1 Public Key Cryptography

For minimizing the interdomain handoff delay, this study provides a solution by reducing the transmission time in the authentication process at the expense of longer computation latency for public key processing. Under the PKI, the trust relationship can be initiated at the RB and transferred to all the involved parties including every MAP, WISP, and MU, where an interdomain handoff is simply treated as a cross-WISP transaction through issuing and reception of D-coin. As before, the most common criticism on using PKC in wireless environments lies in the unacceptable computational complexity and communication overhead. However, recent rapid developments in improving the calculation speed and shortening the overhead of PKC have made it much more friendly in the application scenarios, where some well-known notoriously expensive cryptographic operations can be performed efficiently, such as pairing, which used to take over 1 s to calculate when it was first invented. Today, the hardware acceleration technique can deal with it within 1.3 ms [29]. Therefore, the PKC-based UAB scheme can be a practical distributed security management solution in the application scenario of wireless MANs with through a WMN.

22.6.2 Distributed Security Management of WMN

Another unique feature brought by the UAB is its distributed security management. Although a centralized security management framework is still recommended by the IEEE (e.g., RADIUS) due to its highest security assurance, many academic researchers have argued that such a centralized scheme is not efficient and scalable when the network size is getting large [15, 30]. To achieve performance requirement, developing a security scheme that can initiate a graceful compromise between the performance and security assurance is highly desired and contributive. The proposed UAB is a fully distributed one, where the RB delegates its roaming functionalities to every WISP, which in turn delegates its security capabilities to every MAP in the whole metropolitan-area WMN domain. Under a distributed security architecture, the bottleneck problem can be well resolved while an extremely high level of security guarantee can be achieved by way of a voting and threshold mechanism.

22.7 CONCLUSIONS

We have proposed a novel universal authentication and billing (UAB) architecture for wireless MANs. The new architecture can successfully tackle the challenging tasks such as security guarantee and performance improvement in terms of system compromise resilience capability, interdomain handoff authentication latency, and roaming broker's workload. We have also demonstrated the practicality and feasibility of the UAB in a real-world application scenario.

REFERENCES

1. C. d. Laat, G. Gross, and L. Gommans, "Generic AAA architecture (RFC 2903)," available: http://www.ietf.org/rfc/rfc2903.txt, 2000.
2. B. Anton, B. Bullock, and J. Short, "Best current practices for wireless Internet service provider (WISP) roaming," available: http://www.weca.net/OpenSection/wispr.asp, 2003.
3. S. Pack and Y. Choi, "Fast handoff scheme based on mobility prediction in public wireless LAN systems," *IEE Commun.* 151(5), 489–495 (2004).
4. M. Long, C. H. Wu, and J. D. Irwin, "Localised authentication for inter network roaming across wireless LANs," *IEE Commun.* 151(5), 496–500 (2004).
5. A. Mishra, M. H. Shin, N. L. Petroni, J. T. Clancy, and W. A. Arbauch, "Proactive key distribution using neighbor graphs," *IEEE Wireless Commun.* 11(1), 26–36, (2004).
6. C. Chou and K. G. Shin, "An enhanced inter-access point protocol for uniform intra and intersubnet handoffs," *IEEE Trans. Mobile Comput.* 4(4), 321–334 (2005).
7. Y. Zhang and Y. Fang, "ARSA: An attack-resilient security architecture for multihop wireless mesh networks," *IEEE J. Sel. Areas Commun.* 24(10), 1916–1928, (2006).
8. J. Leu, R. Lai, H. Lin, and W. Shih, "Running Cellular/PWLAN services: Practical considerations for Cellular/PWLAN architecture supporting interoperator roaming," *IEEE Commun. Mag.* 44(2), 73–84 (2006).
9. G. Ateniese, A. Herzberg, H. Krawczyk, and G. Tsudik, "Untraceable mobility or how to travel incognito," *Computer Networks* 310(8), 871–884 (1999).
10. B. Schilit, J. Hong, and M. Gruteser, "Wireless location privacy protection," *Computer* 32(12), 135–137 (2003).
11. H. Zhu, X. Lin, P.-H. Ho, X. Shen, and M. Shi, "TTP based privacy preserving inter-WISP roaming architecture for wireless metropolitan area networks," in *Proceedings of the IEEE Wireless Communications and Networking Conference (WCNC'07)*, Hong Kong, China, Mar. 2007.

12. H. Zhu, X. Lin, R. Lu, P.-H. Ho, and X. Shen. "SLAB: Secure localized authentication and billing scheme for wireless mesh networks," Technical Report, University of Waterloo, Ontario, Canada, Apr. 2007.
13. J. W. Wong, L. Mirlas, W. Kou, and X. Lin, "Credit card based secure online payment and a payment protocol using a trusted third party," in W. Kou (Ed.), *Payment Technologies for E-commerce*, Springer-Verlag, New York, 2003, pp. 227–243.
14. Y. Tsiounis, A. Kiayias, and A. Karygiannis, "A solution for wireless privacy and payments based on E-cash," *in Proceedings of the IEEE International Conference on Security and Privacy for Emerging Areas in Communication Networks (Secure-Comm'05)*, Athens, Greece, Sept. 2005.
15. N. B. Salem and J.-P. Hubaux, "Securing wireless mesh networks," *IEEE Wireless Commun.* 13(2), 50–55 (2006).
16. D. Boneh, B. Lynn, and H. Shacham, "Short signatures from the weil pairing," *J. Crypto.* 17(4), 297–319 (2004).
17. D. Chaum, "Blind signatures for untraceable payments," in *Advances in Cryptography* (*CRYPTO'82*), Springer-Verlag, Santa Barbara, CA, Aug., 1982, pp. 199–204.
18. J. Camenisch, J.-M. Piveteau, and M. Stadler, "Blind signatures based on the discrete logarithm problem," in *Advances in Cryptography* (*EUROCRYPT'94*), Springer-Verlag, Perugia, May 1994, pp. 428–432.
19. M. Abe and E. Fujisaki, "How to date blind signatures," in *Advances in Cryptography* (*ASIACRYPT'96*), Springer-Verlag, Kyongju, Korea, Nov. 1996, pp. 244–251.
20. M. Abe and T. Okamoto, "Provably secure partially blind signatures, in *Advances in Cryptography* (*CRYPTO'97*), Springer-Verlag, Santa Barbara, CA, Aug. 1997, pp. 271–286.
21. M. Zhao, S. W. Smith, and D. M. Nicol, "Aggregated path authentication for efficient BGP security," in *Proceedings of the ACM Conference on Computer and Communications Security* (*CCS'05*), Alexandria, VA, Nov. 2005, pp. 128–138.
22. L. Lamport, "Password authentication with insecure communication," *Commun. ACM* 24(11), 770–772 (1981).
23. D. Boneh and M. Franklin, "Identity-based encryption from the Weil pairing," in *Advances in Cryptography* (*CRYPTO'01*), Springer-Verlag, Santa Barbara, CA, Aug. 2001, pp. 213–229.
24. R. P. Minch, "Privacy issues in location-aware mobile devices," in *Proceedings of the IEEE Annual Hawaii International Conference on System Sciences* (*HICSS'04*), Hawaii, Jan. 2004, pp. 1–10.
25. W. Liang and W. Wang, "A quantitative study of authentication and QoS in wireless IP networks," in *Proceedings of the IEEE Annual Joint Conference of the IEEE Computer and Communications Societies* (*INFOCOM'05*), Miami, FL, Mar. 2005, pp. 1478–1489.
26. F. Hess, "Efficient identity based signature schemes based on pairings," in *Proceedings of Selected Areas in Cryptography* (*SAC'02*), St. John's, Newfoundland, Canada, Aug. 2002, pp. 310–324.
27. S. Baek, S. Pack, T. Kwon, and Y. Choi, "A localized authentication, authorization, and accounting (AAA) protocol for mobile hotspots," in *Proceedings of the Annual*

Conference on Wireless On-demand Network Systems and Services (*WONS'06*), Les Ménuires, France, Jan. 2006, pp. 144–153.

28. *Multiprecision Integer and Rational Arithmetic C/C++ Library* (*MIRACL*), available: http://www.shamus.ie/.
29. T. Kerins, W. P. Marnane, E. M. Popvici, and P. S. L. M. Barreto, "Efficient hardware for the Tate pairing calculation in characteristic three," in *Proceedings of the Workshop on Cryptographic Hardware and Embedded Systems* (*CHES'05*), Edinburgh, Scotland, Aug. 2005, pp. 398–411.
30. I. Akyildiz, X. Wang, and W. Wang, "Wireless mesh networks: A survey," *Computer Networks* 47(4), 445–487 (2005).
31. R. Lu and Z. Cao, "Efficient remote user authentication scheme using smart card," *Computer Networks* 49(4), 535–540 (2005).
32. M. Shi, H. Rutagemwa, X. Shen, J. W. Mark, and A. Saleh "A ticket ID system for service agent based authentication in WLAN/Cellular integrated networks," in *Proceedings of the IEEE Wireless Communications and Networking Conference* (*WCNC'07*), Hong Kong, China, Mar. 2007.
33. R. Lu, Z. Cao, and Y. Zhou, "Proxy blind multi-signature scheme without a secure channel," *Appl. Math. Comput.* 164(1), 179–187 (2005).
34. X. Lin and Y. Yang, "A blind signature scheme based on Lucas sequence," *J. Beijing Univ. Posts Telecommun.* 21(1), 88–91 (1998).

CHAPTER 23

SCHEDULING ALGORITHMS FOR WiMAX NETWORKS: SIMULATOR DEVELOPMENT AND PERFORMANCE STUDY

SAI SUHAS KOLUKULA, M. SAI RUPAK, K. S. SRIDHARAN, and KRISHNA M. SIVALINGAM[1]

23.1 INTRODUCTION

The need for wireless data access to the Internet from "anywhere, anytime" has become part of the everyday life in most parts of the world. The different flavors of the IEEE 802.11 wireless local area network (LAN) standard (also termed as WiFi) has provided part of the access solution and operate at 11–100 Mbps. However, the coverage region is limited to short distances from the wireless access points. On the other hand, cellular networks provide data access covering large areas, on the order of square miles, per wireless cellular base station. However, the data rates of cellular data networks are limited to a few tens of kilobytes per second, although third-generation (3G) technologies promise higher rates up to 1–2 Mbps.

The IEEE 802.16 broadband wireless access standard [1, 2], typically called WiMAX (for worldwide interoperability for microwave access), attempts to bridge the gap by defining technology that can support high bandwidth rates (tens of megabytes per second) covering large regions (on the order of square miles). The standard defines wireless metropolitan area network (MAN) technology with the goal of connecting fixed and mobile users. There are

[1] Figure 23.2 of this chapter is reprinted with permission from IEEE 802.16-2004—IEEE Standard for Local and Metropolitan Area Networks, Part 16: Air Interface for Fixed Broadband Wireless Access Systems, Copyright 2004, by IEEE. The IEEE disclaims any responsibility or liability resulting from the placement and use in the described manner.

Emerging Wireless LANs, Wireless PANs, and Wireless MANs. Edited by Y. Xiao and Y. Pan

several possible applications of WiMAX including broadband access and WiFi HotSpot interconnection. In addition to the IEEE standards, there exists the WiMAX Forum [3], an industry consortium that aims to streamline commercialization, product certification, and adoption of WiMAX standards.

In parallel, ETSI developed the HiperMAN standard for wireless metropolitan area networks [4]. The HiperMAN is designed to interoperate with a subset of the IEEE 802.16 standard. This chapter will primarily focus on the IEEE standard.

There are ongoing installations and planned deployments of WiMAX and mobile WiMAX worldwide including Australia, France, Finland, India, Japan, Pakistan, Poland, Saudi Arabia, South Korea, Taiwan and United States [5]. In the United States, Sprint is planning a nationwide WiMAX network, AT&T is deploying in Nevada, and Clearwire is offering services in 12 states. WiMAX equipment is available from leading vendors including Adaptix, Alcatel, Alvarion, Airspan, Aperta, Intel, Navini, Proxim, Siemens, Telsim, and Wavesat.

This chapter presents an overview of the IEEE 802.16 standard including the physical (PHY) and media access control (MAC) layers. A generic quality of service (QoS) framework that is integrated with the 802.16 MAC layer is presented. The MAC layer and QoS framework is implemented in the popular ns2 simulator [6]. Performance results from the simulator are then presented to understand the protocol's QoS related performance. Additional details about this work may be found in [7, 8].

23.2 IEEE 802.16 STANDARD

This section presents information about IEEE 802.16 standard history and the PHY and MAC layer network protocols.

23.2.1 History

The first version of the IEEE 802.16 standard was approved in December 2001. It defined a wireless MAN air interface for fixed point-to-multipoint broadband wireless access (BWA) systems that are capable of providing multiple services. The communication frequencies were in the 10- to 66-GHz range, with support for line-of-sight (LOS) communications only.

The IEEE 802.16a standard, approved in January 2003, covered transmission frequencies in the 2- to 11-GHz range and included support for non-LOS (NLOS) communications. One of the new features of the IEEE 802.16a was the addition of orthogonal frequency division multiplexing (OFDM), which was selected for its multipath delay tolerance in non-LOS communications. The standard also defined flexible channel bandwidth for better support of low-rate users and adaptive modulation to better handle changing channel conditions.

The standards groups subsequently consolidated standards IEEE 802.16, IEEE 802.16a, and IEEE 802.16c into IEEE 802.16-2004, which was approved in 2004. An amendment denoted as IEEE 802.16e, which primarily included support for mobile users, was approved in December 2005 [9].

WiMAX can be used for broadband access, WiFi coverage extension, metropolitan area wide data-centric (including mobility) connectivity, and fixed broadband access in suburban/rural areas and low density areas. For additional description of the standard and its applications, the reader is referred to [10].

The standard primarily defines the PHY and MAC layers for point-to-multipoint and mesh network modes of operation. In this chapter, we do not discuss the mesh network topology and concentrate the discussion on the point-to-multipoint topology. The following sections describe the PHY and MAC layers in more detail.

23.2.2 Network Architecture

The fixed network architecture consists of a IEEE 802.16–capable base station (BS) (Fig. 23.1) that serves fixed users, called subscriber stations (SS), spread over a range up to a few miles (typically up to 3–10 miles). Each user, typically a home or a building, has a roof-top IEEE 802.16 transceiver that is used for communicating with the base station. In December 2005, IEEE also approved the IEEE 802.16e standard, which supports mobile users. The support of mobility is significant, and, if successful, this standard will provide serious competition to cellular network data users.

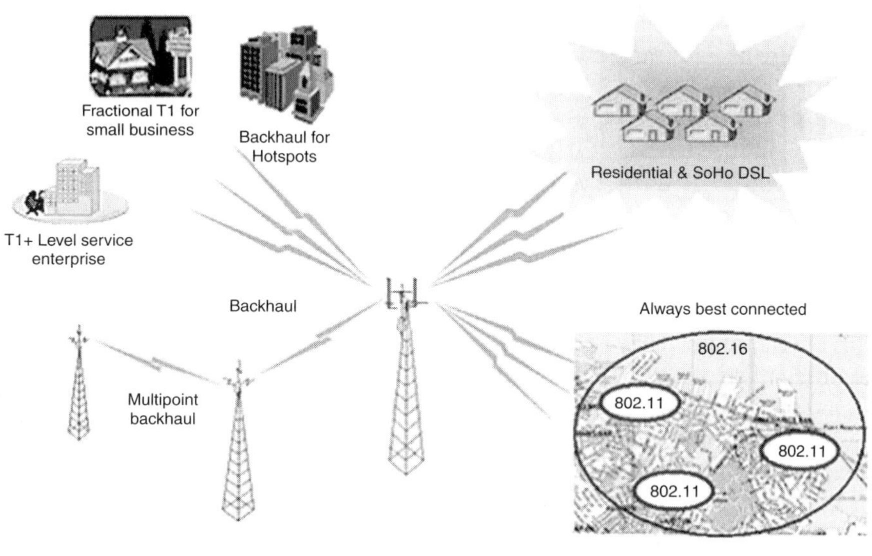

FIGURE 23.1 Typical IEEE 802.16 network. (Copyright Worldwide Interperability for Microwave Access Forum.)

23.2.2.1 Network Protocol Stack. The 802.16 protocol stack's physical layer consists of two sublayers: the physical medium-dependent sublayer that deals with the actual transmission and the transmission convergence (TC) sublayer that hides the different transmission technologies from the MAC layer.

The MAC layer consists of three sublayers as follows:

- The lowest MAC sublayer is the privacy sublayer, which deals with privacy and security. This sublayer provides services including authentication, secure communication, and key exchange. Since encryption applies only to the payload, the headers are transmitted in the clear, allowing for traffic analysis and perhaps network attacks.
- The next sublayer is the common part sublayer (CPS). This is the protocol-independent core, which deals with channel management and slot allocation to stations. The 802.16 MAC is connection oriented and supports continuous and bursty traffic, such as constant bit rate and real-time variable bit rate.
- The service-specific convergence sublayer (CS) provides the interface to the network layer above the MAC layer and is similar to the logical link sublayer in other 802 protocols. Its function is to map transport layer traffic streams to the 802.16 MAC layer connections.

23.2.3 IEEE 802.16 PHY Layer

The key IEEE 802.16 PHY layer technologies include: orthogonal frequency division multiplexing (OFDM) and scalable orthogonal frequency division multiple access (OFDMA), adaptive modulation and coding (AMC), flexible channel bandwidth allocation (1.25–20 MHz), support for adaptive antennas (smart, directional), and support for multiple-input multiple-output (MIMO).

23.2.3.1 10- to 66-GHz Systems. The IEEE 802.16-2001 standard defines operation in the 10- to 66-GHz frequency range, which supports LOS operation with a range of few kilometers and rates up to 120 Mbps. The channel bandwidths expected to be either 25 or 28 MHz. Millimeter waves in this frequency range travel in a straight line, as a result of which the BS can have multiple antennas, each pointing at a different sector. Each sector has its own users and is independent of the adjoining ones. Millimeter wave communications are based on electro-magnetic transmissions with operating wavelength of the order of one to ten millimetres.

Due to a sharp decline in signal strength of millimeter waves with distance from the BS, the signal-to-noise ratio (SNR) also drops very fast. For this reason, 802.16 uses three different modulation schemes with forward error correction (FEC). The modulation schemes are: quadrature phase shift keying (QPSK), which offers 2 bits/baud and is used by subscribers located far away from the BS; 16-bit quadrature amplitude modulation (QAM-16), which offers 4 bits/baud and is used by subscribers located at an intermediate distance from the BS; and 64-bit quadrature amplitude modulation (QAM-64), which offers 6 bits/baud and is used by subscribers located near the BS.

The 802.16 PHY layer allows transmission parameters such as modulation and coding schemes to be adjusted for each node, for every frame. Channels are separated in time using a framing mechanisms, that is, each 28-MHz carrier is subdivided into frames that are repeated continuously. Duration of each frame can be 0.5, 1, or 2 ms. A frame is subdivided in into physical slots. The number of physical slots in a frame is a function of symbol rate and frame duration. Each frame is also divided into two logical channels: downlink and uplink.

The downlink channel is a broadcast channel used by the BS for transmitting downlink data and control information to the SSs. The BS is completely in control for the downlink direction. The BS allocates the slots on the downlink channel based on a suitable scheduling mechanism. The uplink channel is time shared among all SSs. The BS is responsible for granting bandwidth to individual SSs in the uplink direction using demand assigned multiple access time division multiple access (DAMA–TDMA). The BS first allocates bandwidth to each SS to enable it to send requests for bandwidth needed to transmit uplink data. The BS then assigns a variable number of physical slots to each SS for uplink data transmissions according to their bandwidth demand. This information is sent to all SSs through the uplink control message.

IEEE 802.16 supports time division duplexing (TDD) and frequency division duplexing (FDD). In TDD, uplink and downlink channels may share the same frequency channel with each TDD frame having one downlink subframe followed by an uplink subframe. Physical slots allocated to each subframe may vary dynamically according to bandwidth need in each direction. In FDD, uplink and downlink channels operate on separate frequencies. Downlink transmissions occur concurrently with uplink transmissions. Therefore, the duration of a subframe (downlink or uplink) is the same as the frame duration. On the downlink, both full-duplex and half-duplex SSs are supported simultaneously.

23.2.3.2 2- to 11-GHz Systems. The 802.16-2004 standard included support for the 2- to 11-GHz frequency range. This range allows NLOS, wall penetration, communication ranges up to 30 miles, besides other advantages. The standard specifies that OFDM with a 256-point transform may also be used. OFDM has been selected due to its several benefits, including efficient spectrum usage, better protection from radio frequency (RF) interference, and lower multipath effects.

23.2.3.3 Uplink and Downlink Framing. Transmission in the IEEE 802.16 network consists of a downlink frame and uplink frame. The frame consists of a preamble followed by a downlink map (DL-MAP) message and an uplink map (UL-MAP) message. The DL-MAP message contains information on PHY synchronization, a downlink channel descriptor message (DCD), and the

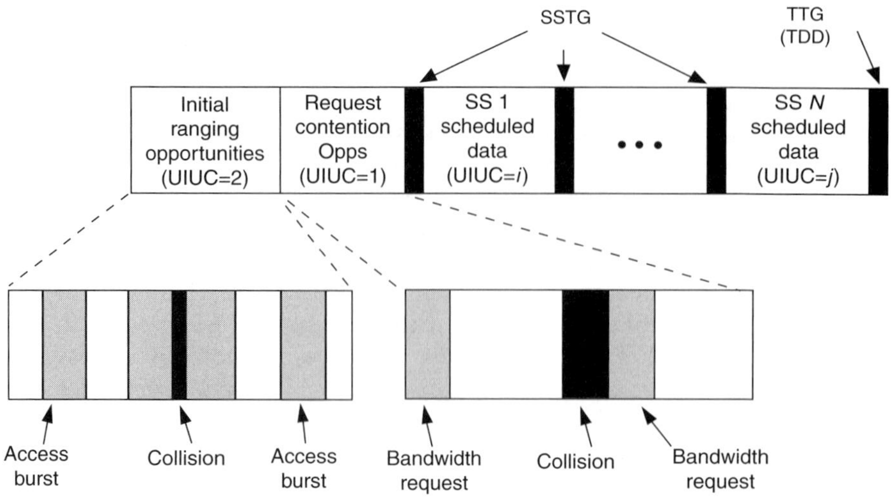

FIGURE 23.2 TDD uplink subframe structure. (From IEEE 802.16-2004, IEEE standard for local and metropolitan area networks, Part 16: Air interface for fixed broadband wireless access systems. Copyright 2004 by IEEE.)

number of downlink slots. The uplink subframe structure (Fig. 23.2) provides information about uplink slot allocation.

The following are the different burst classes sent by an SS:

- Transmitted in contention opportunities reserved for initial ranging
- Transmitted in contention opportunities defined by request intervals, which are reserved in response to polling by the BS
- Transmitted in intervals defined by data grant information elements (IEs), allotted to the SSs

The number and order of these bursts may vary within the frame, as determined by the BS uplink scheduler.

23.2.3.4 Mobility. The IEEE 802.16e-2005 amendment includes support for mobile users. It is based on scalable OFDMA, which allows varying levels of bandwidth allocation (from 1.25- to 20-MHz channels). However, supporting mobility requires significant work at the networking level for tasks such as roaming, handoffs, QoS, and security mobility. The WiMAX Forum Network Working Group (NWG) is working on networking-level issues [11]. This chapter concentrates on fixed wireless systems, and mobile WiMAX is not considered further.

23.2.4 MAC Protocol and Scheduling

The role of the MAC layer is to efficiently share the channel. The MAC layer is designed to support QoS for different applications including voice and video. The 802.16 MAC common part sublayer (CPS) specifies the mechanism to efficiently access the shared medium. On the downlink, the BS is the only central entity transmitting to the SSs. As a result, it does not need to coordinate with other stations. Transmissions are separated into different sectors using suitable sectorized antennas at the BS. Messages sent by the BS may be unicast to a particular SS, multicast to a group of SSs, or broadcast to all SSs.

The IEEE 802.16 MAC is connection oriented. Connections are used to transport a given flow's traffic and is unidirectional. A given MAC-level connection can be shared among multiple higher level flows, including services such as user datagram protocol (UDP), which does not require connections. The connections are identified using per-node unique 16-bit connection ID (CID), thus leading to 64,000 connections per node.

Three different bidirectional management connections are established by the BS, when a new SS joins the network. These include: (i) the basic connection for short, time-critical MAC messages; (ii) primary management connection for messages that can be longer and tolerate longer delay, and (iii) secondary management connection for standards-based messages including dynamic host configuration protocol (DHCP) and simple network management protocol (SNMP) [1]. In addition to the management connections, transport connections are set up based on QoS and traffic requirements and services supported. As mentioned earlier, a transport connection may be shared by multiple higher layer applications, providing that the applications have the same QoS and other requirements.

23.3 QoS PROVISIONS IN IEEE 802.16

The IEEE 802.16 standard supports many traffic types (data, voice, video) with different QoS requirements. In this context, the MAC layer defines QoS signaling mechanisms and functions for data control transmissions between the BS and the SSs. The MAC layer defines the framework to support QoS for both uplink and downlink traffic. As mentioned earlier, the concept of MAC-level connections, mapped to application-level flows, is used to support QoS.

23.3.1 Bandwidth Reservations

Quality of service support is provided using a request and grant protocol. The BS collects the bandwidth requests, computes the transmission schedule, and announces the same to the SSs, using the UL-MAP.

There are two types of SS nodes: grant per connection (GPC), where the SS handles bandwidth requests on a connection basis, and grant per SS (GPSS) basis, where the SS aggregates bandwidth requests for all its connections. GPC systems provide higher level granularity in terms of bandwidth request but are limited in terms of the slot allocation policy followed by the BS. In GPSS, the SS has additional flexibility in assigning the overall transmission slots to its connections. Thus, the GPSS might temporarily allocate a connection more than its requested bandwidth at the expense of another connection, if traffic conditions require such an action.

Bandwidth requests can be done by the SS using bandwidth request management packets or by the BS using polling. With polling, the BS uses an UL-MAP information element (IE) to notify the SS of bandwidth request opportunities. The polling can be done in a unicast, multicast, or broadcast manner. The SS can also utilize a *poll-me bit* in the grant management subheader to request the BS to issue a poll.

With contention-based requests, collisions may occur during contention slots in which all SSs are allowed to transmit their respective ranging or bandwidth requests. A truncated binary exponential backoff mechanism is used. A contending SS selects a random slot within the contention window and then transmits the request till the selected slot. If the transmission was successful and a data grant was received, the node transmits in the next alloted uplink frame slot. The bandwidth requests are not acknowledged by the BS. If an SS does not receive an allocation for a previous request, it will retransmit the request during a later frame.

23.3.2 Service Types

The IEEE 802.16-2004 specifies the following service types, for the network flows.

1. *Unsolicited Grant Service* (*UGS*). This type is used for periodic and fixed packet-size traffic sources similar to the asynchronous transmission mode (ATM) constant bit rate (CBR) service. The BS assigns a fixed number of slots to the connection, based upon the initial request. These services may not request the use of the contention slots for additional bandwidth requests but can use the management packet to indicate lagging bandwidth needs.
2. *Real-Time Polling Service* (*rtPS*). This type is used for periodic and variable bit rate services. The BS assigns periodic bandwidth request opportunity slots to each connection that will be used by the service to indicate bandwidth requests. These services may not request the use of the contention slots for additional bandwidth requests.
3. *Non-Real-Time Polling Service* (*nrtPS*). This is similar to rtPS. The connection is granted a minimal set of bandwidth request opportunities and is expected to use the contention slots for additional bandwidth requests. This service is useful for data transmissions such as file transpoint protocol (FTP).

4. *Best Effort Service* (*BES*). This type is are used for services that do not need any bandwidth or other guarantees. They use the contention slots for bandwidth requests.

Additional details such as the mandatory QoS parameters for these services may be found in [1].

23.4 GENERIC QoS ARCHITECTURE

In order to meet the QoS requirements specified in the IEEE 802.16 standard, we present a generic QoS architecture, as shown in Fig. 23.3. Other similar architectures may be found in [12, 13]. The QoS scheduling architecture is distributed, implementing GPC mode for granting bandwidth (BW) to SSs. A similar architecture supporting GPSS mode is presented in [8] and is not described here due to space constraints.

The following sections describe the QoS components at the BS and the SS.

23.4.1 QoS Components at the SS

The QoS architectural components at the subscriber station (operating in GPC mode) are described in the following sections.

23.4.1.1 SS UL Data Classifier. This component classifies each packet coming from the upper layer into one of the uplink traffic queues. In our implementation of GPC, the SS maintains a separate queue for each of its connections along with their QoS parameters. It uses the CID associated with each packet to classify packets into separate queues. The functioning of the BS data classifier is also similar.

23.4.1.2 SS BW Request Generator. This component is responsible for generating BW requests to be sent to the BS as and when required. It uses the unicast poll opportunities, piggybacking, and the contention slots for this purpose. For each connection, an *aggregate* request is generated. The aggregate request for a connection i is equal to the current queue length for that connection.

23.4.1.3 SS Coordinator. This component coordinates the entire UL transmission process. It reads the UL-MAP sent by the BS and identifies the slots for each of its connections. It coordinates the transmission of data packets for each connection in those slots along with the BW requests that are generated by the BW request generator. It also manages the extra slots given by the BS.

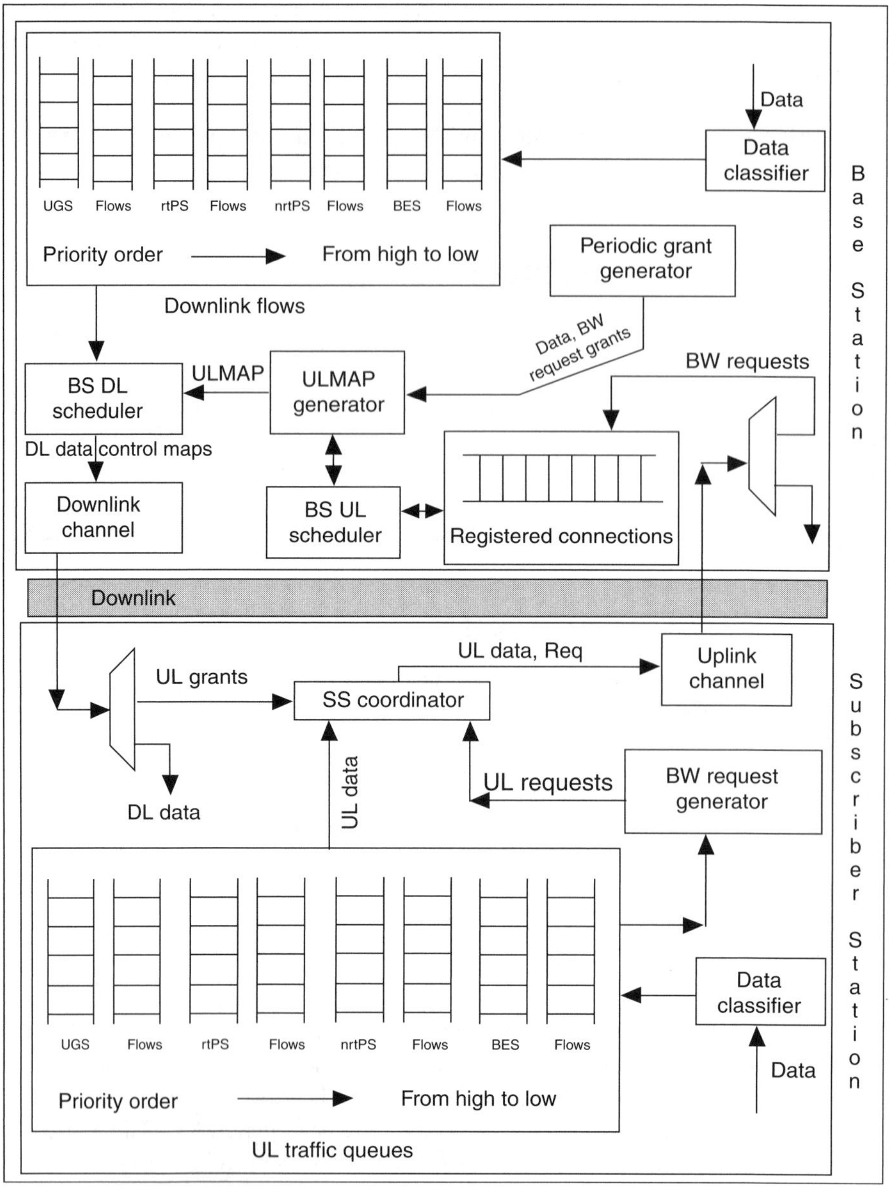

FIGURE 23.3 Proposed QoS architecture for IEEE 802.16.

23.4.2 QoS Components at the BS

The QoS architectural components at the BS are described in the following section.

23.4.2.1 BS Periodic Grant Generator. The periodic data and request grant generator generates periodic data grants for UGS connections and periodic/regular BW request opportunities for the rtps/nrtPS flows as per the QoS parameters agreed upon at connection setup time. These grants are used by the UL-MAP generator while generating the UL-MAP.

23.4.2.2 BS UL-MAP Generator. The UL-MAP generator component is responsible for generating the UL-MAP for the upcoming frame. It uses the grants supplied by the Grant Generator and the BW requirements for each connection. It uses the BS UL scheduler that has the actual logic for the UL scheduling. The generated UL-MAP is given to the BS DL scheduler that schedules the transmission of the UL-MAP in the subsequent frame.

23.4.2.3 BS DL Scheduler. The BS DL scheduler schedules the DL traffic from the DL queues while maintaining fairness and priorities among them. It also schedules for transmission the acknowledgments generated for various connections at the BS. It generates the DL-MAP and schedules for transmission the DL-MAP and the UL-MAP.

23.4.2.4 BS UL Scheduler. This component is responsible for allocating bandwidth to each SS for uplink transmission. The scheduler attempts to meet delay and bandwidth guarantees, while maintaining fairness among the different flows and realizing high channel utilization.

For the specific sequence of MAC-level scheduling related events that occur at the SS and the BS, the reader is referred to [12].

23.4.3 Scheduling Algorithms

We had implemented four different scheduling algorithms in the BS UL scheduler for the purpose of performing a comparative study. They are listed below:

1. Max-min fair allocation
2. Deficit fair priority queue
3. Weighted fair queuing
4. Weighted fair priority queuing

The first two algorithms are from [12] and [14], respectively, modified to suit our requirements. Other related work on QoS and scheduling for WiMAX systems may be found in [15–20].

23.4.3.1 Max-Min Fair Allocation. In [12], the author has proposed and implemented max-min fair allocation for GPSS mode of operation, which we have modified for GPC. Input to this algorithm is total bytes or slots requested for each flow (service type) by all the connections. BS calculates total slots requested for each flow by all connections by examining the BW requests of each connection. Uplink bandwidth is distributed among various SSs in two stages using max-min fair allocation strategy [12].

In the first stage, uplink bandwidth is distributed among the uplink flows. In the first round, each uplink flow is allocated its percentage of bandwidth, normalized by its weight (UGS is given maximum weight followed by rtPS, followed by nrtPS, followed by BES). This ensures that high priority reserved flows are always satisfied before low priority flows. In the second round, excess bandwidth allocated to any flow is distributed among unsatisfied flows in proportion to their weight. This process continues until either all four uplink flows are satisfied or no bandwidth is available.

In the second stage, bandwidth allocated to each flow is distributed among all connections. In the first round, bandwidth allocated to UGS flows is equally distributed among all UGS connections. In the second round, excess bandwidth allocated to any connection more than its requirement for UGS flows is evenly distributed among unsatisfied connections. This process continues until either the UGS requirements of all connections are satisfied or no bandwidth for UGS flows is available. The above process is repeated for rtPS, nrtPS, and BES flows too. For a more detailed algorithm please refer to [12].

23.4.3.2 Deficit Fair Priority Queuing (DFPQ). In DFPQ [14], there is an *active list* maintained in the BS. We only schedule the bandwidth application services in the active list. If the queue for the application is not empty, it will stay in the active list and removed otherwise. The service flows in the active list are queued by strict priority in the following order: UGS, rtPS, nrtPS, and BES. In each round, the highest priority flow will always be served first.

Two variables called D, denoting the DeficitCounter, and *Quantum* are maintained by the BS. The scheduler steps are as follows: (i) visit every nonempty queue and note the pending requests; (ii) let D += Quantum for each visit; and (iii) let S denote the request size (in bits) of the request packet at the head of the queue. If $S \leq D$, decrement D by S and schedule the packet. The process is repeated until either the DeficitCounter is no more greater than zero or the queue is empty, in which case DeficitCounter is reset to zero. The scheduler then evaluates the next backlogged priority queue. If all the queues are satisfied, the algorithm stops; otherwise, if there are more slots left, another pass is done among the unsatisfied flows in the priority order. If there are no more UL slots, the algorithm halts.

In [14], the quantum value for the ith service flow, that is, Quantum [i] is chosen as:

$$\text{Quantum}[i] = \sum_{j=0}^{J_i} r_{\max}(i,j)$$

where r_{max} is the maximum sustained traffic rate and J_i is the total connections for the ith class of service flow. The UGS flows are served whatever they request. So the whole scheduling is only for DL and UL rtPS, nrtPS, and BES with DL queues getting priority over UL in each service type for avoiding buffer overflows and to guarantee latency requirements.

For simplicity, we choose the Quantum value as 30% of slots (left after servicing UGS flows) for rtPS, 20% of slots for nrtPS, and 10% for BES. The remaining 40% is used in the subsequent rounds. In the case of DFPQ, we had also studied a variation by allocating slots on a per-flow basis rather than on a per-connection basis. In other words, during each pass, we allocate Quantum[i] amount of slots for the ith service flow rather than for one connection of the ith service flow. We then divide the slots allocated for each service flow among all connections of that flow. This will result in more fairness among connections of the same service flow especially under heavy loads. We have used this method for comparing with the other algorithms.

23.4.3.3 Weighted Fair Queuing (WFQ). This is a direct implementation of weighted scheduling where the scheduling decisions are based on the amount of BW requested by each connection. The UGS connections are alloted the requested BW (at connection setup time) in each frame. Following this, each connection is assigned a weight depending upon the amount of BW requested by that connection. The weight for a connection is calculated as follows:

$$\text{Weight}[i] = \frac{\text{BWReq}[i]}{\sum_{i=0}^{N} \text{BWReq}[i]}$$

where BWReq[i] is the BW requested by connection i, and N is the total number of registered connections at the BS. Bandwidth is allocated to each connection depending upon its weight. Bandwidth allocated to a connection i, BWAlloc[i] is calculated as follows.

$$\text{BWAlloc}[i] = \text{weight}[i] * \text{TotUL}$$

where TotUL is the total UL bandwidth available (after satisfying the bandwidth requirements for UGS flows). This algorithm is used as a baseline for verification and validation of the other algorithms since it does not take into account the service types and their associated priorities.

23.4.3.4 Weighted Fair Priority Queuing (WFPQ). This is an improvement over the previous algorithm in that we take the priorities of the various service flows into account. As before, UGS connections get the requested BW (at connection setup time) in each frame. For the rest, we allocate BW on a per-flow basis rather than on a per-connection basis. The BS calculates the number of slots required for each service flow by examining the BW requests for each connection.

After satisfying the UGS connections, out of the remaining slots, 50% is assigned to the rtPS flows, 30% to nrtPS, and 20% to BES. The excess bandwidth allocated to all flows (TotExcess) is collected and distributed among the unsatisfied flows using WFQ. Each unsatisfied flow is allocated additional BW (from TotExcess) depending upon its additional bandwidth requirement. Thus, we calculate ExcessAlloc for the *i*th service flow as:

$$\text{ExcessAlloc}[i] = \frac{\text{AdditionalReq}[i]}{\text{TotExcess}}$$

where AdditionalReq[*i*] is the yet unsatisfied bandwidth requirement of the *i*th flow. The bandwidth allocated to each flow is distributed among all connections of that service flow depending on their requests.

23.4.3.5 Spare Slot Allocation. In a particular frame, if the BW requests of all the connections are satisfied and the BS is still left with some more unallocated UL slots, instead of letting them go as waste, it can allocate them to the various connections. These extra slots act as buffer in the presence of changed BW requirements at each SS. This can happen when additional packets arrive at the MAC layer after the BW request has been sent in the previous frame (if any). Hence by using these extra slots, the SS can tide over the sudden additional BW requirements for its connections, especially its rtPS connections. But the BS has no way to anticipate the additional requirements of each connection.

This problem does not arise in the case of GPSS since the BW is allocated for the entire SS and not to the individual connections of that SS. So, the BS can divide the additional slots among the SSs and let them handle the extra slots as they wish. So, if the SS finds that there is an additional requirement for a particular connection, more than what was requested for in the previous frame, it can intelligently reshuffle the extra slots.

In our case, the least we can do is to give priority to different connections when dividing the extra slots, by allocating extra slots to each connection depending on its service type. Thus, the BS can divide the extra slots to each connection depending on the service type for that connection. In other words, UGS connections can be allocated the maximum (say, 40%) of the extra slots, and BES connections can be allocated the minimum (say, 10%) of the extra slots. In this way, higher priority connections have a better chance of utilizing the extra slots to meet their additional requirements and meet their QoS guarantees.

Although this above method will at least give an opportunity for the connections to utilize the spare slots and prevent them from being wasted, in some cases we cannot prevent the wastage. For example, UGS connections need a fixed size grant in each frame. Allocating extra slots to them will result in wastage. But the average delay will reduce since they can send some packets that were meant to be sent in the next frame in the current frame itself if they have already arrived. If there are no packets to be sent, those slots will get wasted. In the case of UGS, we can take the decision that no UGS connections will be allocated extra slots (at the cost of increased average delay) since they will always get wasted. But the same is not the case with other service types since we cannot predict their requirements accurately.

Thus, we propose to use a slightly different strategy for allocation of slots by the BS for the GPC mode of operation. In this strategy, the BS allocates the slots requested by the connections as usual and then allocates the spare slots also to connections in the manner discussed above, that is, allocate about 40% of the spare slots for UGS connections, 30% for rtPS, 20% for nrtPS, and the remaining 10% for BES connections and divide them equally among connections of each type. However, the BS gives the SS the flexibility to use the spare slots in a manner different than that specified by itself. That is, if the SS finds more slots allotted to connection A, which does not need those many extra slots, but there exists another connection B that can use those extra slots, the SS can transfer the extra slots from A to B. In other words, we can think of the above process as the BS allocating the spare slots to an SS rather than to any particular connection of that SS and allowing the SS to use the spare slots as it sees fit. In this way, we are not changing the scheduling decisions of the BS. At the same time, we can improve the BW utilization and achieve better performance.

This may not make much difference if the network load is so high that there are hardly any spare slots left. At lower loads, this can make a difference. We have compared the regular method to ours and present the results in Section 23.5. We have chosen to always allow higher priority connections the chance to use a spare slot not utilized by another connection. In this way we preserve the priority among the connections.

23.5 SIMULATOR AND PERFORMANCE ANALYSIS

In this section, we present the results of the simulation analysis and show that our architecture meets the QoS guarantees of the various service flows. The analysis is done for the different scheduling algorithms for both GPC and GPSS modes.

23.5.1 Simulation Environment

The simulation topology consists of one BS and a number of SSs, a typical example of which is shown in Fig. 23.4. This network has one BS and three SS nodes, with each SS having one UGS, one rtPS, one nrtPS, and one BES connection. Channel bandwidth is 20 Mbps unless otherwise specified.

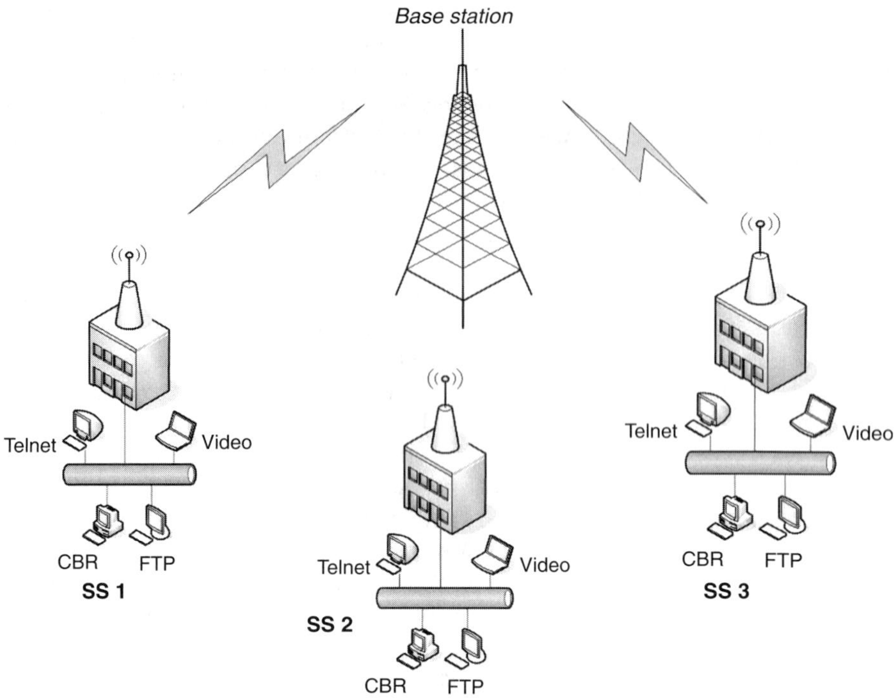

FIGURE 23.4 Simulation setup.

As mentioned earlier, we assume that the channel is error free so that each packet is successfully received at the destination. Also, we do not dynamically change the number of SSs during the simulation. Each SS has a combination of various application flows such as CBR traffic, video traffic, FTP, Telnet, and the like. Each of these application-level flows is mapped to one of the uplink scheduling flows.

The performance metrics considered are: average delay of uplink flows at SSs, average throughput, and number of packets transmitted. Delay is calculated as the amount of time taken for a packet to reach an application level [transmission control protocol (TCP)/user datagram protocol (UDP)] sink from the time that it was sent by a TCP/UDP agent. We can calculate this from trace file that *ns* generates. For the calculation of throughput, we consider only the time used for uplink transmission and also assume fixed size packets, occupying one time slot. Thus, we calculate throughput as the ratio of the number of packets sent and total UL time.

The parameters varied include number of connections, application types, and loads (amount of traffic) offered by each application. For all our experiments, we consider three different scenarios. In the first, the load offered by all four service flows is equal. In the second scenario, the load offered by higher priority flows is more than that offered by the lower priority flows. The total load on the system is 95% of the total channel bandwidth, of which 40% is offered by UGS flows, 30% by rtPS flows, 20% by nrtPS flows, and 10% by BES flows.

In the third scenario, the load offered by lower priority flows is more than that offered by the higher priority flows. The total load on the system is 95% of the total channel bandwidth, of which 40% is offered by BES flows, 30% by nrtPS flows, 20% by rtPS flows, and 10% by UGS flows. Also, unless otherwise specified, each simulation is run for 150 s. The trends did not vary much even if the simulation is run for longer durations.

23.5.2 Spare Slot Allocation Comparison

Here, we present the comparison of results obtained when the BS allows spare slot reallocation by the SS as against the regular method of slot allocation. We have measured the number of packets received, the average delay, and the average throughput for each service flow in the three different scenarios mentioned above. Table 23.1 shows a summary of the results obtained when

TABLE 23.1 Performance with GPC Mode, with and without Extra Slot Adjustment by the SS

	Received Pkts.		Avg. Throughput (kbps)		Avg. Delay (ms)	
	Plain	Extra Slots Adjusted	Plain	Extra Slots Adjusted	Plain	Extra Slots Adjusted
			Equal Loads			
UGS	27987	27987	4078.25	4078.25	50.32	21.88
rtPS	26511	26511	3863.17	3863.17	76.33	26.44
nrtPS	27381	27381	3989.95	3989.95	121.26	33.96
BES	26165	26165	3812.75	3812.75	164.97	46.56
			More Higher Priority Loads			
UGS	41976	41976	6116.72	6116.72	54.1	25.46
rtPS	42201	42225	6149.51	6153.01	176.33	55.24
nrtPS	25058	25058	3651.44	3651.44	197.97	61.21
BES	19783	19650	2882.77	2863.39	233.46	108.62
			More Lower Priority Loads			
UGS	8400	8400	1224.04	1224.04	56.15	22.73
rtPS	28002	28002	4080.44	4080.44	124.95	57.38
nrtPS	34844	55206	5077.45	8044.59	179.8	85.6
BES	35537	43278	5178.43	6306.45	199.18	138.99

the BS scheduler uses the WFPQ scheduling algorithm. The results obtained for the other algorithms were similar. For the regular method of slot allocation, the UGS connections are not given any extra slots, keeping in view that they will get wasted eventually—though the instantaneous delays will reduce. The spare slots are divided by weight among the other three flows with 50% of them being given to rtPS flows, 40% to nrtPS, and the rest 10% to BES.

We observe that in the first scenario, the number of packets sent (and thus the throughput) is the same for both methods of slot allocation. The difference, however, is in the delays. With our method, the delays are reduced drastically. This is because many packets are sent ahead of their time and thus they spend less time in the queue. Thus, the spare slots are utilized more efficiently.

In the second scenario, the regular method actually has more packets received for BES connections than our method. This is because in our method, we always give the opportunity for the higher priority flows to use the spare slots, and, since the load for them is higher, BES connections may lose out on some spare slots. They get an opportunity to send only in their own slots; they do not get any spare slots from the higher priority flows. But the delays are still much lower than the regular method.

In the last scenario, since the UGS and rtPS load is lower and they are also given more spare slots, many of these spare slots are used by nrtPS and BES flows. Thus, the number of packets received for them is much higher than with the regular method. The delays are also much lower.

23.5.3 Delay and Throughput Analysis

With the aim of showing that our architecture meets delay guarantees of real-time applications and maintains fairness among flows in accordance to their priority, we have run our simulations with the three different scenarios.

Scenario 1. In this scenario, as mentioned before, each service flow offers equal load on the system. The topology is as specified in the previous section. We have measured the number of packets sent, average delay, and average throughput for each service flow for the different scheduling algorithms. Table 23.2 gives a summary of the results obtained. It also shows the results obtained for the GPSS mode of operation for the same scheduling algorithms as presented in [8].

The salient points to note from the Table 23.2 are:

- All the algorithms produce similar and predictable results. The number of packets sent and thus the throughput is the same for all the four algorithms. There is only a marginal difference in the average delays for each service flow among the different algorithms. Thus, we could say that in this scenario, any algorithm can be used in the BS scheduler with the same results.

TABLE 23.2 Performance Comparison of GPC and GPSS for all Alogarithms under Equal Loads

	Received Pkts.		Avg. Throughput (kbps)		Avg. Delay (ms)	
	GPC	GPSS	GPC	GPSS	GPC	GPSS
			Weighted Fair Priority Queue			
UGS	27987	27990	4078.25	4078.69	21.88	21.67
rtPS	26511	26511	3863.17	3863.17	26.44	26.91
nrtPS	27381	27381	3989.95	3989.95	33.96	34.21
BES	26165	26165	3812.75	3812.75	46.56	46.92
			Weighted Fair Queue			
UGS	27987	27990	4078.25	4078.69	21.88	21.67
rtPS	26511	26511	3863.17	3863.17	26.44	26.91
nrtPS	27381	27381	3989.95	3989.95	33.96	34.21
BES	26165	26165	3812.75	3812.75	46.56	46.92
			Max-Min Fair Allocation			
UGS	27987	27990	4078.25	4078.69	21.91	21.67
rtPS	26511	26511	3863.17	3863.17	26.46	26.91
nrtPS	27381	27381	3989.95	3989.95	33.85	34.21
BES	26165	26165	3812.75	3812.75	46.62	46.92
			Deficit Fair Priority Queue			
UGS	27987	27990	4078.25	4078.69	21.94	21.67
rtPS	26511	26511	3863.17	3863.17	26.54	26.91
nrtPS	27381	27381	3989.95	3989.95	33.84	34.21
BES	26165	26165	3812.75	3812.75	46.46	46.92

- Though the same load is offered by all the service flows, the priority ordering between them is maintained. This is seen from the fact that UGS connections experience the least delay and BES connections the maximum. Thus, we have shown that the delay guarantees of higher priority flows are not affected by the amount of lower priority load.
- We also see that with the spare slot allocation done by SS, GPC produces delays that are slightly lower than GPSS for all service flows except UGS, thus achieving better performance than GPSS in this scenario.

Figure 23.5*a* shows the plot of instantaneous delays for the various service flows with time (for the WFPQ algorithm). We see that the delay for higher

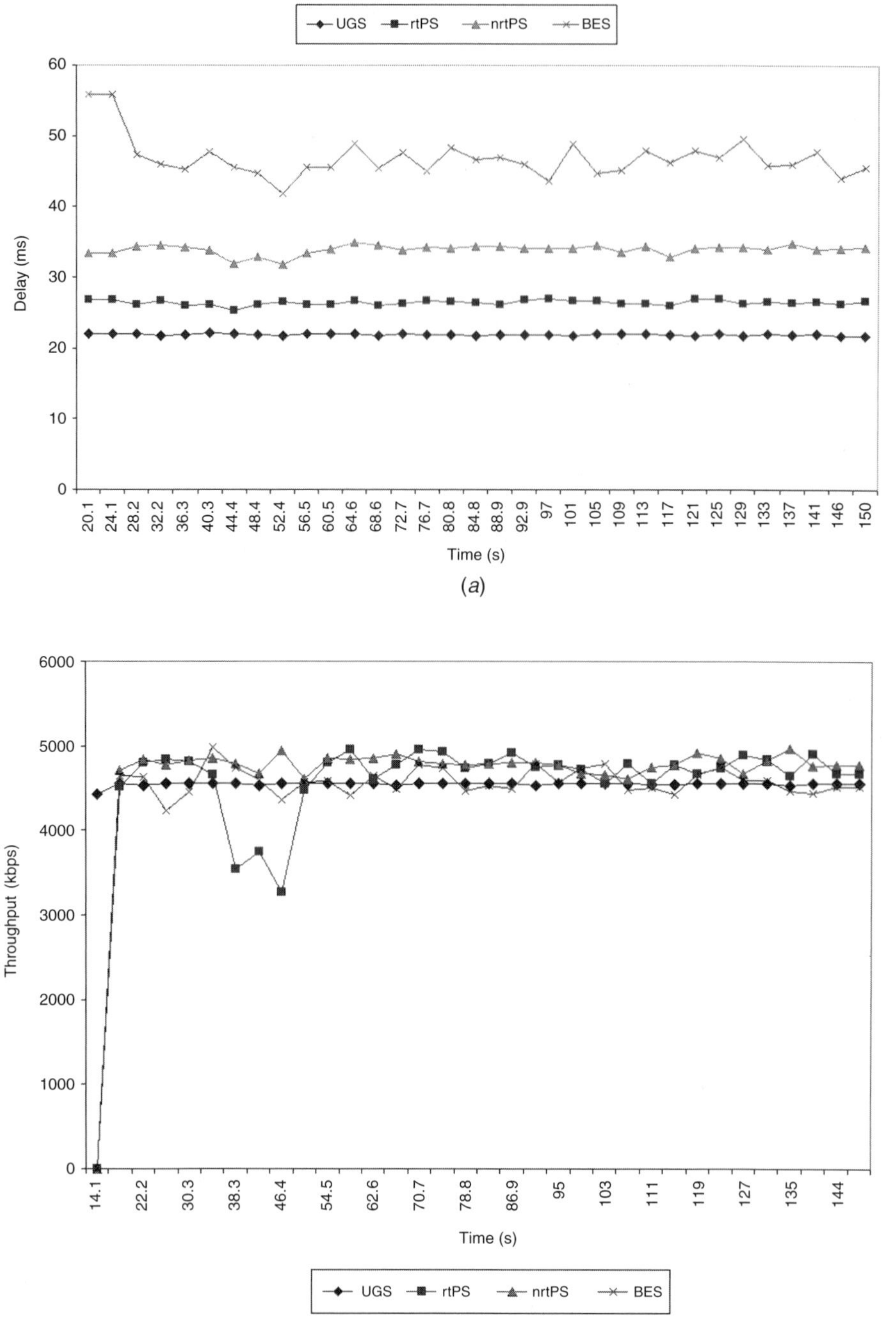

FIGURE 23.5 Instantaneous delay and throughput with GPC mode.

priority flows is almost constant with time, thus maintaining their QoS guarantees. Figure 23.5*b* shows the plot of the instantaneous throughput with time for different service flows. Here we note that since the load offered is the same, throughput is the same in time for all flows.

Scenario 2. Table 23.3 presents the summary of the results obtained in this scenario.

- Of the four algorithms, WFQ performs best in this scenario with the least overall delay of 61.85 ms followed by max-min fair allocation with 61.92 ms, followed by DFPQ with 62.12 ms, and WFPQ with 62.63 ms.

TABLE 23.3 Performance Comparison of GPC and GPSS for all Alogarithms under Heavy Higher Priority Loads

	Received Pkts.		Avg. Throughput (kbps)		Avg. Delay (ms)	
	GPC	GPSS	GPC	GPSS	GPC	GPSS
			Weighted Fair Priority Queue			
UGS	41976	41985	6116.72	6118.03	25.46	25.51
rtPS	42225	42225	6153.01	6153.01	55.24	49.59
nrtPS	25058	25058	3651.44	3651.44	61.21	52.96
BES	19650	19838	2863.39	2890.78	108.62	149.74
			Weighted Fair Queue			
UGS	41976	41985	6116.72	6118.03	25.34	25.51
rtPS	42225	42225	6153.01	6153.01	55.59	49.59
nrtPS	25058	25058	3651.44	3651.44	59.17	52.96
BES	19674	19838	2866.89	2890.78	107.3	149.74
			Max-Min Fair Allocation			
UGS	41976	41985	6116.72	6118.03	25.42	25.51
rtPS	42225	42225	6153.01	6153.01	55.43	49.59
nrtPS	25058	25058	3651.44	3651.44	59.6	52.96
BES	19674	19838	2862.95	2890.78	107.22	149.74
			Deficit Fair Priority Queue			
UGS	41976	41985	6116.72	6118.03	25.44	25.51
rtPS	42225	42225	6153.01	6153.01	55.51	49.59
nrtPS	25058	25058	3651.44	3651.44	59.39	52.96
BES	19634	19838	2861.06	2890.78	108.09	149.74

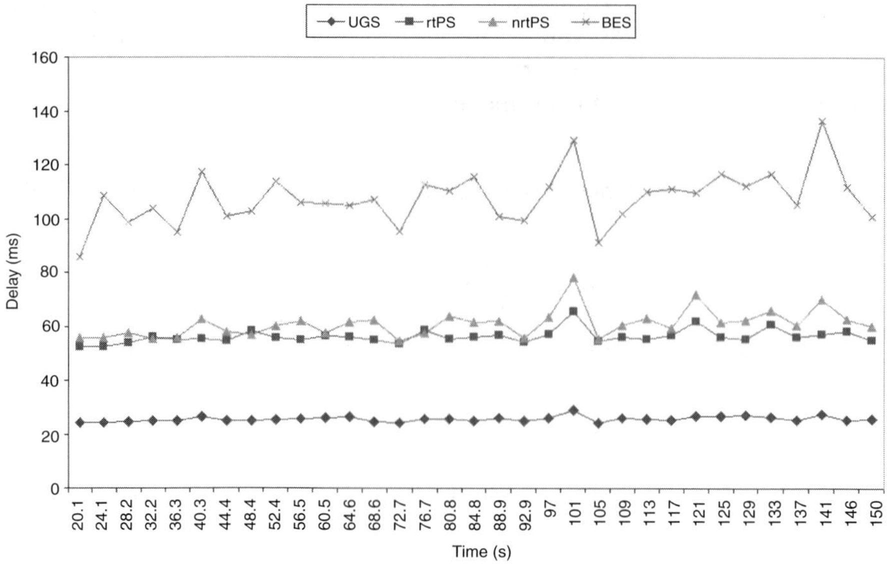

(*a*)

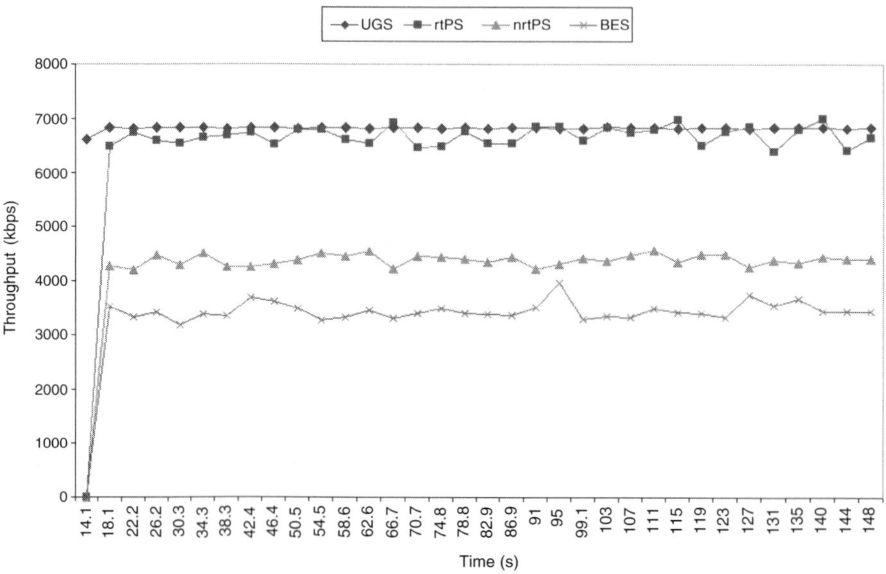

(*b*)

FIGURE 23.6 GPC: Instantaneous delay and throughput under heavy higher priority loads. (*a*) GPC: Delay Under Heavy Higher Priority Loads. (*b*) GPC: Throughput under heavy higher priority loads. (*c*) GPC: Throughput under heavy higher priority loads.

WFQ has the least delay among the four algorithms for UGS, nrtPS, and BES. It also has the maximum number of BES packets transmitted.

- Though the load offered by higher priority connections is more, they still have delays comparatively lower than the lower priority connections. The throughput of the higher priority connections is also higher, keeping in view their higher loads. We also note that though nrtPS and BES connections experience higher delays, it is not unreasonably high, keeping in view the total load on the system.
- Comparing with results obtained for GPSS, we note that for UGS, GPC has slightly lower delay. For rtPS and nrtPS, GPSS performs better, which is to be expected since in the GPSS mode, the SS can better handle the

TABLE 23.4 Performance Comparison of GPC and GPSS for all Alogarithms under Heavy Lower Priority Loads

	Received Pkts.		Avg. Throughput (kbps)		Avg. Delay (ms)	
	GPS	GPSS	GPC	GPSS	GPC	GPSS
			Weighted Fair Priority Queue			
UGS	8400	8400	1224.04	1224.04	22.73	22.44
rtPS	28002	28002	4080.44	4080.44	57.38	41.28
nrtPS	55206	57006	8044.59	8306.89	85.6	79.54
BES	43278	41937	6306.45	6111.04	138.99	144.11
			Weighted Fair Queue			
UGS	8400	8400	1224.04	1224.04	22.77	22.44
rtPS	28002	28002	4080.44	4080.44	58.76	41.28
nrtPS	55271	57006	8054.06	8306.89	85.58	79.54
BES	43223	41937	6298.43	6111.04	139.09	144.11
			Max-Min Fair Allocation			
UGS	8400	8400	1224.04	1224.04	22.9	22.44
rtPS	28002	28002	4080.44	4080.44	58.16	41.28
nrtPS	55242	57006	8049.84	8306.89	85.58	79.54
BES	43206	41937	6295.96	6111.04	139.23	144.11
			Deficit Fair Priority Queue			
UGC	8400	8400	1224.04	1224.04	22.88	22.44
rtPC	28002	28002	4080.44	4080.44	58.54	41.28
nrtPS	55440	57006	8078.69	8306.89	85.2	79.54
BES	43360	41937	6318.4	6111.04	138.35	144.11

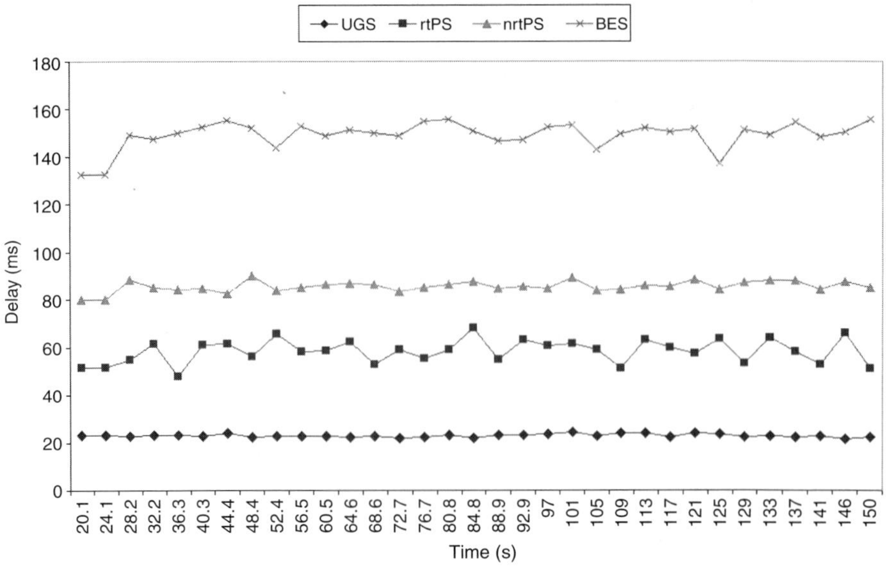

(*a*)

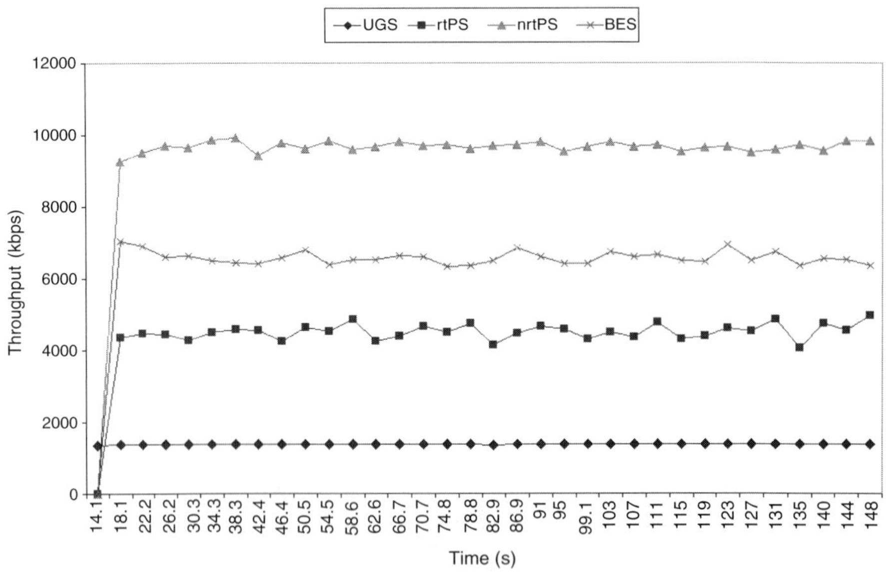

(*b*)

FIGURE 23.7 GPC: Instantaneous delay and throughput under heavy lower priority loads. (*a*) GPC: delay under heavy lower priority loads. (*b*) GPC: throughput under heavy lower priority loads.

changed BW requirements of real-time applications. At the same time, we also note that the BES delay for GPC is significantly lower than for GPSS. Thus, in this case, GPC is more fair to BES. Also, the average delay for all flows taken together is lower for GPC than for GPSS.

From Fig. 23.6*a*, we see that the instantaneous delays for higher priority flows is always lower than for the others, even though they offer more load and it also doesn't fluctuate as much. From Fig. 23.6*b*, we see that the instantaneous throughput for UGS and rtPS is higher than nrtPS or BES.

Scenario 3. Table 23.4 gives the summary of the results obtained in this scenario.

- Among the four algorithms, WFPQ has the least overall delay (76.16 ms), followed by DFPQ (76.24 ms), followed by max-min fair allocation (76.47 ms), and WFQ (76.54 ms). DFPQ has the maximum throughput (maximum number of received packets) and the least delay in the case of BES and nrtPS. Thus, in this scenario, we can say that DFPQ performs best.
- The throughput for the nrtPS and BES flows keeps in view the higher load offered by them. Yet, UGS has the least delay and BES the maximum. Thus, the heavy load from the lower priority traffic does not effect the delays of higher priority traffic.
- Comparing with GPSS, we find that UGS delays are almost the same. GPSS has lower delays for rtPS and nrtPS, whereas GPC has lower delay for BES. Also, throughput for UGS and RTPS is the same. GPSS has more throughput (number of packets received) for nrtPS, whereas GPC has more throughput for BES.

From Fig. 23.7*a*, we see that the instantaneous delays for higher priority flows is always lower than for the others, though the instantaneous throughput for nrtPS and BES is higher than UGS and rtPS (Fig. 23.7*b*).

23.6 CONCLUSIONS

In this chapter, we have presented an overview of the PHY and MAC layers for the IEEE 802.16 standard. We also present a QoS scheduling architecture. We have implemented the core features of IEEE 802.16 MAC with our architecture in the *ns2* simulator framework. We have implemented four different scheduling algorithms (WFQ, WFPQ, max-min fair allocation, and DFPQ) for the BS scheduler and have found that under equal loads from all services, all perform well. We have also concluded that, though WFQ may perform well (as it does in the case of more higher priority load), it is not suitable for use in IEEE 802.16 since it does not take priority of services into account. DFPQ performs well under heavy lower priority loads as does also WFPQ. If the SS is allowed (in the GPC mode) to reallocate the spare slots allocated to its connections, then the network performance can improve. In such a case, it even compares favorably with GPSS in many cases. It is evident through the simulation results that lower priority traffic does not affect QoS guarantees of high priority real-time traffic, that is, fairness is maintained among flows at an SS and across different SS nodes.

Future work on this can include the design of suitable admission control mechanisms, dynamic allocation of variable bandwidth for DL and UL transmissions, and integrated studies with higher level network protocols.

REFERENCES

1. IEEE 802.16-2004, "IEEE standard for local and metropolitan area networks, Part 16: Air interface for fixed broadband wireless access systems," available: http://standards.ieee.org/getieee802/index.html, Oct. 2004.
2. A. Ghosh, D. Wolter, J. Andrews, and R. Chen, "Broadband wireless access with WiMax/802.16: Current performance benchmarks and future potential," *IEEE Commun. Mag*. 43(2), 129–136 (2005).
3. Wimax Forum, "Promoting interoperability standards for broadband wireless access," available: www.wimaxforum.org, 2007.
4. ETSI HiperMAN; "High performance metropolitan area networks," available: http://portal.etsi.org/radio/HiperMAN/HiperMAN.asp, Mar., 2007.
5. "Mobile WiMAX: the best personal broadband experience," available: http://www.wimaxforum.org/technology/downloads/MobileWiMAX_PersonalBroadband.pdf, June 2006.
6. "The Network Simulator," available: ns-2. http://nsnam.isi.edu/nsnam/index.php/Main_Page, 2007.
7. K. Sai Suhas, "IEEE 802.16 WiMAX Protocol: ns2 simulator based implementation and performance evaluation for GPC, "Master's thesis, Sri Sathya Sai University, Puttaparthi, India, Feb. 2006.
8. M. Sai Rupak, "IEEE 802.16 WiMAX : ns2 simulator based implementation and performance analysis for GPSS," Master's thesis, Sri Sathya Sai University, Puttaparthi, India, Feb. 2006.
9. IEEE 802.16e-2005, "IEEE standard for local and metropolitan area networks Part 16: Air interface for fixed and mobile broadband wireless access systems amendment for physical and medium access control layers for combined fixed and mobile operation in licensed bands, available: http://standards.ieee.org/getieee802/download/802.16e-2005.pdf, 2005.
10. R. J. Guice and R. J. Munoz, "IEEE 802.16 commercial off the shelf (COTS) technologies as a compliment to ship to objective maneuver (STOM) communications," Master's thesis, Naval Postgraduate School, Monterey, CA, Sept. 2004.

11. WiMAX Forum Network Working Group, "WiMAX end-to-end network systems architecture," http://www.wimaxforum.org/technology/documents/WiMAX_End_to_End_Network_Systems_Architecture.zip, Dec. 2005.
12. S. Maheshwari, An efficient QoS scheduling architecture for IEEE 802.16 wireless MANs," Master's thesis, Indian Institute of Technology, Mumbai, India, 2005.
13. M. Hawa, Stochastic evaluation of fair scheduling with applications to quality-of-service in broadband wireless access networks," PhD thesis, University of Kansas, Lawrence, KS, Aug. 2003.
14. J. Chen, W. Jiao, and H. Wang, "A service flow management strategy for IEEE 802.16 broadband wireless access systems in TDD mode," in *Proceedings of the International Conference on Communications* (*ICC*), Seoul, Korea, May 2005, pp. 3422–3426.
15. K. Wongthavarawat and A. Ganz, "IEEE 802.16 based last mile broadband wireless military networks with quality of service support, "in *Proceedings of the IEEE MILCOM*, Monterey, CA, Oct. 2003, pp. 779–784.
16. A. Sayenko, O. Alanen, J. Karhula, and T. Hamalainen, "Ensuring the QoS requirements in 802.16 scheduling," in MSWiM '06: Proceedings of the 9th ACM International Symposium on Modeling Analysis and Simulation of Wireless and Mobile Systems, ACM Press, 2006, pp. 108–117.
17. C. Cicconetti, A. Erta, L. Lenzini, and E. Mingozzi, "Performance evaluation of the IEEE 802.16 MAC for QoS Support," *IEEE Trans. Mobile Comput.* 6(1), 26–38 (2007).
18. F. D. Pellegrini, D. Miorandi, E. Salvadori, and N. Scalabrino, "Qos support In WiMAX networks: Issues and experimental measurements, "Technical Report N. 200600009, CreateNet, available: http://www.create-net.org/~dmiorandi/CN_techrep_200600009.pdf, June 2006.
19. O. Yang and J. Lu, "Call admission control and scheduling schemes with QoS support for real-time video applications in IEEE 802.16 networks," *J. Multimedia* 1(2), 21–29 (2006).
20. D. Niyato and E. Hossain, "Queue-aware uplink bandwidth allocation and rate control for polling service in IEEE 802.16 broadband wireless networks," *IEEE Trans. Mobile Comput.* 5(6), 668–679 (2006).

INDEX

ABOUT THE EDITORS

Dr. Yang Xiao worked in industry as a MAC (medium access control) architect involving IEEE 802.11 standard enhancement before he joined the Department of Computer Science at the University of Memphis in 2002. He is currently with the Department of Computer Science at the University of Alabama. He was a voting member of the IEEE 802.11 working group from 2001 to 2004. He is an IEEE senior member. He is a member of the American Telemedicine Association. He currently serves as editor-in-chief for the *International Journal of Security and Networks* (IJSN), *International Journal of Sensor Networks* (IJSNet), and *International Journal of Telemedicine and Applications* (IJTA). He serves as a referee/reviewer for many funding agencies and is a panelist for the U.S. National Science Foundation (NSF) and a member of the Canada Foundation for Innovation (CFI)'s Telecommunications expert committee. He has served on TPC for more than 100 conferences such as INFOCOM, ICDCS, MOBIHOC, ICC, GLOBECOM, and WCNC. He is an associate editor for several journals, e.g., *IEEE Transactions on Vehicular Technology*. His research areas are security, telemedicine, sensor networks, and wireless networks. He has published more than 300 papers in major journals, refereed conference proceedings, and written book chapters related to these research areas. Dr. Xiao's research has been supported by the NSF and U.S. Army Research.

Yi Pan received his BE and ME degrees in computer engineering from Tsinghua University, China, in 1982 and 1984, respectively, and his PhD degree in computer science from the University of Pittsburgh, Pittsburgh, Pennsylvania, in 1991. Currently, he is the chair and a full professor in the Department of Computer Science at Georgia State University, Atlanta, Georgia. Dr. Pan's research interests include high performance computing, networking, and bioinformatics. Dr. Pan has published more than 100 journal papers with over 30 papers published in various IEEE journals. In addition, he has published over 150 papers in refereed conferences. He has also authored/edited over 33 books (including proceedings). Dr. Pan has served as an editor-in-chief or editorial board member for 15 journals, including 5 IEEE transactions. Dr. Pan has delivered over 10 keynote speeches at international conferences.